Interest Rate Modeling
Theory and Practice
Second Edition

D0144098

CHAPMAN & HALL/CRC
Financial Mathematics Series

Aims and scope:
The field of financial mathematics forms an ever-expanding slice of the financial sector. This series aims to capture new developments and summarize what is known over the whole spectrum of this field. It will include a broad range of textbooks, reference works and handbooks that are meant to appeal to both academics and practitioners. The inclusion of numerical code and concrete real-world examples is highly encouraged.

Series Editors

M.A.H. Dempster
Centre for Financial Research
Department of Pure Mathematics and Statistics
University of Cambridge

Dilip B. Madan
Robert H. Smith School of Business
University of Maryland

Rama Cont
Department of Mathematics
Imperial College

Equity-Linked Life Insurance

Partial Hedging Methods

Alexander Melnikov, Amir Nosrati

High-Performance Computing in Finance

Problems, Methods, and Solutions

M.A.H. Dempster, Juho Kanniainen, John Keane, Erik Vynckier

An Introduction to Computational Risk Management of Equity-Linked Insurance

Runhuan Feng

Derivative Pricing

A Problem-Based Primer

Ambrose Lo

Portfolio Rebalancing

Edward E. Qian

Interest Rate Modeling

Theory and Practice, Second Edition

Lixin Wu

For more information about this series please visit: *https://www.crcpress.com/Chapman-and-HallCRC-Financial-Mathematics-Series/book-series/CHFINANCMTH*

Interest Rate Modeling
Theory and Practice
Second Edition

Lixin Wu

CRC Press
Taylor & Francis Group
Boca Raton London New York

CRC Press is an imprint of the
Taylor & Francis Group, an **informa** business

A CHAPMAN & HALL BOOK

CRC Press
Taylor & Francis Group
6000 Broken Sound Parkway NW, Suite 300
Boca Raton, FL 33487-2742

First issued in paperback 2020

ISBN 13: 978-0-367-65655-3 (pbk)
ISBN 13: 978-0-8153-7891-4 (hbk)

Library of Congress Cataloging-in-Publication Data

Names: Wu, Lixin, 1961- author.
Title: Interest rate modeling : theory and practice / Lixin Wu.
Description: 2nd edition. | Boca Raton, Florida : CRC Press, [2019] | Includes bibliographical references and index.
Identifiers: LCCN 2018050904| ISBN 9780815378914 (hardback : alk. paper)| ISBN 9781351227421 (ebook : alk. paper)
Subjects: LCSH: Interest rates--Mathematical models. | Interest rate futures--Mathematical models.
Classification: LCC HG6024.5 .W82 2019 | DDC 332.801/5195--dc23
LC record available at https://lccn.loc.gov/2018050904

To my parents,
To Molly,
Dorothy and Derek

Contents

Preface to the First Edition

Motivations

This book was motivated by my teaching of the subject to graduate students at the Hong Kong University of Science and Technology (HKUST). My interest-rate class usually consists of students working toward both research degrees and professional degrees; their interests and levels of mathematical sophistication vary quite a bit. To meet the needs of the students with diverse backgrounds, I must choose materials that are interesting to the majority of them, and strike a balance between theory and application when delivering the course materials. These considerations, together with my own preferences, have shaped a coherent course curriculum that seems to work well. Given this success, I decided to write a book based on that curriculum.

Interest-rate modeling has long been at the core of financial derivatives theory. There are already quite a number of monographs and textbooks on interest-rate models. It is a good idea to write another book on the subject only if it will contribute significant added value to the literature. This is why I thought about this book. This book portrays the theory of interest-rate modeling as a three-dimensional object of finance, mathematics, and computation. In this book, all models are introduced with financial and economical justifications; options are modeled along the so-called martingale approach; and option evaluations are handled with fine numerical methods. With this book, the reader may capture the interdisciplinary nature of the field of interest-rate (or fixed-income) modeling, and understand what it takes to be a competent quantitative analyst in today's market.

The book takes the top-down approach to introducing interest-rate models. The framework for no-arbitrage models is first established, and then the story evolves around three representative types of models, namely, the Hull–White model, the market model, and affine models. Relating individual models to the arbitrage framework helps to achieve better appreciation of the motivations behind each model, as well as better understanding of the interconnections among different models. Note that these three types of models coexist in the market. The adoption of any of these models or their variants may often be determined by products or sectors rather than by the subjective will. Hence, a quant must have flexibility in adopting models. The premise, of course, is a thorough understanding of the models. It is hoped that, through the top-down approach, readers will get a clear picture of the status of this important subject, without being overwhelmed by too many specific models.

This book can serve as a textbook. Inherited from my lecture notes for a diverse pool of students, the book is not written in a strict mathematical style. Were that the case, there would be a lot more lemmas and theorems in the text. But efforts were indeed made to make the book self-contained in mathematics, and rigorous justifications are given for almost all results. There are quite a number of examples in the text; many of them are based on real market data. Exercise sets are provided for all but one chapter. These exercises often require computer implementation. Students not only learn the martingale approach for interest-rate modeling, but they also learn how to implement various models on computers. The adoption of materials is influenced by my experiences as a consultant and a lecturer for industrial courses. My early students often noted that materials in the course were directly relevant to their work in institutions. Those materials are included here.

To a large extent, this book can also serve as a research monograph. It contains many results that are either new or exist only in recent research articles, including, as only a few examples, adaptive Hull–White lattice trees, market model calibration by quadratic programming, correlation adjustment, and swaption pricing under affine term structure models. Many of the numerical methods or schemes are very efficient or even optimized, owing to my original background as a numerical analyst. In addition, notes are given at the end of most chapters to comment on cutting-edge research, which includes volatility-smile modeling, convexity adjustment, and so on.

Study Guide

When used as a textbook, this book can be covered in two 14-week semesters. For a one-semester course, I recommend the coverage of Chapters 1–4, half of Chapter 5 on lattice trees, and Chapter 6. As a reference book for self-learning, readers should study short-rate models, market models, and affine models in Chapters 5, 6, and 9, respectively. For applications, Chapter 8 is also very useful. Chapter 7, meanwhile, is special in this book, as it is particularly prepared for those readers who are interested in market model calibration.

Chapters 1 and 2 contain the mathematical foundations for interest-rate modeling, where we introduce Ito's calculus and the martingale representation theorem. The presentation of the theories is largely self-contained, except for some omissions in the proof of the martingale representation theorem, which I think is technically too demanding for this type of book.

In Chapter 3, I introduce bonds and bond yields, which constitute the underlying securities or quantities of the interest-rate derivatives markets. I also discuss the composition of bond markets and how they function. For completeness, I include the classical theory of risk management that is based on parallel yield changes.

Chapter 4 is a cornerstone of the book, as it introduces the Heath–Jarrow–Morton (HJM) model, the framework for no-arbitrage pricing models. With market data, we demonstrate the estimation of the HJM model. Forward measures, which are important devices for interest-rate options pricing, are introduced. Change of measures, a very useful technique for option pricing, is discussed in general.

Chapter 5 consists of two parts: a theoretical part and a numerical part. The theoretical part focuses on the issue of when the HJM model implies a Markovian short-rate model, and the numerical part is about the construction and calibration of short-rate lattice models. A very efficient methodology to construct and calibrate a truncated and adaptive lattice is presented with the Hull–White model, which, after slight modifications, is applicable to general Markovian models with the feature of mean reversion.

Chapter 6 is another cornerstone of the book, where I introduce the LIBOR market and the LIBOR market model. After the derivation of the market model, I draw the connection between the model and the no-arbitrage framework of HJM. This chapter contains perhaps the simplest yet most robust formula for swaption pricing in the literature. Moreover, with the pricing of Bermudan swaptions, I give an enlightening introduction to the popular Longstaff–Schwartz method for pricing American options in the context of Monte Carlo simulations.

Chapter 7 discusses an important aspect of model applications in the markets—model calibration. Model calibration is a procedure to fix the parameters of a model based on observed information from the derivatives market. This issue is rarely dealt with in academic or theoretical literature, but in the real world it cannot be ignored. With the LIBOR market model, I show how a problem of calibration can be set up and solved.

In Chapter 8, I address two intriguing industrial issues, namely, volatility and correlation adjustments. Mathematically, these issues are about computing the expectation of a financial quantity under a non-martingale measure. With unprecedented generality and clarity, I offer analytical formulae for these evaluation problems. The adjustment formulae have widespread applications in pricing futures, non-vanilla swaps, and swaptions.

Finally, in Chapter 9, I introduce the class of affine term structure models for interest rates. Rooted in general equilibrium theory for asset pricing, the affine term structure models are favored by many people, particularly those in academic finance. These models are parsimonious in parameterization, and they have a high degree of analytical tractability. The construction of the models is demonstrated, followed by their applications to pricing options on bonds and interest rates.

Preface to the Second Edition

It has been almost ten years since the publication of the first edition of this book. In responses to the 2008 financial crisis, major changes have taken place over the past ten years in financial markets, from changes in regulations to the practice of derivatives pricing and risk management. Regulators have been pushing OTC trades to go through central counterparty clearing houses, which are subject to initial margin (IM) and variable margin (VM). For the remaining OTC trades, collaterals have become market standard, in addition to risk capital requirements. The funding costs for IM, VM, collaterals, and risk capital have become a burden to many firms. How to take into account the funding costs in trade prices has been a central issue to the practitioners, regulators, and researchers. The current solution is to make various valuation adjustments, so-called xVA, to either the trade prices or accounting books, which has been controversial and is still debated today.

Major changes also occurred to the modeling of the interest rate derivatives. Pre-crisis term structure models, which were based on a single forward rate curve – so-called single curve modeling – were replaced by multiple curve models, which simultaneously model multiple forward rate curves. Yet, most multi-curve models are at odds with the basis swap curves, which suggest that the forward rate curves cannot evolve separately in any usual ways. There is an affine solution of multi-curve modeling which is compatible with the basis curves, but such a model is very different from models quants are used to, such as the SABR-LIBOR market model (SABR-LMM) that is popular owing to its capacity to manage volatility smile risks.

In the second edition, we will offer our solutions to xVA and post-crisis interest rate modeling. Specifically, we want to achieve three objectives. First, we will introduce the theories of major smile models for interest rate derivatives, and then adapt the most important one, the SABR-LMM model, to the post-crisis markets. Second, we will introduce models for inflation rate derivatives and credit derivatives. Third, we will introduce our solution to the issue of xVA. Altogether, six new chapters will be added. With exception of the last chapter on xVA, all new chapters will be developed around the central theme: the LIBOR market model (this is a distinguished feature of the second edition).

In Chapter 10, we will introduce the SABR model and the Heston's type LMM model that feature the role of stochastic volatility in the formation of volatility smiles. Through these two models, we try to demonstrate the methodologies and techniques of smile models that are based on stochastic volatilities.

In Chapter 11, we will introduce the Lévy market model, a framework of models that captures volatility smiles based on the dynamics of jumps and diffusions. Although this topic has theoretically been complex, we will offer a simple exposition of model construction and pricing.

In Chapter 12, we take readers to inflation derivatives modeling and pricing, a two-decade-old theoretical subject not yet fully understood. Our aim is to first build a solid foundation, and then on top of that develop the inflation market model to justify the current market practice in pricing and hedging inflation derivatives.

In Chapter 13, we deal with single-name credit derivatives, with the intention of pricing credit instruments, bonds, credit default swaps (CDS), CDS options (or credit swaption), and even collateralized debt obligation (CDO) using an LMM type model. We will redefine risky zero-coupon bonds using tradable securities, and then risky forward rates, and eventually the credit market model. This model allows us to price all instruments except CDOs, for which we need additional tools like copulas to model correlated defaults.

In Chapter 14, we will rebuild the foundation, namely, the risky zero-coupon bonds, for the post-crisis interest rate derivative markets. Based on the new foundation, we redefine LIBOR in the presence of credit risk of LIBOR panel banks, and demonstrate that such a risk is responsible for the emergence of the basis curves. We then define a dual-curve LMM and, more notably, the dual-curve SABR-LMM. A large portion of the chapter is then devoted to the pricing of caplets and swaptions under the dual-curve SABR-LMM, along the approach of the heat kernel expansion method of Henry-Labordère for the SABR-LMM model.

Finally, in Chapter 15, we present an xVA theory, which is applicable to general derivatives pricing, including interest rate derivatives. We will prove that the bilateral credit valuation adjustment is part of the fair price, and demonstrate how funding costs enter the P&L of trades. We show that only the market funding liquidity risk premium can enter into pricing, otherwise price asymmetry will occur.

Acknowledgments to the Second Edition

First I want to thank Mr. Sarfraz Khan, the editor at Taylor & Francis who took the initiative to contract with me for the second edition as otherwise the second edition would not have reached readers in 2019.

Three chapters are based on the joint publications that I worked on with my former PhD students. I want to thank Ho Siu Lam for the joint work on credit derivatives, Frederic Zhang for the joint work on xVA, and Shidong Cui for the joint work on the dual-curve SABR-LMM model. I would like to especially mention Shidong for his work with the very complex results of caplet and swaption pricing; his ability to manage the details is truly amazing.

I also want to thank my wife, Molly, for her support throughout the writing process of the second edition.

Finally, I want to thank my daughter, Dorothy, who helped me to proofread all six new chapters. Her corrections and suggestions have definitely made this book better.

Lixin Wu
Hong Kong

Author

Lixin Wu earned his PhD in applied mathematics from UCLA in 1991. Originally a specialist in numerical analysis, he switched his area of focus to financial mathematics in 1996. Since then, he has made notable contributions to the area. He co-developed the PDE model for soft barrier options and the finite-state Markov chain model for credit contagion. He is, perhaps, best known in the financial engineering community for a series of works on market models, including an optimal calibration methodology for the standard market model, a market model with square-root volatility, a market model for credit derivatives, a market model for inflation derivatives, and a dual-curve SABR market model for post-crisis derivatives markets. He also has made valuable contributions to the topic of xVA. Over the years, Dr. Wu has been a consultant for financial institutions and a lecturer for Risk Euromoney and Marco Evans, two professional education agencies. He is currently a full professor at the Hong Kong University of Science and Technology.

Chapter 1

The Basics of Stochastic Calculus

The seemingly random fluctuation in stock prices is the most distinguishing feature of financial markets and it creates both risks and opportunities for investors. To model random fluctuations in stock prices as well as in other financial time series data (indexes, interest rates, exchange rates, etc.) has long been a central issue in the discipline of quantitative finance. In mathematical terms, a financial time series is a stochastic process. In quantitative modeling, a financial time series is treated as a function of other standardized stochastic processes, and these stochastic processes serve as the engines for the random evolution of financial time series. With these models, we can assess risk, value risky securities, design hedging strategies, and make decisions on asset allocations in a scientific way. Among the many standardized stochastic time series that are available, the so-called Brownian motion is no doubt the most basic yet the most important one. This chapter is devoted to describing Brownian motion and its calculus.

1.1 Brownian Motion

Financial time series data actually exist in discrete form, for example, tick data, daily data, weekly data, and so on. It may be intuitive to describe these data with discrete time series models. Alternatively, we can also describe them with continuous time series models, and then take discrete steps in time. It turns out that working with continuous-time financial time series is a lot more efficient than working with discrete time series models, due to the existence of an arsenal of stochastic analysis tools in continuous time. The theory of Brownian motion is the single most important building block of continuous-time financial time series. We proceed by introducing Brownian motion through a limiting process, starting with simple random walks.

1.1.1   Simple Random Walks

Simple random walks are discrete time series, $\{X_i\}$, defined as

$$X_0 = 0,$$

$$X_{n+1} = \begin{cases} X_n - \sqrt{\Delta t}, & p = \dfrac{1}{2} \\ X_n + \sqrt{\Delta t}, & 1 - p = \dfrac{1}{2} \end{cases}, \quad n = 0, 1, 2, \ldots, \tag{1.1}$$

where $\Delta t > 0$ stands for the interval of time for stepping forward. One can verify that $\{X_i\}$ have the following properties:

1. The increment of $X_{n+1} - X_n$ is independent of $\{X_i\}, \forall i \leq n$.

2. $E[X_n \mid X_m] = X_m, m \leq n$.

An interesting feature of the simple random walk is the linearity of X_i's variance in time: given X_0, the variance of X_i is equal to $i\Delta t$, the time it takes the time series to evolve from X_0 to X_i.

Out of the simple Brownian random walk, we can construct a continuous-time process through linear interpolation:

$$\bar{X}(t) = X_i + \frac{t - i\Delta t}{\Delta t}(X_{i+1} - X_i), \quad t \in [i\Delta t, (i+1)\Delta t]. \tag{1.2}$$

We are interested in the limiting process of $\bar{X}(t)$ as $\Delta t \to 0$, in the hope that the limit remains a meaningful stochastic process. The next theorem confirms just that.

Theorem 1.1.1 (The Lundeberg–Levi Central Limit Theorem). *For the continuous process, $\bar{X}(t)$, there is*

$$\lim_{\Delta t \to 0} P\left\{\bar{X}(s+t) - \bar{X}(s) \leq x\right\} = \frac{1}{\sqrt{2\pi t}} \int_{-\infty}^{x} \exp\left(-\frac{u^2}{2t}\right) du. \tag{1.3}$$

Proof: The proof is a matter of applying the central limit theorem. Without loss of generality, we let $s = 0$ and take $\Delta t = t/n$. Apparently, there are $\bar{X}(0) = X_0, \bar{X}(t) = X_n$, and

$$P\{X_n - X_0 \leq x\} = P\left\{\sum_{i=1}^{n}(X_i - X_{i-1}) \leq x\right\}$$

$$= P\left\{\frac{1/n \sum_{i=1}^{n}(X_i - X_{i-1}) - 0}{\sqrt{\Delta t/n}} \leq \frac{x}{\sqrt{n\Delta t}}\right\}$$

$$= P\left\{\frac{1/n \sum_{i=1}^{n}(X_i - X_{i-1}) - 0}{\sqrt{\Delta t/n}} \leq \frac{x}{\sqrt{t}}\right\}. \tag{1.4}$$

According to the central limit theorem, we have

$$\lim_{\Delta t \to 0} P \left\{ \frac{1/n \sum_{i=1}^{n} (X_i - X_{i-1}) - 0}{\sqrt{\Delta t/n}} \leq \frac{x}{\sqrt{t}} \right\}$$

$$= P \left\{ \varepsilon \leq \frac{x}{\sqrt{t}} \right\}$$

$$= \int_{-\infty}^{x/\sqrt{t}} \frac{1}{\sqrt{2\pi}} \exp\left(-\frac{v^2}{2}\right) dv$$

$$\left(\text{let } u = \sqrt{t}v \right) = \frac{1}{\sqrt{2\pi t}} \int_{-\infty}^{x} \exp\left(-\frac{u^2}{2t}\right) du, \qquad (1.5)$$

where ε is a standard normal random variable. This completes the proof. □

We call the limited process,

$$W(t) = \lim_{\Delta t \to 0} \bar{X}(t), \quad 0 \leq t \leq \infty, \qquad (1.6)$$

a Wiener process in honor of Norbert Wiener, a pioneer in stochastic control theory. It is also called Brownian motion, after Robert Brown, a Scottish botanist who in 1827 observed that pollen grains suspended in liquid moved irregularly in motions that resembled simple random walks. The formal definition of Brownian motion is given in the next section.

1.1.2   Brownian Motion

A continuous stochastic process is a collection of real-valued random variables, $\{X(t, \omega), 0 \leq t \leq T\}$ or $\{X_t(\omega), 0 \leq t \leq T\}$, that are defined on a probability space $(\Omega, \mathcal{F}, \mathbb{P})$. Here Ω is the collection of all ωs, which are so-called sample points, $\mathcal{F}$ the smallest σ-algebra that contains Ωs, and $\mathbb{P}$ a probability measure on Ω. Each random outcome, $\omega \in \Omega$, corresponds to an entire time series

$$t \to X_t(\omega), \quad t \in T, \qquad (1.7)$$

which is called a path of X_t. In view of Equation 1.7, we can regard $X_t(\omega)$ as a function of two variables, ω and t. For notational simplicity, however, we often suppress the ω variable when its explicit appearance is not necessary.

In the context of financial modeling, we are particularly interested in the Brownian motion introduced earlier. Its formal definition is given below.

Definition 1.1.1. *A Brownian motion or a Wiener process is a real-value stochastic process, W_t or $W(t), 0 \leq t \leq \infty$, that has the following properties:*

1. $W(0) = 0$.

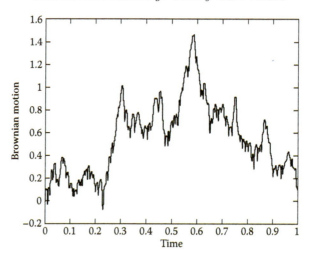

FIGURE 1.1: A sample path of a Brownian motion.

2. $W(t + s) - W(t)$ *is independent of* $\{W(u),\ 0 \le u \le t\}$.

3. *For* $t \ge 0$ *and* $s > 0$, *the increment* $W(t + s) - W(t) \sim N(0, s)$.

4. $W(t)$ *is continuous almost surely (a.s.).*

Here $N(0, s)$ stands for a normal distribution with mean zero and variance s. Note that in some literature, property 4 is not part of the definition, as it can be proved to be implied by the first three properties (Varadhan, 1980a or Ikeda and Watanabe, 1989). A sample path of $W(t)$ is shown in Figure 1.1, which is generated with a step size of $\Delta t = 2^{-10}$.

Brownian motion plays a major role in continuous-time stochastic modeling in physics, engineering and finance. In finance, it has been used to model the random behavior of asset returns. Several major properties of Brownian motion are listed below.

Lemma 1.1.1. *A Brownian motion,* $W(t)$, *has the following properties:*

1. *Self-similarity:* $\forall \lambda > 0\ W(t) \mapsto (1/\sqrt{\lambda})W(\lambda t) = \tilde{W}(t)$ *is also a Brownian motion.*

2. *Unbounded variation: for any* $T \ge 0$, $\lim_{\Delta t \to 0} \sum_{j,t_j \le T} |\Delta W_j| = \infty$, *where* $\Delta W_j = W(t_{j+1}) - W(t_j)$.

3. *Non-differentiability:* $W(t)$ *is not differentiable at any* $t \ge 0$.

The self-similarity property implies that $W(t)$ is a fractal object. This can be proved straightforwardly and the proof is left to readers. We will see that unbounded variation implies non-differentiability. To prove unbounded variation, we will need the following lemma.

Lemma 1.1.2. *Let* $0 = t_0 < t_1 < \cdots < t_n = T$ *represent an arbitrary partition of time interval* $[0,T]$, $\Delta t = \max_j (t_{j+1} - t_j)$ *and* $\Delta W_j = W(t_{j+1}) - W(t_j)$. *Then* $\lim_{\Delta t \to 0} \sum_{j=0}^{n-1} (\Delta W_j)^2 = T$ *almost surely.*

Proof: Let $\Delta t_j = t_{j+1} - t_j, \forall j$. Then we can write

$$\Delta W_j = \varepsilon_j \sqrt{\Delta t_j}, \quad \varepsilon_j \sim N(0,1) \text{ iid.} \tag{1.8}$$

Here, iid stands for "independent with identical distribution." By Kolmogorov's large number theorem,

$$\sum_{j=0}^{n-1} (\Delta W_j)^2 = \sum_{j=0}^{n-1} \Delta t_j \varepsilon_j^2 \xrightarrow{\text{a.s.}} \sum_{j=0}^{n-1} \Delta t_j E\left[\varepsilon_j^2\right] = T, \tag{1.9}$$

since $E[\varepsilon_j^2] = 1, \forall j$. $\qquad\square$

[**Proof** of unbounded variation and non-differentiability]
    We will do the proof by the method of contradiction. Suppose that $W(t)$ has bounded variation over a finite interval $[0,T]$ such that

$$\lim_{\Delta t \to 0} \sum_{j, t_j \leq T} |\Delta W_j| = C < \infty \tag{1.10}$$

for some finite constant C. Since a continuous function is uniformly continuous over a finite interval, we have

$$|\Delta W_j| = o(1) \quad \text{as} \quad \Delta t \to 0. \tag{1.11}$$

Using Lemma 1.2 and Equation 1.11, we then arrive at

$$T \approx \sum_{j, t_j \leq T} |\Delta W_j|^2 \leq (\max |\Delta W_j|) \sum_{j, t_j \leq T} |\Delta W_j| \leq C \cdot o(1) \xrightarrow{\Delta t \to 0} 0, \tag{1.12}$$

which is a contradiction, so the variation must be unbounded.
    The property of non-differentiability follows from the unbounded variation property. In fact, if $W'(t)$ existed and was finite over any interval, say $[0,T]$, then there would be

$$\lim_{\Delta t \to 0} \sum_{j, t_j \leq T} |\Delta W_j| \stackrel{\Delta}{=} \int_0^T |W'(t)| \, dt < \infty, \tag{1.13}$$

where "$\stackrel{\Delta}{=}$" means "is defined as." Equation 1.13 contradicts to the property of unbounded variation we have just proved. $\qquad\square$

1.1.3 Adaptive and Non-Adaptive Functions

We now define the class of functions of stochastic processes such that their values at time t can be determined based on available information up to time t. Formally, we introduce the notion of filtration.

Definition 1.1.2. *Let $\mathcal{F}_t$ denote the smallest σ-algebra containing all sets of the form*

$$\{\omega; W_{t_1}(\omega) \in B_1, \ldots, W_{t_k}(\omega) \in B_k\} \subset \Omega, \tag{1.14}$$

where $k = 1, 2, \ldots, t_j \leq t$ and $B_j \subset \mathbf{R}$ are Borel sets, where $\mathbf{R}$ stands for the set of real numbers. Denote the σ-algebra as $\mathcal{F}_t = \sigma(W(s), 0 \leq s \leq t)$; we call the collection of $(\mathcal{F}_t)_{t \geq 0}$ a Brownian filtration.

For applications in mathematical finance, it suffices to think of $\mathcal{F}_t$ as "information up to time t" or "history of W_s up to time t." According to the definition, $\mathcal{F}_s \subset \mathcal{F}_t$ for $s \leq t$, meaning that a filtration is an increasing stream of information. Readers can find thorough discussions of Brownian filtration in many previous works, for example, Øksendal (1992).

Definition 1.1.3. *A function, $f(t)$, is said to be $\mathcal{F}_t$-adaptive if*

$$f(t) = \tilde{f}(\{W(s), 0 \leq s \leq t\}, t), \quad \forall t, \tag{1.15}$$

that is, the value of the function at time t depends only on the path history up to time t.

Adaptive functions[1] are natural candidates to work with in finance. Suppose that the value of a function represents a decision in investment. Then, such a decision has to be made based on the available information up to the moment of making the decision. The next example gives a good idea of what kind of function is or is not an $\mathcal{F}_t$-adaptive function.

Example 1.1 Function

$$f(t) = \begin{cases} 0 & \min_{0 \leq s \leq t} W(s) < 2 \\ 1 & \min_{0 \leq s \leq t} W(s) \geq 2 \end{cases} \tag{1.16}$$

is $\mathcal{F}_t$-adaptive, whereas

$$f(t) = \begin{cases} 0 & \min_{0 \leq s \leq 1} W(s) < 2 \\ 1 & \min_{0 \leq s \leq 1} W(s) \geq 2 \end{cases} \tag{1.17}$$

is not $\mathcal{F}_t$-adaptive, because $f(t)$ cannot be determined at any time $t < 1$.

---

[1] They are also called non-predictable or non-anticipative functions.

1.2 Stochastic Integrals

Stochastic calculus considers the integration and differentiation of general $\mathcal{F}_t$-adaptive functions. The purpose of developing such a stochastic calculus is to model financial time series (with random dynamics) with either integral or differential equations. According to Lemma 1.1, a Brownian motion, $W(t)$, is nowhere differentiable in the usual sense of differentiation for deterministic functions. To define differentials of stochastic processes in a proper sense, we must first study the notion of stochastic integrals.

Stochastic integrals can be defined for functions in the square-integrable space, $H^2[0,T] = L^2(\Omega \times [0,T], d\mathbb{P} \times dt)$, which is defined to be the collection of functions satisfying

$$E\left[\int_0^T |f(t,\omega)|^2 \, dt\right] < \infty. \tag{1.18}$$

Note that, without indicated otherwise, $E[\cdot]$ means $E^{\mathbb{P}}[\cdot]$, the unconditional expectation under $\mathbb{P}$. The definition consists of a three-step procedure. First, we make the definition for elementary or piecewise constant functions in an intuitive way. Second, we define the integrals of a bounded continuous function as a limit of integrals of elementary functions. Finally, we define the integral of a general square-integrable function as a limit of integrals of bounded continuous functions. The key in this three-step procedure is of course to ensure the convergence of the limits in $L^2(\Omega, \mathcal{F}, \mathbb{P})$, the Hilbert space of random variables satisfying

$$E\left[X^2(\omega)\right] < \infty.$$

This definition approach is taken by Øksendal (1992). Alternative treatments of course also exist; see, for example, Mikosch (1998).

An elementary function has the form

$$\varphi(t,\omega) = \sum_j c_j(\omega)\chi_{(t_j, t_{j+1}]}(t), \tag{1.19}$$

where $\chi_A(t)$ is the indicator function such that $\chi_A(t) = 1$ if $t \in A$, or otherwise $\chi_A(t) = 0$, and $c_j(\omega)$ is adapted to $\mathcal{F}_{t_j}$. For the elementary function, the integral is defined in a rather natural way:

$$\int_0^T \varphi(t,\omega) \, dW(t,\omega) = \sum_j c_j(\omega)\Delta W(t_j, \omega), \tag{1.20}$$

where $\Delta W(t_j, \omega) = W(t_{j+1}, \omega) - W(t_j, \omega)$ is the increment of the over time interval $(t_j, t_{j+1}]$. The next result plays a crucial role in defining the stochastic integral for general functions.

Lemma 1.2.1 (Ito isometry). *If $\varphi(t, \omega)$ is a bounded elementary function, then*

$$E\left[\left(\int_0^T \varphi(t, \omega)\, dW(t, \omega)\right)^2\right] = \int_0^T E\left[\varphi^2(t, \omega)\right] dt. \qquad (1.21)$$

Proof: We proceed straightforwardly:

$$E\left[\left(\int_0^T \varphi(t, \omega)\, dW(t, \omega)\right)^2\right] = E\left[\sum_{i,j=0}^{n-1} c_i c_j \Delta W(t_i, \omega) \Delta W(t_j, \omega)\right]$$

$$= E\left[\sum_{j=0}^{n-1} c_j^2 (\Delta W(t_j, \omega))^2\right]$$

$$= \sum_{j=0}^{n-1} E\left[c_j^2\right] \Delta t_j$$

$$= \int_0^T E[\varphi^2(t, \omega)]\, dt.$$

This completes the proof. $\qquad \qquad \square$

For bounded continuous functions, we can define the stochastic integral intuitively through a limiting process:

$$\int_0^T f(t)\, dW(t) \triangleq \lim_{\Delta t \to 0} \sum_{j=0}^{n-1} f(t_j) \Delta W(t_j). \qquad (1.22)$$

The summation in Equation 1.22 is called a Riemann–Stieltjes sum. The existence and uniqueness of the limit is assured by the next lemma.

Lemma 1.2.1. *Let $f(t)$ be a bounded continuous and $\mathcal{F}_t$-adaptive function. Then, given any partition of $[0, T]$ with $\Delta t = \max_j(t_{j+1} - t_j)$, the limit of the Riemann–Stieltjes sum in Equation 1.23 exists and is unique in $L^2(\Omega, \mathcal{F}, \mathbb{P})$. Moreover, Ito's isometry holds for bounded continuous functions:*

$$E\left[\left(\int_0^T f(t)\, dW(t, \omega)\right)^2\right] = \int_0^T E[f^2(t)]\, dt. \qquad (1.23)$$

Proof: Construct an elementary function:

$$\varphi_n(t, \omega) = \sum_{j=0}^{n-1} f(t_j, \omega) \chi_{(t_j, t_{j+1}]}(t). \qquad (1.24)$$

Note that $\varphi_n(t)$ satisfies Equation 1.21, the Ito's isometry. Due to the continuity of $f(t)$, there is

$$\int_0^T E[(\, f(t) - \varphi_n(t))^2]\, dt \xrightarrow{\Delta t \to 0} 0, \qquad (1.25)$$

which implies that $\varphi_n(t)$ is a Cauchy sequence in $L^2(\Omega, \mathcal{F}, \mathbb{P})$. By making use again of Ito's isometry for $\varphi_n - \varphi_m$, we can see that

$$\int_0^T \varphi_n(t, \omega) \, dW_t(\omega) \tag{1.26}$$

is a Cauchy sequence in $L^2(\Omega, \mathcal{F}, \mathbb{P})$. Hence its limit exists as an element in the space, which, also by Ito isometry, is independent of the partition $\{t_i\}$. We denote the limit as

$$\int_0^T f(t) \, dW(t).$$

Finally, by taking limit $n \to \infty$ for the equality

$$E\left[\left(\int_0^T \varphi_n(t, \omega) \, dW(t, \omega)\right)^2\right] = \int_0^T E\left[\varphi_n^2(t, \omega)\right] dt,$$

we will arrive at Equation 1.24. Hence, Ito isometry holds for continuous functions as well. □

For a general function in $L^2(\Omega, \mathcal{F}, \mathbb{P})$, definition (Equation 1.22) for stochastic integrals is no longer valid. Nonetheless, we can approximate a general function of $L^2(\Omega, \mathcal{F}, \mathbb{P})$ by a sequence of bounded continuous functions in the sense that

$$\int_0^T E\left[(f(t) - f_n(t))^2\right] dt \to 0 \quad \text{as } n \to \infty, \tag{1.27}$$

and thus we define the stochastic integral or Ito's integral for $f(t)$ as the limit

$$\int_0^T f(t) \, dW(t) \triangleq \lim_{n \to \infty} \int_0^T f_n(t) \, dW(t). \tag{1.28}$$

Furthermore, by Ito's isometry for continuous functions, Equation 1.24, we assert that $\int_0^T f_n(t) \, dW(t)$ is a Cauchy sequence in $L^2(\Omega, \mathcal{F}, \mathbb{P})$, and thus its limit exists. The details of the justification are found in Øksendal (1992).

We finish this section by presenting additional properties of Ito's integrals. The proofs are straightforward and thus omitted for brevity.

Properties of Stochastic Integrals:
For an $\mathcal{F}_t$-adaptive function, $f \in L^2(\Omega, \mathcal{F}, \mathbb{P})$,

1. $E\left[\int_t^T f(s) \, dW(s) \Big| \mathcal{F}_t\right] = 0$, where $E[.|\mathcal{F}_t]$ stands for the expectation conditional on $\mathcal{F}_t$.

2. Continuity: $\int_0^t f(s) \, dW(s)$ is also an $\mathcal{F}_t$-adaptive function and is continuous almost surely.

1.2.1 Evaluation of Stochastic Integrals

We now consider the evaluation of stochastic integrals. Suppose that we know the anti-derivative of a function, $f(t)$, such that

$$\frac{dF(t)}{dt} = f(t). \tag{1.29}$$

Could there be

$$\int_0^t f(W(s))\,dW(s) = F(W(t)) - F(W(0))? \tag{1.30}$$

The answer to this question is no. Consider, for example, $f(t) = W(t)$. If Equation 1.30 was correct, then there would be

$$\int_0^t W(s)\,dW(s) = \frac{1}{2}\left[W^2(t) - W^2(0)\right] = \frac{1}{2}W^2(t). \tag{1.31}$$

Taking expectations on both sides and applying the first property of the stochastic integrals, we would obtain

$$0 = E\left[\int_0^t W(s)\,dW(s)\right] = E\left[\frac{1}{2}W^2(t)\right] = \frac{1}{2}t, \tag{1.32}$$

which is a contradiction. This result suggests that general rules in deterministic calculus is not applicable to stochastic integrals.

As a showcase of integral evaluation, we try to work out the integral of $f(t) = W(t)$ according to its definition. Let $t_j = jt/n$ and denote W_j for $W(t_j)$, $j = 0, \ldots, n$. Start from the partial sum as follows:

$$
\begin{aligned}
S_n &= \sum_{j=0}^{n-1} W(t_j)\Delta W(t_j) = \sum_{j=0}^{n-1} W_j(W_{j+1} - W_j) \\
&= \sum_{j=0}^{n-1} W_j W_{j+1} - W_j^2 \\
&= \sum_{j=0}^{n-1} -W_{j+1}^2 + 2W_{j+1}W_j - W_j^2 + W_{j+1}^2 - W_{j+1}W_j \\
&= \sum_{j=0}^{n-1} -(W_{j+1} - W_j)^2 + W_{j+1}^2 - W_j^2 - W_j\Delta W_j \\
&= -\left[\sum_{j=0}^{n-1} (\Delta W_j)^2\right] + W_n^2 - W_0^2 - S_n.
\end{aligned}
\tag{1.33}
$$

We then have, by Kolmogorov's large number theorem,

$$S_n = \frac{1}{2}\left(W_n^2 - W_0^2\right) - \frac{1}{2}\sum_{j=0}^{n-1}(\Delta W_j)^2$$

$$\xrightarrow{\Delta t \to 0} \frac{1}{2}\left[W^2(t) - W^2(0)\right] - \frac{1}{2}t; \qquad (1.34)$$

that is,

$$\int_0^t W(s)\,dW(s) = \frac{1}{2}\left[W^2(t) - W^2(0)\right] - \frac{1}{2}t. \qquad (1.35)$$

Now, both sides of Equation 1.36 vanish under expectations. Compared with deterministic calculus, there is an additional term, $-t/2$, in Equation 1.35. The above procedure of integral valuation, however, is inefficient and cumbersome for general functions. In the next section, we introduce the theory for efficient evaluation of stochastic integrals.

Similar to deterministic calculus, an integral equation implies a corresponding differential equation. To see that, we differentiate Equation 1.36 with respect to t, obtaining

$$W(t)\,dW(t) = \frac{1}{2}d\left[W^2(t) - t\right], \qquad (1.36)$$

or

$$dW^2(t) = 2W(t)\,dW(t) + dt. \qquad (1.37)$$

Equation 1.38 is the first *stochastic differential equation* (SDE) to appear in this book; it relates the differential of $f(t) = W^2(t)$ to the differential of $W(t)$. Knowing Equation 1.38, we can calculate the stochastic integral $\int_0^t W(s)\,dW(s)$ easily, without going through the procedure from Equations 1.34 through 1.36. In the next section, we study the dynamics of general functions in a broader context.

1.3 Stochastic Differentials and Ito's Lemma

In this section, we study the differentials of functions of other stochastic processes. In stochastic calculus, the so-called Ito's process is most often used as the basic stochastic process.

Definition 1.3.1. *Ito's process is a continuous stochastic process of the form:*

$$X(t) = X_0 + \int_0^t \sigma(s)\,dW(s) + \int_0^t \mu(s)\,ds, \qquad (1.38)$$

where $\sigma(s)$ and $\mu(s)$ are adaptive functions satisfying

$$E\left[\int_0^t \left(\sigma^2(s) + |\mu(s)|\right) ds\right] < \infty, \quad \forall t. \tag{1.39}$$

The corresponding differential of Ito's process is

$$dX(t) = \sigma(t) \, dW(t) + \mu(t) \, dt. \tag{1.40}$$

We call $\sigma(t)$ and $\mu(t)$ the volatility and drift of the SDE, respectively.

We now consider a function of $X(t)$, $Y(t) = F(X(t), t)$. The next lemma describes the SDE satisfied by $Y(t)$.

Lemma 1.3.1 (Ito's Lemma). *Let $X(t)$ be Ito's process with drift $\mu(t)$ and volatility $\sigma(t)$, and let $F(x, t)$ be a smooth function with bounded second-order derivatives. Then $Y(t) = F(X(t), t)$ is also Ito's process with drift*

$$N(t) = \frac{\partial F}{\partial t} + \frac{1}{2}\sigma^2(t)\frac{\partial^2 F}{\partial x^2} + \mu(t)\frac{\partial F}{\partial x}, \tag{1.41}$$

and volatility

$$\Sigma(t) = \sigma(t)\frac{\partial F}{\partial x}. \tag{1.42}$$

Proof: By Taylor's expansion,

$$\begin{aligned}
\Delta Y(t_i) &= F(X(t_i + \Delta t), t_i + \Delta t) - F(X(t_i), t_i) \\
&= F_x \Delta X + F_t \Delta t + \frac{1}{2}F_{xx}(\Delta X)^2 + F_{xt}\Delta X \Delta t + \frac{1}{2}F_{tt}(\Delta t)^2 \\
&\quad + \text{higher order terms.}
\end{aligned} \tag{1.43}$$

Because

$$\Delta W(t) = \sqrt{\Delta t} \cdot \varepsilon, \quad \varepsilon \sim N(0, 1), \tag{1.44}$$

we generally have

$$E\left[|\Delta W|^p \, \Delta t^q\right] \propto \Delta t^{(p/2)+q}. \tag{1.45}$$

Here "$\propto$" means "of the order of." Based on Equation 1.46, we know that the order of magnitude of both the cross term and the higher-order terms in Equation 1.45 is $O(\Delta t^{3/2})$, and thus we can rewrite Equation 1.43 as

$$\begin{aligned}
\Delta Y(t_i) &= F_x \Delta X + F_t \Delta t + \frac{1}{2}F_{xx}\sigma^2(t_i)(\Delta W(t_i))^2 + O(\Delta t^{3/2}) \\
&= F_x \Delta X + F_t \Delta t + \frac{1}{2}F_{xx}\sigma^2(t_i)\Delta t \\
&\quad + \frac{1}{2}F_{xx}\sigma^2(t_i)\left(\Delta W^2(t_i) - \Delta t\right) + O(\Delta t^{3/2}).
\end{aligned} \tag{1.46}$$

Now, we sum up the increment for $i = 0, 1, \ldots, n-1$, obtaining

$$Y(t) - Y(0) = \sum_{i=0}^{n-1} F_x(X_i, t_i) \Delta X_i + F_t(X_i, t_i) \Delta t_i + \frac{1}{2} F_{xx}(X_i, t_i) \sigma^2(t_i) \Delta t_i$$

$$+ \frac{1}{2} \sum_{i=0}^{n-1} F_{xx}(X_i, t_i) \sigma^2(t_i)[(\Delta W_i)^2 - \Delta t_i] + O(\Delta t^{\frac{1}{2}}), \quad (1.47)$$

where $X_i = X(t_i)$ and $\Delta X_i = X(t_{i+1}) - X(t_i)$. Next, we will show that the fourth term on the right-hand side converges to zero in $L^2(\Omega, \mathcal{F}, \mathbb{P})$ as the partition of time shrinks to zero. Clearly, the mean of the fourth term is

$$E\left[\sum_{i=0}^{n-1} F_{xx}(X_i, t_i) \sigma^2(t_i)(\Delta W_i^2 - \Delta t_i)\right]$$

$$= \sum_{i=0}^{n-1} E\left[F_{xx}(X_i, t_i) \sigma^2(t_i)\right] \left[E\left(\Delta W_i^2\right) - \Delta t_i\right] = 0. \quad (1.48)$$

Here, we have used the fact that $F(X_t, t)$ and $\sigma(t)$ are both $\mathcal{F}_t$-adaptive functions. What is left is to show that the variance of the term converges to zero. For notational simplicity, we denote

$$a_i = F_{xx}(X_i, t_i) \sigma^2(t_i). \quad (1.49)$$

We then have

$$\text{Var}\left[\sum_{i=0}^{n-1} a_i \left((\Delta W_i)^2 - \Delta t_i\right)\right]$$

$$= E\left[\left(\sum_{i=0}^{n-1} a_i \left((\Delta W_i)^2 - \Delta t_i\right)\right)^2\right]$$

$$= E\left[\sum_{i=0}^{n-1} a_i a_j \left((\Delta W_i)^2 - \Delta t_i\right)\left((\Delta W_j)^2 - \Delta t_j\right)\right]$$

$$= E\left[\sum_{i=0}^{n-1} a_i^2 \left((\Delta W_i)^2 - \Delta t_i\right)^2\right]$$

$$= E\left[\sum_{i=0}^{n-1} a_i^2 \left((\Delta W_i)^4 - 2\Delta t_i(\Delta W_i)^2 + (\Delta t_i)^2\right)\right]$$

$$= \sum_{i=0}^{n} E\left[a_i^2\right] \cdot \left\{E\left[(\Delta W_i)^4\right] - \Delta t_i^2\right\}$$

$$= \sum_{i=0}^{n} E\left[a_i^2\right] \cdot \left(3(\Delta t_i)^2 - \Delta t_i^2\right)$$

$$\leq 2\left(\max_j \Delta t_j\right) \cdot \sum_{i=0}^{n-1} E\left[a_i^2\right] \Delta t_i$$

$$\approx 2\left(\max_j \Delta t_j\right) \cdot \int_0^T E[a^2]\, dt \to 0 \quad \text{as} \quad \max_j \Delta t_j \to 0. \quad (1.50)$$

Based on Equations 1.48 and 1.50, we conclude that the fourth term of Equation 1.48 vanishes in $L^2(\Omega, \mathcal{F}, \mathbb{P})$. Hence, as $\Delta t = \max_j \Delta t_j \to 0$, Equation 1.47 becomes

$$Y(t) - Y(0) = \int_0^t F_x \mathrm{d}X(s) + F_t \mathrm{d}s + \frac{1}{2} F_{xx} \sigma^2(s) \mathrm{d}s. \tag{1.51}$$

In differential form, the above equation becomes

$$\mathrm{d}Y(t) = F_x \mathrm{d}X(t) + F_t \mathrm{d}t + \frac{1}{2} F_{xx} \sigma^2(t) \mathrm{d}t$$

$$= \left(F_t + \mu(t) F_x + \frac{1}{2} \sigma^2(t) F_{xx} \right) \mathrm{d}t + \sigma(t) F_x \mathrm{d}W(t). \tag{1.52}$$

This finishes the proof. $\square$

Next, we study the application of Ito's lemma with two examples.

Example 1.2 Consider again the differential of the function $f(t) = W_t^2$. We have

$$\frac{\partial f}{\partial t} = 0, \qquad \frac{\partial f}{\partial W} = 2W_t, \quad \text{and} \quad \frac{\partial^2 f}{\partial W^2} = 2. \tag{1.53}$$

According to Ito's lemma, we have

$$\mathrm{d}f = 2W_t \mathrm{d}W_t + \mathrm{d}t, \tag{1.54}$$

which reproduces Equation 1.38.

Example 1.3 (A lognormal process). A lognormal model is perhaps the most authoritative model for asset prices in financial studies. This model is based on the assumption that the return on an asset over a fixed horizon obeys a normal distribution. Let $S(t)$ denote the price of an asset at time t. Then, the return over a horizon, $(t, t + \Delta t)$, is defined as

$$\ln \frac{S_{t+\Delta t}}{S_t} = \mu_t \Delta t + \sigma_t \varepsilon \sqrt{\Delta t} \sim N(\mu_t \Delta t, \sigma_t^2 \Delta t). \tag{1.55}$$

Note that the random term is proportional to $\sqrt{\Delta t}$, whereas the drift term is proportional to Δt. Hence, for a small Δt, the random term dominates the deterministic term. This is consistent to the widely held belief that, in an efficient market, asset price movements over a short horizon are pretty much random. Taking the limit $\Delta t \to \mathrm{d}t$, we arrive at the model with a differential equation:

$$\mathrm{d}(\ln S_t) = \mu_t \, \mathrm{d}t + \sigma_t \, \mathrm{d}W_t. \tag{1.56}$$

By integrating the above equation over the interval $(0, t)$, we then obtain

$$\ln S_t = \ln S_0 + \int_0^t \mu_s \, \mathrm{d}s + \sigma_s \, \mathrm{d}W_s, \tag{1.57}$$

or

$$S_t = S_0 \mathrm{e}^{\int_0^t \mu_s \mathrm{d}s + \sigma_s \mathrm{d}W_s}. \tag{1.58}$$

To derive the differential equation for S_t, we let $X_t = \ln S_t/S_0$ and write $S_t = S_0 e^{X_t}$. Obviously, there are

$$\frac{\partial S_t}{\partial t} = 0, \qquad \frac{\partial S_t}{\partial X} = \frac{\partial^2 S_t}{\partial X^2} = S_0 e^{X_t} = S_t. \tag{1.59}$$

According to Ito's lemma,

$$dS_t = \left(\mu_t + \frac{1}{2}\sigma_t^2\right) S_t\, dt + \sigma S_t\, dW_t. \tag{1.60}$$

This is the very popular lognormal process or geometric Brownian motion for asset price modeling.

Remark 1.3.1. *In formalism, Equation 1.50 may imply*

$$\int_0^t a(s)\,(dW_s)^2 = \int_0^t a(s)\,ds \tag{1.61}$$

for any $\mathcal{F}_t$-adaptive square-integrable function, $a(s)$. Based on Equation 1.62, we define the following operational rules:

$$(dW_t)^2 = dt \tag{1.62}$$

and

$$dW_t\, dt = 0 \quad and \quad dt\, dt = 0 \tag{1.63}$$

for the Brownian motion. Note that Equation 1.62 is obviously at odds with $dW_t = \varepsilon\sqrt{\Delta t}$, $\varepsilon \sim N(0,1)$, so it has to be interpreted according to Equation 1.61.

The above operation rules offer a great deal of convenience to stochastic differentiations. Taking the derivation of Ito's lemma for example, we may now proceed as

$$dY(t) = \frac{\partial F}{\partial t}dt + \frac{\partial F}{\partial x}dX(t) + \frac{1}{2}\frac{\partial^2 F}{\partial x^2}\,(dX(t))^2$$

$$= \frac{\partial F}{\partial t}\,dt + \frac{\partial F}{\partial x}\,(\sigma(t)\,dW_t + \mu(t)\,dt) + \frac{1}{2}\frac{\partial^2 F}{\partial x^2}\sigma^2(t)dt \tag{1.64}$$

$$= \left(\frac{\partial F}{\partial t} + \mu(t)\frac{\partial F}{\partial x} + \frac{1}{2}\sigma^2(t)\frac{\partial^2 F}{\partial x^2}\right)dt + \sigma(t)\frac{\partial F}{\partial x}dW_t.$$

Equation 1.64 carries an important insight for stochastic differentiation: when calculating the differential of a general function, we must retain the second-order terms in stochastic variables.

1.4   Multi-Factor Extensions

The need to extend Ito process and Ito's lemma to multiple dimensions stems from the fact that the values of both a single asset and a portfolio of assets are affected by multiple sources of risks. This is obvious to see for a portfolio. For an asset, we sometimes need to distinguish idiosyncratic risk from *systematic* risk. Idiosyncratic risk is asset-specific, whereas systematic risk is market-wide risk, or risk of a macroeconomic nature.

1.4.1   Multi-Factor Ito's Process

A multiple-factor Ito's process takes the form

$$dX_i(t) = \mu_i(t)\, dt + \sum_{j=1}^{n} \sigma_{ij}(t)\, dW_j(t)$$
$$= \mu_i(t)\, dt + \boldsymbol{\sigma}_i^{\mathrm{T}}(t)\, d\mathbf{W}_t,$$

where

$$\mathbf{W}(t) = \begin{pmatrix} W_1(t) \\ W_2(t) \\ \vdots \\ W_n(t) \end{pmatrix}$$

is a vector of independent Brownian motion, and

$$\boldsymbol{\sigma}_i(t) = \begin{pmatrix} \sigma_{i1}(t) \\ \sigma_{i2}(t) \\ \vdots \\ \sigma_{in}(t) \end{pmatrix}$$

is called the volatility vector. Let $\mathbf{X}(t) = (X_1(t), X_2(t), \ldots, X_n(t))^{\mathrm{T}}$. In integral form, the multi-factor Ito's process is

$$\mathbf{X}_t = \mathbf{X}_0 + \int_0^t \boldsymbol{\mu}_s\, ds + \int_0^t \boldsymbol{\Sigma}(s)\, d\mathbf{W}_s,$$

where

$$\boldsymbol{\mu}_t = \begin{pmatrix} \mu_1(t) \\ \mu_2(t) \\ \vdots \\ \mu_n(t) \end{pmatrix}$$

is the vector of drifts, and

$$\mathbf{\Sigma}(t) = \begin{pmatrix} \boldsymbol{\sigma}_1^{\mathrm{T}}(t) \\ \boldsymbol{\sigma}_2^{\mathrm{T}}(t) \\ \vdots \\ \boldsymbol{\sigma}_n^{\mathrm{T}}(t) \end{pmatrix}$$

is the volatility matrix. Note that both $\boldsymbol{\mu}(t)$ and $\boldsymbol{\sigma}_i(t)$, $i = 1, \ldots, n$ are $\mathcal{F}_t$-adaptive processes, and they satisfy

$$E\left[\int_0^t \left(\sum_{j=1}^n \|\boldsymbol{\sigma}_j(s)\|^2 + |\boldsymbol{\mu}_s|_1 \right) \mathrm{d}s\right] < \infty, \quad \forall t.$$

Namely, $\boldsymbol{\sigma}_j(s), j = 1, \ldots, n$ are square integrable and $\boldsymbol{\mu}_s$ has bounded variation.

1.4.2   Ito's Lemma

The one-factor Ito's lemma can be generalized directly to a multi-factor situation. A proof parallel to that of Lemma 1.2 can be assembled by using vector notations.

Lemma 1.4.1 (Ito's Lemma). *Let f be a deterministic twice continuous differentiable function, and let $\mathbf{X}_t$ be an n-factor Ito's process. Then $Y_t = f(\mathbf{X}_t, t)$ is also an n-factor Ito's process, and it satisfies*

$$dY_t = \left(\frac{\partial f}{\partial t} + \sum_{j=1}^n \mu_j(t)\frac{\partial f}{\partial X_j} + \frac{1}{2}\sum_{i,j=1}^n \boldsymbol{\sigma}_i^{\mathrm{T}}(t)\boldsymbol{\sigma}_j(t)\frac{\partial^2 f}{\partial X_i \partial X_j}\right) dt$$

$$+ \sum_{j=1}^n \frac{\partial f}{\partial X_j}\boldsymbol{\sigma}_j^{\mathrm{T}}(t)\, d\mathbf{W}_t.$$

1.4.3   Correlated Brownian Motions

Let $W(t)$ and $\tilde{W}(t)$ be two Brownian motions under the probability space $(\Omega, \mathcal{F}, \mathbb{P})$. We say that $W(t)$ and $\tilde{W}(t)$ are correlated if

$$\mathrm{Cov}\left[\Delta W(t), \Delta \tilde{W}(t)\right] = \rho \Delta t, \quad \rho \neq 0. \tag{1.65}$$

Equivalently, we can write

$$\begin{aligned} W(t) &= W_1(t), \\ \tilde{W}(t) &= \rho W_1(t) + \sqrt{1 - \rho^2}W_2(t), \end{aligned} \tag{1.66}$$

where $W_1(t)$ and $W_2(t)$ are independent Brownian motions. We have the following additional operation rule for correlated Brownian motions:

$$dW(t)\, d\tilde{W}(t) = \rho\, dt. \tag{1.67}$$

With the help of the above operation rule, we can derive the processes of the product and quotient of two Ito's processes. The following results are very useful for financial modeling, and for this reason we call them the *product rule* and the *quotient rule*, respectively.

Product rule: Let $X(t)$ and $Y(t)$ be two Ito's processes such that

$$
\begin{aligned}
dX(t) &= \sigma_X(t)\, dW(t) + u_X(t)\, dt,\\
dY(t) &= \sigma_Y(t)\, d\tilde{W}(t) + u_Y(t)\, dt,
\end{aligned}
\tag{1.68}
$$

where $dW(t)\, d\tilde{W}(t) = \rho\, dt$. Then,

$$
\begin{aligned}
d\left(X(t)Y(t)\right) &= X(t)\, dY(t) + Y(t)\, dX(t) + dX(t)\, dY(t)\\
&= X(t)\, dY(t) + Y(t)\, dX(t) + \sigma_X(t)\sigma_Y(t)\rho\, dt.
\end{aligned}
\tag{1.69}
$$

Quotient rule: Let $X(t)$ and $Y(t)$ be two Ito's processes. Then,

$$d\left(\frac{X(t)}{Y(t)}\right) = \frac{dX(t)}{Y(t)} - \frac{X(t)\, dY(t)}{(Y(t))^2} - \frac{dX(t)\, dY(t)}{(Y(t))^2} + \frac{X(t)\, (dY(t))^2}{(Y(t))^3}. \tag{1.70}$$

The proofs for both rules are left as exercises.

1.4.4 The Multi-Factor Lognormal Model

As an important area of application for the multi-factor Ito's lemma, we now introduce the classic model of a financial market with multiple assets. This financial market consists of a money market account (also called a savings account), B_t, and n risky assets, $\left\{S_t^i\right\}_{i=1}^n$. The price evolutions of these $n+1$ assets are governed by the following equations:

$$
\begin{aligned}
dB_t &= r_t B_t dt\\
dS_t^i &= S_t^i(\mu_t^i dt + \boldsymbol{\sigma}_i^{\mathrm{T}}(s)\, d\mathbf{W}_t), \quad i = 1, 2, \ldots, n.
\end{aligned}
$$

Here r_t is the risk-free interest rate, μ_t^i and $\boldsymbol{\sigma}_i(s)$ the rate of return and volatility of the ith asset, and

$$
\boldsymbol{\sigma}_i(t) = \begin{pmatrix} \sigma_{i1}(t)\\ \sigma_{i2}(t)\\ \vdots\\ \sigma_{in}(t) \end{pmatrix} \quad \text{and} \quad \mathbf{W}_t = \begin{pmatrix} W_1(t)\\ W_2(t)\\ \vdots\\ W_n(t) \end{pmatrix}.
$$

Driving the market are n independent Brownian motions. We therefore call the above model an n-factor model. Note that the savings account is considered a riskless asset so that it is not driven by any Brownian motion.

By the multi-factor Ito's lemma, we can derive the equations for the log of asset prices:

$$\mathrm{d}\ln S_t^i = \left(\mu_t^i - \frac{1}{2}\|\boldsymbol{\sigma}_i(t)\|^2\right)\mathrm{d}t + \boldsymbol{\sigma}_i^{\mathrm{T}}(t)\,\mathrm{d}\mathbf{W}_t.$$

The above equation readily allows us to solve for the asset price:

$$S_t^i = S_0^i \exp\left(\int_0^t \boldsymbol{\sigma}_i^{\mathrm{T}}(s)\,\mathrm{d}\mathbf{W}_s + \left(\mu_s^i - \frac{1}{2}\|\boldsymbol{\sigma}_i(s)\|^2\right)\mathrm{d}s\right),$$

for $i = 1, 2, \ldots, n$. The value of the money market account, meanwhile, is simply

$$B_t = \exp\left(\int_0^t r_s\,\mathrm{d}s\right).$$

1.5 Martingales

We finish this chapter with the introduction of martingales, which is a key concept in derivatives modeling. The definition is given below.

Definition 1.5.1. *A stochastic process, M_t, is called a $\mathbb{P}$-martingale if and only if it has the following properties:*

1. *$E^{\mathbb{P}}\left[|M_t|\right] < \infty, \quad \forall t.$*

2. *$E^{\mathbb{P}}\left[M_t|\mathcal{F}_s\right] = M_s, \quad \forall s \leq t.$*

The martingale properties are associated with fair games in investments or speculations. Let us think of $M_t - M_s$ as the profit or loss (P&L) of a gamble between two parties over the time period (s, t). Then the game is considered fair if the expected P&L is zero. Daily life examples of fair games include the coin tossing game and futures investments in financial markets. In mathematics, there are plenty of examples as well. In fact, we have already seen several of them so far, of which we remind readers below.

Example 1.4

1. The simple random walk, X_n, is a martingale because $E[|X_n|] < n\sqrt{\Delta t}$ and $E[X_n \mid \mathcal{F}_m] = X_m$, $m \leq n$.

2. A $\mathbb{P}$-Brownian motion, W_t, is a martingale by definition.

3. The stochastic integral $X_t = \int_0^t f(u)\,\mathrm{d}W_u$ is a martingale, since

$$E^{\mathbb{P}}\left[X_t \,|\, \mathcal{F}_s\right] = E^{\mathbb{P}}\left[\int_0^s + \int_s^t f(u)\,\mathrm{d}W_u \,\middle|\, \mathcal{F}_s\right]$$
$$= \int_0^s f(u)\,\mathrm{d}W_u = X_s, \quad \forall s \leq t. \tag{1.71}$$

Here, we have applied the first property of stochastic integrals (see page 11).

4. The process $M_t = \exp\left(\int_0^t \sigma_s\mathrm{d}W_s - \dfrac{1}{2}\sigma_s^2\mathrm{d}s\right)$ is an exponential martingale. In fact, using the Ito's lemma, we can show that

$$\mathrm{d}M_t = \sigma_t M_t\,\mathrm{d}W_t, \tag{1.72}$$

which is an Ito's process without drift. It follows that

$$M_t = M_s + \int_s^t M_u \sigma_u\,\mathrm{d}W_u. \tag{1.73}$$

Based on the conclusion of the last example, we know that M_t is a martingale.

We emphasize here that an Ito's process is a martingale process if and only if its drift term is zero. Finally, we present two additional examples.

5. $M_t = W_t^2 - t$ is a martingale. Here is the verification: for $s \leq t$,

$$E^{\mathbb{P}}\left[W_t^2 - t \,\middle|\, \mathcal{F}_s\right] = E^{\mathbb{P}}\left[(W_t - W_s + W_s)^2 - t \,\middle|\, \mathcal{F}_s\right]$$
$$= E^{\mathbb{P}}\left[(W_t - W_s)^2 + 2W_s(W_t - W_s)\right.$$
$$\left. + W_s^2 - t \,\middle|\, \mathcal{F}_s\right]$$
$$= (t - s) + 0 + W_s^2 - t = W_s^2 - s. \tag{1.74}$$

6. Let X_T be a contingent claim depending on information up to time T. Define

$$N_t = E^{\mathbb{P}}[X_T \,|\, F_t].$$

Then N_t is a $\mathbb{P}$-martingale. In fact, for any $s \leq t$, we have

$$E^{\mathbb{P}}[N_t \,|\, \mathcal{F}_s] = E^{\mathbb{P}}\left[E^{\mathbb{P}}[X_T|\mathcal{F}_t] \,\middle|\, \mathcal{F}_s\right]$$
$$= E^{\mathbb{P}}[X_T \,|\, \mathcal{F}_s]$$
$$= N_s. \tag{1.75}$$

Equation (9.77) demonstrates the so-called *tower law*, namely, an expectation conditioned first on history up to time t and then on history up to an earlier time, s, is the same as that conditioned on history up to the earlier time, s.

Exercises

1. Solve subproblem (a) with the help of Excel® or MATLAB®.

   (a) Draw N standard normal random numbers, $X_i, i = 1, \ldots, N$ and verify that

   $$\frac{1}{N} \sum_{i=1}^{N} X_i^2 \to 1 \quad \text{as } N \to \infty,$$

   by increasing N from 10 to 1000. Plot the above average value versus N.

   (b) Let $W(t)$ be a Brownian motion. For fixed t, prove that

   $$\sum_{i=1}^{N} (\Delta W(i\Delta t))^2 \to t \quad \text{a.s.}$$

   Here $\Delta t = t/N$.

2. Prove directly from the definition of Ito's integrals that

   $$\int_0^t s \, dW(s) = tW(t) - \int_0^t W(s) \, ds.$$

   $\left(\text{Hint: } \sum_j \Delta(s_j W_j) = \sum_j s_j \Delta W_j + \sum_j W_j \Delta s_j.\right)$

3. Justify the following equality with and without using Ito's lemma.

   $$\int_0^t W^2(s) \, dW(s) = \frac{1}{3} W^3(t) - \int_0^t W(s) \, ds.$$

4. Let $f(t)$ be a continuous and adaptive function. Prove that

   $$E\left[\int_0^T f(t) \, dW(t)\right] = 0,$$

   $$E\left[\left(\int_0^T f(t) \, dW(t)\right)^2\right] = E\left[\int_0^T |f(t)|^2 \, dt\right].$$

5. Let X_t and Y_t be Ito's processes and $f(x, y, t)$ be a twice continuous differentiable function of x, y and t. Use a two-dimensional Ito's lemma of the form

$$df = \frac{\partial f}{\partial t} dt + \frac{\partial f}{\partial x} dX_t + \frac{\partial f}{\partial y} dY_t$$
$$+ \frac{1}{2} \left(\frac{\partial^2 f}{\partial x^2} (dX_t)^2 + 2 \frac{\partial^2 f}{\partial x \partial y} (dX_t dY_t) + \frac{\partial^2 f}{\partial y^2} (dY_t)^2 \right)$$

to prove the *Quotient* rule:

$$d\left(\frac{X_t}{Y_t} \right) = \frac{dX_t}{Y_t} - \frac{X_t dY_t}{Y_t^2} - \frac{dX_t dY_t}{Y_t^2} + \frac{X_t (dY_t)^2}{Y_t^3}.$$

Chapter 2

The Martingale Representation Theorem

In this chapter, we introduce the martingale approach to derivatives pricing. This approach consists of two major steps: the derivation of the martingale probability measure and the construction of the replication strategy. The mathematical foundation of this approach is formed by two theorems, namely, the Cameron–Martin–Girsanov (CMG) theorem and the martingale representation theorem. The derivation of the martingale probability measure is achieved by using the CMG theorem, while the construction of the replication strategy is based on the martingale representation theorem. A significant portion of this chapter is devoted to establishing these two theorems. We motivate our discussion with a simple binomial model for tradable assets and then proceed to establish the two theorems in continuous time for a complete market with multiple underlying securities. Once the pricing formula for general options has been established, we price call options as an important example and derive the famous Black–Scholes formulae. Until up to Chapter 9, we limit ourselves to a discussion of a complete market for which every source of risk can be traded. The proof of the martingale representation theorem is provided in this chapter's Appendix. At the end of the chapter, we give some references on derivative pricing in incomplete markets.

2.1 Changing Measures with Binomial Models

2.1.1 A Motivating Example

Consider the simplest option-pricing model with an underlying asset following a one-period binomial process, as depicted in Figure 2.1. In Figure 2.1, $0 \leq p \leq 1$ and $\bar{p} = 1 - p$. The option's payoffs at time 1, $f(S_u)$ and $f(S_d)$, are given explicitly, and we want to determine $f(S)$, the value of the option at time 0. Without loss of generality, we assume that there is a zero interest rate in the model. To avoid arbitrage, we must impose the order $S_d \leq S \leq S_u$. We call $\mathbb{P} = \{p, \bar{p}\}$ the objective measure of the underlying process.

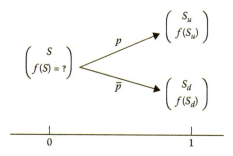

FIGURE 2.1: A binomial model for option pricing.

It may be tempted to price the option by expectation under $\mathbb{P}$:

$$f(S) = E^{\mathbb{P}}[f(S_1)]$$
$$= p\,f(S_u) + \bar{p}\,f(S_d). \tag{2.1}$$

However, except for a special p, the above price generates arbitrage and thus is wrong. To see that, we replicate the payoff of the option at time 1 using a portfolio of the underlying asset and a cash bond, with respective numbers of units, α and β, such that, at time 1,

$$\alpha S_u + \beta = f(S_u),$$
$$\alpha S_d + \beta = f(S_d). \tag{2.2}$$

Solving for α and β, we obtain

$$\alpha = \frac{f(S_u) - f(S_d)}{S_u - S_d},$$
$$\beta = \frac{S_u f(S_d) - S_d f(S_u)}{S_u - S_d}. \tag{2.3}$$

Equation 2.2 implies that the time-1 values of the portfolio and option are identical. To avoid arbitrage, their values at time 0 must be identical as well[1] which yields the arbitrage price of the option at time 0:

$$f(S) = \alpha S + \beta$$
$$= q\,f(S_u) + \bar{q}\,f(S_d)$$
$$= E^{\mathbb{Q}}[f(S_1)], \tag{2.4}$$

where $\mathbb{Q} = \{q, \bar{q}\}$, and

$$q = \frac{S - S_d}{S_u - S_d}, \quad \bar{q} = 1 - q \tag{2.5}$$

---

[1]Such an argument is also called the dominance principle.

is a different set of probabilities. Note that Equation 2.4 gives the no-arbitrage price of the option. Any other price will induce arbitrage to the market. Hence, the expectation price, in Equation 2.1, is correct only if $p = q$. In fact, $\{q, \bar{q}\}$ is the only set of probabilities that satisfies

$$S = qS_u + \bar{q}S_d = E^{\mathbb{Q}}(S_1). \tag{2.6}$$

The price formulae, Equations 2.4 and 2.6, have rather general implications. First, the price of the option can be expressed as an expectation of the payoff under a special probability distribution. Second, this special probability distribution is nothing else but the "martingale measure" for the underlying asset. As a result, the original objective measure plays no role in derivatives pricing and it needs to be changed into the "martingale measure" for such a purpose. If we introduce a stochastic variable, ζ, such that

$$\zeta = \begin{cases} \zeta_u = \dfrac{q}{p}, & \text{if } S_1 = S_u, \\[2ex] \zeta_d = \dfrac{\bar{q}}{\bar{p}}, & \text{if } S_1 = S_d, \end{cases} \tag{2.7}$$

we then can rewrite the price formula as

$$\begin{aligned} E^{\mathbb{Q}}[\,f(S_1)] &= p\zeta_u f(S_u) + \bar{p}\zeta_d f(S_d) \\ &= E^{\mathbb{P}}[\zeta f(S_1)]\,. \end{aligned} \tag{2.8}$$

In the finance literature, ζ is called a pricing kernel. Hence, finding the pricing measure is equivalent to finding the pricing kernel. In the next section, we elaborate the pricing kernel with a multi-period binomial tree.

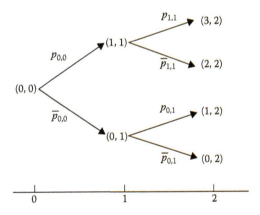

FIGURE 2.2: A two-period binomial tree.

2.1.2 Binomial Trees and Path Probabilities

Let us move one step forward and consider the binomial tree model up to two time steps, as shown in Figure 2.2, where each pair of numbers represents a state (which can be associated with the price of an asset if necessary). Out of each state at time j, two possible states are generated at time $j + 1$. Hence, we have 2^j states at time j, starting with a single state at time 0. The branching probabilities for reaching the next two states from one state, (i, j), are $p_{i,j} \in [0, 1]$ and $\bar{p}_{i,j} = 1 - p_{i,j}$, respectively. The collection of branching probabilities, $\mathbb{P} = \{p_{i,j}, \bar{p}_{i,j}\}$, is again called a measure. As is shown in Figure 2.2, there are two paths over the time horizon from 0 to 1, whereas there are four paths over the time horizon from 0 to 2. The corresponding path probabilities for the horizon from 0 to 1 are

$$\pi_{0,1} = \bar{p}_{0,0} \quad \text{and} \quad \pi_{1,1} = p_{0,0}, \tag{2.9}$$

whereas for the horizon from 0 to 2, they are

$$\pi_{0,2} = \bar{p}_{0,0}\bar{p}_{0,1}, \pi_{1,2} = \bar{p}_{0,0}p_{0,1}, \pi_{2,2} = p_{0,0}\bar{p}_{1,1}, \text{and } \pi_{3,2} = p_{0,0}p_{1,1}. \tag{2.10}$$

The path probabilities can also be marked in a binomial tree as is shown in Figure 2.3.

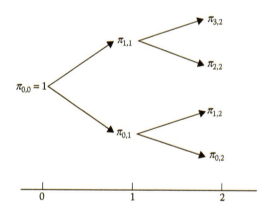

FIGURE 2.3: A path probability tree under $\mathbb{P}$.

Consider now another set of branching probabilities, $\mathbb{Q} = \{q_{i,j}, \bar{q}_{i,j} = 1 - q_{i,j}\}$, for the same tree. The corresponding path probabilities are

$$\pi'_{0,1} = \bar{q}_{0,0} \quad \text{and} \quad \pi'_{1,1} = q_{0,0} \tag{2.11}$$

up to time 1, and

$$\pi'_{0,2} = \bar{q}_{0,0}\bar{q}_{0,1}, \pi'_{1,2} = \bar{q}_{0,0}q_{0,1}, \pi'_{2,2} = q_{0,0}\bar{q}_{1,1}, \text{ and } \pi'_{3,2} = q_{0,0}q_{1,1} \tag{2.12}$$

up to time 2. Suppose that the $\mathbb{P}$-probability of paths $\pi_{i,j} \neq 0$ for all i,j. We then can define the ratio of path probabilities as follows:

$$\zeta_{i,j} = \frac{\pi'_{i,j}}{\pi_{i,j}}. \tag{2.13}$$

These ratios can then be equipped with the original measure, $\mathbb{P}$, and thus be treated as possible values of a random process, ζ, as is shown in Figure 2.4.

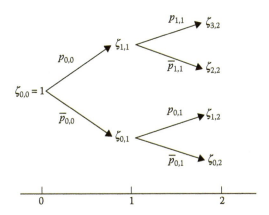

FIGURE 2.4: A Radon–Nikodym random process for the two-period tree.

The purpose of introducing ζ is to allow a change of measures. Consider pricing a contingent claim with maturity j that pays $x_{i,j}$ if the (i,j) state is realized. For $j = 1$, the expectation price of the claim under $\mathbb{Q}$ is

$$E^{\mathbb{Q}}[X_1 \mid \mathcal{F}_0] = \sum_{i=0}^{1} \pi'_{i,1} x_{i,1} = \sum_{i=0}^{1} \pi_{i,1}\left(\frac{\pi'_{i,1}}{\pi_{i,1}}\right)x_{i,1} = E^{\mathbb{P}}[\zeta_1 X_1 \mid \mathcal{F}_0]. \tag{2.14}$$

For $j = 2$, the expectation price is

$$E^{\mathbb{Q}}[X_2 \mid \mathcal{F}_0] = \sum_{i=0}^{3} \pi'_{i,2} x_{i,2} = \sum_{i=0}^{3} \pi_{i,2}\left(\frac{\pi'_{i,2}}{\pi_{i,2}}\right)x_{i,2} = E^{\mathbb{P}}[\zeta_2 X_2 \mid \mathcal{F}_0]. \tag{2.15}$$

The role of ζ is self-explanatory in Equations 2.14 and 2.15: if $\mathbb{Q}$ is the pricing measure, then ζ is the pricing kernel. In the mathematical literature, the kernel is often called a Radon–Nikodym derivative between the two measures, and it is denoted as

$$\zeta \triangleq \frac{d\mathbb{Q}}{d\mathbb{P}}. \tag{2.16}$$

An alternative viewpoint is to treat the derivative ζ as a mapping of the filtration, $\mathcal{F}_t$, to a number:

$$\left.\frac{d\mathbb{Q}}{d\mathbb{P}}\right|_{\mathcal{F}_t} = \zeta_t. \tag{2.17}$$

Intuitively, Equations 2.14 and 2.15 can be generalized to binomial processes with multiple time steps.

Next, we consider the change of measures under filtrations larger than $\mathcal{F}_0$. According to the definition of $\zeta_t, t \geq 0$, the ratio of path probabilities over (t, T) is

$$\frac{\zeta_T}{\zeta_t}. \tag{2.18}$$

Conditional on the filtration, $\mathcal{F}_t$, ζ_t is known for certainty and thus we obtain the formula

$$E^{\mathbb{Q}}[X_T \mid \mathcal{F}_t] = E^{\mathbb{P}}\left[\frac{\zeta_T}{\zeta_t} X_T \middle| \mathcal{F}_t \right] = \zeta_t^{-1} E^{\mathbb{P}}[\zeta_T X_T \mid \mathcal{F}_t]. \tag{2.19}$$

Let us take the binomial model in Figure 2.2 as an example of Equation 2.19. Suppose that, at $t = 1$, we have reached the lower state, $(0, 1)$, and we thus have filtration $\mathcal{F}_1 = \{(0, 0), (0, 1)\}$. Then,

$$\begin{aligned}
E^{\mathbb{Q}}[X_2 \mid \mathcal{F}_1] &= \bar{q}_{0,1} x_{0,2} + q_{0,1} x_{1,2} \\
&= \left(\frac{\bar{q}_{0,0}}{\bar{p}_{0,0}} \right)^{-1} \left(\bar{p}_{0,1} \frac{\bar{q}_{0,0} \bar{q}_{0,1}}{\bar{p}_{0,0} \bar{p}_{0,1}} x_{0,2} + p_{0,1} \frac{\bar{q}_{0,0} q_{0,1}}{\bar{p}_{0,0} p_{0,1}} x_{1,2} \right) \\
&= \left(\frac{\pi'_{0,1}}{\pi_{0,1}} \right)^{-1} \left(\bar{p}_{0,1} \frac{\pi'_{0,2}}{\pi_{0,2}} x_{0,2} + p_{0,1} \frac{\pi'_{1,2}}{\pi_{1,2}} x_{1,2} \right) \\
&= \zeta_1^{-1} E^{\mathbb{P}}[\zeta_T X_T \mid \mathcal{F}_1].
\end{aligned} \tag{2.20}$$

We next make a remark on the martingale property of ζ_t. By substituting $X_T = 1$ in Equation 2.19, we obtain

$$1 = \zeta_t^{-1} E^{\mathbb{P}}[\zeta_T \mid \mathcal{F}_t], \quad \forall t \leq T, \tag{2.21}$$

or

$$\zeta_t = E^{\mathbb{P}}[\zeta_T \mid \mathcal{F}_t], \tag{2.22}$$

which indicates that the Radon–Nikodym derivative, ζ_t, is a martingale process under the measure $\mathbb{P}$.

Finally, we discuss the existence of the Radon–Nikodym derivative, ζ_t. In our binomial tree model in Figure 2.2, the Radon–Nikodym derivatives can obviously be defined if $\pi_{i,j} \neq 0$ for all i, j. Even if $\pi_{i,j} = 0$, the derivative can still be defined if we simultaneously have $\pi'_{i,j} = 0$. In fact, with the following definition for the Radon–Nikodym derivative:

$$\zeta_j = \begin{cases} \dfrac{\pi'_{i,j}}{\pi_{i,j}}, & \text{if the } i\text{th path is taken and } \pi_{i,j} \neq 0, \\ 0, & \text{if the } i\text{th path is taken and } \pi_{i,j} = \pi'_{i,j} = 0, \end{cases} \tag{2.23}$$

previous arguments on the change of measures remain valid. In general, a Radon–Nikodym derivative, $d\mathbb{Q}/d\mathbb{P}$, exists as long as $\mathbb{Q}$ is said to be absolutely continuous with respect to $\mathbb{P}$.

Definition 2.1.1. *Measures* $\mathbb{P}$ *and* $\mathbb{Q}$ *operate on the same sample space,* S. *We say that* $\mathbb{Q}$ *is absolutely continuous with respect to* $\mathbb{P}$ *if, for any subset* $A \subseteq S$,

$$\mathbb{P}(A) = 0 \Longrightarrow \mathbb{Q}(A) = 0, \tag{2.24}$$

and it is denoted as $\mathbb{Q} \ll \mathbb{P}$.

For later use, we also introduce the concept of equivalence between two measures.

Definition 2.1.2. *Measures* $\mathbb{P}$ *and* $\mathbb{Q}$ *operate in the same sample space,* S. *If* $\mathbb{Q} \ll \mathbb{P}$ *and* $\mathbb{P} \ll \mathbb{Q}$ *hold simultaneously, we say that the two measures are equivalent and we write* $\mathbb{Q} \approx \mathbb{P}$.
An identical definition for equivalence is

$$\mathbb{P}(A) = 0 \Longleftrightarrow \mathbb{Q}(A) = 0. \tag{2.25}$$

That is, a $\mathbb{P}$-*null set is also a* $\mathbb{Q}$-*null set, and vice versa.*

2.2 Change of Measures under Brownian Filtration

2.2.1 The Radon–Nikodym Derivative of a Brownian Path

Consider a path of $\mathbb{P}$-Brownian motion over $(0, t)$ with discrete time stepping,

$$\{W(0) = 0, W(\Delta t), W(2\Delta t), \dots, W(n\Delta t)\}, \tag{2.26}$$

where $\Delta t = t/n$. With the probability ratio in mind, our immediate question is what the path probability is. The answer, unfortunately, is zero. The implication that we cannot define the notion of the probability ratio given that the same path is realized under two different probability measures. To circumvent this problem, we first seek to calculate the probability for the Brownian motion to travel in a corridor (the so-called corridor probability), as is shown in Figure 2.5, and then we define the ratio of the corridor probabilities. The ratio of the path probabilities is finally defined through a limiting procedure. The corridor can be represented by the intervals $A_i = (x_i - (\Delta x/2), x_i + (\Delta x/2)), i = 1, 2, \dots, n$, where $x_i = W(i\Delta t)$ and $\Delta x > 0$ is a small number.

For a Brownian motion, the marginal distribution at $t_i = i\Delta t$ is known to be

$$f_{\mathbb{P}}(x) = \frac{1}{\sqrt{2\pi\Delta t}} e^{-(1/2)[(x-x_i)^2/\Delta t]} \sim N(x_i, \Delta t).$$

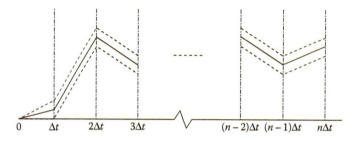

FIGURE 2.5: A corridor for Brownian motion with step size Δt.

Hence, the probability for the next step to fall in A_{i+1} is

$$\text{Prob}_{\mathbb{P}}(A_{i+1}) = \int_{x_{i+1}-\Delta x/2}^{x_{i+1}+\Delta x/2} f_{\mathbb{P}}(x)\mathrm{d}x$$

$$\approx f_{\mathbb{P}}(x_{i+1})\Delta x = \frac{\Delta x}{\sqrt{2\pi\Delta t}}e^{-(1/2)[(x_{i+1}-x_i)^2/\Delta t]}. \qquad (2.27)$$

Approximately, we can define the corridor probability to be

$$\prod_{i=1}^{n}\text{Prob}_{\mathbb{P}}(A_i) = \left(\frac{\Delta x}{\sqrt{2\pi\Delta t}}\right)^n e^{-(1/2\Delta t)\sum_{i=0}^{n-1}(x_{i+1}-x_i)^2}. \qquad (2.28)$$

Next, suppose that the same path is realized under a different marginal probability,

$$f_{\mathbb{Q}}(x) = \frac{1}{\sqrt{2\pi\Delta t}}e^{-(1/2)[(x-x_i+\gamma\Delta t)^2/\Delta t]} \sim N(x_i - \gamma\Delta t, \Delta t), \quad \forall i, \qquad (2.29)$$

where γ is taken to be constant for simplicity. Then the corresponding corridor probability can be similarly obtained to be

$$\prod_{i=1}^{n}\text{Prob}_{\mathbb{Q}}(A_i) = \left(\frac{\Delta x}{\sqrt{2\pi\Delta t}}\right)^n e^{-(1/2\Delta t)\sum_{i=0}^{n-1}(x_{i+1}-x_i+\gamma\Delta t)^2}. \qquad (2.30)$$

It follows that the ratio of the two corridor probabilities is

$$\zeta_t = \exp\left(-\frac{1}{2\Delta t}\sum_{i=0}^{n-1}\left[(x_{i+1}-x_i+\gamma\Delta t)^2 - (x_{i+1}-x_i)^2\right]\right)$$

$$= \exp\left(-\frac{1}{2\Delta t}\sum_{i=0}^{n-1}\left[2(x_{i+1}-x_i)\cdot\gamma\Delta t + \gamma^2\Delta t^2\right]\right)$$

$$= \exp\left(-\gamma\sum_{i=0}^{n-1}(x_{i+1}-x_i) - \frac{1}{2}\gamma^2\Delta t\cdot n\right)$$

$$\qquad (2.31)$$

$$= \exp\left(-\gamma(x_n - x_0) - \frac{1}{2}\gamma^2 t\right)$$

$$= e^{-\gamma W_t - (1/2)\gamma^2 t},$$

which depends neither on the path history nor on Δt, but only on the terminal point, W_t, of the Brownian motion. Hence, if we take the limit, $\Delta t \to 0$, for the corridor probabilities, the result will remain unchanged. Note that ζ_t is an exponential martingale, and can be used to define a new measure, $\mathbb{Q}$, such that

$$\left.\frac{d\mathbb{Q}}{d\mathbb{P}}\right|_{\mathcal{F}_t} = \zeta_t. \tag{2.32}$$

Now let us consider the following question: what is the distribution of the $\mathbb{P}$-Brownian motion, W_t, under the new measure, $\mathbb{Q}$? To answer this question, we evaluate the moment-generating function of W_t under $\mathbb{Q}$:

$$\begin{aligned} E^{\mathbb{Q}}\left[e^{\lambda W_t}\right] &= E^{\mathbb{P}}\left[\zeta_t e^{\lambda W_t}\right] \\ &= E^{\mathbb{P}}\left[e^{-\gamma W_t - (1/2)\gamma^2 t + \lambda W_t}\right] \\ &= E^{\mathbb{P}}\left[e^{(\lambda - \gamma)W_t}\right] e^{-(1/2)\gamma^2 t} \\ &= e^{(1/2)(\lambda - \gamma)^2 t - (1/2)\gamma^2 t} \\ &= e^{(1/2)\lambda^2 t - \lambda \gamma t}. \end{aligned} \tag{2.33}$$

The above result indicates that, under $\mathbb{Q}$,

$$W_t \sim N(-\gamma t, t), \tag{2.34}$$

or, equivalently, we can say that $\tilde{W}_t \triangleq W_t + \gamma t$ is a $\mathbb{Q}$-Brownian motion. A formal statement on the relationship between the measures and Brownian motions is given in the following section.

2.2.2 The CMG Theorem

The following theorem states that a Brownian motion with a drift is in fact a standard Brownian motion under a different measure. The Radon–Nikodym derivative can be expressed in terms of the drift, which can be a time-dependent adaptive function.

Theorem 2.2.1 (The CMG Theorem). *Let W_t be a $\mathbb{P}$-Brownian motion and γ_t be an $\mathcal{F}_t$-adaptive process satisfying the Novikov condition,*

$$E^{\mathbb{P}}\left[\exp\left(\frac{1}{2}\int_0^T \gamma_t^2\, dt\right)\right] < \infty. \tag{2.35}$$

Define a new measure, $\mathbb{Q}$, as

$$\left.\frac{d\mathbb{Q}}{d\mathbb{P}}\right|_{\mathcal{F}_t} = \exp\left(\int_0^t -\gamma_s\, dW_s - \frac{1}{2}\gamma_s^2\, ds\right). \tag{2.36}$$

Then $\mathbb{Q}$ is equivalent to $\mathbb{P}$ and

$$\tilde{W}_t = W_t + \int_0^t \gamma_s \, ds \qquad (2.37)$$

is a $\mathbb{Q}$-Brownian motion.

Proof: The equivalence is because the Radon–Nikodym derivative is positive. To prove the remaining parts of the theorem, we need to show that $\tilde{W}_t$ satisfies the following three properties under $\mathbb{Q}$:

1. $\tilde{W}_0 = 0$;

2. $\tilde{W}_{t+s} - \tilde{W}_t$ is independent of $\{\tilde{W}_u, 0 \le u \le t\}$; and

3. $\tilde{W}_{t+s} - \tilde{W}_t \sim N(0, s)$.

Properties 1 and 2 are easy to show and thus their proofs are omitted. Let us proceed to establishing the third property by checking the moment-generating function of the increment. For any real constant, λ, we have

$$E^{\mathbb{Q}}\left[e^{\lambda(\tilde{W}_{t+s} - \tilde{W}_t)} \Big| \mathcal{F}_t \right]$$

$$= E_t^{\mathbb{P}}\left[\frac{d\mathbb{Q}}{d\mathbb{P}} e^{\lambda(\tilde{W}_{t+s} - \tilde{W}_t)} \right]$$

$$= E_t^{\mathbb{P}}\left[\exp\left(\int_t^{t+s} -\gamma_u dW_u - \frac{1}{2}\gamma_u^2 du + \lambda\left(\tilde{W}_{t+s} - \tilde{W}_t \right) \right) \right]$$

$$= E_t^{\mathbb{P}}\left[\exp\left(\int_t^{t+s} -\gamma_u dW_u - \frac{1}{2}\gamma_u^2 du \right.\right.$$
$$\left.\left. + \lambda\left(W_{t+s} - W_t \right) + \lambda \int_t^{t+s} \gamma_u du \right) \right]$$

$$= e^{(1/2)\lambda^2 s} E_t^{\mathbb{P}}\left[\exp\left(\int_t^{t+s} -(\gamma_u - \lambda) \, dW_u - \frac{1}{2}(\gamma_u - \lambda)^2 \, du \right) \right]$$

$$= e^{(1/2)\lambda^2 s}. \qquad (2.38)$$

Here $E_t^{\mathbb{P}}[\cdot]$ is the short form of $E^{\mathbb{P}}[\cdot|\mathcal{F}_t]$. Based on the property of independent increments of Brownian motion, we can also show that Equation 2.38 holds for any filtration, $\mathcal{F}_u, u \le t$. The theorem is thus proved. □

We remark here that, conventionally, $E_0^{\mathbb{P}}[\cdot]$ or $E^{\mathbb{P}}[\cdot|\mathcal{F}_0]$ is considered an unconditional expectation, and is simply written as $E^{\mathbb{P}}[\cdot]$.

2.3 The Martingale Representation Theorem

The martingale representation theorem plays a critical role in the so-called martingale approach to derivatives pricing. This theorem has two important consequences. First, it leads to a general principle for derivatives pricing. Second, it implies a replication or hedging strategy of a derivative using its underlying security. We first present a simple version of the theorem based on a single Brownian filtration, $\mathcal{F}_t = \sigma(W_s, 0 \leq s \leq t)$. We begin with a martingale process, M_t, such that

$$dM_t = \sigma_t dW_t, \tag{2.39}$$

and we call σ_t the volatility of M_t.

Theorem 2.3.1 (The Martingale Representation Theorem). *Suppose that N_t is a $\mathbb{Q}$-martingale process that is adaptive to $\mathcal{F}_t$ and satisfies $E^{\mathbb{Q}}[N_T^2] < \infty$ for some T. If the volatility of M_t is non-zero almost surely, then there exists a unique $\mathcal{F}_t$-adaptive process, φ_t, such that $E^{\mathbb{Q}}[\int_0^T \varphi_t^2 \sigma_t^2 \, dt] < \infty$ almost surely, and*

$$N_t = N_0 + \int_0^t \varphi_s \, dM_s, \quad t \leq T, \tag{2.40}$$

or, in differential form,

$$dN_t = \varphi_t \, dM_t. \tag{2.41}$$

The proof combining the techniques of Steele (2000) and Øksendal (2003) is provided in the appendix of this chapter. A different proof can be found in Korn and Korn (2000).

2.4 A Complete Market with Two Securities

We consider the first "complete market" in continuous time, which consists of a money market account and a risky security. The price processes for the two securities, B_t and S_t, are assumed to be

$$dB_t = r_t B_t \, dt, \qquad B_0 = 1,$$
$$dS_t = S_t(\mu_t \, dt + \sigma_t \, dW_t), \quad S_0 = S_0.$$

Here, the volatility of the risky asset is $\sigma_t \neq 0$ almost surely, and the short rate, r_t, can be stochastic. Denote the discounted price of the risky asset as $Z_t = B_t^{-1} S_t$, which can be shown to follow the process

$$dZ_t = Z_t \left((\mu_t - r_t)dt + \sigma_t dW_t \right)$$
$$= Z_t \sigma_t d \left(W_t + \int_0^t \frac{(\mu_s - r_s)}{\sigma_s} ds \right). \tag{2.42}$$

By introducing

$$\gamma_t = \frac{\mu_t - r_t}{\sigma_t}, \tag{2.43}$$

which is $\mathcal{F}_t$-adaptive, and by defining a new measure, $\mathbb{Q}$, according to Equation 2.36, we have

$$\tilde{W}_t = W_t + \int_0^t \gamma_s \, ds,$$

which is a $\mathbb{Q}$-Brownian motion. In terms of $\tilde{W}_t$, Z_t satisfies

$$dZ_t = \sigma_t Z_t \, d\tilde{W}_t, \tag{2.44}$$

which is a lognormal $\mathbb{Q}$-martingale. Recall that in the binomial model for option pricing, we also derived the martingale measure for the underlying security.

2.5   Replicating and Pricing of Contingent Claims

Let X_T be a contingent claim (or option) with payoff day or maturity T. The claim is an $\mathcal{F}_T$-adaptive function whose value depends on $\{S_t, 0 \leq t \leq T\}$. Define first a $\mathbb{Q}$-martingale with the discounted payoff:

$$N_t = E^{\mathbb{Q}}(B_T^{-1} X_T \mid \mathcal{F}_t).$$

Without loss of generality, we assume that $E^{\mathbb{Q}}[N_t^2] < \infty$. According to the martingale representation theorem, there exists an $\mathcal{F}_t$-adaptive function, φ_t, such that

$$dN_t = \varphi_t \, dZ_t, \tag{2.45}$$

where Z_t, defined in the last section, is the discounted price of S_t. Next, we define

$$\psi_t = N_t - \varphi_t Z_t. \tag{2.46}$$

Consider now the portfolio with φ_t units of the stock and ψ_t units of the money market account, denoted as (φ_t, ψ_t). According to the definition of ψ_t, the discount value of the replication portfolio is

$$\tilde{V}_t = \varphi_t Z_t + \psi_t = N_t. \tag{2.47}$$

This portfolio has two important properties. First, at time T, when the option matures,

$$\tilde{V}_T = N_T = B_T^{-1} X_T, \tag{2.48}$$

which suggests that the (discounted) value of the portfolio equals that of the option. In other words, the portfolio replicates the payoff of the contingent claim. Second, the replicating portfolio is a *self-financing* one, meaning that it can track the asset allocation, (φ_t, ψ_t), without the need for either capital infusion or capital withdrawal. In fact, based on Equations 2.45 and 2.47, we have

$$d\tilde{V}_t = dN_t = \varphi_t dZ_t. \tag{2.49}$$

In terms of the spot value, B_t and S_t, Equation 2.49 becomes

$$
\begin{aligned}
dV_t &= d(\tilde{V}_t B_t) \\
&= B_t d\tilde{V}_t + \tilde{V}_t dB_t \\
&= B_t \varphi_t dZ_t + (\varphi_t Z_t + \psi_t) dB_t \\
&= \varphi_t (B_t dZ_t + Z_t dB_t) + \psi_t dB_t \\
&= \varphi_t d(B_t Z_t) + \psi_t dB_t \\
&= \varphi_t dS_t + \psi_t dB_t.
\end{aligned} \tag{2.50}
$$

A direct consequence of the above equation is the equality

$$
\begin{aligned}
\varphi_{t+dt} S_{t+dt} + \psi_{t+dt} B_{t+dt} &= \varphi_t S_t + \psi_t B_t + \varphi_t dS_t + \psi_t dB_t \\
&= \varphi_t S_{t+dt} + \psi_t B_{t+dt},
\end{aligned} \tag{2.51}
$$

which says that the values of the portfolio before and after rebalancing at time $t + dt$ are equal, and no cash flow is generated when we update the asset allocation from (φ_t, ψ_t) to $(\varphi_{t+dt}, \psi_{t+dt})$. This is what we mean by a self-financing portfolio. By the dominance principle, the value of the contingent claim is nothing but the value of the replication portfolio:

$$V_t = B_t E^{\mathbb{Q}}\left[B_T^{-1} X_T \,|\, \mathcal{F}_t\right] = E^{\mathbb{Q}}\left[e^{-\int_t^T r_s ds} X_T \,|\, \mathcal{F}_t\right]. \tag{2.52}$$

By already knowing the pricing measure, $\mathbb{Q}$, we have thus established a general pricing principle or pricing formula for contingent claims.

An important implication of the above pricing approach is hedging. A contingent claim can be replicated dynamically by its underlying risky asset and the money market account, or it can be dynamically hedged by the underlying security. The hedge ratio is φ_t, about which we know only its existence and its uniqueness based on the martingale representation theorem. Fortunately, Ito's lemma tells us more about φ_t. We understand that the price of the contingent claim is a function of the underlying security and the short rate. Assume, for simplicity, that the short rate is a deterministic function and the value function, V_t, has continuous second-order partial derivatives. Then, by Ito's lemma, we have the following expression of the price process of the contingent claim:

$$dV_t = \frac{\partial V_t}{\partial S_t} dS_t + \left(\frac{\partial V_t}{\partial t} + \frac{1}{2}\sigma_t^2 S_t^2 \frac{\partial^2 V_t}{\partial S_t^2}\right) dt. \tag{2.53}$$

Through comparison with Equation 2.50, we obtain

$$\varphi_t = \frac{\partial V_t}{\partial S_t}, \tag{2.54}$$

due to the uniqueness of φ_t. Hence, the hedge ratio is just the rate of change of the contingent claim with respect to the underlying asset.

We can also produce interesting results by comparing the drift terms of Equations 2.50 and 2.53, yielding

$$\frac{\partial V_t}{\partial t} + \frac{1}{2}\sigma_t^2 S_t^2 \frac{\partial^2 V_t}{\partial S_t^2} = r_t B_t \psi_t. \tag{2.55}$$

According to Equation 2.47,

$$\begin{aligned} B_t \psi_t &= B_t(\tilde{V}_t - \varphi_t Z_t) \\ &= V_t - \varphi_t S_t \\ &= V_t - S_t \frac{\partial V_t}{\partial S_t}. \end{aligned} \tag{2.56}$$

By substituting the right-hand side of Equation 2.56 into 2.55 and rearranging the terms, we end up with

$$\frac{\partial V_t}{\partial t} + \frac{1}{2}\sigma_t^2 S_t^2 \frac{\partial^2 V_t}{\partial S_t^2} + r_t S_t \frac{\partial V_t}{\partial S_t} - r_t V_t = 0, \tag{2.57}$$

which is the celebrated Black–Scholes–Merton equation (Black and Scholes, 1973; Merton, 1973). To evaluate V_t, we may solve Equation 2.57 with the terminal condition

$$V(S_T, T) = X_T. \tag{2.58}$$

We emphasize here that Equation 2.57 is valid when the short rate is deterministic. Moreover, for standard European call or put options, it is more convenient to calculate the expectation (Equation 2.52) for their values, which will be demonstrated in Section 2.8.

2.6   Multi-Factor Extensions

In derivatives pricing, we often need to model simultaneously the dynamics of multiple risky securities, using multiple risk factors. Because of that, we must extend several major results established so far to the setting of multiple risk sources or assets. These results include the CMG theorem, the martingale representation theorem, and the option pricing formula, as in Equation 2.52. The proofs are parallel to those for the one-dimensional case and thus are omitted for brevity. Hereafter, we use a superscript "T" to denote the transposition of a matrix.

Theorem 2.6.1 (The CMG Theorem). *Let* $\mathbf{W}_t = (W_1(t), W_2(t), \ldots,$

$W_n(t))^{\mathrm{T}}$ *be an n-dimensional* $\mathbb{P}$-*Brownian motion, and let* $\boldsymbol{\gamma}_t = (\gamma_1(t),$ $\gamma_2(t), \ldots, \gamma_n(t))^{\mathrm{T}}$ *be an n-dimensional* $\mathcal{F}_t$-*adaptive process, such that*

$$E^{\mathbb{P}}\left[\exp\left(\frac{1}{2}\int_0^T \|\boldsymbol{\gamma}_t\|_2^2 \, dt\right)\right] < \infty.$$

Define a new measure, $\mathbb{Q}$, *with a Radon–Nikodym derivative*

$$\left.\frac{d\mathbb{Q}}{d\mathbb{P}}\right|_{\mathcal{F}_t} = \exp\left(\int_0^t -\boldsymbol{\gamma}_s^{\mathrm{T}}\, d\mathbf{W}_s - \frac{1}{2}\int_0^t \|\boldsymbol{\gamma}_s\|_2^2 \, ds\right). \qquad (2.59)$$

Then $\mathbb{Q}$ *is equivalent to* $\mathbb{P}$, *and*

$$\tilde{\mathbf{W}}_t = \mathbf{W}_t + \int_0^t \boldsymbol{\gamma}_s \, ds$$

is an n-dimensional $\mathbb{Q}$-*Brownian motion.*

Theorem 2.6.2 (The Martingale Representation Theorem). *Let* $\mathbf{W}_t$ *be an n-dimensional Brownian motion and suppose that* $\mathbf{M}_t$ *is an n-dimensional* $\mathbb{Q}$-*martingale process,* $\mathbf{M}_t = (M_1(t), M_2(t), \ldots, M_n(t))^{\mathrm{T}}$, *such that*

$$dM_i(t) = \sum_{j=1}^n a_{ij}(t) \, dW_j(t).$$

Let $\mathbf{A} = (a_{ij})$ *be a non-singular matrix. If* N_t *is any one-dimensional* $\mathbb{Q}$-*martingale with* $E^{\mathbb{Q}}[N_t^2] < \infty$, *there exists an n-dimensional* $\mathcal{F}_t$-*adaptive process,* $\boldsymbol{\Phi}_t = (\varphi_1(t), \varphi_2(t), \ldots, \varphi_n(t))^{\mathrm{T}}$, *such that*

$$E^{\mathbb{Q}}\left[\int_0^t \left(\sum_j a_{ij}^2(s)\varphi_j^2(s)\,ds\right)\right] < \infty, \quad \forall i, \qquad (2.60)$$

and

$$N_t = N_0 + \sum_{j=1}^n \int_0^t \varphi_j(s)\,dM_j(s)$$

$$\triangleq N_0 + \int_0^t \boldsymbol{\Phi}^{\mathrm{T}}(s)\,d\mathbf{M}(s).$$

2.7 A Complete Market with Multiple Securities

With the preparations of the previous sections, we are ready to address the pricing of a contingent claim whose value depends on the values of n risky securities. Similar to the pricing of options on a single asset, the pricing of options on multiple assets consists of two steps: the construction of a martingale measure for the assets and the construction of the replication strategy.

2.7.1   Existence of a Martingale Measure

We consider a standard model of a complete financial market with a money market account and n risky securities. Let the time t prices be B_t and S_t^i, $1 \le i \le n$, respectively. We assume lognormal price processes for all assets:

$$\mathrm{d}B_t = r_t B_t \mathrm{d}t$$

$$\mathrm{d}S_t^i = S_t^i \left(\mu_t^i \mathrm{d}t + \sum_{j=1}^{n} \sigma_{ij} \mathrm{d}W_j(t) \right) \qquad (2.61)$$

$$= S_t^i \left(\mu_t^i \mathrm{d}t + \boldsymbol{\sigma}_i^{\mathrm{T}}(t) \mathrm{d}\mathbf{W}_t \right), \quad i = 1, 2, \ldots, n.$$

Here,

$$\boldsymbol{\sigma}_i^{\mathrm{T}}(t) = (\sigma_{i,1}, \sigma_{i,2}, \ldots, \sigma_{i,n}).$$

Let $Z_t^i = B_t^{-1} S_t^i$ denote the discounted asset price of the ith asset. It then follows that

$$\mathrm{d}Z_t^i = Z_t^i \left[\boldsymbol{\sigma}_i^{\mathrm{T}}(t) \mathrm{d}\mathbf{W}_t + \left(\mu_t^i - r_t \right) \mathrm{d}t \right], \quad i = 1, 2, \ldots, n. \qquad (2.62)$$

To construct a martingale measure for $Z_t^i, \forall i$, we must "absorb" the drift terms in Equation 2.62 into the Brownian motion. For that reason, we define an $\mathcal{F}_t$-adaptive function, $\boldsymbol{\gamma}_t$, via the following equations:

$$\boldsymbol{\sigma}_i^{\mathrm{T}}(t)\boldsymbol{\gamma}_t = \mu_t^i - r_t, \quad i = 1, 2, \ldots, n. \qquad (2.63)$$

Suppose that $\boldsymbol{\gamma}_t$, the solution to Equation 2.63, exists and satisfies

$$E^{\mathbb{P}} \left[\exp \left(\int_0^T \|\boldsymbol{\gamma}_t\|^2 \mathrm{d}t \right) \right] < \infty \qquad (2.64)$$

for some $T > 0$. We then can define a new measure, $\mathbb{Q}$, according to Equation 2.59. Under this newly defined $\mathbb{Q}$,

$$\tilde{\mathbf{W}}_t = \mathbf{W}_t + \int_0^t \boldsymbol{\gamma}_s \, \mathrm{d}s$$

is a multi-dimensional Brownian motion, with which we can rewrite the price processes for the discounted assets into

$$\mathrm{d}Z_t^i = Z_t^i \boldsymbol{\sigma}_i^{\mathrm{T}}(t) \mathrm{d}\tilde{\mathbf{W}}_t, \quad i = 1, 2, \ldots, n,$$

and $Z_t^i, i = 1, 2, \ldots, n$ are lognormal $\mathbb{Q}$-martingales.

We now study the existence of $\boldsymbol{\gamma}_t$ and condition 2.64. In matrix form, Equation 2.63 can be recast into

$$\boldsymbol{\Sigma}\boldsymbol{\gamma}_t = \boldsymbol{\mu}_t - r_t \mathbf{I}, \qquad (2.65)$$

where

$$\Sigma = \begin{pmatrix} \sigma_1^T \\ \sigma_2^T \\ \vdots \\ \sigma_n^T \end{pmatrix}, \quad \mu_t = \begin{pmatrix} \mu_t^1 \\ \mu_t^2 \\ \vdots \\ \mu_t^n \end{pmatrix}, \quad \text{and} \quad I = \begin{pmatrix} 1 \\ 1 \\ \vdots \\ 1 \end{pmatrix}.$$

Consider first the case when the inverse of the volatility matrix, Σ^{-1}, exists and is bounded. Then we can determine γ_t uniquely as

$$\gamma_t = \Sigma^{-1}(\mu_t - r_t I),$$

and the boundedness condition 2.64 is obviously satisfied as well. The condition of non-singularity of Σ is, however, sufficient but not necessary for the existence of γ_t.

Next, we show that the absence of arbitrage is a sufficient condition for the existence of γ_t as a solution to Equation 2.63. We may interpret the notion of the absence of arbitrage as follows: any riskless portfolios will earn a return equal to the risk-free rate. Let $\{\theta_i\}$ be a portfolio, where θ_i is the number of units in the ith asset. If the portfolio is a riskless one, then, by the absence of arbitrage, the return of the portfolio is zero; that is,

$$d\left(\sum_i \theta_i Z_t^i\right) = \left(\sum_i \theta_i Z_t^i (\mu_t^i - r_t)\right) dt + \left(\sum_i \theta_i Z_t^i \sigma_i^T\right) dW_t = 0.$$

The last equation implies that, as long as

$$(\theta_1 Z_t^1, \dots, \theta_n Z_t^n) \begin{pmatrix} \sigma_1^T \\ \vdots \\ \sigma_n^T \end{pmatrix} = (\theta_1 Z_t^1, \dots, \theta_n Z_t^n) \Sigma = (0, \dots, 0),$$

there will be

$$(\theta_1 Z_t^1, \dots, \theta_n Z_t^n) \begin{pmatrix} \mu_t^1 - r_t \\ \vdots \\ \mu_t^n - r_t \end{pmatrix} = 0.$$

The statements above imply that $(\mu_t^1 - r_t, \dots, \mu_t^n - r_t)^T$ must lie in the linear space spanned by the columns of Σ, meaning that there exists a coefficient vector, $\gamma_t = (\gamma_t^1, \dots, \gamma_t^n)$, such that

$$\mu_t - r_t I = \Sigma \gamma_t. \tag{2.66}$$

In terms of components, Equation 2.66 reads as follows:

$$\mu_t^i - r_t = \sum_{j=1}^n \sigma_{ij} \gamma_t^j, \quad 1 \le i \le n. \tag{2.67}$$

The above arguments justify the existence of γ_t in the absence of arbitrage.

The solution, γ_t, to Equation 2.65 is not unique unless $\boldsymbol{\Sigma}$ is nonsingular almost surely. Regardless of if $\boldsymbol{\Sigma}$ is singular, we can always find a bounded solution of γ_t provided that all non-zero singular values of $\boldsymbol{\Sigma}$ stay away from zero, thus satisfying the Novikov condition, Equation 2.64. When $\boldsymbol{\Sigma}$ is non-singular, we say that the market is non-degenerate. In a financial market, non-degeneracy means that none of the securities can be dynamically replicated by other securities, and thus none of them is redundant. Note that when there is only one risky security, $n = 1$, Equation 2.67 reduces to Equation 2.43.

In the finance literature, the components of γ_t are called the market prices of risks and each of these components can be interpreted as the excess of returns per unit of risk. Under the objective measure, $\mathbb{P}$, the discounted price process of an asset is not necessarily a martingale process. This means that trading an asset at the market price may not be a fair game in the sense that the expected return does not equal the risk-free rate. In fact, empirical studies often suggest $\mu_t^i > r_t, \forall i$, which reflects an important reality of our financial markets where typical investors are risk averse and demand a premium for taking risk.

We finish this section with the remark that the market prices of risks of any two risky assets depending on a single risk source are equal. To see this, we let the prices of two tradable assets, S_t^1 and S_t^2, be driven by the same Brownian motion:

$$\mathrm{d}S_t^i = S_t^i(\mu_t^i\mathrm{d}t + \sigma_t^i\mathrm{d}W_t), \quad i = 1, 2.$$

According to the no-arbitrage principle, there is an $\mathcal{F}_t$-adaptive process, γ_t, such that

$$\mu_t^i - r_t = \sigma_t^i\gamma_t, \quad i = 1, 2. \tag{2.68}$$

It follows that

$$\frac{\mu_t^1 - r_t}{\sigma_t^1} = \frac{\mu_t^2 - r_t}{\sigma_t^2} = \gamma_t. \tag{2.69}$$

In much of the literature, Equation 2.69 is used as a starting point for the derivation of the Black–Scholes–Merton equation.

2.7.2 Pricing Contingent Claims

Now we are ready to address the pricing of a contingent claim depending on the prices of multiple underlying securities. Having found the martingale measure, $\mathbb{Q}$, for the underlying securities, we define a $\mathbb{Q}$-martingale as

$$N_t = E^{\mathbb{Q}}(B_T^{-1}X_T \mid \mathcal{F}_t),$$

using the discounted value of X_T, the payoff function of the claim at time T. Without loss of generality, we assume that the volatility matrix of the underlying risky securities, $\boldsymbol{\Sigma}$, is non-singular.[2] According to the

---

[2]Otherwise, the market is degenerate and the replication portfolio is not unique.

martingale representation theorem, there exists an $\mathcal{F}_t$-adaptive function, $\Phi_t = (\varphi_1(t), \ldots, \varphi_n(t))^{\mathrm{T}}$, such that

$$dN_t = \Phi_t^{\mathrm{T}} d\mathbf{Z}_t,$$

where $\mathbf{Z}_t$ is the vector of the discounted prices. We now define another process,

$$\psi_t = N_t - \Phi_t^{\mathrm{T}} \mathbf{Z}_t,$$

and form a portfolio with ψ_t units of the money market account and $\phi_i(t)$ units of the ith risky security, $i = 1, \ldots, n$. The discounted value of the portfolio is

$$\tilde{V}_t = \Phi_t^{\mathrm{T}} \mathbf{Z}_t + \psi_t = N_t. \tag{2.70}$$

The last equation implies replication of the payoff of the contingent portfolio. Furthermore, from Equation 2.70, we can derive

$$dV_t = \Phi_t^{\mathrm{T}} d\mathbf{S}_t + \psi_t dB_t,$$

which implies that the replication strategy is a self-financing one. So, we conclude that the value of the contingent claim equals that of the portfolio and thus is given by

$$V_t = B_t E^{\mathbb{Q}} \left[B_T^{-1} X_T \mid \mathcal{F}_t \right] = E^{\mathbb{Q}} \left[e^{-\int_t^T r_s ds} X_T \mid \mathcal{F}_t \right]. \tag{2.71}$$

Formally, Equation 2.71 is identical to Equation 2.52, the formula for options on a single underlying security.

2.8 The Black–Scholes Formula

Consider the pricing of a European call option on an asset, S_t, which has the payoff

$$C_T = (S_T - K)^+ \tag{2.72}$$

at time T. Assume that the short rate is a constant, $r_t = r$. According to the Black–Scholes–Merton Equation 2.57 and the terminal condition 2.58, the value of the call option satisfies

$$\begin{cases} \dfrac{\partial C_t}{\partial t} + \dfrac{1}{2}\sigma_t^2 S^2 \dfrac{\partial^2 C_t}{\partial S^2} + rS\dfrac{\partial C_t}{\partial S} - rC_t = 0 \\ C_T = (S - K)^+ \end{cases}. \tag{2.73}$$

By solving this terminal-value problem of the partial differential equation (PDE), we can obtain the price of the option.

Alternatively, we can derive the formula for the call options by working out the expectation in Equation 2.71 directly. We write

$$C_t = e^{-r(T-t)} E_t^{\mathbb{Q}} \left[(S_T - K)^+ \right]$$

$$= e^{-r(T-t)} \left(E_t^{\mathbb{Q}}[S_T 1_{S_T > K}] - K E_t^{\mathbb{Q}}[1_{S_T > K}] \right). \quad (2.74)$$

Since

$$S_T = S_t \exp \left[\left(r - \frac{1}{2} \bar{\sigma}^2 \right) \tau + \bar{\sigma} \sqrt{\tau} \cdot \varepsilon \right], \quad \varepsilon \sim N(0, 1), \quad (2.75)$$

where $\tau = T - t$ and $\bar{\sigma}$ is the mean volatility,

$$\bar{\sigma} = \sqrt{\frac{1}{\tau} \int_0^\tau \sigma_s^2 \mathrm{d}s}, \quad (2.76)$$

we have

$$E_t^{\mathbb{Q}}[1_{S_T > K}] = \mathrm{Prob} \left(\varepsilon > -\frac{\ln(S_t/K) + \left(r - (1/2) \bar{\sigma}^2 \right) \tau}{\bar{\sigma} \sqrt{\tau}} \right) = \Phi(d_2), \quad (2.77)$$

with

$$d_2 = \frac{\ln(S_t/K) + \left(r - (1/2) \bar{\sigma}^2 \right) \tau}{\bar{\sigma} \sqrt{\tau}}. \quad (2.78)$$

Meanwhile,

$$E_t^{\mathbb{Q}}[S_T 1_{S_T > K}] = \frac{1}{\sqrt{2\pi}} \int_{-d_2}^\infty S_t \exp \left[\left(r - \frac{1}{2} \bar{\sigma}^2 \right) \tau + \bar{\sigma} \sqrt{\tau} x - \frac{1}{2} x^2 \right] \mathrm{d}x$$

$$= \frac{S_t e^{r\tau}}{\sqrt{2\pi}} \int_{-d_2 - \bar{\sigma}\sqrt{\tau}}^\infty \exp \left(-\frac{1}{2} y^2 \right) \mathrm{d}y$$

$$= S_t e^{r\tau} \Phi(d_1), \quad (2.79)$$

where

$$d_1 = d_2 + \bar{\sigma} \sqrt{\tau}. \quad (2.80)$$

By substituting Equations 2.77 and 2.79 into 2.74, we arrive at the celebrated Black–Scholes formula:

$$C_t = S_t \Phi(d_1) - e^{-r(T-t)} K \Phi(d_2). \quad (2.81)$$

By direct verification, we can show that the hedge ratio, φ_t, is

$$\frac{\partial C_t}{\partial S_t} = \Phi(d_1). \quad (2.82)$$

Next, we proceed to derive the formula for a put option, which has the payoff function

$$P_T = (K - S_T)^+. \quad (2.83)$$

Instead of pricing the put option by taking the expectation, we make use of the so-called *call–put parity*: a long call and a short put are equivalent to a forward contract, provided that they have the same strike. As a formula, it is

$$C_t - P_t = S_t - e^{-r(T-t)} K. \tag{2.84}$$

Equality (Equation 2.84) implies the formula for the put option:

$$\begin{aligned}
P_t &= C_t - \left(S_t - e^{-r(T-t)} K\right) \\
&= S_t \left(\Phi(d_1) - 1\right) - e^{-r(T-t)} K \left(\Phi(d_2) - 1\right) \\
&= e^{-r(T-t)} K \Phi(-d_2) - S_t \Phi(-d_1).
\end{aligned} \tag{2.85}$$

The hedge ratio, φ_t, can analogously be derived to be

$$\frac{\partial P_t}{\partial S_t} = -\Phi(-d_1). \tag{2.86}$$

The negative number means that the hedger sells short the underlying asset.

2.9 Notes

This chapter provides a rather comprehensive introduction to arbitrage pricing theory in a complete market, a market where the prices of risky assets are driven by lognormal processes. We have shown that the existence of a unique martingale measure that is equivalent to the physical measure is both necessary and sufficient for the absence of arbitrage in the complete market. We emphasize that the mathematical underpinning underlying the arbitrage pricing is the martingale representation theory. For a discussion of arbitrage pricing theory in incomplete markets, we refer readers to Harrison and Kreps (1979) and Harrison and Pliska (1981).

Exercises

1. Check if the following stochastic processes are $\mathbb{P}$-martingales.

$$X_t = W_t^2 - t,$$

$$N(t) = W_t^3 - 3tW_t,$$

$$S_t = S_0 e^{\gamma W_t - (1/2)\gamma^2 t},$$

$$S_t = S_0 e^{\int_0^t \sigma(s)dW_s - (1/2)\sigma^2(s)ds}.$$

Here W_t is a $\mathbb{P}$-Brownian motion.

2. Numerically verify the CMG theorem. First, simulate 1000 terminal values of W_t for $t = 1$, denote the simulated values as $W^{(i)}, i = 1, \ldots, 1000$, and then do the following:

   (a) Compute the mean and variance of $\{W^{(i)}\}_{i=1}^{1000}$ using the uniform probability (or weight), $1/1000$.

   (b) For $\gamma = 0.2$, compute the mean and variance of the Brownian motion with a drift, $\{W^{(i)} + \gamma\}_{i=1}^{1000}$, with non-uniform weights $1/1000 \times \exp\{-\gamma W^{(i)} - \frac{1}{2}\gamma^2\}$.

   Discuss your results.

3. The moment-generating function of a normal random variable, $X \sim N(0, \sigma^2)$, is

$$E\left[e^{\lambda X}\right] = e^{(1/2)\lambda^2\sigma^2},$$

   where λ is a constant. Let W_t be a $\mathbb{P}$-Brownian motion. Prove that

$$X_t = \int_0^t \sigma(s)\, dW_s \sim N\left(0, \int_0^t \sigma^2(s)\, ds\right)$$

   by checking its moment-generating function.

4. Let W_t be a $\mathbb{P}$-Brownian motion and γ_t be a $\mathcal{F}_t$-adaptive process, such that

$$E^{\mathbb{P}}\left[\exp\left(\int_0^t \gamma_s^2\, ds\right)\right] < \infty, \quad \forall t > 0.$$

   Define a new measure, $\mathbb{Q}$, such that

$$\left.\frac{d\mathbb{Q}}{d\mathbb{P}}\right|_{\mathcal{F}_t} = \exp\left(\int_0^t -\gamma_s\, dW_s - \frac{1}{2}\gamma_s^2\, ds\right).$$

   Let

$$\tilde{W}_t = W_t + \int_0^t \gamma_s\, ds.$$

   Prove that, under $\mathbb{Q}$,

   (a) for any $t \geq 0$ and $a \geq 0$, $\tilde{W}_{t+a} - \tilde{W}_t$ is independent of $\left\{\tilde{W}_s, s \leq t\right\}$;

   (b) $\tilde{W}_{t+a} - \tilde{W}_t \sim N(0, a)$ by checking the moment-generating function.

5. Let $\tilde{X}_t = X_t/B_t$ and $\tilde{S}_t = S_t/B_t$ be the discounted price of two assets. If

$$d\tilde{X}_t = \phi_t\, d\tilde{S}_t,$$

   prove that

$$dX_t = \phi_t\, dS_t + \psi_t\, dB_t,$$

   where $\psi_t = X_t - \phi_t S_t$.

6. Let $\mathbf{Z}(t) = (Z_1(t), Z_2(t), \ldots, Z_n(t))^{\mathrm{T}} \in \mathbf{R}^n$ be an independent $\mathbb{P}$-Brownian motion and $\mathbf{f}(t) = (f_1(t), f_2(t), \ldots, f_n(t))^{\mathrm{T}} \in \mathbf{R}^n$ be a $\mathcal{F}_t$-adaptive function with respect to $\mathbf{Z}(t)$. Define the stochastic integral,

$$\int_0^t \mathbf{f}^{\mathrm{T}}(s)\,\mathrm{d}\mathbf{Z}(s) = \sum_{i=1}^n \int_0^t f_i(s)\,\mathrm{d}Z_i(s).$$

Prove that

$$E^{\mathbb{P}}\left[\int_0^t \mathbf{f}^{\mathrm{T}}(s)\,\mathrm{d}\mathbf{Z}(s)\right] = 0,$$

$$E^{\mathbb{P}}\left[\left(\int_0^t \mathbf{f}^{\mathrm{T}}(s)\,\mathrm{d}\mathbf{Z}(s)\right)^2\right] = E^{\mathbb{P}}\left[\int_0^t \|\mathbf{f}(s)\|_2^2\,\mathrm{d}s\right].$$

7. For $\mathcal{F}_t$-adaptive functions, $c(t), \boldsymbol{\alpha}(t) = (\alpha_1(t), \ldots, \alpha_n(t))^{\mathrm{T}}$, and independent $\mathbb{P}$-Brownian motion, $\mathbf{Z}(t) = (Z_1(t), Z_2(t), \ldots, Z_n(t))^{\mathrm{T}} \in \mathbf{R}^n$, define

$$X_t = \exp\left(\int_0^t c(s)\,\mathrm{d}s + \boldsymbol{\alpha}^{\mathrm{T}}(s)\,\mathrm{d}\mathbf{Z}(s)\right).$$

Prove that

$$\mathrm{d}X_t = \left(c(t) + \frac{1}{2}\|\boldsymbol{\alpha}(t)\|_2^2\right)X_t\mathrm{d}t + X_t\boldsymbol{\alpha}^{\mathrm{T}}(t)\,\mathrm{d}\mathbf{Z}(t).$$

For what function of $c(t)$ is the random process X_t an exponential martingale? Why?

8. (CMG theorem in n-space.) Consider a probability measure, $\mathbb{P}$, on the space of paths $\mathbf{Z}(t) = (Z_1(t), Z_2(t), \ldots, Z_n(t))^{\mathrm{T}} \in \mathbf{R}^n, t \leq T$, such that $\mathbf{Z}(t)$ is a vector of an independent Brownian motion. Assume that $\boldsymbol{\lambda}(t)$ is a vector of $\mathcal{F}_t$-adaptive functions and set

$$M(t) = \exp\left(\int_0^t \boldsymbol{\lambda}^{\mathrm{T}}(s)\,\mathrm{d}\mathbf{Z}(s) - \frac{1}{2}\|\boldsymbol{\lambda}(s)\|_2^2\,\mathrm{d}s\right), \quad \forall t \leq T.$$

Define a new measure, $\mathbb{Q}$, such that

$$\left.\frac{\mathrm{d}\mathbb{Q}}{\mathrm{d}\mathbb{P}}\right|_{\mathcal{F}_t} = M(t), \quad \forall t \geq 0.$$

Prove that the random processes

$$W_j(t) = Z_j(t) - \int_0^t \lambda_j(s)\,\mathrm{d}s, \quad 1 \leq j \leq n$$

are independent Brownian motions under the measure $\mathbb{Q}$.

Appendix: The Martingale Representation Theorem

We first prove a slightly more restricted version of the martingale representation theorem, where one of the martingale processes is a Brownian motion. We combine and modify the approaches of Steele (2000) and Øksendal (2003) to yield a simpler proof. To proceed, we define a functional space:

$$H^2[0,T] = \left\{ f \left| E^P \left[\int_0^T f^2(\omega, t)dt \right] < \infty \right. \right\}.$$

Theorem 2.9.1. *Suppose that N_t is a $\mathbb{P}$-martingale such that $E^\mathbb{P}[N_t^2] < \infty, t \leq T$ and $N_0 = 0$. Then, there is a unique $\mathcal{F}_t$-adaptive function $\varphi(\omega, t) \in H^2[0,T]$, such that*

$$N_t = \int_0^t \varphi(\omega, s)\, dW_s, \quad \forall 0 \leq t \leq T. \tag{2.87}$$

To prove this theorem, we need a series of lemmas. But, first of all, we prove the uniqueness ahead of the existence.

Suppose that N_t can be expressed in terms of another function, $\psi(\omega, t)$. Then, there will be

$$0 = \int_0^t (\varphi(\omega, s) - \psi(\omega, s))\, dW_s, \quad \forall 0 \leq t \leq T. \tag{2.88}$$

By Ito's isometry, there is

$$0 = E^\mathbb{P} \left[\left(\int_0^t (\varphi(\omega, s) - \psi(\omega, s))\, dW_s \right)^2 \right]$$

$$= \int_0^t E^\mathbb{P} \left[(\varphi(\omega, s) - \psi(\omega, s))^2 \right] dt. \tag{2.89}$$

Hence,

$$\varphi(\omega, s) = \psi(\omega, s) \quad \text{in } H^2[0,T]. \tag{2.90}$$

The uniqueness is thus proved.

Next, we state a well-known result without giving proof, for which we refer to Øksendal (2003).

Lemma 2.9.1. *Fix $T > 0$. The set of random variables*

$$\{\phi(\Delta W_{t_0}, \ldots, \Delta W_{t_n-1}); \ t_i \in [0,T], \phi \in C_0^\infty(\mathbf{R}^n), n = 1, 2 \ldots\}$$

is dense in $L^2(\mathcal{F}_T, P)$.

Lemma 2.9.2. *For a Brownian motion, W_t, we have the following integral representation:*

$$\exp\left[\frac{\theta^2 t}{2} + i\theta\left(W_{t+s} - W_t\right)\right]$$

$$= 1 + \int_s^{s+t} i\theta \exp\left(\frac{\theta^2(u-s)}{2} + i\theta\left(W_u - W_s\right)\right) dW_u. \tag{2.91}$$

Proof: It is well known that, for a fixed s, the exponential martingale,

$$X_t = \exp\left(-\frac{\sigma^2 t}{2} + \sigma\left(W_{t+s} - W_s\right)\right), \tag{2.92}$$

satisfies the integral equation

$$X_t = 1 + \int_0^t \sigma X_u \, dW_{u+s}. \tag{2.93}$$

Substituting Equation 2.92 into 2.93, we obtain

$$\exp\left(-\frac{\sigma^2 t}{2} + \sigma\left(W_{t+s} - W_s\right)\right)$$

$$= 1 + \int_0^t \sigma \exp\left(-\frac{\sigma^2 u}{2} + \sigma\left(W_u - W_s\right)\right) dW_{u+s}$$

$$= 1 + \int_s^{t+s} \sigma \exp\left(-\frac{\sigma^2(u-s)}{2} + \sigma\left(W_u - W_s\right)\right) dW_u$$

$$= 1 + \int_0^{t+s} 1_{\{u \geq s\}} \sigma \exp\left(\frac{-\sigma^2(u-s)}{2} + \sigma\left(W_u - W_s\right)\right) dW_u. \tag{2.94}$$

By replacing σ with $i\theta$, we obtain the desired result. $\qquad\square$

Next, we will show that a class of random variables constructed using complex exponential functions has the desired integral representations.

Lemma 2.9.3. *For any $0 = t_0 < t_1 < \cdots < t_n = T$, the random variable*

$$M_t = \prod_{j=0}^{n-1} \exp\left[i\theta_j\left(W_{t_{j+1}} - W_{t_j}\right)\right] \tag{2.95}$$

has the following integral representation:

$$M_T = E^{\mathbb{P}}[M_T] + \int_0^T \varphi(\omega, t) \, dW_t \tag{2.96}$$

for some $\varphi \in H^2[0, T]$.

Proof: We use the method of induction. Let $M(0) = 1$ and suppose that we already have

$$M(t_{n-1}) = E^{\mathbb{P}}[M(t_{n-1})] + \int_0^{t_{n-1}} \varphi_{n-1}(\omega, t) \, dW_t \qquad (2.97)$$

for some $\varphi_{n-1}(\omega, t) \in H^2[0, T]$. Then,

$$
\begin{aligned}
M(t_n) &= M(t_{n-1}) \exp\left(i\theta_{n-1}\left(W_{t_n} - W_{t_{n-1}}\right)\right) \\
&= M(t_{n-1})\left(\exp\left(\frac{-\theta_{n-1}^2 (t_n - t_{n-1})}{2}\right)\right. \\
&\quad \left. + \int_{t_{n-1}}^{t_n} i\theta_{n-1} \exp\left(\frac{\theta_{n-1}^2 (t - t_n)}{2} + i\theta_{n-1}\left(W_t - W_{t_{n-1}}\right)\right) dt\right) \\
&= \left(E\left[M(t_{n-1})\right] + \int_0^{t_{n-1}} \varphi_{n-1}(\omega, t) \, dW_t\right) \exp\left(\frac{-\theta_{n-1}^2 (t_n - t_{n-1})}{2}\right) \\
&\quad + \int_{t_{n-1}}^{t_n} i\theta_{n-1} M(t_{n-1}) \exp\left(\frac{\theta_{n-1}^2 (t - t_n)}{2} + i\theta_{n-1}\left(W_t - W_{t_{n-1}}\right)\right) dt \\
&= E^{\mathbb{P}}[M(t_n)] + \int_0^{t_n} \varphi_n(\omega, t) \, dW_t, \qquad (2.98)
\end{aligned}
$$

where

$$
\begin{aligned}
\varphi_n(\omega, t) &= \varphi_{n-1}(\omega, t) \exp\left(\frac{\theta_{n-1}^2 (t_n - t_{n-1})}{2}\right) + 1_{t \geq t_{n-1}} i\theta_{n-1} M(t_{n-1}) \\
&\quad \times \exp\left(\frac{\theta_{n-1}^2 (t - t_n)}{2} + i\theta_{n-1}\left(W_t - W_{t_{n-1}}\right)\right). \qquad (2.99)
\end{aligned}
$$

Apparently,

$$
\begin{aligned}
E^{\mathbb{P}}[M(t_n)] &= E[M(t_{n-1})] \exp\left(\frac{\theta_{n-1}^2 (t_n - t_{n-1})}{2}\right) \\
&= \prod_{j=0}^{n-1} \exp\left(\frac{\theta_j^2 (t_{j+1} - t_j)}{2}\right) \qquad (2.100)
\end{aligned}
$$

and $\varphi_n \in H^2[0, T]$ provided $\varphi_{n-1} \in H^2[0, T]$. $\qquad \square$

Lemma 2.9.4. *If S is the linear span of the set of random variables of the form*

$$\prod_{j=0}^{n-1} \exp\left(i\theta_j \left(W_{t_{j+1}} - W_{t_j}\right)\right) \qquad (2.101)$$

over all $n \geq 1, 0 = t_0 < t_1 < \cdots < t_n = T$, and $\theta_j \in \mathbb{R}$, then S is dense in the space of square-integrable complex-valued random variables, $L^2(\Omega, \mathcal{F}, \mathbb{P})$.

Proof: Suppose $g \in L^2(\mathcal{F}_T, P)$ is orthogonal to all functions of the form (2.101), then there will be

$$G(\theta) = \int_{\Omega} \exp\left\{i\theta_1 \Delta W_{t_0} + \ldots + i\theta_n \Delta W_{t_{n-1}}\right\} g(\omega) dP(\omega) = 0 \qquad (2.102)$$

for all $\theta = (\theta_1, \ldots, \theta_n) \in \mathbf{R}^n$ and all $t_0, \ldots, t_{n-1} \in [0, T]$. Then, for $\phi \in C_0^{\infty}(\mathbf{R}^n)$,

$$\int_{\Omega} \phi(\Delta W_{t_0}, \ldots, \Delta W_{t_{n-1}}) g(\omega) dP(\omega)$$

$$= \int_{\Omega} (2\pi)^{-n/2} \left(\int_{\mathbf{R}^n} \hat{\phi}(\theta) e^{i\theta_1 \Delta W_{t_0} + \ldots + i\theta_n \Delta W_{t_{n-1}}} g(\omega) dP(\omega) \right) d\theta$$

$$= (2\pi)^{-n/2} \int_{\mathbf{R}^n} \hat{\phi}(\theta) \left(\int_{\mathbf{R}^n} \hat{\phi}(\theta) e^{i\theta_1 \Delta W_{t_0} + \ldots + i\theta_n \Delta W_{t_{n-1}}} g(\omega) dP(\omega) \right) d\theta$$

$$= (2\pi)^{-n/2} \int_{\mathbf{R}^n} \hat{\phi}(\theta) G(\theta) dy = 0,$$

$$(2.103)$$

where

$$\hat{\phi}(y) = (2\pi)^{-n/2} \int_{\mathbf{R}^n} \phi(x) e^{ix \cdot y} dx$$

is the Fourier transform of ϕ and we have used the inverse Fourier transform theorem

$$\phi(x) = (2\pi)^{-n/2} \int_{\mathbf{R}^n} \hat{\phi}(y) e^{-ix \cdot y} dy.$$

By (2.103) and Lemma 2.9.1 g is orthogonal to a dense subset of $L^2(\mathcal{F}_T, P)$, so we conclude that $g = 0$. Therefore the linear span of the function in (2.101) must be dense in $L^2(\mathcal{F}_T, P)$ as claimed. □

Lemma 2.9.5. *Suppose that $X_n, n = 1, 2, \ldots$, is a sequence of random variables that have a martingale representation of the form*

$$X_n = E^{\mathbb{P}}[X_n] + \int_0^T \varphi_n(\omega, t) \, dW_t \quad \text{with } \varphi_n \in H^2[0, T]. \qquad (2.104)$$

If $X_n \to X$ in $L^2(\Omega, \mathcal{F}, \mathbb{P})$, then there is a $\varphi \in H^2[0, T]$, such that $\varphi_n \to \varphi$ in $H^2[0, T]$ and

$$X = E^{\mathbb{P}}[X] + \int_0^T \varphi(\omega, t) \, dW_t. \qquad (2.105)$$

Proof: If $X_n \to X$ in $L^2(\Omega, \mathcal{F}, \mathbb{P})$, then $E^{\mathbb{P}}[X_n] \to E^{\mathbb{P}}[X]$, and $X_n - E^{\mathbb{P}}[X_n]$ is a Cauchy sequence in $L^2(\Omega, \mathcal{F}, \mathbb{P})$. This implies that the corresponding φ_n for the representation of X_n is also a Cauchy sequence in $H^2[0, T]$, and the limit exists as $n \to \infty$. By Ito's isometry, the integral in Equation 2.102 also converges to the integral in Equation 2.103. The lemma is thus proved. □

Combining Lemmas 2.2 through 2.4, we then arrive at the proof for Theorem 2.5. Finally, based on Theorem 2.5, the proof for Theorem 2.4 is trivial.

[**Proof** for Theorem 2.4] According to the statements of the theorem, there are $\mathcal{F}_t$-adaptive functions, σ and σ_N, in $H^2[0,T]$, such that

$$\begin{aligned} \mathrm{d}M_t &= \sigma \, \mathrm{d}W_t \\ \mathrm{d}N_t &= \sigma_N \, \mathrm{d}W_t. \end{aligned} \tag{2.106}$$

If $\sigma \neq 0$ almost surely, we let $\varphi = \sigma_N/\sigma$ and then write

$$\mathrm{d}N_t = \varphi\sigma \, \mathrm{d}W_t = \varphi \, \mathrm{d}M_t. \tag{2.107}$$

By definition, $\varphi\sigma = \sigma_N \in H^2[0,T]$. The proof is completed. □

The following lemma is quite useful in finding the representation. The proof is straightforward.

Lemma 2.9.6. *Suppose a martingale M_t for $t \in [0,T]$ has the representation*

$$M_t = M_0 + \int_0^t \phi_s dW_s,$$

then ϕ_t is given by

$$\phi_t = \frac{d\langle M_t, W_t \rangle}{dt} \tag{2.108}$$

Chapter 3

Interest Rates and Bonds

To the general public, the term "interest rate" likely means the instantaneous rate that is used for interest accrual in a savings account. But to the participants in fixed-income markets, this term has a much broader meaning, and it encompasses an entire class of *rates of return* that are associated with various fixed-income instruments. The interest rates most frequently referred to include money market rates, U.S. Treasury yields, zero-coupon yields, forward rates, and swap rates. In this chapter, we introduce the money market and U.S. Treasury instruments and then define various interest rates or yields as the rates of returns for investing in these instruments. A bootstrapping technique for calculating zero-coupon yields is also described. To prepare for our study of interest rate modeling in Chapter 4, forward rates are also introduced in conjunction with zero-coupon bonds. The chapter ends with a discussion of classic techniques for risk management of a portfolio of interest-rate-sensitive instruments.

3.1 Interest Rates and Fixed-Income Instruments

The bulk of fixed-income instruments are coupon bonds. But the most popular fixed-income investment among the general public, although many may not even know it, is an investment in the money market through a savings account. Our introduction to interest rates and fixed-income instruments begins with the savings accounts.

3.1.1 Short Rate and Money Market Accounts

The short rate is associated with a savings account in a bank. The short rate at time t is conventionally denoted as r_t. Interest on a savings account is accrued daily, using the actual/365 convention. Let B_t denote the account balance at time (or date) t, and let $\Delta t = 1$ day $= 1/365$ year. Then the new balance the next day at $t + \Delta t$ is

$$B_{t+\Delta t} = B_t(1 + r_t \Delta t). \tag{3.1}$$

Because $\Delta t \ll 1$, daily compounding is very well approximated by *continuous compounding*: in the limit of $\Delta t \to 0$, Equation 3.1 becomes

$$dB_t = r_t B_t dt. \tag{3.2}$$

Because r_t is applied to $(t, t + dt)$, an infinitesimal interval of time, it is also called the instantaneous interest rate. As a mathematical approximation and idealization, continuous compounding is necessary to continuous-time finance. Suppose that a sum of money is deposited at $t = 0$ into a savings account and that there has not been a deposit or withdrawal since. Then the balance at a later time, t, is

$$B_t = B_0 e^{\int_0^t r_s ds}. \tag{3.3}$$

In the real world, the balance, B_t, is not known in advance due to the stochastic nature of the short rate. Nonetheless, the deposit in the savings account is considered a risk-free security, and its return is used as a benchmark to measure the profits and losses of other investments.

In reality, savings accounts for institutions and for individuals offer different interest rates, which reflect different overhead management costs for institutional and individual clients. To distinguish from an individual's account, we call the savings account for an institution a *money market account*. Note that this is somewhat an abuse of terminology. In the United States, a money market account is also a type of savings account for retail customers, which offers higher interest rates under some restrictions, including minimum balances and limited numbers of monthly withdrawals. Its compounding rule is also different from continuous compounding. Hence, we need to emphasize here that, in fixed-income modeling, a money market account means a savings account for institutions that compounds continuously. Such a money market account plays an important role in continuous-time modeling of finance.

3.1.2   Term Rates and Certificates of Deposit

Term rates are associated to certificates of deposit (CD). A CD is a deposit that is committed for a fixed period of time, and the interest rate applied to the CD is called a term rate. For retail customers, the available terms are typically one month, three months, six months, and one year. Usually, the longer the term, the higher the term rate, as investors are awarded a higher premium for committing their money for a longer period of time. The interest payments of CDs use simple compounding. Let $r_{t,\Delta t}$ be the interest rate for the term Δt and I_t be the value of the deposit at time t. Then the balance at the maturity of the CD is

$$I_{t+\Delta t} = I_t(1 + r_{t,\Delta t}\Delta t). \tag{3.4}$$

Investors of CDs often roll over their CDs, meaning that after a CD matures, the entire amount (principal plus interest) is deposited into another CD with the same terms but with the prevailing term rate at the time when the rolling over takes place. Suppose that a CD is rolled over n times. Then the terminal balance at time $t + n\Delta t$ is

$$I_{t+n\Delta t} = I_t \cdot \prod_{i=1}^{n} (1 + r_{t+(i-1)\Delta t, \Delta t} \Delta t). \qquad (3.5)$$

If the Δt term rate remains unchanged over the investment horizon, that is, $r_{t+(i-1)\Delta t, \Delta t} = r_{t,\Delta t}$, $i = 1, \ldots, n$, then there is

$$I_{t+n\Delta T} = I_t (1 + r_{t,\Delta t} \Delta t)^n, \qquad (3.6)$$

and we say that the deposit is compounded n times with interest rate $r_{t,\Delta t}$. We call $\omega = 1/\Delta t$ the compounding frequency, which is the number of compoundings per year. For example, when $\Delta t = 3$ months or 0.25 year, we have $\omega = 1/\Delta t = 4$, corresponding to the so-called quarterly compounding. By the way, a savings account is compounded daily, corresponding to $\omega = 365$.

Different term rates often mean different rates of return. One way to compare CDs of different terms is to check their *effective annual yields* (EAY), defined as the dollar-value return over a year for a $1 initial investment:

$$\text{EAY} = (1 + r_{t,\Delta t} \Delta t)^{1/\Delta t} - 1. \qquad (3.7)$$

Should interest rates stay constant over the investment horizon, then a higher EAY gives a higher return in value. In reality, term rates change in a correlated yet random way. Hence, for any fixed investment horizon when rolling over is needed, it is difficult to judge in advance which term is optimal to an investor. In fact, investors often choose terms based on cash flow considerations.

3.1.3 Bonds and Bond Markets

The bulk of fixed-income investment has to do with bonds. A bond is a financial contract that promises to pay a stream of cash flows over a certain time horizon. The cash flows consist of payments of interest and a payment of the principal. The interest payments are made periodically until the maturity of the bond, and the principal is, out of which interest payments are generated, typically paid back at the maturity date. The interest payments are also called *coupons*. Figure 3.1 illustrates the cash flows of a typical coupon bond, where c and Pr stand for the coupon rate and the principal value for the bond, respectively, and ΔT is the gap of time, measured in units of a year, between two consecutive coupon payments. The principal value is also called the face value or par value.

The above cash flow structure is considered standard for bonds, and we call bonds with such a cash flow pattern bullet bonds or straight bonds. There are a variety of cash flow patterns or structures for non-bullet bonds. One important example of a non-bullet bond is the class of floating-rate bonds, where the coupon rates are not fixed, but are indexed to three- or six-month CD rates. Comprehensive discussions of the floating-rate bonds are given in Chapter 6.

Bonds are tools to raise capital for their issuers, which include central governments, local governments, and industrial corporations. In most developed countries, bond markets are mature and have notional values that are often larger than those of their corresponding equity markets. The major reasons for the attractiveness of the bond markets are the stable returns and lower risk (i.e., smaller price fluctuation) of bonds. With the globalization of the capital markets, investors now have many types of bonds to choose from. They can invest in domestic bonds or international bonds, in government or corporate bonds. The profiles of the risks and returns of the investment alternatives vary. For a comprehensive discussion, please see other works, for example, Fabozzi (2003).

Let us take a quick look at the U.S. domestic bond markets. The U.S. domestic bond markets mainly consist of three sectors:

1. *Government bonds:* Bonds issued by the Treasury Department of the U.S. government and local governments.

2. *Agency bonds:* Bonds issued by certain agencies of the U.S. government and guaranteed or sponsored by the U.S. government.

3. *Corporate bonds:* Bonds issued by U.S. corporations.

The U.S. Treasury Department is the largest single issuer of debt in the world. Bonds in this sector are deemed riskless as they are guaranteed by the full faith of the U.S. government. This sector is also distinguished by the large volume of total debt and the large size of any single issue, two factors that have contributed to making the Treasuries market the most active as well as most liquid market in the world.

Treasury securities include Treasury bills, Treasury notes, Treasury bonds, and Treasury Inflation-Protected Securities (TIPS). The Treasury bills are bonds without coupons (the so-called zero-coupon bonds). Upon issuance, the maturities of zero-coupon bonds vary from 4 weeks (one month) to 13 weeks (three months) and 26 weeks (six months). The Treasury notes are coupon bonds with maturities between 2 and 10 years upon issuance. The Treasury

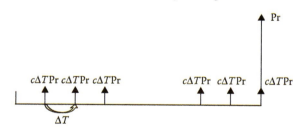

FIGURE 3.1: Cash flows of a coupon bond.

bonds have original maturities greater than 10 years. TIPS are issued with maturities of 5, 10, and 20 years. Thirty-year TIPS have also been issued occasionally. TIPS are issued with fixed coupon rates. The principal of a TIPS is adjusted semiannually according to the inflation rate.

The Treasury securities are issued through an auction process that takes place in a regular cycle of time. Four-week bills are auctioned every Tuesday, whereas 13- and 26-week bills are auctioned every Monday. Treasury notes are issued much less frequently. Two- and five-year notes are offered at the end of each month, and 3- and 10-year notes on the 15th of February, May, August, and November. The 30-year bonds are auctioned in February and August. TIPS are auctioned in different cycles. Five-year TIPS are generally auctioned in the last week of April; 10-year TIPS are generally auctioned in the second week of January and July; and 20-year TIPS are generally auctioned in the last week of January.

In terms of notional value, the Treasury bond sector is not the largest sector of bonds, and it trails behind the agency bond sector. The agency bond sector includes fixed-income securities backed by student loans, public power systems, and, in particular, mortgages. Most agency bonds are not guaranteed, except the mortgage-backed bonds. Other agency bonds, therefore, are actually subject to credit risk. The corporate bond market is also called a credit market, where the uncertainty of credit worthiness is the major source of risk and returns. The modeling of credit risk is a very big subject of its own and will be dealt with in Chapter 13.

3.1.4   Quotation and Interest Accrual

The price of zero-coupon bonds and coupon bonds are quoted using different conventions. The former is quoted using a discount yield, which translates conveniently into a dollar price, whereas the latter is quoted as a percentage of the principal value, using a price tick of 1/32nd of a percentage point. As examples, the quotes of newly issued treasury securities are given in Table 3.1.

TABLE 3.1: Quotes for U.S. Treasuries as of July 3, 2008

U.S. Treasuries Bills		Maturity Date	Discount/Yield	Discount/ΔY
3-month		6/5/2008	1.42/1.44	−0.01/0.087
6-month		9/4/2008	1.51/1.54	0.03/−0.031
Notes/Bonds	Coupon	Maturity Date	Price/Yield	Price/ΔY
2-year	2	2/28/2010	100-29$\frac{3}{4}$/1.52	−0-00$\frac{3}{4}$/0.012
3-year	4.75	3/31/2011	103-21$\frac{3}{4}$/1.42	−0-02/0.018
5-year	2.75	2/28/2013	101-16/2.43	0-06/−0.040
10-year	3.5	2/15/2018	99-23+/3.53	0-14/−0.053
30-year	4.375	2/15/2038	97-08$\frac{3}{4}$/4.54	0-08+/−0.017

Available at http://www.bloomberg.com/markets/rates/index.html.

Some explanations are needed on how to figure out the dollar values of Treasury bills and notes/bonds from the quotes. We begin with Treasury notes/bonds. One may notice that a "+" appears in the quotation of 10-year notes. Such a "+" stands for 0.5, or half a tick. The dollar values of the five Treasury notes and bonds are calculated as follows:

$$100\text{-}29\tfrac{3}{4} = 100 + \frac{29.75}{32} = 100.9297,$$

$$103\text{-}21\tfrac{1}{4} = 103 + \frac{21.25}{32} = 103.6641,$$

$$101\text{-}16 = 101 + \frac{16}{32} = 101.5, \qquad (3.8)$$

$$99\text{-}23+ = 99 + \frac{23.5}{32} = 99.7344,$$

$$97\text{-}08\tfrac{1}{2} = 97 + \frac{8.5}{32} = 97.2656.$$

The above dollar prices are not the prices for transactions, however. For transactions, we must add the interest values accrued since the last coupon payment, which are calculated as follows. Consider a coupon bond with coupon rate c and coupon dates $\{T_i\}$. Suppose that we are at a moment, t, between two consecutive coupon dates, T_j and T_{j+1} (e.g., $T_j < t \leq T_{j+1}$). Then the accrued interest is

$$\mathrm{AI}(t, T_j) = \Delta T \cdot c \cdot \mathrm{Pr} \cdot q,$$

where q is the fraction of time elapsed since the last coupon date over the current coupon period:

$$q = \frac{t - T_j}{T_{j+1} - T_j}. \qquad (3.9)$$

The transaction price of the bond is simply the sum of a quote price plus the accrued interest:

$$\text{Transaction price} = \text{Quote price} + \mathrm{AI}(t, T_j).$$

Note that the industrial jargon for quote prices is clean prices, whereas the transaction prices are called dirty prices or full prices.

Let us look at an example of how to calculate accrued interest.

Example 3.1.1. *Consider a 10-year Treasury note maturing on February 15, 2018. On March 7, 2008, the bond quote is 99-23+ (= 99.7344). We need to compute the accrued interest and then the dirty price.*

The coupon dates are February 15 and August 15. There are 182 days between the coupon dates, and on March 7, 21 days have elapsed since February 15, the last coupon date. Hence,

$$q = \frac{21}{182} = 0.115385,$$

$$\mathrm{AI} = 0.5 \times 3.5\% \times 100 \times 0.115385 = 0.2019.$$

Then the full price is

$$B^c = 99.7344 + 0.2019 = 99.9363. \tag{3.10}$$

The dollar value of a Treasury bill is calculated using the discount yield according to the formula

$$V = \Pr \cdot \left(1 - \frac{\tau}{360} Y_d\right), \tag{3.11}$$

where τ is the number of days remaining to maturity. Suppose, for instance, that the six-month Treasury bill has a time to maturity of $\tau = 100$ days. Then its price is

$$P = 100 \times \left(1 - \frac{100}{360} \times 1.51\%\right) = \$99.5806.$$

Note that the discount yield is a quoting mechanism rather than a good measure of returns on an investment in a Treasury bill. There are two reasons for this. First, the yield is not calculated as a return for the initial amount of investment, and, second, the yield is annualized according to a 360-day year instead of 365-day year. In the next section, we discuss the measure of returns for bond investments.

3.2 Yields

3.2.1 Yield to Maturity

In a free market, the price of a bond is determined by supply and demand. Due to discounting, the full price is normally smaller than the total notional value of coupons plus the principal. Denote the full price of a bullet bond as B^c. Suppose that all cash flows are discounted by a uniform rate, y, of compounding frequency ω. Then y should satisfy the following equation:

$$B^c = \Pr \cdot \left(\sum_{i=1}^{n} \frac{c\Delta T}{(1 + y\Delta t)^{i\Delta T/\Delta t}} + \frac{1}{(1 + y\Delta t)^{n\Delta T/\Delta t}}\right), \tag{3.12}$$

where n is the number of coupons and $\Delta t = 1/\omega$. In bond mathematics, the compounding frequency is taken to be $\omega = 1/\Delta T$ by default, when there is $\Delta t = \Delta T$. This discount rate, which can be easily solved by a trial-and-error procedure using Equation 3.12, is defined to be the *yield to maturity* (YTM), as well as the *internal rate of return* (IRR) of the bond, and it is often simply called the *bond yield*.

As the function of the yield (for $\omega = 1/\Delta T$), the formula for a general time, $t \leq T$, is

$$B_t^c = \Pr \cdot \left(\sum_{i;i\Delta T > t}^{n} \frac{c\Delta T}{(1 + y\Delta T)^{(i\Delta T - t)/\Delta T}} + \frac{1}{(1 + y\Delta T)^{(n\Delta T - t)/\Delta T}}\right). \tag{3.13}$$

Assuming that $t \in (T_j, T_{j+1}]$, and introducing

$$q = \frac{t - T_j}{T_{j+1} - T_j} = \frac{t - T_j}{\Delta T},$$

we then can write

$$t = T_j + \Delta T q = (j + q)\Delta T \text{ and } i\Delta T - t = (i - j - q)\Delta T, \quad \forall i.$$

It follows that

$$B_t^c = \text{Pr} \cdot \left(\sum_{i=j+1}^{n} \frac{c\Delta T}{(1 + y\Delta T)^{i-j-q}} + \frac{1}{(1 + y\Delta T)^{n-j-q}} \right) \tag{3.14}$$

$$= \text{Pr} \cdot (1 + y\Delta T)^q \left(\sum_{i=1}^{n-j} \frac{c\Delta T}{(1 + y\Delta T)^i} + \frac{1}{(1 + y\Delta T)^{n-j}} \right).$$

Given the bond price at any time, t, the bond yield is implied by Equation 3.14. A rough way to compare the relative cheapness/richness of two bonds with the same coupon frequency is to compare their yields. Intuitively, a bond with a higher yield is cheaper and thus may be more attractive.

There is a one-to-one price–yield relationship, as shown in Figure 3.2. Because of this relationship, a bond price is also quoted using its yield in the industry. As we can see in Figure 3.2, a bond price is a convex function of the yield. Such a feature will be used later for convexity adjustment related to futures trading.

The price–yield relationship of a zero-coupon bond simplifies to

$$P = \text{Pr} \cdot (1 + y\Delta T)^{-(T-t)/\Delta T}.$$

Although simple, the above equation does not lend itself to a manual solution of the yield for a general maturity, $T - t$. To derive an approximate value of the yield, we consider the following approximation of the "return on the investment":

$$\frac{\text{Pr} - P}{P} = (1 + y\Delta T)^{(T-t)/\Delta T} - 1 \approx y \times (T - t), \tag{3.15}$$

using the Taylor expansion. Equation 3.15 gives rise to an approximate yield-to-maturity:

$$y \approx \frac{1}{(T - t)} \frac{\text{Pr} - P}{P},$$

which is also called the bond equivalent yield. Note that for Treasury zero-coupon bonds, the year has 365 days, meaning that

$$T - t \approx \frac{\tau}{365}, \tag{3.16}$$

where τ is the number of days to maturity. In Table 3.1, part of the third column shows the discount yields and bond equivalent yields for three- and six-month Treasury bills.

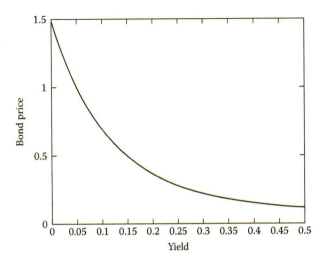

FIGURE 3.2: The price–yield relationship for a 10-year bond with $c = 0.05$, $\Delta T = 0.5$.

3.2.2   Par Bonds, Par Yields, and the Par Yield Curve

The summation in Equation 3.12 can be worked out so that

$$B^c = \Delta T \cdot c \cdot \Pr \sum_{i=1}^{n} (1 + y\Delta T)^{-i} + \Pr(1 + y\Delta T)^{-n}$$
$$= \Pr \left[1 - \left(1 - \frac{c}{y} \right) \left(1 - \frac{1}{(1 + y\Delta T)^n} \right) \right]. \tag{3.17}$$

From the above expression, we can tell when the price is smaller, equal to, or larger than the principal value.

1. When $c < y$, $B^c < \Pr$. In such a case, we say that the bond is sold at discount (of the par value).

2. When the coupon rate is $c = y$, then $B^c = \Pr$, that is, the bond price equals the par value of the bond. In such a case, we call the bond a *par bond*, and the corresponding coupon rate a *par yield*.

3. When $c > y$, $B^c > \Pr$. In such a case, we call the bond a premium bond (it is traded at a premium to par).

Par yields play an important role in today's interest-rate derivatives market. As we shall see later, there are many derivatives based on the par yields.

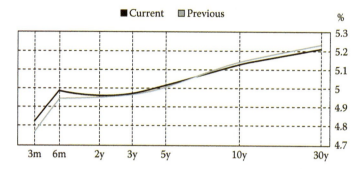

FIGURE 3.3: The U.S. Treasury yield curve for April 28, 2006 (gray) and May 1, 2006 (black) (data from http://www.bloomberg.com/markets/rates/index.html).

3.2.3   Yield Curves for U.S. Treasuries

A bond issuer may routinely issue bonds of various maturities, and, in a market, there can be many bonds of the same issuer being traded. For various reasons, some bonds are more liquid than others. The most liquid ones are often called benchmark bonds for the issuer. Their yields reflect the level of borrowing costs the market demands from the issuer. Moreover, the prices of the benchmark bonds imply a discount curve for cash flows from the issuer, and the discount curve can be used to gauge the relative cheapness/expensiveness of the issuer's other bonds. If a relatively cheaper or more expensive bond is found, one may trade against this bond using the benchmark bonds and thus take an arbitrage profit. Hence, the prices or yields of the benchmark bonds carry essential information for the arbitrage pricing of the issuer's other bonds, and they are treated as a summary of the status quo of all bonds offered by the same issuer.

In the U.S. Treasury market, newly issued bills and notes/bonds are called on-the-run Treasury securities. Traditionally, the on-the-run issues enjoy higher liquidity and are thus treated as benchmarks. Table 3.1 provides the closing price quotes of the on-the-run issues for July 3, 2008. As can be seen in the table, the on-the-run issues have maturities of 3 months, 6 months, 2 years, 3 years, 5 years, 10 years, and 30 years. When we connect the yields of the benchmark bonds through interpolation, we obtain a so-called yield curve. Since bond yields vary from day-to-day so does the yield curve. Figure 3.3 shows the yield curves for the U.S. Treasuries constructed by linear interpolation for April 28 and May 1, 2006, two consecutive trading days.

A yield curve is constructed based on yields of on-the-run issues using the interpolation technique. It provides a rough idea of the level of yields for various maturities. Further, the Treasury yield curve implies a discount curve, namely, the collection of prices of all zero-coupon bonds. The discount curve is used for pricing off-the-run Treasury securities, or marking to market

Treasury portfolios. Moreover, the discount curve is also essential for pricing future cash flows of any security, either deterministic or stochastic. To price a portfolio of interest-rate derivatives, we may model the dynamics of the entire yield curve, in contrast to modeling the dynamics of a stock price for stock options. In the next section, we describe the technique for "backing" out the discount curve from the yield curve.

3.3   Zero-Coupon Bonds and Zero-Coupon Yields

3.3.1   Zero-Coupon Bonds

Let us refer to the cash flow of the coupon bond shown in Figure 3.1 again. When there is no coupon, $c = 0$, the principal is the only cash flow, and the coupon bond is reduced to a zero-coupon bond. The corresponding yield is called a zero-coupon yield. As a convention, the time-t price of a zero-coupon bond maturing at time T into a par value of one dollar is denoted as $P(t, T)$ or P_t^T. In terms of its yield, the price of the zero-coupon bond is

$$P_t^T = \frac{1}{(1 + y\Delta t)^{(T-t)/\Delta t}}, \tag{3.18}$$

where the time to maturity, $T - t$, does not have to be a multiple of Δt. The collection of P_t^T for $T \geq t$ is called a discount curve.

With the discount curve, one can price any bond portfolio with deterministic cash flows. This is because any such portfolio can be treated as a portfolio of zero-coupon bonds. For example, we can express the price of the coupon bond in terms of those of zero-coupon bonds:

$$B^c(0) = \sum_{i=1}^{n} c \cdot \Delta T \Pr \cdot P_0^{i\Delta T} + \Pr \cdot P_0^{n\Delta T}. \tag{3.19}$$

In continuous-time finance, it is often favorable to work with continuous compounding, that is, by letting the term $\Delta t \to 0$ in Equation 3.18. At this limit, we have

$$P_t^T = e^{-y \times (T-t)}.$$

Given P_t^T, the corresponding zero-coupon yield can be calculated from the last equation:

$$y_{T-t} = -\frac{1}{T-t} \ln P_t^T.$$

In most fixed-income markets, zero-coupon bonds are rare, with the exception of the U.S. Treasury markets, where they are actually quoted and traded, although they occupy only a small fraction of the daily turnover of the

Treasury bond market. Zero-coupon bonds first appeared in 1982, when both Merrill Lynch and Salomon Brothers created synthetic zero-coupon Treasury receipts whose cash flows were backed by Treasury coupon bonds. In 1985, the U.S. Treasury launched its Separate Trading of Registered Interest and Principal of Securities (STRIPS) program. This program facilitated the stripping of designated Treasury securities. Today, all Treasury notes and bonds, both fixed-principal and inflation-indexed principal, are eligible for stripping. In principle, one can replicate Treasury coupon bonds with STRIPS, but the relatively low liquidity of STRIPS makes such replication not always practical. Because of this, prices of STRIPS are not taken as literally as the discount bond prices. The market convention is to "back out" the discount curve from the liquid Treasury securities through a bootstrapping process. The prices of the STRIPS are actually subordinated to the discount curve so obtained.

3.3.2   Bootstrapping the Zero-Coupon Yields

The determination of the zero-coupon yield curve (or discount curve) based on the yields of the on-the-run issues is an under-determined problem: we need to solve for infinitely many unknowns based on a few inputs. To define a meaningful solution, one must parameterize the zero-coupon yield curve. The simplest parameterization that is financially acceptable is to assume piecewise constant functional forms for the zero-coupon yield curve. Under such a parameterization, the zero-coupon yield curve can be derived sequentially. Such a procedure is often called bootstrapping in finance. Next, we describe the bootstrapping procedure with the construction of the zero-coupon yield curve for U.S. Treasuries.

Let $\{B_j^c, T_j\}_{j=1}^7$ be the prices and maturities of the seven on-the-run issues. Let $T_0 = 0$ and $\Delta T = 0.5$. We assume that the zero-coupon yield for maturities between $[T_0, T_7]$ is a piecewise linear function. The determination of the YTMs is done sequentially. Because the first two on-the-run issues are zero-coupon bonds, we first back out $y(0.25)$ and $y(0.5)$, the zero yields for $(0, T_1]$ and $(T_1, T_2]$, using formula 3.18. This will require a root-finding procedure. Once $y(0.5)$ is found, we proceed to determining $y(i\Delta T), i = 2, 3, 4$ from the following equation:

$$B_3^c = \frac{c_3 \Delta T}{(1 + y(\Delta T)\Delta T)^i} + \sum_{i=2}^4 \frac{c_3 \Delta T}{(1 + y(i\Delta T)\Delta T)^i} + \frac{1}{(1 + y(4\Delta T)\Delta T)^4},$$

$$(3.20)$$

where

$$y(i\Delta T) = y(0.5) + \alpha \times (i\Delta T - 0.5), \quad i = 2, 3, 4.$$

So, our zero-coupon yield is a linear function over $T \in [T_2, T_3]$. Equation 9.20 become the equation for α, which can be determined through a root-finding procedure. This procedure can continue all the way to $j = 7$. The entire zero-coupon yield curve for maturity $T \leq 30$ so-determined is displayed in Figure 3.4.

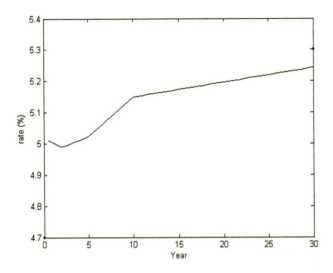

FIGURE 3.4: The zero-coupon yield curve of U.S. Treasuries on May 1, 2006.

A zero-coupon yield curve implies a discount curve. Suppose that the y_T is the zero-coupon yield for maturity T. Then the corresponding zero-coupon bond price is calculated according to Equation 3.18. With discount bond prices, we can value any coupon bond using Equation 3.19.

3.3.2.1 Future Value and Present Value

Owing to the (mostly) positive interest rates, the notional value of I_t today will become $I_{t+T}(>I_t)$ T years later. In terms of zero-coupon bond prices, we can define the present value of a future cash flow, I_{t+T}, to be

$$I_t = I_{t+T} P(t, t+T).$$
(3.21)

Conversely, we can rewrite Equation 3.21 as

$$I_{t+T} = \frac{I_t}{P(t, t+T)},$$
(3.22)

and call I_{t+T} the future value of I_t.

3.4   Forward Rates and Forward-Rate Agreements

A forward-rate agreement (FRA) is a contract between two parties to lend and borrow a certain amount of money for some future period of time with a pre-specified interest rate. The agreement is so structured that neither party needs to make an upfront payment. This is equivalent to saying that, as a financial instrument, the value of the contract is zero when the agreement is entered. The key to such a contract lies in the lending rate that should be fair to both parties. Fortunately, this fair rate can be determined through arbitrage arguments.

Let the time now be t and the fair lending rate for a future period, $[T, T + \Delta T]$, be $f_{\Delta T}(t, T)$. To finance the lending, the lender may short $P(t, T)/P(t, T + \Delta T)$ units of the $(T + \Delta T)$-maturity zero-coupon bond, and then long one unit of the T-maturity zero-coupon bond. At time T, the proceeds from the T-maturity zero are lent out for a period of ΔT with the interest rate $f_{\Delta T}(t, T)$. At time $T + \Delta T$, the loan is paid back from the borrower and the short position of $(T + \Delta T)$-maturity zero-coupon bond (which just matures) is covered, yielding a net cash flow of

$$V = 1 + \Delta T f_{\Delta T}(t, T) - \frac{P(t, T)}{P(t, T + \Delta T)}.$$

Because this is a set of zero net transactions initially, in the absence of arbitrage, V must be zero, which leads to the following expression of the fair lending rate:

$$f_{\Delta T}(t, T) = \frac{1}{\Delta T} \left(\frac{P(t, T)}{P(t, T + \Delta T)} - 1 \right).$$

Hence, the arbitrage free or fair forward lending rate is totally determined by the prices of zero-coupon bonds. We call $f_{\Delta T}(t, T)$ the simple forward rate for the period $(T, T + \Delta T)$ seen at time t, or simply a forward rate.

Consider now the limiting case, $\Delta T \to 0$, for the forward rate. There is

$$
\begin{aligned}
f(t, T) &\triangleq \lim_{\Delta T \to 0} f_{\Delta T}(t, T) \\
&= \lim_{\Delta T \to 0} \frac{1}{\Delta T} \left(\frac{P(t, T)}{P(t, T + \Delta T)} - 1 \right) \\
&= \frac{-1}{P(t, T)} \frac{\partial P(t, T)}{\partial T} \\
&= -\frac{\partial \ln P(t, T)}{\partial T}.
\end{aligned}
\tag{3.23}
$$

We call $f(t, T)$ an *instantaneous forward rate*. According to Equation 3.23, we can express the price of a T-maturity zero-coupon bond in terms of $f(t, s)$, $t \leq s \leq T$:

$$P(t, T) = e^{-\int_t^T f(t,s)\, ds}.$$

Note that once we have a model for the forward rates, we also have a model for the discount bonds. Because of the above relationship, the instantaneous forward rates are treated as potential candidates of state variables for interest-rate modeling.

3.5 Yield-Based Bond Risk Management

In previous sections, we have completely characterized the price–yield relationship of bonds. Bond price fluctuations can be attributed to yield fluctuations, and price risks can be treated as yield risks. It has become a commonplace that yields to maturities of all bonds from the same issuer are highly correlated: yields (of different maturities) often move in the same direction with comparable magnitudes. This reality makes it possible to hedge against the price change of one bond with another. In this section, we introduce a yield-based theory of bond risk management, which is about hedging the bond price risk against the parallel shift of the yield curve. The theory was initially developed for bonds. Yet, as we shall see, it also applies to managing the yield risk of other interest-rate-sensitive instruments.

3.5.1 Duration and Convexity

In the bond market, bond prices change unpredictably on a daily basis. The changes in bond prices can be interpreted as the consequence of unpredictable changes in yields. The duration of a bond is a measure of risk exposure with respect to a possible change in the bond yield. It has been observed that the prices of long-maturity bonds are more sensitive to change in yields than are the prices of short-maturity bonds, and the impact of yield changes on bond prices seems proportional to the cash flow dates of the bonds. Intuitively, Macaulay (1938) introduced the weighted average of the cash flow dates as a measure of price sensitivity with respect to the bond yield:

$$
D_{\mathrm{mac}} = \frac{\mathrm{Pr}}{B_t^c} \left[\sum_{i,T_i>t}^{n} \Delta T \cdot c(1+y\Delta T)^{-(T_i-t)/\Delta T}(T_i-t) \right.
$$

$$
\left. + (1+y\Delta T)^{-(T_n-t)/\Delta T}(T_n-t) \right].
$$

This measure is called the *Macaulay duration* in the bond market. Note that, for a zero-coupon bond, the duration is simply its maturity. It was later understood that the Macaulay duration is closely related to the derivative of the bond price with respect to its yield. In fact, differentiating Equation 3.13 with respect to y yields

$$\frac{\mathrm{d}B_t^c}{\mathrm{d}y} = -\frac{\mathrm{Pr}}{1 + y\Delta T}\left[\sum_{i;T_i > t}^{n} \Delta T \cdot c(1 + y\Delta T)^{-(T_i - t)/\Delta T}(T_i - t)\right.$$

$$\left. + (1 + y\Delta T)^{-(T_n - t)/\Delta T}(T_n - t)\right]. \tag{3.24}$$

In terms of D_{mac}, the Macaulay duration just defined, we have

$$\frac{\mathrm{d}B_t^c}{B_t^c} = -\frac{D_{\mathrm{mac}}}{1 + y\Delta T}\mathrm{d}y \quad \text{or} \quad \frac{1}{B_t^c}\frac{\mathrm{d}B_t^c}{\mathrm{d}y} = -\frac{D_{\mathrm{mac}}}{1 + y\Delta T}. \tag{3.25}$$

According to Equation 3.25, the Macaulay duration is essentially the rate of change with respect to the yield for each dollar of market value of the bond. After multiplying by the change in the yield, the Macaulay duration gives the percentage change in the value of the bond. For convenience, we define

$$D_{\mathrm{mod}} = \frac{D_{\mathrm{mac}}}{1 + y\Delta T},$$

and call it the *modified duration*. Then the first equation of Equation 3.25 can be written in the following simple form:

$$\frac{\mathrm{d}B_t^c}{B_t^c} = -D_{\mathrm{mod}}\,\mathrm{d}y. \tag{3.26}$$

Both D_{mac} and D_{mod} are called duration measures of bonds.

By using Equation 3.17, the succinct bond formula, we can obtain the following formula for the modified duration:

$$D_{\mathrm{mod}} = \frac{\mathrm{Pr}}{B^c}\left[\frac{c}{y^2}\left(1 - \frac{1}{(1 + y\Delta T)^n}\right) + \left(1 - \frac{c}{y}\right)\frac{n\Delta T}{(1 + y\Delta T)^{n+1}}\right]. \tag{3.27}$$

The above expression is simplified for par bonds. When $c = y$ and $B^c = \mathrm{Pr}$, we have

$$D_{\mathrm{mod}} = \frac{1}{y}[1 - (1 + y\Delta T)^{-n}]. \tag{3.28}$$

Note that Treasury bonds are quoted in yields and that recent issues are usually traded close to par, so Equation 3.28 gives us an approximate value of the durations for bonds being traded close to par.

The next example shows how much the dollar value of a bond changes given its duration.

Example 3.5.1. *Given a 30-year Treasury yielding 5% and trading at par, we can calculate the modified duration using Equation 3.28 and obtain $D_{\mathrm{mod}} = 15.45$ years. This means that a one basis point variation in the yield will cause a change of 15.45 cents per 100 dollars.*

From a mathematical viewpoint, the estimation of price changes using duration is equivalent to estimating functional value using a linear approximation. The accuracy of such an approximation becomes poorer when the change of the yield becomes larger. To see that, we consider the second-order expansion of the bond price change with respect to the yield:

$$\Delta B_t^c = \frac{\mathrm{d} B_t^c}{\mathrm{d} y} \Delta y + \frac{1}{2} \frac{\mathrm{d}^2 B_t^c}{\mathrm{d} y^2} (\Delta y)^2. \tag{3.29}$$

We also introduce the *convexity measure*:

$$
\begin{aligned}
C = \frac{1}{B_t^c} \frac{\mathrm{d}^2 B_t^c}{\mathrm{d} y^2} &= \frac{\mathrm{Pr}}{B_t^c} \left[\sum_{i; T_i > t}^{N} \Delta T \cdot c (T_i - t)(T_{i+1} - t) \right. \\
&\times (1 + y \Delta T)^{-((T_{i+1}-t)/\Delta T)-1} \\
&\left. + (T_N - t)(T_{N+1} - t)(1 + y \Delta T)^{-((T_{N+1}-t)/\Delta T)-1} \right].
\end{aligned} \tag{3.30}
$$

Then we can rewrite Equation 3.29 as

$$\frac{\Delta B_t^c}{B_t^c} = -D_{\mathrm{mod}} \Delta y + \frac{1}{2} C \Delta y^2. \tag{3.31}$$

Note that C captures the convexity or curvature of the bond–price curve. Given the duration measure, D_{mod}, and the convexity measure, C, we can readily calculate the percentage change in the price with Equation 3.31. When $\Delta y^2 \ll \Delta y$, we can neglect the second-order term in Equation 3.31 and thus return to Equation 3.26. However, for relatively large Δy, the inclusion of the convexity term in Equation 3.31 will be necessary to produce a good approximation of the percentage change. The curves of the exact price, the linear approximation, and the quadratic approximation are shown in Figure 3.5.

We remark here that although the duration measure and the convexity measure are introduced here for bonds, their applications are not restricted to bonds. We can apply them to any interest-rate-sensitive instruments.

3.5.2 Portfolio Risk Management

We can also calculate the duration and convexity of a portfolio of fixed-income instruments. Consider a portfolio of N instruments, with n_i units and price B_i^c for the ith instrument. Then, the absolute change in the portfolio value upon a parallel yield shift is given by

$$\mathrm{d} V = \sum_i n_i \mathrm{d} B_i^c = \sum_i n_i B_i^c \cdot \left(-D_{\mathrm{mod}}^i \mathrm{d} y + \frac{1}{2} C^i \mathrm{d} y^2 \right).$$

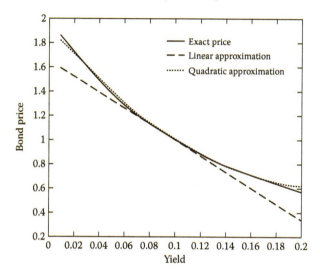

FIGURE 3.5: Linear and quadratic approximations of bond prices.

The percentage change is then

$$\frac{\mathrm{d}V}{V} = -\left(\sum_i x_i D^i_{\mathrm{mod}}\right)\mathrm{d}y + \frac{1}{2}\left(\sum_i x_i C^i\right)\mathrm{d}y^2, \qquad (3.32)$$

where $x_i = n_i B^c_i / V$ is the percentage of the value in the ith instrument. Equation 3.32 indicates that the duration and convexity of a portfolio are the weighted average of the duration and convexity of its components, respectively.

In classical risk management, a portfolio manager can limit his/her exposure to interest-rate risk by reducing the duration while increasing the convexity of the portfolio. To avoid possible losses in case of large yield moves, the manager usually will not tolerate negative net convexity. A portfolio with very small duration is called a duration-neutral portfolio. Practically, interest-rate futures and swaps are often used as hedging instruments for duration management.

The basic premise of the duration and convexity technology is that the yield curves shift in parallel, either upward or downward by the same amount. This is, however, a very crude assumption about the yield curve movement as, in reality, points in a yield curve do not often shift by the same amount and sometimes they do not even move in the same direction. For a more elaborate model of yield curve dynamics, we will have to resort to stochastic calculus in a multi-factor setting.

Exercises

1. Compute the Macaulay duration of a 30-year par bond with 6% coupons paid semiannually. If the yield decreases by one basis point, by how much will the value of the bond change?

2. Let t_j^+ be the moment immediately after a coupon payment. Prove that, at $t = t_j^+$, the price of a coupon bond is

$$B = \frac{c \cdot \text{Pr}}{y} \left(1 - \frac{1}{(1 + y\Delta t)^{N-j}} \right) + \frac{\text{Pr}}{(1 + y\Delta t)^{N-j}}.$$

Furthermore, do the following:

(a) Given $c = 5\%$, $j = 0$, $N = 20$, and $\Delta t = 0.5$, plot B against y for y varying from 0 to 0.5.

(b) Given, in addition, $B = 102\text{-}24$, calculate y.

3. Use formula (Equation 3.17) to prove that at $t = t_j^+$, the modified duration of a coupon bond trading at par and having $\nu = N - j$ coupons remaining, is given by

$$D_{\text{mod}} = \frac{1}{y} \left(1 - \frac{1}{(1 + y\Delta t)^{\nu}} \right).$$

4. A coupon-paying bond will mature in 15 and a quarter years and the coupons are paid semiannually. Suppose that the principal is 100, the coupon rate is 7%, and the dirty price is 102. Compute the yield, modified duration, and convexity of the bond.

5. Assume continuous compounding. Prove the yield–forward rate relation

$$f_{\Delta T}(0, T_1) = \frac{T_2 y(T_2) - T_1 y(T_1)}{T_2 - T_1}, \quad T_2 > T_1,$$

where $\Delta T = T_2 - T_1$, and

$$f(0; T) = y(T) + T y_T(T).$$

Here, $y(T)$ is the T-maturity term rate that relates to the zero-coupon price by $P(0, T) = e^{-T y(T)}$.

Chapter 4

The Heath–Jarrow–Morton Model

Interest-rate models are necessary for pricing most interest-rate derivatives and for gauging the risk of general interest-rate instruments. The natural candidate for the state variable of interest-rate models seems to be the short rate, which in fact is the only state variable in many early models, like the well-known Vasicek (1977) model or the Cox–Ingersoll–Ross (CIR) (1985) model. Early models are largely based on macro-economical arguments, and they are thus called equilibrium models. Equilibrium models may have a sound financial economical background, but they do not automatically reproduce the market price of benchmark bonds, or reproduce the yield curve, unless a calibration procedure has been carried through. It is known that such a procedure may result in a model that is twisted and sometimes at odds with financial intuition. In fact, as models based on a single-state variable, short-rate models seem to lack sufficient capacity to describe the dynamic features of the entire yield curve.

To exclude arbitrage, an interest-rate model should take the entire yield curve as an input instead of output. This consideration has catalyzed the development of the so-called arbitrage pricing models. A basic feature of arbitrage pricing models of interest rates is that the prices of a whole class of basic securities, for example, the prices of zero-coupon bonds of all maturities, are taken as primitive state variables. Equivalently, an arbitrage pricing model can also take some kind of yield curve, discrete or continuous, as the state variables. The zero-coupon yield curve and the par-yield curve are two examples. It turns out that (the continuous or discrete compounding) forward rates are often the best choice of state variables for many applications. With the continuous compounding forward rates, Heath, Jarrow, and Morton (1992) developed an arbitrage framework for interest-rate modeling, such that any specific arbitrage-free model may be fitted to the framework as a special case. This chapter describes the derivation, estimation, and application of the Heath–Jarrow–Morton (HJM) model.

With historical data of U.S. Treasury yields, we demonstrate the estimation of model parameters of the HJM model, using an important technique called principal component analysis (PCA). We then study in depth two classic special cases under the HJM framework, namely, the Ho–Lee model (1986) and the Hull–White (1989) model, which played important roles both in the development of interest-rate models and in practical applications. Through our analysis here, we should have a better idea of what we expect from an interest-rate model.

Derivatives pricing under the HJM model is another focus of this chapter. As a Markovian model in forward rates, the HJM model can be conveniently implemented through Monte Carlo simulation methods. Without loss of generality, we demonstrate the Monte Carlo pricing of bond options with a succinct algorithm under a two-factor HJM model. But this is not our focus. Our focus instead is on the use of forward measures, a kind of powerful device for pricing interest-rate derivatives. The forward measures are the martingale measures with zero-coupon bonds to be the numeraires. Pricing under an appropriate forward measure can make both pricing and hedging transparent. By changing from the risk-neutral measure to forward measures, we highlight the importance of making the right choice of numeraire assets, which is at the disposal of quantitative analysts or traders, in derivatives pricing. We will discuss the change of measure corresponding to the change of numeraire assets in general. Finally, we will introduce the linear Gaussian model, an arbitrage-free model which enjoys certain popularity in the market due to its flexibility and simplicity.

We need to point out that there does not exist a framework yet out of which any arbitrage pricing model can be derived. The HJM model may be better treated as a necessary condition for a large class of no-arbitrage models. It is not yet known in general how to specify the HJM model in order to avoid negative interest rates, which is certainly a kind of arbitrage to the markets. On the other hand, it is very difficult, if not impossible, to interpret positive interest-rate models in the HJM context. At the end of this chapter, we offer more comments on general positive interest-rate models.

4.1 Lognormal Model: The Starting Point

The theoretical basis of this chapter starts from the usual assumption of lognormal asset dynamics for zero-coupon bonds of all maturities:

$$dP(t,T) = P(t,T) \left[\mu(t,T)\, dt + \mathbf{\Sigma}^{\mathrm{T}}(t,T)\, d\mathbf{W}_t \right], \qquad (4.1)$$

under the physical measure, $\mathbb{P}$. Here $\mu(t,T)$ is a scalar function of t and T, $\mathbf{\Sigma}(t,T)$ is a column vector,

$$\mathbf{\Sigma}(t,T) = \left(\Sigma_1(t,T), \Sigma_2(t,T), \ldots, \Sigma_n(t,T) \right)^{\mathrm{T}},$$

and $\mathbf{W}_t$ is an n-dimensional $\mathbb{P}$-Brownian motion,

$$\mathbf{W}_t = \left(W_1(t), W_2(t), \ldots, W_n(t) \right)^{\mathrm{T}}.$$

In principle, the coefficients in Equation 4.1 can be estimated from time series data of zero-coupon bonds, yet it is not guaranteed that Equation 4.1 with estimated drift and volatility functions can exclude arbitrage. For the time being, we assume that both $\mu(t,T)$ and $\boldsymbol{\Sigma}(t,T)$ are sufficiently regular deterministic functions of t, so that the SDE (Equation 4.1) admits a unique strong solution.

The purpose of a model like Equation 4.1 is to price derivatives depending on (a portfolio of) $P(t,T), \forall T$ and $t \leq T$. For this purpose, we need to find a martingale measure for zero-coupon bonds of all maturities. Similar to our discussions on the multiple-asset market, we define an $\mathcal{F}_t$-adaptive process, $\boldsymbol{\gamma}_t$, that satisfies the following equation:

$$\boldsymbol{\Sigma}^{\mathrm{T}}(t,T)\boldsymbol{\gamma}_t = \boldsymbol{\mu}(t,T) - r_t\mathbf{I}.$$

Suppose that such a $\boldsymbol{\gamma}_t$ exists, is independent of T, and satisfies the Novikov condition. We can define a measure, $\mathbb{Q}$, as

$$\left.\frac{\mathrm{d}\mathbb{Q}}{\mathrm{d}\mathbb{P}}\right|_{\mathcal{F}_t} = \exp\left(\int_0^t -\boldsymbol{\gamma}_s^{\mathrm{T}}\mathrm{d}\mathbf{W}_s - \frac{1}{2}\|\boldsymbol{\gamma}_s\|_2^2\,\mathrm{d}s\right). \tag{4.2}$$

Then, by the CMG theorem, the process

$$\tilde{\mathbf{W}}_t = \mathbf{W}_t + \int_0^t \boldsymbol{\gamma}_s\,\mathrm{d}s \tag{4.3}$$

is a $\mathbb{Q}$-Brownian motion, and, in terms of $\tilde{\mathbf{W}}_t$, we can rewrite Equation 4.1 as

$$\begin{aligned}
\mathrm{d}P(t,T) &= P(t,T)\left[r_t\,\mathrm{d}t + \boldsymbol{\Sigma}^{\mathrm{T}}(t,T)\left(\mathrm{d}\mathbf{W}_t + \boldsymbol{\gamma}_t\,\mathrm{d}t\right)\right] \\
&= P(t,T)\left[r_t\,\mathrm{d}t + \boldsymbol{\Sigma}^{\mathrm{T}}(t,T)\,\mathrm{d}\tilde{\mathbf{W}}_t\right].
\end{aligned}$$

It then follows that the discounted prices of all maturities, $B_t^{-1}P(t,T)$, are $\mathbb{Q}$-martingales.

Now let us address the existence of such a $\boldsymbol{\gamma}_t$. Without loss of generality, we assume that the market of zero-coupon bonds is non-degenerate. That is, there exist at least n distinct zero-coupon bonds such that their volatility vectors constitute a non-singular matrix. Let $\{T_i\}_{i=1}^n$ be the maturities of the n bonds such that $T_i < T_{i+1}$, and let $\{\boldsymbol{\Sigma}(t,T_i)\}_{i=1}^n$ be the column vectors of their volatilities. By introducing matrices

$$\mathbf{A} = \begin{pmatrix} \boldsymbol{\Sigma}^{\mathrm{T}}(t,T_1) \\ \boldsymbol{\Sigma}^{\mathrm{T}}(t,T_2) \\ \vdots \\ \boldsymbol{\Sigma}^{\mathrm{T}}(t,T_n) \end{pmatrix}, \quad \boldsymbol{\mu}_t = \begin{pmatrix} \mu(t,T_1) \\ \mu(t,T_2) \\ \vdots \\ \mu(t,T_n) \end{pmatrix}, \quad \text{and} \quad \mathbf{I} = \begin{pmatrix} 1 \\ 1 \\ \vdots \\ 1 \end{pmatrix}, \tag{4.4}$$

we then define $\boldsymbol{\gamma}_t$ as the solution to the linear system

$$\mathbf{A}\boldsymbol{\gamma}_t = \boldsymbol{\mu}_t - r_t\mathbf{I}. \tag{4.5}$$

This solution is unique provided that $\mathbf{A}$ is non-singular. Such a solution, however, appears to depend on $T_i, 1 \leq i \leq n$, the input maturities. With such a

γ_t, we can define a new measure, $\mathbb{Q}$, from Equation 4.2, and under which the discounted prices of those n zero-coupon bonds, $B_t^{-1}P(t, T_i), i = 1, \ldots, n$, are martingales.

Next, we will show that the solution to Equation 4.5, γ_t, also satisfies

$$\mathbf{\Sigma}^{\mathrm{T}}(t, T)\gamma_t = \mu(t, T) - r_t, \quad \forall T \leq T_n. \tag{4.6}$$

The implication is that the new measure, $\mathbb{Q}$, defined using γ_t is a martingale measure for zero-coupon bonds of all maturities $T \leq T_n$. To show that, we define

$$N_t = E^{\mathbb{Q}}\left[B_T^{-1} \mid \mathcal{F}_t\right], \quad \forall T \leq T_n, \tag{4.7}$$

which is a $\mathbb{Q}$-martingale by definition. According to the (multi-factor version of the) martingale representation theorem, we know that N_t is the value of a self-financing portfolio consisting of the money market account and the n zero-coupon bonds of maturities, $\{T_i\}_{i=1}^n$. The value of the portfolio at time T is nothing else but one, which is identical to the value of the T-maturity zero-coupon bond at maturity. By the dominance principle, there must be $B_t^{-1}P(t, T) = N_t$, so it follows that $B_t^{-1}P(t, T)$ is also a $\mathbb{Q}$-martingale. In terms of the $\tilde{\mathbf{W}}_t$ defined in Equation 4.3, the price process of $P(t, T)$ can be written as

$$\mathrm{d}\left(\frac{P(t, T)}{B(t)}\right) = \frac{P(t, T)}{B(t)}\left[\left(\mu(t, T) - r_t - \mathbf{\Sigma}^{\mathrm{T}}(t, T)\gamma_t\right)\mathrm{d}t + \mathbf{\Sigma}^{\mathrm{T}}(t, T)\,\mathrm{d}\tilde{\mathbf{W}}_t\right].$$

$$\tag{4.8}$$

The martingale property of $B_t^{-1}P(t, T)$ dictates that the drift term must vanish, yielding Equation 4.6.

The equality in Equation 4.6 can also be justified without using the martingale representation theorem. Because $\mathbf{A}$ is non-singular, for any T between 0 and T_n, there exists a unique vector, $\boldsymbol{\theta} = (\theta_1, \theta_2, \ldots, \theta_n)$, such that

$$(\theta_1 P_1, \theta_2 P_2, \ldots, \theta_n P_n)\mathbf{A} = P(t, T)\mathbf{\Sigma}^{\mathrm{T}}(t, T), \tag{4.9}$$

where we have denoted $P_i = P(t, T_i)$ for notational simplicity. Consider the portfolio

$$V_t = P(t, T) - \sum_{i=1}^n \theta_i P(t, T_i). \tag{4.10}$$

The value process of the portfolio is

$$\begin{aligned}
\mathrm{d}V_t &= \mathrm{d}P(t, T) - \sum_{i=1}^n \theta_i \,\mathrm{d}P(t, T_i) \\
&= \left[P(t, T)\mu(t, T) - \sum_{i=1}^n \theta_i P(t, T_i)\mu(t, T_i)\right]\mathrm{d}t \\
&\quad + \left[P(t, T)\mathbf{\Sigma}^{\mathrm{T}}(t, T) - \sum_{i=1}^n \theta_i P(t, T_i)\mathbf{\Sigma}^{\mathrm{T}}(t, T_i)\right]\mathrm{d}\tilde{\mathbf{W}}_t. \tag{4.11}
\end{aligned}$$

According to the definition of θ_i, the volatility coefficient of Equation 4.11 is zero, implying that V_t is a riskless portfolio. In the absence of arbitrage, the riskless portfolio must earn a return rate equal to the risk-free rate, meaning

$$P(t,T)\mu(t,T) - \sum_{i=1}^{n} \theta_i P(t,T_i)\mu(t,T_i) = r_t \left[P(t,T) - \sum_{i=1}^{n} \theta_i P(t,T_i) \right]. \quad (4.12)$$

By making use of Equation 4.5, we can rewrite Equation 4.12 as

$$\mu(t,T) - r_t = \sum_{i=1}^{n} \theta_i \frac{P(t,T_i)}{P(t,T)} (\mu(t,T_i) - r_t)$$

$$= \sum_{i=1}^{n} \theta_i \frac{P(t,T_i)}{P(t,T)} \mathbf{\Sigma}^{\mathrm{T}}(t,T_i)\boldsymbol{\gamma}_t$$

$$= \mathbf{\Sigma}^{\mathrm{T}}(t,T)\boldsymbol{\gamma}_t,$$

which is exactly Equation 4.6.

We comment here that the components of $\boldsymbol{\gamma}_t$ are considered to be the market prices of risks for zero-coupon bonds, $P(t,T), \forall T$. Hence, if zero-coupon bonds of all maturities are driven by a finite number of Brownian motions, then they share the same set of market prices of risks.

4.2 The HJM Model

Under the martingale measure, $\mathbb{Q}$, the price process of a zero-coupon bond becomes

$$\mathrm{d}P(t,T) = P(t,T) \left[r_t \, \mathrm{d}t + \mathbf{\Sigma}^{\mathrm{T}}(t,T) \, \mathrm{d}\tilde{\mathbf{W}}_t \right]. \quad (4.13)$$

For the purpose of derivatives pricing, $\mathbf{\Sigma}(t,T)$ must satisfy at least the following additional conditions: (1) $\mathbf{\Sigma}(t,t) = 0, \forall t$; and (2) $P(t,t) = 1, \forall t$. These two conditions reflect only one fact: at maturity, the price of the zero-coupon bond equals its par value and thus has no volatility.

The specification of $\mathbf{\Sigma}(t,T)$ is a difficult job if we work directly with the process of $P(t,T)$. But the job will become quite amenable if we work with the process of forward rates. By Ito's lemma, there is

$$\mathrm{d}\ln P(t,T) = \left[r_t - \frac{1}{2}\mathbf{\Sigma}^{\mathrm{T}}(t,T)\mathbf{\Sigma}(t,T) \right] \mathrm{d}t + \mathbf{\Sigma}^{\mathrm{T}}(t,T) \, \mathrm{d}\tilde{\mathbf{W}}_t. \quad (4.14)$$

Assume, moreover, that $\mathbf{\Sigma}_T(t,T) = \partial\mathbf{\Sigma}(t,T)/\partial T$ exists and $\int_0^T \|\mathbf{\Sigma}_T(t,T)\|^2 \, \mathrm{d}t < \infty$. By differentiating Equation 4.14 with respect to T and recalling that

$$f(t,T) = -\frac{\partial \ln P(t,T)}{\partial T}, \quad (4.15)$$

we obtain the process of forward rates under the $\mathbb{Q}$-measure,

$$\mathrm{d}f(t,T) = \mathbf{\Sigma}_T^{\mathrm{T}}\mathbf{\Sigma}\,\mathrm{d}t - \mathbf{\Sigma}_T^{\mathrm{T}}\,\mathrm{d}\tilde{\mathbf{W}}_t. \tag{4.16}$$

Here the arguments of $\mathbf{\Sigma}$ are omitted for simplicity. We consider $-\mathbf{\Sigma}_T(t,T)$ to be the volatility of the forward rate, and retag it as

$$\boldsymbol{\sigma}(t,T) = -\mathbf{\Sigma}_T(t,T). \tag{4.17}$$

By integrating the above equation with respect to T, we then obtain the volatility of the zero-coupon bond:

$$\mathbf{\Sigma}(t,T) = -\int_t^T \boldsymbol{\sigma}(t,s)\,\mathrm{d}s. \tag{4.18}$$

Here we have made use of the condition $\mathbf{\Sigma}(t,t) = 0$. Equation 4.18 fully describes the relationship between the volatility of a forward rate and the volatility of its corresponding zero-coupon bond. In terms of $\boldsymbol{\sigma}(t,T)$, we can rewrite this forward-rate process as

$$\mathrm{d}f(t,T) = \left(\boldsymbol{\sigma}^{\mathrm{T}}(t,T)\int_t^T \boldsymbol{\sigma}(t,s)\,\mathrm{d}s\right)\mathrm{d}t + \boldsymbol{\sigma}^{\mathrm{T}}(t,T)\,\mathrm{d}\tilde{\mathbf{W}}_t. \tag{4.19}$$

Equation 4.19 is the famous HJM equation. An important feature of Equation 4.19 is that the drift term of the forward-rate process under the martingale measure, $\mathbb{Q}$, is completely determined by its volatility.

The HJM model lays down the foundation of arbitrage pricing in the context of fixed-income derivatives, and it is considered a milestone of financial derivative theory. Before 1992, fixed-income modeling was dominated by the so-called equilibrium models. Such models are based on macroeconomic arguments. The major limitation of equilibrium models is that these models do not naturally reproduce the market prices of basic instruments, zero-coupon bonds in particular, unless users go through a calibration procedure. With arbitrage pricing models, the prices of the basic instruments are treated as model inputs rather than as outputs, so their prices are naturally reproduced. The arbitrage pricing models are rooted in the efficient market hypothesis, which states that market prices of instruments do not induce any arbitrage opportunity. With an arbitrage model, derivative securities will be priced consistently with the basic instruments in the sense that no arbitrage opportunity would be induced.

There are two ways to specify the forward-rate volatility in the HJM model. The first way is to estimate $\boldsymbol{\sigma}(t,T)$ (together with its dimension) directly from time series data on forward rates of various maturities. Note that $\boldsymbol{\sigma}^{\mathrm{T}}(t,T)\boldsymbol{\sigma}(t,T')$ should reflect the covariance between $f(t,T)$ and $f(t,T')$. We will carry out such an estimation in Section 4.4. The second way is to specify $\boldsymbol{\sigma}(t,T)$ exogenously, using certain parametric functions of t and T. Note that different specifications of $\boldsymbol{\sigma}(t,T)$ generate different concrete models for applications. Because of that, the HJM model is treated not only as a model in its own right, but also as a framework for fixed-income models that are deemed arbitrage-free.

The HJM Equation 4.19 is a necessary condition for no-arbitrage models, but not a sufficient one. For usual specifications of the volatility function, $\boldsymbol{\sigma}(t,T)$, the forward rate has a Gaussian distribution, so it can assume negative values with a positive possibility. It is not yet clear under what conditions for $\boldsymbol{\sigma}(t,T)$ we can ensure that the forward rate stays positive. On the other hand, it can also be very difficult to identify the corresponding forward-rate volatility functions for some existing short-rate models, particularly the short-rate models that guarantee positive interest rates. One objective of Chapter 5 is to derive the corresponding forward-rate volatility function for short-rate models.

The HJM model also implies the dynamics of the short rate. By integrating Equation 4.16 from 0 to t, we obtain the expression of forward rates:

$$f(t,T) = f(0,T) + \int_0^t \boldsymbol{\Sigma}^{\mathrm{T}}(s,T)\frac{\partial \boldsymbol{\Sigma}(s,T)}{\partial T}\,\mathrm{d}s - \frac{\partial \boldsymbol{\Sigma}^{\mathrm{T}}(s,T)}{\partial T}\,\mathrm{d}\tilde{\mathbf{W}}_s. \qquad (4.20)$$

Then by setting $T = t$, we obtain an expression for the short rate:

$$r_t = f(t,t) = f(0,t) + \int_0^t \frac{1}{2}\frac{\partial \|\boldsymbol{\Sigma}(s,t)\|^2}{\partial t}\,\mathrm{d}s - \frac{\partial \boldsymbol{\Sigma}^{\mathrm{T}}(s,t)}{\partial t}\,\mathrm{d}\tilde{\mathbf{W}}_s. \qquad (4.21)$$

In differential form, Equation 4.21 becomes

$$\mathrm{d}r_t = \mathrm{drift} - \left(\int_0^t \frac{\partial^2 \boldsymbol{\Sigma}^{\mathrm{T}}(s,t)}{\partial t^2}\,\mathrm{d}\tilde{\mathbf{W}}_s\right)\mathrm{d}t - \left.\frac{\partial \boldsymbol{\Sigma}^{\mathrm{T}}(s,t)}{\partial t}\right|_{s=t}\mathrm{d}\tilde{\mathbf{W}}_t. \qquad (4.22)$$

By examining Equation 4.22, we understand that, under the HJM framework, the short rate is in general a non-Markovian random variable, unless

$$\int_0^t \frac{\partial^2 \boldsymbol{\Sigma}^{\mathrm{T}}(s,t)}{\partial t^2}\,\mathrm{d}\tilde{\mathbf{W}}_s \qquad (4.23)$$

can be expressed as a function of some Markovian variables. Only in such a situation can we call the short-rate dynamics a Markovian variable. We devote all of Chapter 5 to Markovian short-rate models.

For some subsequent applications, we rewrite Equation 4.21 as

$$r_t = f(0,t) + \frac{\partial}{\partial t}\int_0^t \frac{1}{2}\|\boldsymbol{\Sigma}(s,t)\|^2\mathrm{d}s - \boldsymbol{\Sigma}^{\mathrm{T}}(s,t)\,\mathrm{d}\tilde{\mathbf{W}}_s. \qquad (4.24)$$

Here, we have applied the following stochastic Fubini theorem (see Karatzas and Shreve, 1991) to exchange the order of differentiation and integration:

$$\frac{\partial}{\partial t}\int_0^t \theta(s,t)\,\mathrm{d}\tilde{W}_s = \theta(t,t)\frac{\mathrm{d}\tilde{W}_t}{\mathrm{d}t} + \int_0^t \frac{\partial}{\partial t}\theta(s,t)\,\mathrm{d}\tilde{W}_s,$$

and we have applied the property $\Sigma(t, t) = 0$.

Finally, we note that, under the HJM model, the following price formula for zero-coupon bonds exists:

$$P(t, T) = P(0, T) \exp \left\{ \int_0^t \left(r_s - \frac{1}{2} \Sigma^{\mathrm{T}}(s, T) \Sigma(s, T) \right) \mathrm{d}s + \Sigma^{\mathrm{T}}(s, T) \, \mathrm{d}\tilde{\mathbf{W}}_s \right\},$$

(4.25)

which will be used repeatedly for analyses as well as computations.

4.3 Special Cases of the HJM Model

Since the publication of the HJM model in 1992, arbitrage pricing models have quickly acquired dominant status in fixed-income modeling. Arbitrage pricing models have been generated from the HJM framework by making various specifications of forward-rate volatility. In this section, we study two specifications of the forward-rate volatility that, in terms of the short-rate dynamics, reproduce the popular one-factor models of Ho and Lee (1986) and Hull and White (1989), respectively.

4.3.1 The Ho–Lee Model

The simplest specification of the HJM model is $\sigma = \text{const}$ for $n = 1$, corresponding to the forward-rate equation

$$\mathrm{d}f(t, T) = \sigma \, \mathrm{d}\tilde{W}_t + \sigma^2 (T - t) \, \mathrm{d}t.$$

By integrating the equation over $[0, t]$, we obtain

$$f(t, T) - f(0, T) = \sigma \tilde{W}_t + \frac{1}{2} \sigma^2 t \left(2T - t \right).$$

By making $T = t$, we have the expression for the short rate:

$$r_t = f(t, t) = f(0, t) + \frac{1}{2} \sigma^2 t^2 + \sigma \tilde{W}_t.$$

In differential form, the last equation becomes

$$\mathrm{d}r_t = \left(f_T(0, t) + \sigma^2 t \right) \mathrm{d}t + \sigma \, \mathrm{d}\tilde{W}_t.$$

(4.26)

Equation 4.26 is interpreted as the continuous-time version of the so-called Ho–Lee (1986) model, which was first developed in the context of binomial trees.

Let us take a look at a basic feature of the Ho–Lee model. It is quite obvious to see that

$$E^{\mathbb{Q}}[r_t] = f(0,t) + \frac{1}{2}\sigma^2 t^2,$$
$$\text{Var}[r_t] = \sigma^2 t. \tag{4.27}$$

The two equations of Equation 4.27 suggest that the short rate will fluctuate around a quadratic function of time with increasing variance. This feature is counter to common sense, and it has motivated alternative specifications to make the short rate behave more reasonably.

With the Ho–Lee model, we can obtain the following price formula of zero-coupon bonds:

$$P(t,T) = \exp\left\{ -\int_t^T \left[f(0,s) + \sigma\tilde{W}_t + \frac{\sigma^2}{2}t\,(2s-t) \right] ds \right\}$$
$$= \exp\left\{ -\left[\int_t^T f(0,s)\,ds + \sigma\tilde{W}_t\,(T-t) + \frac{\sigma^2}{2}tT\,(T-t) \right] \right\}$$
$$= \frac{P(0,T)}{P(0,t)} \exp\left\{ -\left[\sigma\tilde{W}_t\,(T-t) + \frac{\sigma^2}{2}tT\,(T-t) \right] \right\}.$$

This formula can be used to, among other applications, price options on zero-coupon bonds.

4.3.2 The Hull–White (or Extended Vasicek) Model

It has been empirically observed that forward-rate volatility decays with time-to-maturity, $T-t$. This motivates the following specification of the volatility:

$$\sigma(t,T) = \sigma e^{-\kappa(T-t)}, \quad \kappa > 0. \tag{4.28}$$

That is, the volatility decays exponentially as time goes forward. The corresponding HJM equation now reads,

$$df(t,T) = \sigma e^{-\kappa(T-t)}d\tilde{W}_t + \left[\sigma e^{-\kappa(T-t)} \int_t^T \sigma e^{-\kappa(s-t)}ds \right] dt$$
$$= \sigma e^{-\kappa(T-t)}d\tilde{W}_t + \sigma e^{-\kappa(T-t)}\frac{\sigma}{\kappa}\left[1 - e^{-\kappa(T-t)} \right] dt$$
$$= \sigma e^{-\kappa(T-t)}d\tilde{W}_t + \frac{\sigma^2}{\kappa}\left[e^{-\kappa(T-t)} - e^{-2\kappa(T-t)} \right] dt.$$

Integrating the above equation over $(0,t)$ yields

$$f(t,T) = f(0,T) + \sigma\int_0^t e^{-\kappa(T-s)}d\tilde{W}_s$$
$$+ \frac{\sigma^2}{2\kappa^2}\left[\left(1 - e^{-\kappa T}\right)^2 - \left(1 - e^{-\kappa(T-t)}\right)^2 \right].$$

By making $T = t$, we obtain the expression for the short rate:

$$r_t = f(0, t) + \sigma \int_0^t e^{-\kappa(t-s)} \mathrm{d}\tilde{W}_s + \frac{\sigma^2}{2\kappa^2} \left(1 - e^{-\kappa t}\right)^2. \qquad (4.29)$$

Yet again, let us check the mean and variance of the short rate, which are

$$E^{\mathbb{Q}}[r_t] = f(0, t) + \frac{\sigma^2}{2\kappa^2} \left(1 - e^{-\kappa t}\right)^2,$$

$$\mathrm{Var}[r_t] = \sigma^2 \int_0^t e^{-2\kappa(t-s)} \mathrm{d}s = \sigma^2 \frac{1}{2\kappa} \left(1 - e^{-2\kappa t}\right) < \frac{\sigma^2}{2\kappa}.$$

The above equations suggest that both the mean and the variance of the short rate stay bounded, a very plausible feature for a short-rate model.

Next, let us study the differential form of Equation 4.29. Denote

$$X_t = \sigma \int_0^t e^{-\kappa(t-s)} \mathrm{d}\tilde{W}_s,$$

which satisfies

$$\mathrm{d}X_t = \sigma \, \mathrm{d}\tilde{W}_t - \sigma\kappa \int_0^t e^{-\kappa(t-s)} \mathrm{d}\tilde{W}_s \, \mathrm{d}t$$

$$= \sigma \, \mathrm{d}\tilde{W}_t - \kappa X_t \, \mathrm{d}t, \qquad (4.30)$$

and relates to the short rate as

$$X_t = r_t - f(0, t) - \frac{\sigma^2}{2\kappa^2} \left(1 - e^{-\kappa t}\right)^2. \qquad (4.31)$$

By differentiating Equation 4.29 and making use of Equation 4.30, we then obtain

$$\mathrm{d}r_t = f_T(0, t) \, \mathrm{d}t + \sigma \, \mathrm{d}\tilde{W}_t - \kappa X_t \, \mathrm{d}t + \frac{\sigma^2}{\kappa} e^{-\kappa t} \left(1 - e^{-\kappa t}\right) \mathrm{d}t$$

$$= \kappa(\theta_t - r_t) \, \mathrm{d}t + \sigma \, \mathrm{d}\tilde{W}_t, \qquad (4.32)$$

where

$$\theta_t \overset{\Delta}{=} f(0, t) + \frac{1}{\kappa} f_T(0, t) + \frac{\sigma^2}{2\kappa^2} \left(1 - e^{-2\kappa t}\right). \qquad (4.33)$$

Equation 4.32 is called the Hull–White (1989) model or sometimes the extended Vasicek model, because Vasicek (1977) was the first to adopt the formalism of Equation 4.32 for short-rate modeling in an equilibrium approach. Note that when $\kappa \to 0$, the Hull–White model reduces to the Ho–Lee model (Equation 4.26).

Let us highlight the so-called mean-reverting feature of the Hull–White model: when $r_t > \theta_t$, the drift is negative; when $r_t < \theta_t$, the drift turns positive. The drift term acts like a force that pushes the short rate toward its mean level, θ_t. The contribution of Hull and White is to identify the level of mean reversion, θ_t, displayed in Equation 4.33, so that zero-coupon bond prices of all maturities are reproduced.

In terms of the short rate, we have the following formula for the zero-coupon bond price (Hull and White, 1989):

$$P(t,T) = A(t,T)e^{-B(t,T)r_t}, \qquad (4.34)$$

where

$$B(t,T) = \frac{1 - e^{-\kappa(T-t)}}{\kappa}, \qquad (4.35)$$

and

$$\ln A(t,T) = \ln \frac{P(0,T)}{P(0,t)} - B(t,T)\frac{\partial \ln P(0,t)}{\partial t}$$
$$- \frac{\sigma^2}{4\kappa^3}\left(e^{-\kappa T} - e^{-\kappa t}\right)^2 \left(e^{2\kappa t} - 1\right). \qquad (4.36)$$

The proof is left as an exercise.

Let us make some additional comments on the Hull–White model from the perspective of applications. Under the Hull–White model, the short rate is a Gaussian random variable and can take a negative value (with a positive probability), which is unrealistic and is considered a major disadvantage of the model. Yet the advantages of this model often outweigh its disadvantages in applications. Due to mean reversion, the probability for negative interest rates is often small. Moreover, since the short rate is a Gaussian variable, the model can be implemented through a lattice tree and thus it is a convenient choice of model to price path-dependent options.

We finish this section with a cautionary note regarding the specification of forward-rate volatility. We have considered constant and exponentially decaying volatilities and thus have reproduced Ho–Lee and Hull–White models. We are, however, unable to go too far in this direction and reproduce many other short-rate models, particularly the ones that guarantee positive short rate. In addition, it has been quite prohibitive to adopt a state-dependent forward-rate volatility. If we do so, we will turn the HJM equation into a non-linear equation for the forward rate, yet the existence of a solution for this kind of non-linear SDE is often problematic. In fact, researchers have tried to make the volatility depend linearly on the forward rate, but only to find out that such a model blows up in finite time. The details are provided in the appendix to this chapter.

4.4   Estimating the HJM Model from Yield Data

We now study the direct specifications for the general HJM model (Equation 4.19) based on the distributional properties of historical data of interest rates. Specifically, we need to specify the dimensions of the model, n, as well as the volatility function of the forward rates, $\sigma(t, T)$, based on time series data of bond yields of various maturities. Note that the obtained HJM model is considered a process under the physical measure instead of the risk-neutral measure. For pricing purposes, we may have to calibrate the model to observed data of benchmark derivatives. This, however, is a very different issue and it is addressed later only in the context of the market model.

4.4.1   From a Yield Curve to a Forward-Rate Curve

The inputs for estimating the HJM model are historical data on U.S. Treasury yields, demonstrated in Figure 4.1 with monthly quotes of yields for a 10-year period, from 1996 to 2006. There are seven curves in the figure, which are dot plots of yield-to-maturities for 3-month, 6-month, 2-year, 3-year, 5-year, 10-year, or 30-year maturity benchmark U.S. Treasury bonds, respectively.

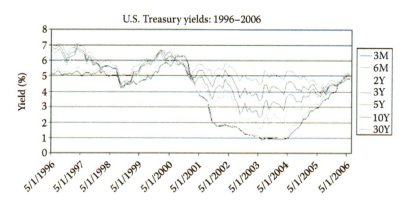

FIGURE 4.1: The monthly quotes of U.S. Treasury benchmark yields. (Adapted from Reuters.)

The first step of our model specification is to estimate the entire forward-rate curve, $f(\tau, \tau + T)$, for each day, τ, in the data set and for the 30-year horizon, $0 \leq T \leq 30$. Since we do not have any detailed information about the benchmark Treasury bonds over this 10-year period, we treat the yields in the input data set for the last five maturities, $T_3 = 2$, $T_4 = 3$, $T_5 = 5$,

$T_6 = 10$, and $T_7 = 30$, as par yields.[1] The instantaneous forward-rate curve for the day is determined by reproducing the value of the Treasury bills,

$$\frac{1}{(1 + y_i \Delta T)^{i/2}} = P(\tau, \tau + T_i), \quad \text{for } i = 1, 2, \tag{4.37}$$

and the value of the par bonds,

$$1 = \sum_{j=1}^{n_i} y_i(\tau) \cdot \Delta T P(\tau, \tau + j\Delta T) + P(\tau, \tau + n_i \Delta T) \tag{4.38}$$

for $i = 3, \ldots, 7$. Here $y_i(\tau)$ is the yield or par yield shown in Figure 4.1, $n_i = T_i/\Delta T$, $i = 1, 2, \ldots, 7$, $\Delta T = 0.5$, and

$$P(\tau, \tau + T) = e^{-\int_\tau^{\tau+T} f(\tau,s)ds}. \tag{4.39}$$

Because there are only a few inputs of yields, interpolation is necessary for constructing the forward rates. For clarity and notational simplicity, we present the algorithm for constructing the forward-rate curve for date $\tau = 0$ only with linear interpolation. The construction method for subsequent dates is identical. Define, in addition, $T_0 = 0$. Using $T_i, i = 0, 1, \ldots, 7$ as knot points, we assume that the forward rate is a continuous linear function between any two adjacent maturities:

$$\begin{aligned} f(0, T) &= r_0, & &\text{for } T_0 \leq T < T_1, \\ f(0, T) &= f(0, T_{i-1}) + \alpha_i(T - T_{i-1}), & &\text{for } T_{i-1} \leq T < T_i, \\ & & & i = 2, \ldots, 7, \end{aligned}$$

where r_0 is, in particular, taken to be the three-month rate for continuous compounding, given by

$$r_0 = \frac{1}{\Delta T} \ln(1 + y_1 \Delta T), \tag{4.40}$$

and $\{\alpha_n\}_2^7$ will be determined from the following procedure of bootstrapping.

1. Determine α_2 by matching $P(0, T_2)$ defined by Equation 4.39 to the bond price given by Equation 4.37.

2. Assume that we already have $\alpha_2, \alpha_3, \ldots, \alpha_i$.

3. To compute α_{i+1}, we use the $(i + 1)$st bond, and decompose the price into

$$1 = B_{i+1}^0 + B_{i+1}^1, \tag{4.41}$$

---

[1]Otherwise we would need coupon rates and exact maturities of the bonds. The 3- and 6-month yields are for zero-coupon bonds.

where the first term represents the present value of the coupons due on or before date T_i,

$$B_{i+1}^0 = \sum_{t_j \le T_i} \Delta T y_{i+1} P_0^{t_j}, \tag{4.42}$$

where y_{i+1} is the coupon rate for the $(i+1)$st bond, and t_j the jth coupon date of the bond. The discounted factor is calculated using the known forward rates, $f(0,T)$, for $T \le T_i$. Assume that t_j lies between T_{k-1} and $T_k, k \le i$, then

$$P_0^{t_j} = P_0^{T_{k-1}} \exp\left\{ -\int_{T_{k-1}}^{t_j} f(0,s)\, ds \right\}$$

$$= P_0^{T_{k-1}} \exp\left\{ -(t_j - T_{k-1}) \left[f(0,T_{k-1}) + \frac{1}{2}\alpha_k(t_j - T_{k-1}) \right] \right\},$$

where

$$P_0^{T_{k-1}} = \exp\left\{ -\int_0^{T_{k-1}} f(0,s)\, ds \right\}$$

$$= \exp\left\{ -\sum_{j=1}^{k-1} \int_{T_{j-1}}^{T_j} [f(0,T_{j-1}) + \alpha_j(s - T_{j-1})]\, ds \right\}$$

$$= \exp\left\{ -\sum_{j=1}^{k-1} (T_j - T_{j-1}) \left[f(0,T_{j-1}) + \frac{1}{2}\alpha_j(T_j - T_{j-1}) \right] \right\}. \tag{4.43}$$

The second term in Equation 4.41 is

$$B_{i+1}^1 = P_0^{T_i} \left(\sum_{T_i < t_j \le T_{i+1}} \Delta T y_{i+1} e^{-(t_j - T_i)(f_i + (1/2)\alpha_{i+1}(t_j - T_i))} \right.$$

$$\left. + e^{-(T_{i+1} - T_i)(f_i + (1/2)\alpha_{i+1}(T_{i+1} - T_i))} \right), \tag{4.44}$$

where $f_i = f(0,T_i)$. By combining Equations 4.41, 4.42, and 4.44, we obtain an equation for α_{i+1}:

$$\frac{1 - \sum_{t_j \le T_i} \Delta T y_{i+1} P_0^{t_j}}{P_0^{T_i}} = \sum_{T_i < t_j \le T_{i+1}} \Delta T y_{i+1} e^{-(t_j - T_i)(f_i + (1/2)\alpha_{i+1}(t_j - T_i))}$$

$$+ e^{-(T_{i+1} - T_i)(f_i + (1/2)\alpha_{i+1}(T_{i+1} - T_i))}.$$

We then solve for α_{i+1} from the above equation by a root-finding algorithm. This process can continue until $i = 7$; then we will obtain all $f(0,T)$ for $T \le 30$.

As an example, we display in Figure 4.2 the instantaneous forward-rate curve for May 31, 1996, the first date in our data, where the "*"s mark the seven benchmark maturities.

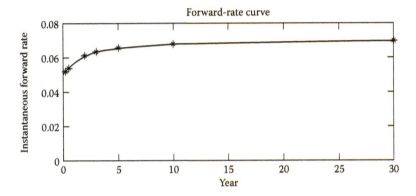

FIGURE 4.2: Instantaneous forward-rate curve of May 31, 1996, by linear interpolation.

There is a slight problem with the use of linear interpolation in the boot-strapping procedure: the curve is not differentiable at the knot points, $\{T_i\}$. If piece-wise constant interpolation was used instead, then we would have in-duced jumps across those maturities, making things even worse. This kind of non-smoothness is artificial and financially not justifiable, and it can cause potential problems. According to the HJM equation, such non-smoothness propagates over time, causing non-differentiability (or jumps if piece-wise con-stant interpolation is used) in the short rate across certain dates in the future. When the non-smooth forward-rate curve is applied to derivatives pricing, the non-smoothness may translate into extra volatility and thus cause mispricing.

The solution for better smoothness, fortunately, is simple: we can adopt spline interpolation instead of linear interpolation. A spline is a piece-wise cubic polynomial that has a continuous second-order derivative. The forward-rate curve so constructed is shown in Figure 4.3.

For each date in the data set, we can construct one forward-rate curve with the above procedure. On the forward-rate curve of each date, we pick seven points, as are marked in Figure 4.4, that are corresponding to the value of forward rates, $f(\tau, \tau + T_i)$, of the seven maturities (3-month, 6-month, 2-year, 3-year, 5-year, 10-year, or 30-year). By plotting $f(\tau, \tau + T_i), i = 1, \ldots, 7$ against τ in the data set, we obtain the time series data for the forward rates of the seven maturities, shown in Figure 4.4, where the time gap for the plots is one month, $\Delta\tau = 1/12$.

Before we move on to analyze the covariance of the forward rates, we make a comment on the construction of the forward-rate curves. Sometimes, in applications, we may need to enhance the smoothness or reduce the curva-ture of the forward-rate curve. These objectives can be achieved by adopting

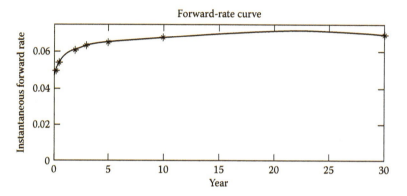

FIGURE 4.3: The instantaneous forward-rate curve for May 31, 1996, by spline interpolation.

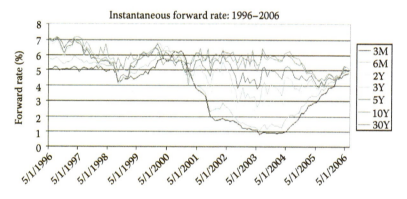

FIGURE 4.4: Forward-rate curve for the period 1996–2006.

appropriate regularizations. Popular regularization functions include smoothness regularization,

$$\int_0^T \left| \frac{\partial f(0,s)}{\partial s} \right|^2 ds, \tag{4.45}$$

and curvature regularization,

$$\int_0^T \left| \frac{\partial^2 f(0,s)}{\partial s^2} \right|^2 ds. \tag{4.46}$$

Note that there also exist other parameterizations. For example, one may consider the following parameterization of the forward-rate function

$$f(0,T) = \sum_{j=0}^N \theta_j \left(\frac{T}{1 + \theta_{N+1} T} \right)^j$$

suggested in Avellaneda and Laurence (1999).

4.4.2 Principal Component Analysis

Having constructed the time series data of forward rates of the seven maturities, shown in Figure 4.4, we now proceed to the estimation of covariance among those forward rates. We then perform PCA with the covariance matrix. The results will shed light on the proper number of random factors that drive the evolution of the forward-rate curve, so that we can determine n and subsequently $\sigma(t, T)$ for the HJM equation.

Let $f(n\Delta\tau, n\Delta\tau + T_i), n = 0, 1, \ldots, N$ be the forward rates for the seven maturities, $T_i, i = 1, 2, \ldots, 7$, where $\Delta\tau = 1/12$ represents the observation interval of one month, and N the total number of months. For forward rates of each maturity, $T_i, i = 1, 2, \ldots, 7$, we calculate the change over $\Delta\tau$:

$$\Delta f_{n,i} = f((n+1)\Delta\tau, (n+1)\Delta\tau + T_i)$$
$$- f(n\Delta\tau, n\Delta\tau + T_i), \quad n = 0, 1, \ldots, N - 1. \qquad (4.47)$$

The empirical covariance between $\Delta f_{\cdot,i}$ and $\Delta f_{\cdot,j}$ is, straightforwardly,

$$\hat{c}_{ij} = \frac{1}{N} \sum_{n=0}^{N-1} (\Delta f_{n,i} - \overline{\Delta f_i})(\Delta f_{n,j} - \overline{\Delta f_j}),$$

where

$$\overline{\Delta f_i} = \frac{1}{N} \sum_{n=0}^{N-1} \Delta f_{n,i}. \qquad (4.48)$$

By performing eigenvalue decomposition on the covariance matrix, $\hat{C} = (\hat{c}_{ij})$, we obtain

$$\hat{C} = V \Lambda V^T = \sum_{k=1}^{7} \lambda_k \Delta\tau \mathbf{v}_k \mathbf{v}_k^T, \qquad (4.49)$$

where $\Lambda = \Delta\tau \mathrm{diag}(\lambda_1, \lambda_2, \ldots, \lambda_7)$ and $V = (\mathbf{v}_1, \mathbf{v}_2, \ldots, \mathbf{v}_7)$ are eigenvalue and eigenvector matrices, respectively, the λs are put in descending order, that is, $\lambda_1 \geq \lambda_2 \geq \cdots \geq \lambda_7$, and the $\mathbf{v}_k$s are normalized, $\|\mathbf{v}_k\|_2 = 1$, $k = 1, \ldots, 7$. These eigenvectors $\{\mathbf{v}_1, \mathbf{v}_2, \ldots, \mathbf{v}_7\}$ are also called principal components of $\hat{C}$. In terms of components, Equation 4.49 reads as

$$\hat{c}_{ij} = \sum_{k=1}^{7} \lambda_k \Delta\tau v_{ik} v_{jk}. \qquad (4.50)$$

Based on the above eigenvalue decomposition, we can model the random increments of forward rates as follows:

$$\Delta \mathbf{f}_n = \overline{\Delta \mathbf{f}} + \sum_{k=1}^{7} \sqrt{\lambda_k \Delta\tau} \, \mathbf{v}_k \xi_{k,n}. \qquad (4.51)$$

Here,

$$\Delta \mathbf{f}_n = \begin{pmatrix} \Delta f_{n,1} \\ \Delta f_{n,2} \\ \vdots \\ \Delta f_{n,7} \end{pmatrix}, \quad \overline{\Delta \mathbf{f}} = \begin{pmatrix} \overline{\Delta f_1} \\ \overline{\Delta f_2} \\ \vdots \\ \overline{\Delta f_7} \end{pmatrix},$$

and $\{\xi_{k,n}\}$ are independent random variables with mean and variance equal to zero and one, respectively. We say that the forward-rate curve is driven by these seven random factors. By generating seven of these random variables and applying them to Equation 4.51, we simulate the change of all forward rates over one time step, $\Delta\tau$.

The importance of the seven factors, however, is not the same. Because $\lambda_i \geq \lambda_{i+1}$, $\mathbf{v}_i$ is considered more important than $\mathbf{v}_{i+1}$ in shaping the forward-rate curve. Moreover, suppose that, for some $\nu < 7$, there is

$$\lambda_1 + \lambda_2 + \cdots + \lambda_\nu \gg \lambda_{\nu+1} + \cdots + \lambda_7. \tag{4.52}$$

We then can ignore those λ_k for $k > \nu$ and approximate $\hat{C}$ by

$$\hat{C} \approx (\mathbf{v}_1 \cdots \mathbf{v}_\nu) \begin{pmatrix} \lambda_1 & & \\ & \ddots & \\ & & \lambda_\nu \end{pmatrix} \begin{pmatrix} \mathbf{v}_1^T \\ \vdots \\ \mathbf{v}_\nu^T \end{pmatrix}.$$

Accordingly, we can discard the less important factors in modeling forward-rate increments:

$$\Delta \mathbf{f}_n \approx \overline{\Delta \mathbf{f}} + \sum_{k=1}^{\nu} \sqrt{\lambda_k \Delta\tau} \mathbf{v}_k \xi_{k,n}. \tag{4.53}$$

Let us take a look at the weights of various factors that drive the Treasury forward-rate curve. The eigenvalues of the covariance matrix of forward rates calculated based on the monthly observed U.S. Treasury yields are listed in Table 4.1. The second column shows their weights. We can see that the first three principal components carry an aggregated weight of over 93%.

Consider taking the first three principal components, that is, $\nu = 3$, in modeling the forward-rate increments by Equation 4.53. A random shock, $\xi_{k,n}$, will deform forward rates of the seven maturities by $\sqrt{\lambda_k \Delta\tau} v_{i,k} \xi_{k,n}, i = 1, \ldots, 7$. To understand the impacts of the random shocks on the forward rates, we look at the first three principal components (i.e., the eigenvectors of the first three largest eigenvalues) in Figure 4.5.

We want to highlight some stylized facts here about the PCA of yield curves. As we can see from Figure 4.5, the leading component is relatively flat, with all elements positive, which will cause roughly a parallel move of yields in response to a random shock. The second principal component tilts: its elements change from positive values to negative values. Given a positive

TABLE 4.1: Eigenvalues and Weights for the Covariance Matrix

Eigenvalue, λ_k	$\lambda_k/\sum_j \lambda_j$ (%)
0.000514	64.09
0.000152	18.91
8.32E−05	10.39
3.97E−05	4.95
1.25E−05	1.56
6.35E−07	0.08
1.78E−07	0.02

For the covariance matrix of U.S. dollar (USD) forward rates under continuous compounding.

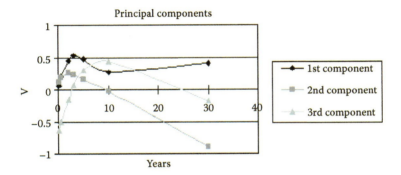

FIGURE 4.5: The first three principal components of USD forward rates.

shock, short maturity yields will increase, while long maturity yields will decrease. The third principal component bends: it starts in the negative territory, goes to the positive territory, and ends up in the negative territory again. A positive shock to this component will move both short- and long-term yields lower, yet such a shock will move the median-term yields higher, thus "bending" the yield curve. Knowing that the aggregated weight of the first three eigenvalues is 93%, we may say that the deformation of the forward-rate curve is caused by parallel shifts, tilting, and bending, corresponding to the shapes of the three leading eigenvectors, 93% of the time. The eigenvalues, meanwhile, differentiate the magnitudes of the impacts on average. These stylized facts were first discovered by Litterman and Scheinkman (1991). With Treasury yield data prior to 1991, they obtained an aggregated weight for the first three eigenvalues to be 98%, higher than the 93% obtained using the data between 1996 and 2006. The difference can be explained by the decorrelation of the Treasury yields that had taken place during the period from 2000 to 2006. Over this period, the short rates first decreased and then increased, while the long rates did not change by much.

Another feature of the leading principal components that is worth mentioning is that the variability concentrates in the short-term ranges, which simply reflects the fact that we witness more variation of the yield curve in short maturities.

Now we are ready to build the HJM model using the results from PCA. The process consists of the following steps:

1. Interpolate $\{v_{i,k}, i = 1, 2, \ldots, 7\}$ and $\overline{\Delta f_i}$ to get $\{v_k(T)\}$ and $\overline{\Delta f}(T)$, respectively, using a method such as continuous spline interpolation.

2. Define $\Delta f(t, t + T) = \sum_{k=1}^{\nu} v_k(T)\sqrt{\lambda_k \Delta t}\xi_k(t) + \frac{\overline{\Delta f}(T)}{\Delta \tau}\Delta t$ for general t and T, where $\xi_k(t)$ is a standard normal random variable.

3. Through comparing to the HJM equation, we define

$$df(t, t + T) = \boldsymbol{\sigma}^{\mathrm{T}}(T)\, d\tilde{\mathbf{W}}_t + \mu(T)\, dt, \qquad (4.54)$$

where $\sigma_k(T) = v_k(T)\sqrt{\lambda_k}$ and $\mu(T) = \overline{\Delta f}(T)/\Delta \tau$. Note that σ_k and μ depend on the time to maturity only.

4. By replacing $t + T$ by T, we finally have

$$df(t, T) = \boldsymbol{\sigma}^{\mathrm{T}}(T - t)\, d\tilde{\mathbf{W}}_t + \mu(T - t)\, dt. \qquad (4.55)$$

For the risk-neutral process of the forward rate, the volatility function is all we need. The risk-neutral drift is calculated according to

$$\mu(t, T) = \boldsymbol{\sigma}^{\mathrm{T}}(T - t) \int_t^T \boldsymbol{\sigma}(s - t)\, ds. \qquad (4.56)$$

For a three-factor HJM model, the forward-rate volatility components estimated using yield data from 1996 to 2006 are displayed in Figure 4.6. In terms of the magnitude, we can see that there is a descending order: $\|\sigma_1(\cdot)\|_2 > \|\sigma_2(\cdot)\|_2 > \|\sigma_3(\cdot)\|_2$.

We finish this section with additional remarks on PCA analysis and HJM model estimation. In much of the literature (e.g., Litterman and Schainkman, 1991; Avellaneda and Laurence, 1999), PCA is carried out with data on yields to maturities or zero-coupon yields. We take the PCA analysis of zero-coupon yields as an example. Out of the YTM data of 1996–2006 displayed in Figure 4.1, we can bootstrap and obtain the zero-coupon yields for the same period, which are displayed in Figure 4.7. PCA analysis yields principal components for the zero-coupon yields, displayed in Figure 4.8. Yet again, we see the stylized features of the first three principal components, namely, being flat, tilted, and bent. The principal components give rise to the volatility of the zero-coupon yields.

Some studies have suggested constructing the forward-rate volatilities from the volatilities of zero-coupon yields. This approach, however, is numerically

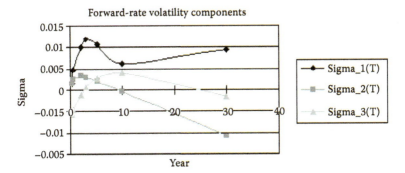

FIGURE 4.6: The three components of the forward-rate volatility, $\sigma_i(T)$, $i = 1, 2, 3$.

FIGURE 4.7: Zero-coupon yields obtained through bootstrapping.

unstable. Note that the forward-rate volatility relates the zero-coupon yield volatility, $\sigma_y(t, T)$, through the relationship

$$\int_t^T \sigma(t, s)\, \mathrm{d}s = (T - t)\sigma_y(t, T). \qquad (4.57)$$

It follows that

$$\sigma(t, T) = \sigma_y(t, T) + (T - t)\frac{\partial \sigma_y(t, T)}{\partial T}. \qquad (4.58)$$

The above expression contains a numerical differentiation that is then multiplied by a factor, $T - t$, which can reach 30 (years). Numerical studies have shown that forward-rate volatilities so generated are unrealistically larger than the observed volatilities. Hence, such an approach is not feasible.

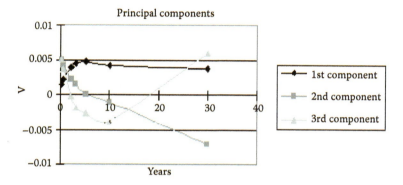

FIGURE 4.8: The first three principal components of USD zero yields.

4.5   A Case Study with a Two-Factor Model

Yields or forward rates of different maturities are not perfectly correlated, which is made evident in the last section with historical data. In this section, we demonstrate how to parameterize the forward-rate volatility to capture the stylized features of the forward-rate curves that are shaped by principal components.

We consider a two-factor HJM model ($\nu = 2$) with the following forward-rate volatility components:

$$
\begin{aligned}
\sigma_1(T) &= ae^{-k_1 T}, \\
\sigma_2(T) &= b(1 - 2e^{-k_2 T}),
\end{aligned}
\tag{4.59}
$$

where a, b, k_1, and k_2 are constants. To get a "flat" $\sigma_1(T)$ and a "tilted" $\sigma_2(T)$, we choose k_1 and k_2 such that $0 \leq k_1 \ll 1, k_1 \ll k_2$. Similar to Avellaneda and Laurence (1999), we consider the following choice of parameters

$$
a = 0.008, \quad b = 0.003, \quad k_1 = 0.0, \quad \text{and} \quad k_2 = 0.35.
$$

Note that if we take $b = 0$, this two-factor model reduces to the Hull–White model, under which forward rates of all maturities are perfectly correlated.

Let us examine the correlation between the 3-month (i.e., short-term) and 30-year (long-term) forward rates. In general, the covariance between forward rates of two maturities, T and T', is calculated according to

$$
c(T, T') = \sigma_1(T)\sigma_1(T') + \sigma_2(T)\sigma_2(T').
\tag{4.60}
$$

The correlation between forward rates of two maturities is thus

$$
\rho(T, T') = \frac{c(T, T')}{\sqrt{c(T, T)}\sqrt{c(T', T')}}.
\tag{4.61}
$$

Taking $T = 0.25$ (3-month) and $T' = 30$, we have

$$c(0.25, 0.25) = 7.024 \times 10^{-5},$$
$$c(0.25, 30) = 5.651 \times 10^{-5},$$
$$c(30, 30) = 7.300 \times 10^{-5}.$$

The correlation coefficient between the 3-month and 30-year forward rates is

$$\rho(0.25, 30) = \frac{c(0.25, 30)}{\sqrt{c(0.25, 0.25)}\sqrt{c(30, 30)}} = 79\%,$$

which indicates a high level of correlation between the two rates. The eigenvalues and normalized eigenvectors of the covariance matrix, $C_{2\times2}$, are listed in Table 4.2.

The relative importance of the two modes is reflected by the weights

$$\frac{\lambda_1}{\lambda_1 + \lambda_2} = 89.46\% \quad \text{and} \quad \frac{\lambda_2}{\lambda_1 + \lambda_2} = 10.54\%,$$

which are consistent with the PCA of U.S. Treasury data by Litterman and Scheinkman (1991).

Finally, we comment that the choices of the parameters correspond to reasonable short-rate volatility, which is calculated according to

$$\sigma_r = \sqrt{c(0,0)} = \sqrt{\sigma_1^2(0) + \sigma_2^2(0)} = \sqrt{a^2 + b^2} = 0.85\%.$$

Suppose the short rate is $r(0) = 5\%$, then the percentage of volatility of the short rate is $\sigma_r/r(0) = (0.85/5)\% = 17\%$. The interval of one standard deviation is

$$((5 - 0.85)\%, (5 + 0.85)\%) = (4.15\%, 5.85\%),$$

into which the short rate will fall with 67% probability.

4.6 Monte Carlo Implementations

We now consider the application of the HJM model to derivatives pricing. As a demonstration, we consider the pricing of a bond option that matures at

TABLE 4.2: PCA Analysis of Short and Long Rates

$\lambda_1 = 1.281 \times 10^{-4}$	$\lambda_2 = 1.51 \times 10^{-5}$
$v_1 = \begin{pmatrix} 0.6984 \\ 0.7157 \end{pmatrix}$	$v_2 = \begin{pmatrix} -0.7157 \\ 0.6984 \end{pmatrix}$
(for parallel shifts)	(for tilt moves)

T_0 with payoff

$$X_{T_0} = \left(\sum_{i=1}^{n} \Delta T \cdot c \cdot P_{T_0}^{T_i} + P_{T_0}^{T_n} - K \right)^{+}.$$

Here c is the coupon rate of the bond, K the strike price of the option, and $T_i = T_0 + i\Delta T$ the cash flow date of the ith coupon of the underlying bond. We call $T_n - T_0$, the life of the underlying bond beyond T_0, the tenor of the bond. The value of the option is given by

$$V_0 = E^{\mathbb{Q}} \left[\frac{1}{B_{T_0}} \left(\sum_{i=1}^{n} \Delta T \cdot c \cdot P_{T_0}^{T_i} + P_{T_0}^{T_n} - K \right)^{+} \Bigg| \mathcal{F}_0 \right]$$

$$= E^{\mathbb{Q}} \left[\left(\sum_{i=1}^{n} \Delta T c \cdot \frac{P_{T_0}^{T_i}}{B_{T_0}} + \frac{P_{T_0}^{T_n}}{B_{T_0}} - \frac{K}{B_{T_0}} \right)^{+} \Bigg| \mathcal{F}_0 \right], \qquad (4.62)$$

where $\mathbb{Q}$ is the risk-neutral measure. Based on Equation 4.25, we have the following expression for the discounted value of zero-coupon bonds:

$$\frac{P_{T_0}^{T_i}}{B_{T_0}} = P_0^{T_i} \exp \left(\int_0^{T_0} -\frac{1}{2} \|\mathbf{\Sigma}(t, T_i)\|^2 \, dt + \mathbf{\Sigma}^{\mathrm{T}}(t, T_i) \, d\tilde{\mathbf{W}}_t \right), \qquad (4.63)$$

for $i = 0, 1, \ldots, n$. Taking $i = 0$, in particular, we obtain the expression for the reciprocal of the money market account:

$$\frac{1}{B_{T_0}} = P_0^{T_0} \exp \left(\int_0^{T_0} -\frac{1}{2} \|\mathbf{\Sigma}(t, T_0)\|^2 \, dt + \mathbf{\Sigma}^{\mathrm{T}}(t, T_0) \, d\tilde{\mathbf{W}}_t \right). \qquad (4.64)$$

Equations 4.63 and 4.64 allow us to calculate the option's payoff. The expectation in Equation 4.62 can be evaluated by the following Monte Carlo simulation algorithm:

1. Construct the discount curve, P_0^T, $\forall T$, with the U.S. Treasury data at $t = 0$.

2. Simulate a number of Brownian paths and calculate the discounted prices according to the scheme

$$\frac{P_{t+\Delta t}^{T_i}}{B_{t+\Delta t}} = \frac{P_t^{T_i}}{B_t} \exp \left(-\frac{1}{2} \|\mathbf{\Sigma}(t, T_i)\|^2 \, \Delta t + \mathbf{\Sigma}^{\mathrm{T}}(t, T_i) \Delta \tilde{\mathbf{W}}_t \right) \qquad (4.65)$$

for $i = 0, 1, \ldots, n$, until $t + \Delta t$ equals T_0.

3. Average the payoffs:

$$\left(\sum_{i=1}^{n} \Delta T c \cdot \frac{P_{T_0}^{T_i}}{B_{T_0}} + \frac{P_{T_0}^{T_n}}{B_{T_0}} - \frac{K}{B_{T_0}} \right)^{+}. \qquad (4.66)$$

We will use the Monte Carlo simulation method with the pricing of a set of options, which have the same strike price, $K = 1$, and special coupon rates defined as

$$c = \frac{\left(P_0^{T_0} - P_0^{T_n}\right)}{\sum_{i=1}^{n} \Delta T P_0^{T_i}}. \tag{4.67}$$

As we shall see later, such special bond options are equivalent to swaptions, the options on interest-rate swaps.

To implement the Monte Carlo method, we need to prescribe the volatility functions of zero-coupon bonds. Under the HJM framework, these functions are derived from the volatility functions of forward rates. For demonstration purposes, let us take the two-factor model discussed in the last section, which has a forward-rate volatility prescribed in Equation 4.59. We consider the following two sets of parameterizations:

1. One-factor model: $a = 0.008544$, $b = 0$, $k_1 = 0.0$, $k_2 = 0.35$.

2. Two-factor model: $a = 0.008$, $b = 0.003$, $k_1 = 0.0$, $k_2 = 0.35$.

Note that, for both models, $k_1 = 0$, and the one-factor model is exactly a Ho–Lee model. For comparison, we have chosen $a = 0.008544$ for the one-factor model so that the corresponding short-rate volatilities of the two models, $\sigma_r = \sqrt{a^2 + b^2}$, are identical. The volatility for zero-coupon bond P_t^T is obtained from an integration:

$$\Sigma(t, T) = -\int_t^T \left(b \left(1 - \frac{a}{2} e^{-k_2 s} \right) \right) ds$$

$$= -\left(b \left[(T - t) + (2/k_2) \left(e^{-k_2 T} - e^{-k_2 t} \right) \right] \right). \tag{4.68}$$

The size of each time step for the Monte Carlo simulation is $\Delta t = 0.25$. To construct the discount curve, we use the yield data of March 15, 2007, listed in Table 4.3, and go through the bootstrapping procedure described earlier.

The calculated bond option prices are listed in Table 4.4. As can be seen from the table, the Ho–Lee model consistently produces higher prices, which can be explained as follows: under the one-factor model, the prices of all zero-coupon bonds are perfectly correlated, whereas under the two-factor model, they are not. Because a coupon bond is a portfolio of zero-coupon bonds, its volatility will be larger if all zero-coupon bonds move in the same direction, provided that the short-rate volatilities are the same. The larger volatility leads to a higher option premium.

4.7   Forward Prices

As was seen in the last section, option pricing under the HJM model can be achieved through Monte Carlo simulations. In fact, for interest-rate derivatives pricing under the HJM model, the Monte Carlo method is the only method that has been developed so far. Although this method is flexible and widely applicable, it suffers from slow convergence, and thus it is usually not the choice of market participants for whom pricing in real time is necessary. Hence, fast pricing methods must be developed. An important device for speedy option pricing is by a proper change of measure. As a preparation, we first introduce forward contracts and the notion of forward prices.

Let us begin with forward contracts. Suppose that we want to enter a deal now to purchase an asset at a future time when both payment and delivery take place. What should be taken as the fair price for this transaction? This contract is called a forward contract. We will try to figure out the fair price for the transaction, if there is one, by arbitrage arguments. To ensure delivery, the seller must borrow money now and acquire certain units of the asset. Denote the current price of the asset by S_t, the current time by t, the delivery time by T, and the unknown fair transaction price by F. Assume, at first, for simplicity that the asset pays no dividend. To be able to deliver one unit of the asset, the seller then does the following transactions:

1. Short S_t/P_t^T units of T-maturity zero-coupon bond.

2. Long 1 unit of the asset.

Note that 1 and 2 are a set of zero-net transactions at time t. At the delivery time, T, the seller will deliver the asset to the buyer for the price of F, and thus ends up with the following P&L value,

$$V_T = F - \frac{S_t}{P_t^T}.$$

If arbitrage is not possible, there must be $V_T = 0$, giving the fair transac-

TABLE 4.3: Yields at March 15, 2007

Maturity	Yield
3 months	4.89
6 months	4.88
2 years	4.56
3 years	4.47
5 years	4.44
10 years	4.52
30 years	4.69

TABLE 4.4: Prices of Par-Bond Options in Basis Points

Bond Option		Model	
Maturity, T_0	Tenor, $T_n - T_0$	Ho–Lee	Two-factor
1	0.25	8.23	7.85
2	0.25	11.07	10.34
5	0.25	15.33	14.69
1	5	145.43	140.41
2	5	200.29	191.07
5	5	275.08	272.14
1	10	262.88	254.99
2	10	364.54	349.50
5	10	489.43	487.53

The options with tenor 0.25 are called caplets, while the other are called swaptions. For caplets, $\Delta T = 0.25$, whereas for swaptions, $\Delta T = 0.5$. One basis point corresponds to one cent for the notional of $100.

tion price

$$F = \frac{S_t}{P_t^T}.$$

In an economy where there is no arbitrage, this price is fair and unique.

Next, we derive the fair transaction price for an asset that pays continuous dividends with a dividend yield of $q > 0$. The dividend is calculated as follows: over a short time interval, $(t, t + dt)$, the asset holder receives qdt additional units of the asset, or equivalently, $qS_t dt$ amount of cash. The transactions of the seller will

1. short $S_t \exp(-q(T - t))/P_t^T$ units of T-maturity zero-coupon bonds, and

2. long $\exp(-q(T - t))$ units of the asset, and, before T, the seller will continue to receive the dividend asset.

This is again a set of zero-net transactions. Because of the dividend payment, the second transaction will produce exactly one unit of asset at time T. The net value after delivering the asset for payment at T is

$$V_T = F - \frac{S_t \exp(-q(T - t))}{P_t^T}.$$

In the absence of arbitrage, there must be $V_T = 0$, yielding the fair price

$$F = \frac{S_t \exp(-q(T - t))}{P_t^T}.$$

Finally, we derive the fair transaction price when the asset pays discrete dividends. Assume that, prior to T, the asset pays cash dividend q_i at time $T_i \leq T, i = 1, \ldots, n$, such that $T_{i-1} < T_i$. In such a circumstance, the seller's strategy is to

1. short q_i units of T_i-maturity zero-coupon bonds $(i \leq n)$;

2. short $(S_t - \sum_{t<T_i \leq T} q_i P_t^{T_i})/P_t^T$ units of T-maturity zero-coupon bonds; and

3. long 1 unit of the asset.

The proceeds from shorting are just enough to purchase one unit of the asset. Hence, this is still a set of zero-net transactions. After closing out all positions at time T, the seller ends up with the net value of

$$V_T = F - \frac{\left(S_t - \displaystyle\sum_{t<T_i \leq T} q_i P_t^{T_i} \right)}{P_t^T}.$$

Hence, the fair price for a transaction is

$$F = \frac{\left(S_t - \displaystyle\sum_{t<T_i \leq T} q_i P_t^{T_i} \right)}{P_t^T}. \tag{4.69}$$

For all three cases, we define

$$\hat{S}_t = \begin{cases} S_t, & \text{no dividend} \\ S_t \exp(-q(T-t)), & \text{dividend yield } q \\ S_t - \displaystyle\sum_{t<T_i \leq T} q_i P_t^{T_i}, & \text{discrete dividend } \{q_i\} \end{cases} \tag{4.70}$$

as the *stripped-dividend* price of the asset. In terms of the stripped-dividend prices, we present the following definition.

Definition 4.7.1. *The price of a stripped-dividend asset relative to the T-maturity zero-coupon bond,*

$$F_t^T = \frac{\hat{S}_t}{P_t^T},$$

is called the forward price with delivery at time T.

Like the original asset itself, a stripped-dividend asset is also tradable (or replicable, in principle). Hence, its pricing process under the risk neutral measure $\mathbb{Q}$ is usually assumed to be

$$d\hat{S}_t = \hat{S}_t (r_t \, dt + \boldsymbol{\Sigma}_S^T \, d\tilde{\mathbf{W}}_t).$$

The process for the original asset then follows. In fact, for the asset paying a continuous dividend yield, the price process is

$$dS_t = S_t \left[(r_t - q) \, dt + \boldsymbol{\Sigma}_S^T \, d\tilde{\mathbf{W}}_t \right], \tag{4.71}$$

whereas for the asset paying discrete dividends, the price process is

$$dS_t = \left[r_t S_t - \sum q_i \delta(T_i - t) \right] dt$$
$$+ \left[S_t \mathbf{\Sigma}_S^{\mathrm{T}} - \sum q_i \mathbf{1}_{t \leq T_i} P_t^{T_i} (\mathbf{\Sigma}_S - \mathbf{\Sigma})^{\mathrm{T}} \right] d\tilde{\mathbf{W}}_t, \qquad (4.72)$$

where $\delta(x)$ is the Dirac delta function and $\mathbf{1}_{t \leq T_i}$ the indicator function.

4.8   Forward Measure

Because the price of a zero-coupon bond equals par at maturity, a forward price equals its spot price at the delivery date, that is, $F_T^T = \hat{S}_T = S_T$. As a result, any options written on S_T can equivalently be treated as an option on F_T^T. Next, we will try to price an option on F_T^T. For this purpose, we first need to derive the dynamics that F_t^T follows.

As tradable assets, the price of a stripped-dividend asset and a zero-coupon bond are assumed to be, respectively,

$$d\hat{S}_t = \hat{S}_t \left(r_t \, dt + \mathbf{\Sigma}_S^{\mathrm{T}}(t) \, d\tilde{\mathbf{W}}_t \right),$$
$$dP_t^T = P_t^T \left(r_t \, dt + \mathbf{\Sigma}^{\mathrm{T}}(t, T) \, d\tilde{\mathbf{W}}_t \right). \qquad (4.73)$$

By the quotient rule, the forward price satisfies

$$d\left(\frac{\hat{S}_t}{P_t^T} \right) = \frac{d\hat{S}_t}{P} - \frac{\hat{S}_t dP}{P^2} - \frac{d\hat{S}_t dP}{P^2} + \frac{\hat{S}_t (dP)^2}{P^3}$$

$$= \frac{\hat{S}_t}{P} \left(r_t dt + \mathbf{\Sigma}_S^{\mathrm{T}} d\tilde{\mathbf{W}}_t - r_t \, dt - \mathbf{\Sigma}^{\mathrm{T}} d\tilde{\mathbf{W}}_t - \mathbf{\Sigma}_S^{\mathrm{T}} \mathbf{\Sigma} dt + \mathbf{\Sigma}^{\mathrm{T}} \mathbf{\Sigma} dt \right)$$

$$= \frac{\hat{S}_t}{P} (\mathbf{\Sigma}_S - \mathbf{\Sigma})^{\mathrm{T}} \left(d\tilde{\mathbf{W}}_t - \mathbf{\Sigma} dt \right).$$

Here, on the right-hand side of the equation, we have omitted the sub- and sup-index of P for simplicity. Define now a new measure, $\mathbb{Q}_T$, as

$$\left. \frac{d\mathbb{Q}_T}{d\mathbb{Q}} \right|_{\mathcal{F}_t} = \exp\left(\int_0^t \mathbf{\Sigma}^{\mathrm{T}} d\tilde{\mathbf{W}}_s - \frac{1}{2} \mathbf{\Sigma}^{\mathrm{T}} \mathbf{\Sigma} \, ds \right) = \zeta_t. \qquad (4.74)$$

Then, by the CMG theorem,

$$\hat{\mathbf{W}}_t = \tilde{\mathbf{W}}_t - \int_0^t \mathbf{\Sigma}(s, T) \, ds \qquad (4.75)$$

is a $\mathbb{Q}_T$-Brownian motion. It follows that F_t^T is also a lognormal $\mathbb{Q}_T$-martingale, such that it follows a driftless process,

$$\mathrm{d}F_t^T = F_t^T \boldsymbol{\Sigma}_F^\mathrm{T} \,\mathrm{d}\hat{\mathbf{W}}_t,$$

where

$$\boldsymbol{\Sigma}_F(t) = \boldsymbol{\Sigma}_S(t) - \boldsymbol{\Sigma}(t, T). \tag{4.76}$$

We call $\mathbb{Q}_T$ the forward measure with delivery at T, or simply the T-forward measure. According to the definition, there is a one-to-one correspondence between the T-maturity zero-coupon bond and the T-forward measure.

Based on the understanding that the forward price is a $\mathbb{Q}_T$-martingale, we can derive a general pricing principle of options on F_T^T. Define another $\mathbb{Q}_T$-martingale as

$$N_t = E^{\mathbb{Q}_T}[X_T \mid \mathcal{F}_t]. \tag{4.77}$$

According to the martingale representation theorem, there exists an $\mathcal{F}_t$-adaptive process, φ_t, such that

$$\mathrm{d}N_t = \varphi_t \mathrm{d}F_t^T. \tag{4.78}$$

We now form a portfolio consisting of

   – φ_t units of the underlying asset, and

   – $\psi_t = N_t - \varphi_t F_t^T$ units of the T-maturity zero-coupon bond.

Let $\hat{V}_t$ be the forward price of the portfolio at time t. By definition, $\hat{V}_t = N_t$ for all $t \le T$, and the spot price of the portfolio,

$$V_t = N_t P_t^T. \tag{4.79}$$

At the maturity of the option, there is $V_T = N_T = X_T$, meaning that the portfolio replicates the payoff of the option. In addition, we have

$$\begin{aligned}
\mathrm{d}V_t &= P_t^T \mathrm{d}N_t + N_t \mathrm{d}P_t^T + \mathrm{d}N_t \mathrm{d}P_t^T \\
&= P_t^T \varphi_t \mathrm{d}F_t^T + \left(\varphi_t F_t^T + \psi_t\right) \mathrm{d}P_t^T + \varphi_t \mathrm{d}F_t^T \mathrm{d}P_t^T \\
&= \varphi_t \left(P_t^T \mathrm{d}F_t^T + F_t^T \mathrm{d}P_t^T + \mathrm{d}F_t^T \mathrm{d}P_t^T\right) + \psi_t \mathrm{d}P_t^T \\
&= \varphi_t \mathrm{d}S_t + \psi_t \mathrm{d}P_t^T,
\end{aligned} \tag{4.80}$$

which implies that the portfolio, (φ_t, ψ_t), is a self-financing one. In the absence of arbitrage, the value of the option should be nothing else but that of the replicating portfolio. This yields the general price formula

$$V_t = P_t^T E^{\mathbb{Q}_T}[X_T \mid \mathcal{F}_t] \tag{4.81}$$

for the option. Note that this is the second option price formula in addition to the one developed in Chapter 2,

$$V_t = B_t E^{\mathbb{Q}}[B_T^{-1} X_T \mid \mathcal{F}_t], \tag{4.82}$$

obtained under the risk-neutral measure, $\mathbb{Q}$. Both formulae are obtained through arbitrage arguments, so the values of the two formulae must be identical, or otherwise we would have a big problem.

Mathematically, it remains interesting to verify that the two formulae give an identical price. In fact, we can derive one formula from the other by merely a change of measure. We know that under the risk-neutral measure, $\mathbb{Q}$, P_t^T follows

$$dP_t^T = P_t^T \left(r_t dt + \mathbf{\Sigma}^{\mathrm{T}}(t,T) \, d\tilde{\mathbf{W}}_t \right), \tag{4.83}$$

or

$$d\left(\frac{P_t^T}{B_t}\right) = \left(\frac{P_t^T}{B_t}\right) \mathbf{\Sigma}^{\mathrm{T}}(t,T) \, d\tilde{\mathbf{W}}_t, \tag{4.84}$$

where $\tilde{\mathbf{W}}_t$ is a $\mathbb{Q}$-Brownian motion. By solving the equation, we obtain

$$\frac{P_T^T}{B_T} = \frac{P_t^T}{B_t} \exp\left(\int_t^T -\frac{1}{2} \mathbf{\Sigma}^{\mathrm{T}} \mathbf{\Sigma} \, ds + \mathbf{\Sigma}^{\mathrm{T}} d\tilde{\mathbf{W}}_s \right). \tag{4.85}$$

From this equation, we can express B_t in terms of P_t^T:

$$\frac{B_t}{B_T} = P_t^T \exp\left(\int_t^T -\frac{1}{2} \mathbf{\Sigma}^{\mathrm{T}} \mathbf{\Sigma} \, ds + \mathbf{\Sigma}^{\mathrm{T}} d\tilde{\mathbf{W}}_s \right) = P_t^T \frac{\zeta_T}{\zeta_t}, \tag{4.86}$$

by making use of Equation 4.74. Hence, starting from Equation 4.82, we have

$$\begin{aligned}
V_t &= E^{\mathbb{Q}} \left[\frac{B_t}{B_T} X_T \,\middle|\, \mathcal{F}_t \right] \\
&= E^{\mathbb{Q}} \left[P_t^T \frac{\zeta_T}{\zeta_t} X_T \,\middle|\, \mathcal{F}_t \right] \\
&= P_t^T E^{\mathbb{Q}_T} [X_T \mid \mathcal{F}_t].
\end{aligned} \tag{4.87}$$

This procedure can be reversed to derive Equation 4.82 from Equation 4.81.

For completeness, we finish this section with a lemma on the martingale property of forward rates.

Lemma 4.8.1. *The forward rate $f(t,T)$ is a $\mathbb{Q}_T$-martingale, and it satisfies*

$$df(t,T) = \boldsymbol{\sigma}^{\mathrm{T}}(t,T) \, d\hat{\mathbf{W}}_t, \tag{4.88}$$

where $\hat{\mathbf{W}}_t$ is a $\mathbb{Q}_T$-Brownian motion defined in Equation 4.75.

Proof. Recall the HJM Equation 4.16 for the forward rate. We have

$$df(t,T) = \boldsymbol{\sigma}^{\mathrm{T}}(t,T) \left(d\tilde{\mathbf{W}}_t - \mathbf{\Sigma}(t,T) \, dt \right), \tag{4.89}$$

where $\tilde{\mathbf{W}}_t$ is a Brownian motion under the risk-neutral measure, $\mathbb{Q}$. The conclusion then follows. $\qquad\square$

4.9   Black's Formula for Call and Put Options

In this section, we derive the price formula for both call and put options using the forward price and under its corresponding forward measure. The payoff of a call option on an asset, S_t, is

$$V_T = \max(S_T - K, 0) \triangleq (S_T - K)^+. \tag{4.90}$$

In terms of the forward price, $F_t^T = \hat{S}_t / P_t^T$, we also have

$$V_T = (F_T^T - K)^+. \tag{4.91}$$

Under the T-forward measure, we know that the price is given by

$$V_t = P_t^T E^{\mathbb{Q}_T} \left[\left(F_T^T - K \right)^+ \Big| \mathcal{F}_t \right]. \tag{4.92}$$

The good news here is that F_t^T is a lognormal martingale under $\mathbb{Q}_T$:

$$\mathrm{d}F_t^T = F_t^T \mathbf{\Sigma}_F^{\mathrm{T}} \mathrm{d}\hat{\mathbf{W}}_t, \tag{4.93}$$

where $\mathbf{\Sigma}_F$ is the difference between the volatilities of the asset and the T-maturity zero-coupon bond:

$$\mathbf{\Sigma}_F = \mathbf{\Sigma}_S - \mathbf{\Sigma}(t, T). \tag{4.94}$$

By repeating the procedure to derive the Black–Scholes formula, we obtain

$$V_t = \hat{S}_t \Phi(d_1) - K P_t^T \Phi(d_2), \tag{4.95}$$

where $\Phi(\cdot)$ is the normal accumulative function,

$$\Phi(x) = \frac{1}{\sqrt{2\pi}} \int_{-\infty}^x \exp\left(-\frac{y^2}{2}\right) \mathrm{d}y, \tag{4.96}$$

and

$$d_1 = \frac{\ln\left(\hat{S}_t \Big/ \left(P_t^T K\right)\right) + (1/2)\sigma_F^2 (T - t)}{\sigma_F \sqrt{T - t}},$$

$$d_2 = d_1 - \sigma_F \sqrt{T - t},$$

with

$$\sigma_F^2 = \frac{1}{T - t} \int_t^T \|\mathbf{\Sigma}_F\|^2 \mathrm{d}s = \frac{1}{T - t} \int_t^T \|\mathbf{\Sigma}_S(s) - \mathbf{\Sigma}(s, T)\|^2 \mathrm{d}s. \tag{4.97}$$

The price formula (Equation 4.95) is called Black's formula, in recognition of a similar formula for futures options developed by Black (1976). For later reference, we call σ_F the Black's volatility of the option. Note that when the short rate becomes deterministic, $\Sigma(t,T) = 0$, Black's formula reduces to the Black–Scholes formula.

In addition to the option's price, Black's formula also offers a hedging strategy: we can hedge the option by purchasing $\varphi_t = \Phi(d_1)$ units of underlying asset. Alternatively, we may say that the option can be replicated by a portfolio consisting of

- $\varphi_t = \Phi(d_1)$ units of the underlying asset, and

- $\psi_t = -K\Phi(d_2)$ units of T-maturity zero-coupon bonds.

The price formula for put options can be derived through the *call–put parity*: for the same strike, K, the prices of a call option, a put option, and a forward contract satisfy the relation

$$C(K) - P(K) = \hat{S}_t - P(t,T)K. \tag{4.98}$$

Thus, the formula for a put option follows:

$$\begin{aligned} P(K) &= C(K) - \hat{S}_t + P(t,T)K \\ &= KP_t^T(1 - \Phi(d_2)) - \hat{S}_t(1 - \Phi(d_1)) \\ &= KP_t^T\Phi(-d_2) - \hat{S}_t\Phi(-d_1). \end{aligned} \tag{4.99}$$

The hedging strategy is to short $\Phi(-d_1)$ units of the underlying asset.

4.9.1 Equity Options under the Hull–White Model

To price either a call or a put option by the Black's formula, we need to calculate the Black's volatility of the forward price (Equation 4.97). In applications, asset volatilities are often given in the form of a scalar instead of a vector, and asset correlations are given explicitly. In such a situation, the (square of) Black's volatility, Equation 4.97, takes a different form. In this section, we present Black's volatility for both the Ho–Lee model and the Hull–White model, and we work out two examples of option pricing under these models.

Consider first the pricing of an equity call option under the Ho–Lee model for interest rates. The forward-rate volatility under the Ho–Lee model is σ_0, resulting in the volatility of the zero-coupon bond being $\Sigma(t,T) = -\sigma_0(T-t)$. Assume that the local volatility of the underlying asset is a constant, σ_S, and the correlation between the asset and the zero-coupon bond is ρ. Then, the

Black's volatility of the forward price can be calculated as

$$
\begin{aligned}
\sigma_F^2 &= \frac{1}{T-t} \int_t^T \|\mathbf{\Sigma}_S(u) - \mathbf{\Sigma}(u,T)\|^2 \, du \\
&= \frac{1}{T-t} \int_t^T \left(\|\mathbf{\Sigma}_S(u)\|^2 - 2\mathbf{\Sigma}_S^{\mathrm{T}}(u)\mathbf{\Sigma}(u,T) + \|\mathbf{\Sigma}(u,T)\|^2 \right) du \\
&= \frac{1}{T-t} \int_t^T \left(\sigma_S^2 - 2\rho\sigma_S\sigma_0(T-u) + \sigma_0^2(T-u)^2 \right) du \\
&= \sigma_S^2 - \rho\sigma_S \cdot \sigma_0(T-t) + \frac{1}{3}\sigma_0^2(T-t)^2.
\end{aligned} \tag{4.100}
$$

Note that a positive correlation reduces Black's volatility. Let us witness the effect on the price of asset–interest rate correlation in the following example.

Example 4.9.1. *We examine the price of a call option as a function of the asset–interest rate correlation. The parameters are taken as follows.*

$$
\begin{aligned}
\hat{S}_0 &= 1, \quad \sigma_S = 0.3, \\
f(0,T) &= 0.02 + 0.002T, \quad \forall T; \quad \sigma_0 = 0.002, \\
K &= 1, \\
\rho &= -1 : 0.1 : 1.
\end{aligned} \tag{4.101}
$$

This option is called an at-the-money (ATM) option because the strike price equals the spot price. The price curve of the option against the correlation is presented in Figure 4.9, where one can see that option prices decrease gradually, from 0.1292 to 0.1282. The explanation is intuitive: a positive correlation between the asset and the zero-coupon bond leads to a smaller volatility in the forward price, and hence a lower value for the option.

Let us consider another example with an option on a zero-coupon bond under both the Ho–Lee and the Hull–White models.

Example 4.9.2. *Consider the pricing of a zero-coupon bond option with payoff $V_T = (P(T,\tau) - K)^+$ for $\tau > T$. Let us take $T = 2, \tau = 5$, and $K = P(0,\tau)/P(0,T)$. The initial term structure of interest rates is given by*

$$
f(0,T) = 0.02 + 0.002T. \tag{4.102}
$$

We will price the option first under the Ho–Lee model, with a forward-rate volatility of $\sigma_0 = 0.005$.

To calculate the strike, we need the price of zero-coupon bonds:

$$
\begin{aligned}
P(0,T) &= \exp\left\{ -\int_0^T f(0,u) \, du \right\} \\
&= \exp\left\{ -\int_0^T (0.02 + 0.002u) \, du \right\} \\
&= \exp\left\{ -\left(0.02T + 0.001 \cdot T^2\right) \right\}.
\end{aligned} \tag{4.103}
$$

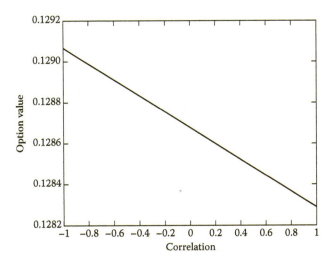

FIGURE 4.9: The price of an ATM option in relation to the correlation.

With the above function, we obtain

$$P(0,2) = 0.9570,$$
$$P(0,5) = 0.8825,$$

(4.104)

and the value of the strike,

$$K = F_0^2 = \frac{P(0,5)}{P(0,2)} = 0.9222.$$

(4.105)

For the Ho–Lee model, the volatility of a zero-coupon bond is $\Sigma(t,T) = -\sigma_0(T-t)$. Thus,

$$\Sigma(t,\tau) - \Sigma(t,T) = -\sigma_0(\tau - T).$$

(4.106)

Black's volatility is then

$$
\begin{aligned}
\sigma_F^2 &= \frac{1}{T} \int_0^T (\Sigma(t,\tau) - \Sigma(t,T))^2 \, dt \\
&= (\sigma_0)^2 (\tau - T)^2 \\
&= (0.005)^2 \times 9 \\
&= 0.000225.
\end{aligned}
$$

(4.107)

It follows that

$$d_1 = 0.0106,$$
$$d_2 = -d_1 = -0.0106,$$

(4.108)

and the value of the option is

$$V_0 = P(0,5)N(d_1) - P(0,2)KN(d_2)$$
$$= P(0,5)(1 - 2N(d_2)) = 0.0075, \tag{4.109}$$

or 75 basis points.

Next, we show the pricing of the same option under the Hull–White model, with a forward-rate volatility of

$$\sigma(t,T) = \sigma_0 e^{-\alpha(T-t)}, \quad \alpha = 0.1. \tag{4.110}$$

The volatility of a zero-coupon bond is then

$$\Sigma(t,T) = -\int_t^T \sigma(t,u)\,du = \frac{\sigma_0}{\alpha}\left[e^{-\alpha(T-t)} - 1\right]. \tag{4.111}$$

It follows that,

$$\Sigma(t,\tau) - \Sigma(t,T) = \frac{\sigma_0}{\alpha}\left[e^{-\alpha(\tau-t)} - e^{-\alpha(T-t)}\right]$$
$$= \frac{\sigma_0}{\alpha}e^{\alpha t}(e^{-\alpha\tau} - e^{-\alpha T}), \tag{4.112}$$

and

$$\sigma_F^2 = \frac{1}{T}\frac{\sigma_0^2}{\alpha^2}\left(e^{-\alpha\tau} - e^{-\alpha T}\right)^2 \int_0^T e^{2\alpha t}\,dt$$
$$= \frac{1}{T}\frac{\sigma_0^2}{\alpha^2}\left(e^{-\alpha\tau} - e^{-\alpha T}\right)^2 \frac{1}{2\alpha}\left(e^{2\alpha T} - 1\right)$$
$$= \frac{1}{2T}\frac{\sigma_0^2}{\alpha^3}\left(e^{-\alpha(\tau-T)} - 1\right)^2\left(1 - e^{-2\alpha T}\right). \tag{4.113}$$

By inserting Black's volatility into Black's formula, we obtain

$$\sigma_F^2 = 0.00013841,$$
$$d_1 = -d_2 = 0.0083, \tag{4.114}$$

and finally

$$V_0 = 0.0059, \tag{4.115}$$

or 59 basis points. The smaller price under the Hull–White model only reflects the fact of the smaller forward-rate volatility as compared to that of the Ho–Lee model.

4.9.2   Options on Coupon Bonds

Options on coupon bonds actually belong to the first generation of fixed-income derivatives. Options on Treasury bonds are liquidly traded.

In Section 4.6, we have studied the pricing of coupon bonds using Monte Carlo simulations. Here, we instead introduce a methodology for approximate pricing of options on coupon bonds.

As was already presented in Section 4.6, the payoffs of call options on coupon bonds take the form

$$V_T = \left(\sum_{i=1}^{N} \Delta T c P(T_0, T_i) + P(T_0, T_N) - K \right)^+ , \qquad (4.116)$$

where T_0 is the maturity of the option. Let B_t^c denote the bond price at time t. Then the T_0-forward price of the coupon bond is

$$
\begin{aligned}
F_t^{T_0} &= \frac{B_t^c}{P(t, T_0)} \\
&= \sum_{i=1}^{N} \Delta T c \frac{P(t, T_i)}{P(t, T_0)} + \frac{P(t, T_N)}{P(t, T_0)} \\
&= \sum_{i=1}^{N} \Delta T c \frac{P(0, T_i)}{P(0, T_0)} M_i(t) + \frac{P(0, T_N)}{P(0, T_0)} M_N(t). \qquad (4.117)
\end{aligned}
$$

Here,

$$
\begin{aligned}
M_i(t) = \exp \bigg(& \int_0^t -\frac{1}{2} \left\| \mathbf{\Sigma}(s, T_i) - \mathbf{\Sigma}(s, T_0) \right\|^2 \, \mathrm{d}s \\
& + \left(\mathbf{\Sigma}(s, T_i) - \mathbf{\Sigma}(s, T_0) \right)^{\mathrm{T}} \, \mathrm{d}\hat{\mathbf{W}}_s \bigg) \qquad (4.118)
\end{aligned}
$$

is a martingale under the T_0-forward measure. For convenience we now rewrite Equation 4.117 as

$$\frac{F_t^{T_0}}{F_0^{T_0}} = \sum_{i=1}^{N} \omega_i M_i(t), \qquad (4.119)$$

where

$$\omega_i = \begin{cases} \dfrac{\Delta T c P(0, T_i)}{B_0^c}, & i < N, \\[2ex] \dfrac{(1 + \Delta T c) P(0, T_N)}{B_0^c}, & i = N. \end{cases} \qquad (4.120)$$

To price the option, we approximate the process of $F_{T_0}^{T_0}$ by a lognormal variable through moment matching:

$$\sum_{i=1}^{N} \omega_i M_i(T_0) \approx \exp \left(-\frac{1}{2} \sigma_B^2 T_0 + \sigma_B \sqrt{T_0} \cdot \varepsilon \right), \qquad (4.121)$$

where $\varepsilon \sim N(0,1)$ under $\mathbb{Q}_{T_0}$, and

$$
\begin{aligned}
\sigma_B^2 &= \frac{1}{T_0} \ln E^{\mathbb{Q}_{T_0}} \left[\left(\sum_{i=1}^{N} \omega_i M_i(T_0) \right)^2 \right] \\
&= \frac{1}{T_0} \ln \left(\sum_{i,j} \omega_i \omega_j E^{\mathbb{Q}_{T_0}} \left[M_i(T_0) M_j(T_0) \right] \right) \\
&= \frac{1}{T_0} \ln \left(\sum_{i,j} \omega_i \omega_j e^{s_i s_j \rho_{ij} T_0} \right),
\end{aligned}
\tag{4.122}
$$

with

$$
\begin{aligned}
s_i^2 &= \frac{1}{T_0} \int_0^{T_0} \| \boldsymbol{\Sigma}(t, T_i) - \boldsymbol{\Sigma}(t, T_0) \|^2 \, dt, \\
\rho_{ij} &= \frac{1}{T_0 s_i s_j} \int_0^{T_0} \left(\boldsymbol{\Sigma}(t, T_i) - \boldsymbol{\Sigma}(t, T_0) \right)^{\mathrm{T}} \left(\boldsymbol{\Sigma}(t, T_j) - \boldsymbol{\Sigma}(t, T_0) \right) dt.
\end{aligned}
\tag{4.123}
$$

Note that when $|s_i s_j \rho_{ij} T_0| \ll 1$ for all i and j, we can approximate σ_B as

$$
\sigma_B \approx \sqrt{\sum_{ij} \omega_i \omega_j s_i s_j \rho_{ij}}.
\tag{4.124}
$$

By utilizing the lognormal approximation of the forward bond price, we have the following Black's formula for the approximate price of bond options:

$$
\begin{aligned}
V_0 &= P(0, T_0) E^{\mathbb{Q}_{T_0}} \left[\left(F_{T_0}^{T_0} - K \right)^+ \Big| \mathcal{F}_0 \right] \\
&\approx P(0, T_0) E^{\mathbb{Q}_{T_0}} \left[\left(F_0^{T_0} e^{-(1/2) \sigma_B^2 T_0 + \sigma_B \sqrt{T_0} \cdot \varepsilon} - K \right)^+ \Big| \mathcal{F}_0 \right] \\
&= P(0, T_0) \left[F_0^{T_0} \Phi(d_1) - K \Phi(d_2) \right] \\
&= B_0^c \Phi(d_1) - K P(0, T_0) \Phi(d_2),
\end{aligned}
\tag{4.125}
$$

where

$$
\begin{aligned}
d_1 &= \frac{\ln\left(F_0^{T_0}/K \right) + (1/2) \sigma_B^2 T_0}{\sigma_B \sqrt{T_0}} = \frac{\ln(B_0^c / K P(0, T_0)) + (1/2) \sigma_B^2 T_0}{\sigma_B \sqrt{T_0}}, \\
d_2 &= d_1 - \sigma_B \sqrt{T_0}.
\end{aligned}
$$

Although appearing rough, the above approximation often works quite well in the marketplace. In fact, it has become a common practice to approximate an addition of lognormal random variables by a single lognormal random variable, as such an approximation seems to be rather accurate when the original random variables are reasonably correlated.

TABLE 4.5: Option Prices by Black's Formula and by the Monte Carlo Simulation Method

Par-Bond Option		Ho–Lee Model		Two-Factor Model	
Maturity, T_0	Tenor $T_n - T_0$	Black	MC	Black	MC
1	0.25	8.12	8.23	7.67	7.85
2	0.25	11.01	11.07	10.31	10.34
5	0.25	15.30	15.33	14.77	14.69
1	5	147.84	145.43	139.40	140.41
2	5	200.20	200.29	190.95	191.07
5	5	276.62	275.08	271.28	272.14
1	10	264.69	262.88	253.35	254.99
2	10	357.57	364.54	345.94	349.50
5	10	492.64	489.43	486.42	487.53

Example 4.9.3. *To demonstrate the accuracy of Black's formula for options on coupon bonds, we reprice the options introduced in Section 4.6, and compare the results with those from Monte Carlo simulations. The results are listed in Table 4.5. One can see that the prices from Black's formula and from Monte Carlo simulations are fairly close. In fact, the root mean squared difference for all price pairs is very close to 1%. Such closeness really does support very positively the lognormal approximation of the price distribution of coupon bonds.*

4.10 Numeraires and Changes of Measure

A major achievement so far in this chapter is to take zero-coupon bonds as numeraires and price options under their corresponding forward measures. Mathematically, this is merely a technique of changing the numeraire asset, followed by taking the expectation of the option payoffs under the martingale measures of the numeraire assets. In this section, we discuss this technique in a general context.

Let $\mathbb{Q}_A$ be the martingale measure associated with reference asset A_t, meaning that, for any traded asset V_t, its price relative to that of asset A_t,

$$\frac{V_t}{A_t},\tag{4.126}$$

is a $\mathbb{Q}_A$-martingale. Consider another asset[2], B_t, and its associated martingale measure, $\mathbb{Q}_B$. According to the one-price principle, the value of any traded asset at time t, V_t, satisfies

$$V_t = A_t E_t^{\mathbb{Q}_A}\left[A_T^{-1}V_T\right] = B_t E_t^{\mathbb{Q}_B}\left[B_T^{-1}V_T\right].\tag{4.127}$$

---

[2] A_t and B_t are either stripped divided or cum-dividend.

From the above equation, we obtain

$$E_t^{\mathbb{Q}_B}\left[B_T^{-1}V_T\right] = \frac{A_t}{B_t}E_t^{\mathbb{Q}_A}\left[\frac{B_T}{A_T}\left(B_T^{-1}V_T\right)\right]. \tag{4.128}$$

Let ζ be the Radon–Nikodym derivative between $\mathbb{Q}_B$ and $\mathbb{Q}_A$:

$$\left.\frac{d\mathbb{Q}_B}{d\mathbb{Q}_A}\right|_{\mathcal{F}_t} = \zeta_t. \tag{4.129}$$

Then,

$$E_t^{\mathbb{Q}_B}\left[B_T^{-1}V_T\right] = \zeta_t^{-1}E_t^{\mathbb{Q}_A}\left[\zeta_T(B_T^{-1}V_T)\right]. \tag{4.130}$$

Subtracting Equation 4.128 from Equation 4.130, we obtain

$$0 = E_t^{\mathbb{Q}_A}\left[\left(\frac{\zeta_T}{\zeta_t} - \frac{B_T/A_T}{B_t/A_t}\right)(B_T^{-1}V_T)\right]. \tag{4.131}$$

Note that Equation 4.131 holds for prices of any tradable assets, so we can argue that

$$\zeta_t = \left.\frac{d\mathbb{Q}_B}{d\mathbb{Q}_A}\right|_{\mathcal{F}_t} = \frac{B_t/A_t}{B_0/A_0} \quad \text{a.s.} \tag{4.132}$$

Example 4.10.1. *We have already established the following correspondence between a numeraire and its martingale measure:*

$$\begin{aligned} B_t &= \exp\left(\int_0^t r_s\,ds\right) \longleftrightarrow \mathbb{Q}, \\ P_t^T &= P_0^T\exp\left(\int_0^t\left(r_s - \frac{1}{2}\mathbf{\Sigma}^{\mathrm{T}}\mathbf{\Sigma}\right)ds + \mathbf{\Sigma}^{\mathrm{T}}d\tilde{\mathbf{W}}_s\right) \longleftrightarrow \mathbb{Q}_T. \end{aligned} \tag{4.133}$$

According to the general formula 4.133, the Radon–Nikodym derivative of $\mathbb{Q}_T$ with respect to $\mathbb{Q}$ is

$$\begin{aligned} \left.\frac{d\mathbb{Q}_T}{d\mathbb{Q}}\right|_{\mathcal{F}_t} &= \frac{P_t^T/P_0^T}{B_t/B_0} \\ &= \exp\left(\int_0^t -\frac{1}{2}\mathbf{\Sigma}^{\mathrm{T}}\mathbf{\Sigma}\,ds + \mathbf{\Sigma}^{\mathrm{T}}d\tilde{\mathbf{W}}_s\right), \end{aligned} \tag{4.134}$$

which is exactly Equation 4.74.

4.11   Linear Gaussian Models

For the purpose of both pricing and hedging, it is not necessary to have a numeraire that is tradable. As it is pointed out in Hagan and Woodward

(1999), a "modern interest rate model consists of three parts: a numeraire, a set of random evolution equations in the risk neutral world, and the martingale pricing formula." The so-called Linear Gaussian Markov Model (LGM) is such an artificial device for interest-rate pricing, which is defined as follows.

1. The numeraire:

$$N_t = \frac{1}{P(0,t)} \exp\left\{ h(t)z_t + \frac{1}{2}h^2(t)\zeta_t \right\}. \tag{4.135}$$

2. The random evolution equation:

$$dz_t = \alpha_t dW_t, \quad z_0 = 0, \quad \text{and} \quad \zeta_t = \int_0^t \alpha^2(s)ds. \tag{4.136}$$

3. The martingale pricing formula:

$$\frac{V_t}{N_t} = E_t^N \left[\frac{V_T}{N_T} \right]. \tag{4.137}$$

where the sup-index N means the martingale measure corresponding to N_t as the numeraire.

Under the LGM model we have closed-form formula for zero-coupon bonds. By putting $V_T = 1$, we can derive the formula for zero-coupon bonds as

$$P(t,T;z_t) = \frac{P(0,T)}{P(0,t)} \exp\left\{ -(h(T) - h(t))z_t - \frac{1}{2}\left(h^2(T) - h^2(t)\right)\zeta_t \right\}. \tag{4.138}$$

One can easily verify that the LGM automatically calibrates to the spot discount curve. Since both the numeraire and the zero-coupon bond prices are driven by the same normal random factor, the model has a high level of analytical tractability. For instance, under LGM swap rates can be expressed in closed form. Under the LGM $h(t)$ is at our disposal, and different choices or features of $h(t)$ will yield different models. In fact, if we take

$$h(t) = \int_0^t e^{-\kappa s} ds \quad \text{for} \quad \kappa > 0,$$

we will reproduce the Hull-White model with "mean reversion speed" κ. LGM can be generalized to multiple driving factors. Before pricing interest rate derivatives, we ought to calibrate α_t to vanilla interest rate derivatives like ccaps, floors and swaptions.

4.12 Notes

We have already witnessed that a conventional specification of forward-rate volatility will result in a Gaussian model for forward rates, which can take negative values and therefore can induce arbitrage. This serious limitation of the HJM framework motivated many studies on frameworks of interest-rate models that maintain positive interest rates. Rogers (1997) takes a fundamental approach and works on a pricing kernel, which is the Radon–Nikodym derivative of the risk-neutral measure with respect to the physical measure. The key is to treat the pricing kernel as a potential,[3] which will result in monotonically decreasing zero-coupon bond price with respect to maturity, and, according to Equation 4.15, thus ensure positive forward rates. Rogers casts the pricing kernel as

$$\zeta_t = E_t^{\mathbb{P}}\left[\int_t^\infty \mu_\zeta(\tau)\,\mathrm{d}\tau\right].\tag{4.139}$$

Then, what is left to do is to describe $\mu_\zeta(\tau)$, the expected rate of change of the pricing kernel.

Rogers' approach includes a slightly earlier model by Flesaker and Hughston (1996) as a special case. But Jin and Glasserman (2001) proved later that these two approaches are equivalent for positive interest-rate models, in the sense that a model under one specification can be converted to that under the other.

There are however outstanding issues in both the theory and application of positive interest-rate models. It is fair to say that the positive interest-rate framework by either Rogers or Flesaker and Hughston is not very convenient for option pricing. On the other hand, the connection between these frameworks and other existing positive interest-rate models, including the CIR (1985) model, the Black–Karasinski (1991) model, general affine term structure models (ATSMs), and the London Interbank Offer Rates (LIBOR) market model, is far from clear. These issues call for further research on interest-rate modeling.

Exercises

1. It is known that zero-coupon bonds relate to forward rates by

$$P(t,T) = \exp\left(-\int_t^T f(t,s)\,\mathrm{d}s\right) \triangleq \mathrm{e}^{M_t}.$$

Given the HJM equation under the physical measure,

$$\mathrm{d}f(t,T) = \mu(t,T)\,\mathrm{d}t + \boldsymbol{\sigma}^{\mathrm{T}}(t,T)\,\mathrm{d}\mathbf{W}_t,$$

prove that the price processes of zero-coupon bonds are

$$\mathrm{d}P(t,T) = P(t,T)\left(r_t\,\mathrm{d}t + \boldsymbol{\Sigma}^{\mathrm{T}}(t,T)\left(\mathrm{d}\mathbf{W}_t + \boldsymbol{\gamma}_t\,\mathrm{d}t\right)\right),$$

---

[3]A potential, say Z_t, is a right-continuous non-negative supermartingale satisfying $E^{\mathbb{P}}[Z_t] \overset{t\to\infty}{\longrightarrow} 0$.

where $\mathbf{\Sigma}(t,T) = -\int_t^T \boldsymbol{\sigma}(t,s)\,\mathrm{d}s$ and $\boldsymbol{\gamma}_t$ satisfies

$$\mu(t,T) + \boldsymbol{\sigma}^{\mathrm{T}}(t,T)\left(\mathbf{\Sigma}(t,T) - \boldsymbol{\gamma}_t\right) = 0.$$

2. With Ho–Lee's model, prove that

$$P_0^t = E^{\mathbb{Q}}\left[\mathrm{e}^{-\int_0^t r_s\,\mathrm{d}s}\right],$$

where $\mathbb{Q}$ stands for the risk-neutral measure.

3. The risk-neutral process for zero-coupon bonds under Ho–Lee's model is

$$\frac{\mathrm{d}P_t^T}{P_t^T} = -(T-t)\sigma\,\mathrm{d}\tilde{W}_t + \left(f(0;t) + \sigma\tilde{W}_t + \frac{1}{2}\sigma^2 t^2\right)\mathrm{d}t.$$

Use the above process to show that $P_T^T = 1$, that is, the price equals par at maturity.

4. Prove that the Hull–White model is mean-reverting by showing that the short rate follows

$$\mathrm{d}r_t = \kappa(\theta_t - r_t)\,\mathrm{d}t + \sigma\,\mathrm{d}\tilde{W}_t,$$

with

$$\theta(t) = f(0,t) + \frac{1}{\kappa}f_t(0,t) + \frac{\sigma^2}{2\kappa^2}\left(1 - \mathrm{e}^{-2\kappa t}\right).$$

5. For the HJM model, prove the following formulae for (discounted) zero-coupon bonds and money market account:

$$\frac{P_{T_0}^{T_j}}{B_{T_0}} = P_0^{T_j}\exp\left(\int_0^{T_0} -\frac{1}{2}\|\mathbf{\Sigma}(t,T_j)\|^2\,\mathrm{d}t + \mathbf{\Sigma}^{\mathrm{T}}(t,T_j)\,\mathrm{d}\tilde{\mathbf{W}}_t\right),$$

$$T_j \geq T_0,$$

$$B_{T_0} = \frac{1}{P_0^{T_0}}\exp\left(\int_0^{T_0}\frac{1}{2}\|\mathbf{\Sigma}(t,T_0)\|^2\,\mathrm{d}t - \mathbf{\Sigma}^{\mathrm{T}}(t,T_0)\,\mathrm{d}\tilde{\mathbf{W}}_t\right),$$

where $\tilde{\mathbf{W}}_t$ is a Brownian motion under the risk-neutral measure.

6. Let a *stripped-dividend price process*, $\hat{S}_t$, follow

$$\mathrm{d}\hat{S}_t = \hat{S}_t\left(r_t\,\mathrm{d}t + \mathbf{\Sigma}_S^{\mathrm{T}}\,\mathrm{d}\tilde{\mathbf{W}}_t\right)$$

under the risk-neutral measure. Derive the risk-neutral process for S_t. Consider both cases of continuous and discrete dividends.

7. Suppose that the risk-neutral process of the short rate follows the *Vasicek* model:

$$dr_t = \kappa(\theta - r_t)\,dt + \sigma\,dW_t,$$

where κ, θ, and σ are constants. Prove the following price formula for zero-coupon bonds:

$$P(t,T) = A(t,T)e^{-B(t,T)r_t},$$

where

$$B(t,T) = \frac{1 - e^{-\kappa(T-t)}}{\kappa},$$

$$A(t,T) = \exp\left\{ \frac{(B(t,T) - T + t)\left(\kappa^2\theta - \sigma^2/2\right)}{\kappa^2} - \frac{\sigma^2 B^2(t,T)}{4\kappa} \right\}.$$

8. Consider the *risk-neutralized* processes of two assets:

$$\frac{dS_1}{S_1} = r_t\,dt + \sigma_1\,dW_1(t),$$

$$\frac{dS_2}{S_2} = r_t\,dt + \sigma_2\,dW_2(t),$$

where $dW_1 dW_2 = \rho\,dt$ and r_t may follow some stochastic process.

   (a) What kind of process does S_1/S_2 follow under the risk-neutral measure?

   (b) Can you define a new measure under which S_1/S_2 is a martingale?

   (c) Express the Radon–Nikodym derivative in terms of the asset prices.

9. Derive the pricing formula for a put option with payoff

$$V_T = (K - S_T)^+$$

under a stochastic interest rate. Assume that the volatility vector of the asset and the T-maturity zero-coupon bond are

$$\Sigma_s = \sigma_s \begin{pmatrix} \rho \\ \sqrt{1 - \rho^2} \end{pmatrix} \quad \text{and} \quad \Sigma_P = \sigma_P \begin{pmatrix} 1 \\ 0 \end{pmatrix},$$

where both σ_s and σ_P are time-dependent scalars.

   (a) Show that the correlation between $d\ln S_t$ and $d\ln P(t,T)$ is ρ.

   (b) Express the option formula in terms of σ_s, σ_P, and ρ.

10. Price a five-year maturity option on a 10-year zero-coupon bond under the one-factor Ho–Lee model. Let the spot forward-rate term structure be

$$f(0,T) = 0.02 + 0.001T,$$

the forward-rate volatility be $\sigma = 0.015$, and the strike price be the forward price of the 10-year zero-coupon bond, $K = P(0,10)/P(0,5)$.

11. Redo the above problem using the Hull–White model: all inputs remain the same except that the forward-rate volatility becomes

$$\sigma(t, T) = 0.015 e^{-0.1(T-t)}.$$

12. Derive the price formula for zero-coupon bonds, Equations 4.34 through 4.36, for the Hull–White model.

13. Price a five-year maturity option on the post-dividend price of a coupon bond under the one-factor Ho-Lee model. The coupon rate and the maturity of the bond are 5.25% and nine and a half years, respectively. The strike price of the option is \$100. Let the spot forward-rate term structure be flat at 5% and the forward-rate volatility be $\sigma = 0.015$.

14. Redo the pricing of the par-bond options in Table 4.4 using T_0-forward measure. Prove first the following formulae for the T_0-forward prices of zero-coupon bonds:

$$\frac{P_t^{T_j}}{P_t^{T_0}} = \frac{P_0^{T_j}}{P_0^{T_0}} \exp\left\{ \int_0^t -\frac{1}{2} \|\Sigma(s, T_j) - \Sigma(s, T_0)\|^2 ds \right.$$
$$\left. + (\Sigma(s, T_j) - \Sigma(s, T_0))^T d\hat{\mathbf{W}}_t \right\}, \quad T_j \geq T_0,$$

where $\hat{\mathbf{W}}_t$ is a Brownian motion under the T_0-forward measure. Then, propose a scheme for Monte Carlo simulations.

15. Prove that the risk-neutral dynamics of an asset with price S_t and stochastic dividend yield q_t remains to be

$$dS_t = S_t(r_t - q_t)dt + \Sigma_S^T d\mathbf{W}_t.$$

Appendix: On the Lognormal Specification of Forward Rates

We now explore the possibility of using the state-dependent volatility function in the HJM model. Without loss of generality, we consider the forward-rate volatility function of the form

$$\sigma(t, T) = \sigma_0(t, T) f^\alpha(t, T), \tag{4.140}$$

where $\sigma_0(t, T)$ is a deterministic function and α a positive exponent. In the special case, $\alpha = 0$, we obtain a Gaussian model.

Similar to Avellaneda and Laurence (1999), we show that the "lognormal" model, corresponding to $\alpha = 1$, blows up in finite time in the sense that a forward rate reaches infinity. This result was first obtained by Morton (1988). One can imagine that similar results may apply to the case of $\alpha > 0$. Hence, volatility specification in the form of Equation 4.136 is denied.

It suffices to show the result with a one-factor model. The no-arbitrage condition dictates that the drift must be

$$\mu(t,T) = f(t,T)\sigma_0(t,T) \int_t^T f(t,s)\sigma_0(t,s)\,\mathrm{d}s, \tag{4.141}$$

which depends on the entire curve of $f(t,s)$, $t \le s \le T$. Consider the simplest specification of $\sigma_0(t,T)$: $\sigma_0(t,T) = \sigma_0 = $ constant. The HJM equation then becomes

$$\frac{\mathrm{d}f(t,T)}{f(t,T)} = \sigma_0\,\mathrm{d}\tilde{W}_t + \left(\sigma_0^2 \int_t^T f(t,s)\,\mathrm{d}s\right)\mathrm{d}t.$$

The formal solution to the above equation is

$$f(t,T) = f(0,T)\exp\left(\sigma_0\tilde{W}_t - \frac{\sigma_0^2}{2}t + \sigma_0^2\int_0^t\left(\int_s^T f(s,u)\,\mathrm{d}u\right)\mathrm{d}s\right)$$

$$= f(0,T)M(t)\exp\left(\sigma_0^2\int_0^t\left(\int_s^T f(s,u)\,\mathrm{d}u\right)\mathrm{d}s\right), \tag{4.142}$$

where $M(t) = \exp(\sigma_0\tilde{W}_t - (\sigma_0^2/2)t)$. Assume for simplicity that the initial term structure is flat, that is, $f(0,T) = f_0 = $ constant. Differentiating both sides of Equation 4.138 with respect to T, we obtain

$$\frac{\partial}{\partial T}f(t,T) = f_0 M(t)\exp\left(\sigma_0^2\int_0^t\left(\int_s^T f(s,u)\,\mathrm{d}u\right)\mathrm{d}s\right)\sigma_0^2\int_0^t f(s,T)\,\mathrm{d}s$$

$$= \sigma_0^2 f(t,T)\int_0^t f(s,T)\,\mathrm{d}s$$

$$= \frac{\sigma_0^2}{2}\frac{\partial}{\partial t}\left(\int_0^t f(s,T)\,\mathrm{d}s\right)^2.$$

Integrating the above equation with respect to t, we then have

$$\frac{\partial}{\partial T}\int_0^t f(s,T)\,\mathrm{d}s = \frac{\sigma_0^2}{2}\left(\int_0^t f(s,T)\,\mathrm{d}s\right)^2. \tag{4.143}$$

Now setting

$$X_t(T) = \int_0^t f(s,T)\,\mathrm{d}s$$

and solving for $X_t(t)$ from Equation 4.139, we obtain

$$X_t(T) = \frac{2X_t(t)}{2 - \sigma_0^2 X_t(t)(T-t)}.$$

It is not hard to see that for any given $X_t(t) > 0$, $X_t(T)$ blows up at

$$T_0 = t + \frac{2}{\sigma_0^2 X_t(t)}.$$

This implies that for any $T \geq T_0$, there will be $f(s, T) = \infty$ for some $s \leq t$, that is, the forward rate blows up in finite time.

Through a proper transformation, we can show that the forward rate also blows up for the volatility specification (Equation 4.136) with $\alpha > 0$. As a result, level-dependent volatilities are ruled out for the HJM model.

Chapter 5

Short-Rate Models and Lattice Implementation

Short-rate models hold a special place in fixed-income modeling: they are the first generation of interest-rate models, and some of them still play active roles in today's applications. Short-rate models remain attractive due to two distinguishing advantages. First, they are intuitive, as many of them were established based on theories from financial economics. Second, as Markovian models of single-state variables, they can be implemented by lattice trees, so they are often used for pricing path-dependent options. One of the two objectives of this chapter is to study the implementation of short-rate models by lattice methods. The other objective is, in continuation of the theoretical analysis of the HJM framework in Chapter 4, to study under what conditions an HJM model implies a Markovian short-rate model.

A necessary step in the applications of short-rate models is to make sure that they are consistent with the current term structure of interest rates. In other words, a short-rate model must price zero-coupon bonds of all relevant maturities correctly. There are two ways to make a short-rate model consistent with the term structure. The first way is to identify the model as a special case of the HJM framework, and thus to derive the parameters for the short-rate model from its corresponding HJM equation. The second way is to calibrate a short-rate model, which is a procedure of numerical fitting. In Section 5.1, we first derive the expression of forward-rate volatility for general short-rate models. Then, using the well-known Cox et al. (1985) model, we show the reality that it can be far from trivial to identify the corresponding forward-rate volatility for production use. The implication is that the first way is not feasible in general. We then move in the opposite direction and address, in Section 5.2, the question of under what conditions an HJM model implies a Markovian short-rate model. A sufficient condition will be obtained, which will allow us to implement the HJM model through a short-rate model, while skipping model calibration.

Calibration has been a key part of implementing short-rate models. In Section 5.3, we present a major improvement to the technology for interest-rate lattice or tree construction. Traditionally, a lattice tree is built first and then it is calibrated afterwards, before being used for pricing derivatives. We however make the calibration part of construction procedure, so that the tree is calibrated once it is built. The calibration requirement affects both branching and branching probabilities, and the resulting tree will automatically be truncated and will evolve along the level of the mean interest rate. The new calibration technology is illustrated with the Hull–White model and it can be applied to a very general class of short-rate models. It also has the flexibility to take variable sizes of time stepping.

5.1   From Short-Rate Models to Forward-Rate Models

Short-rate models dominated fixed-income modeling before the emergence of the no-arbitrage framework of Heath, Jarrow, and Morton (1992), which is based on forward rates. Short-rate models can be made arbitrage free by taking appropriate drift terms, such as the Ho–Lee model and the Hull–White model. But this is not always easy. One way to derive the correct drift term is to identify the corresponding forward-rate volatility and then to solve for the expression of the forward rates, which include the short rate as an extreme case, from the HJM equation. The focus in this section is on how to derive the corresponding forward-rate volatility in order to identify the model as a special case of the HJM framework.

Consider in general an Ito's process for the short rate under the risk-neutral measure, $\mathbb{Q}$,

$$\mathrm{d}r_t = \nu(r_t, t)\,\mathrm{d}t + \rho(r_t, t)\,\mathrm{d}W_t, \tag{5.1}$$

where the drift, $\nu(r_t, t)$, and volatility, $\rho(r_t, t)$, are deterministic functions of their arguments. Note that, for notational simplicity, we hereafter drop "$\sim$" over the $\mathbb{Q}$-Brownian motion, W_t. Define an auxiliary function

$$g(x, t, T) = -\ln E^{\mathbb{Q}}\left[\exp\left(-\int_t^T r_s\,\mathrm{d}s\right)\bigg|\, r_t = x\right]. \tag{5.2}$$

We have the following result (Baxter and Rennie, 1996).

Theorem 5.1.1. *An arbitrage-free short-rate model is an HJM model with forward-rate volatility given by*

$$\sigma(t, T) = \rho(r_t, t)\frac{\partial^2 g}{\partial x \partial T}(r_t, t, T). \tag{5.3}$$

Proof. According to its definition, $g(r_t, t, T) = -\ln P(t, T)$. It follows that

$$f(t, T) = -\frac{\partial \ln P(t, T)}{\partial T} = \frac{\partial g}{\partial T}(r_t, t, T).$$

Hence, the process for the instantaneous forward rate is

$$\mathrm{d}f(t,T) = \frac{\partial f}{\partial x}\mathrm{d}r_t + \left(\frac{\partial f}{\partial t} + \frac{1}{2}\rho^2\frac{\partial^2 f}{\partial x^2}\right)\mathrm{d}t$$

$$= \rho\frac{\partial^2 g}{\partial x\partial T}\mathrm{d}W_t + \text{drift term.}$$

(5.4)

Note that the drift term for the risk-neutral measure is determined by the volatility term. This ends the proof. □

We now try to derive the forward-rate volatility, based on Theorem 5.1, for a few popular short-rate models. We start with the Hull–White model for the short rate,

$$\mathrm{d}r_t = \kappa(\theta_t - r_t)\,\mathrm{d}t + \sigma\,\mathrm{d}W_t. \tag{5.5}$$

As is shown in Equations 4.34 through 4.36, the expectation in Equation 5.2 can be derived as

$$g(r_t, t, T) = B(t,T)r_t - \ln A(t,T), \tag{5.6}$$

where

$$B(t,T) = \frac{1 - \mathrm{e}^{-\kappa(T-t)}}{\kappa}. \tag{5.7}$$

Obviously, there is

$$\frac{\partial^2 g}{\partial x\partial T}(r_t, t, T) = \frac{\partial B(t,T)}{\partial T} = \mathrm{e}^{-\kappa(T-t)}. \tag{5.8}$$

According to Theorem 5.1, the volatility of the forward rate is

$$\sigma(t,T) = \sigma\frac{\partial^2 g}{\partial x\partial T}(r_t, t, T) = \sigma\mathrm{e}^{-\kappa(T-t)}, \tag{5.9}$$

which is already a known result for the Hull–White model. Putting $\kappa = 0$, we also obtain the forward-rate volatility of the Ho–Lee model.

The application of Theorem 5.1 to short-rate models other than Hull–White, however, is often less trivial. Let us consider another short-rate model of the form

$$\mathrm{d}r_t = \sigma_t\sqrt{r_t}\,\mathrm{d}W_t + \kappa_t(\theta_t - r_t)\,\mathrm{d}t, \tag{5.10}$$

where σ_t and $\kappa_t > 0$ are deterministic functions of time, and θ_t the subject to the term structure of interest rates. This is the famous Cox et al. (1985) model, which was proposed as a modification of the Vasicek (1977) model, such that the interest rate remains positive. Intuitively, due to the factor of $\sqrt{r_t}$ in the volatility term, $\mathrm{d}r_t$ is positive at $r_t = 0$, provided that there is $\kappa_t\theta_t > 0$, which pulls the interest rate into the positive territory.

To derive the corresponding forward-rate volatility, we need to evaluate $g(r_t, t, T)$. It will be shown in Chapter 9 that function $g(r_t, t, T)$ takes the following form:

$$g(r, t, T) = rB(t, T) + \int_t^T \theta_s B(s, T) \, ds, \tag{5.11}$$

where $B(t, T)$ satisfies the Riccarti equation,

$$\frac{\partial B}{\partial t} = \frac{1}{2} \sigma_t^2 B^2(t, T) + \kappa_t B(t, T) - 1, \quad B(T, T) = 0. \tag{5.12}$$

For simplicity, we assume constant volatility and reversion strength, that is, $\sigma_t = \sigma$ and $\kappa_t = \kappa$. Then the above Riccarti equation can be solved analytically, such that

$$B(t, T) = \frac{d - \kappa}{\sigma^2} \frac{1 - e^{d(T-t)}}{1 - h e^{d(T-t)}}, \tag{5.13}$$

with

$$d = \sqrt{\kappa^2 + 2\sigma^2}, \qquad h = \frac{d - \kappa}{d + \kappa}. \tag{5.14}$$

Meanwhile, θ_t, the mean of the short rate, is implied by the integral Equation 5.11. The CIR process under the HJM terms is represented by the following volatility function of the forward rate,

$$\sigma(t, T) = \sigma_t \sqrt{r_t} \left(\frac{\partial B(t, T)}{\partial T} + \int_t^T \frac{\partial \theta_s}{\partial r_t} \frac{\partial B(s, T)}{\partial T} \, ds \right). \tag{5.15}$$

There are undesirable features in Equation 5.15. First, θ_t is implied by the integral Equation 5.11, for which a reliable solution is not easily obtainable. Second, $\sigma(t, T)$ depends on r_t, another stochastic process, so the HJM equation is not closed by itself, and r_t is yet another state variable. This feature will be echoed in the next section when we study an alternative formulation of the CIR model. From both analytic and computational point of view, Equation 5.15 is not very useful because we must solve for θ_t numerically in advance, a procedure equivalent to calibration, which should not be necessary in the HJM context. To some extent, the undesirable features in Equation 5.15 demonstrate the limitation of the HJM model as a framework for producing positive interest-rate models.

5.2 General Markovian Models

Existing short-rate models are Markovian models. A no-arbitrage short-rate model should also be derived from the HJM framework. However, this can be quite difficult. In this section, we address the opposite question: under what kind of forward-rate volatility specifications should the resulting short-rate model be a Markovian random variable? Answering this question will help us to calibrate and implement a short-rate model more efficiently.

According to Equation 4.21, the short rate can be expressed as

$$r_t = f(t, t) = f(0, t) + \int_0^t \left[-\boldsymbol{\sigma}^{\mathrm{T}}(s, t)\boldsymbol{\Sigma}(s, t)\, \mathrm{d}s + \boldsymbol{\sigma}^{\mathrm{T}}(s, t)\, \mathrm{d}\mathbf{W}_s \right], \quad (5.16)$$

where $\mathbf{W}_t$ is the n-dimensional Brownian motion under the risk-neutral measure, $\boldsymbol{\sigma}(t, T)$ the forward-rate volatility, and $\boldsymbol{\Sigma}(t, T)$ the volatility of the T-maturity zero-coupon bond, given by $\boldsymbol{\Sigma}(t, T) = -\int_t^T \boldsymbol{\sigma}(t, u)\mathrm{d}u$. The stochastic differentiation of the short rate is

$$\begin{aligned}
\mathrm{d}r_t &= \left[f_t(0, t) + \int_0^t \left(-\frac{\partial}{\partial t}(\boldsymbol{\sigma}^{\mathrm{T}}(s, t)\boldsymbol{\Sigma}(s, t))\, \mathrm{d}s + \frac{\partial \boldsymbol{\sigma}^{\mathrm{T}}(s, t)}{\partial t}\, \mathrm{d}\mathbf{W}_s \right) \right] \mathrm{d}t \\
&\quad + \boldsymbol{\sigma}^{\mathrm{T}}(t, t)\, \mathrm{d}\mathbf{W}_t \\
&= [f_t(t, T)]_{T=t}\, \mathrm{d}t + \boldsymbol{\sigma}^{\mathrm{T}}(t, t)\, \mathrm{d}\mathbf{W}_t.
\end{aligned} \quad (5.17)$$

Based on Equation 5.17 we can make the following judgment: for the short-rate model to be a Markovian process, we need the drift term, $[f_t(t, T)]_{T=t}$, to be a function of a finite set of state variables that are jointly Markovian in their evolution.

To write the short rate as a function of several state variables, we introduce auxiliary functions

$$b_i(t, T) = \sigma_i(t, T) \int_t^T \sigma_i(t, s)\, \mathrm{d}s, \quad i = 1, 2, \ldots, n. \quad (5.18)$$

If we define

$$\chi_i(t) = \int_0^t b_i(s, t)\, \mathrm{d}s + \sigma_i(s, t)\, \mathrm{d}W_i(s), \quad i = 1, 2, \ldots, n, \quad (5.19)$$

we can then write

$$r_t = f(0, t) + \sum_{i=1}^n \chi_i(t). \quad (5.20)$$

If we can find the conditions for $\chi_i(t)$ to be Markovian variables, then, under the same conditions, r_t will also be a Markovian variable.

Define, in addition,

$$\varphi_i(t) = \int_0^t \sigma_i^2(s, t)\, \mathrm{d}s, \quad i = 1, 2, \ldots, n. \quad (5.21)$$

The following theorem presents a sufficient condition under which the pair of functions $\{\chi_i, \varphi_i\}$ are jointly Markovian variables (Ritchken and Sankarasubramanian, 1995; Inui and Kijima, 1998).

Theorem 5.2.1. *Suppose that the forward-rate volatility satisfies*

$$\frac{\partial \sigma_i(t,T)}{\partial T} = -\kappa_i(T)\sigma_i(t,T), \quad i = 1, 2, \ldots, n, \tag{5.22}$$

for some deterministic functions, $\kappa_i(T)$. Then,

$$\begin{aligned}
d\varphi_i(t) &= \left(\sigma_i^2(t,t) - 2\kappa_i(t)\varphi_i(t)\right) dt, \\
d\chi_i(t) &= \left(\varphi_i(t) - \kappa_i(t)\chi_i(t)\right) dt + \sigma_i(t,t)\, dW_i(t),
\end{aligned} \tag{5.23}$$

for $i = 1, \ldots, n$.

Proof. For φ_i, there is

$$\begin{aligned}
d\varphi_i(t) &= \sigma_i^2(t,t)\, dt + \left(\int_0^t 2\sigma_i(s,t)\frac{\partial \sigma_i(s,t)}{\partial t}\, ds\right) dt \\
&= \sigma_i^2(t,t)\, dt - 2\kappa_i(t)\left(\int_0^t \sigma_i^2(s,t)\, ds\right) dt \\
&= \left(\sigma_i^2(t,t) - 2\kappa_i(t)\varphi_i(t)\right) dt,
\end{aligned} \tag{5.24}$$

while for $\chi_i(t)$, we have, noticing that $b_i(t,t) = 0$,

$$d\chi_i(t) = \left(\int_0^t \frac{\partial b_i(s,t)}{\partial t}\, ds + \frac{\partial \sigma_i(s,t)}{\partial t}\, dW_i(s)\right) dt + \sigma_i(t,t)\, dW_i(t). \tag{5.25}$$

Because of Equation 5.22 and

$$\begin{aligned}
\frac{\partial b_i(s,t)}{\partial t} &= \frac{\partial \sigma_i(s,t)}{\partial t}\int_s^t \sigma_i(s,u)\, du + \sigma_i^2(s,t) \\
&= -\kappa_i(t)b_i(s,t) + \sigma_i^2(s,t),
\end{aligned} \tag{5.26}$$

we can rewrite Equation 5.25 as

$$\begin{aligned}
d\chi_i(t) &= -\kappa_i(t)\left(\int_0^t b_i(s,t)\, ds + \sigma_i(s,t)\, dW_i(s)\right) dt \\
&\quad + \left(\int_0^t \sigma_i^2(s,t)\, ds\right) dt + \sigma_i(t,t)\, dW_i(t) \\
&= \left(\varphi_i(t) - \kappa_i(t)\chi_i(t)\right) dt + \sigma_i(t,t)\, dW_i(t).
\end{aligned} \tag{5.27}$$

This completes the proof. $\qquad\square$

The implication of the above theorem is that the HJM model is a Markovian model under condition 5.22. The short-rate process is then

$$dr_t = f_t(0, t)\, dt + \sum_{i=1}^{n} d\chi_i(t)$$

$$= f_t(0, t)\, dt + \sum_{i=1}^{n} (\varphi_i(t) - \kappa_i(t)\chi_i(t))\, dt + \sum_{i=1}^{n} \sigma_i(t, t)\, dW_i(t)$$

$$= \left(f_t(0, t) + \sum_{i=1}^{n} (\varphi_i(t) + (\kappa_n(t) - \kappa_i(t))\,\chi_i(t)) \right) dt$$

$$- \kappa_n(t) \left(\sum_{i=1}^{n} \chi_i(t) \right) dt + \sum_{i=1}^{n} \sigma_i(t, t)\, dW_i(t). \tag{5.28}$$

Denote

$$\Phi(t) = \sum_{i=1}^{n} (\varphi_i(t) + (\kappa_n - \kappa_i)\chi_i(t)), \tag{5.29}$$

and then

$$dr_t = (\kappa_n(t)\, (f(0, t) - r_t) + f_t(0, t) + \Phi(t))\, dt + \sum_{i=1}^{n} \sigma_i(t, t)\, dW_i(t). \tag{5.30}$$

For $\kappa_n(t) > 0$, Equation 5.30 demonstrates the mean-reverting feature for the short-rate process.

Next, we will show that zero-coupon bonds across all maturities can be expressed in terms of Markovian state variables. Under the HJM model, the formula for the price of a zero-coupon bond is

$$P(t, T) = \frac{P(0, T)}{P(0, t)} \exp\left\{ - \int_t^T \left(\sum_{i=1}^{n} \int_0^t b_i(s, u)\, ds + \sigma_i(s, u)\, dW_i(s) \right) du \right\}. \tag{5.31}$$

For the first term in the exponent, we have

$$\int_t^T \int_0^t b_i(s, u)\, ds\, du = \int_0^t \left(\int_t^T b_i(s, u)\, du \right) ds$$

$$= \int_0^t \left(\int_t^T \sigma_i(s, u) \int_s^u \sigma_i(s, v)\, dv\, du \right) ds, \tag{5.32}$$

while the integrand of Equation 5.32 can be written as

$$
\int_t^T \sigma_i(s,u) \int_s^u \sigma_i(s,\nu)\,d\nu\,du
$$

$$
= \sigma_i(s,t) \int_t^T e^{-\int_t^u \kappa_i(x)\,dx} \left[\int_s^t + \int_t^u \sigma_i(s,\nu)\,d\nu \right] du
$$

$$
= b_i(s,t) \int_t^T e^{-\int_t^u \kappa_i(x)\,dx}\,du
$$

$$
+ \sigma_i^2(s,t) \int_t^T e^{-\int_t^u \kappa_i(x)\,dx} \int_t^u e^{-\int_t^\nu \kappa_i(x)\,dx}\,d\nu\,du. \tag{5.33}
$$

Defining

$$
\beta_i(t,T) = \int_t^T e^{-\int_t^u \kappa_i(x)\,dx}\,du, \quad t \le T, \tag{5.34}
$$

and noticing $\beta_i(t,t) = 0$, we have

$$
\int_t^T e^{-\int_t^u \kappa_i(x)\,dx} \int_t^u e^{-\int_t^\nu \kappa_i(x)\,dx}\,d\nu\,du = \int_t^T \frac{\partial \beta_i(t,u)}{\partial u}\beta_i(t,u)\,du
$$

$$
= \frac{1}{2}\beta_i^2(t,T). \tag{5.35}
$$

By combining Equations 5.32 through 5.35, we obtain

$$
\int_t^T b_i(s,u)\,du = \beta_i(t,T)b_i(s,t) + \frac{1}{2}\beta_i^2(t,T)\sigma_i^2(s,t). \tag{5.36}
$$

Now, let us consider the second term in the exponent of Equation 5.31. Equation 5.22 implies that

$$
\int_t^T \sigma_i(s,u)\,du = \beta_i(t,T)\sigma_i(s,t), \tag{5.37}
$$

and it follows that

$$
\int_t^T \int_0^t \left(b_i(s,t)\,ds + \sigma_i(s,t)\,dW_i(s) \right) du
$$

$$
= \beta_i(t,T) \int_0^t \left(b_i(s,t)\,ds + \sigma_i(s,t)\,dW_i(s) \right) + \frac{1}{2}\beta_i^2(t,T) \int_0^t \sigma_i^2(s,t)\,ds
$$

$$
= \beta_i(t,T)\chi_i(t) + \frac{1}{2}\beta_i^2(t,T)\varphi_i(t). \tag{5.38}
$$

The price formula for zero-coupon bonds is thus

$$
P(t,T) = \frac{P(0,T)}{P(0,t)} \exp\left(-\sum_{i=1}^n \beta_i(t,T)\chi_i(t) - \frac{1}{2}\sum_{i=1}^n \beta_i^2(t,T)\varphi_i(t) \right), \quad t \le T. \tag{5.39}
$$

Note that in order to price options on coupon bonds, all we need is the distribution of $\varphi_i(t)$ and $\chi_i(t)$.

Condition 5.22 is sufficient for the short-rate to be a Markovian variable. It becomes a necessary condition if the short-rate volatility is state independent.

Theorem 5.2.2. *Suppose that the short-rate volatility, $\sigma(t,t)$, is a deterministic function of time. Then a necessary condition for the short rate to be Markovian is*

$$\frac{\partial \sigma_i(t,T)}{\partial T} = -\kappa_i(T)\sigma_i(t,T), \qquad (5.40)$$

for some scalar function, $\kappa_i(T), 1 \le i \le n$.

Proof. For the short rate to be Markovian, we require that $r_T - r_t$ depends only on r_t and $\{\mathrm{d}\mathbf{W}_s, s \in (t,T)\}$. In fact, we have

$$
\begin{aligned}
r_T - r_t &= f(0,T) - f(0,t) + \int_0^T \boldsymbol{\sigma}^{\mathrm{T}}(s,T)\boldsymbol{\Sigma}(s,T)\,\mathrm{d}s \\
&\quad - \int_0^t \boldsymbol{\sigma}^{\mathrm{T}}(s,t)\boldsymbol{\Sigma}(s,t)\,\mathrm{d}s + \int_0^T \boldsymbol{\sigma}^{\mathrm{T}}(s,T)\,\mathrm{d}\mathbf{W}_s - \int_0^t \boldsymbol{\sigma}^{\mathrm{T}}(s,t)\,\mathrm{d}\mathbf{W}_s \\
&= f(0,T) - f(0,t) + \int_0^T \boldsymbol{\sigma}^{\mathrm{T}}(s,T)\boldsymbol{\Sigma}(s,T)\,\mathrm{d}s \\
&\quad - \int_0^t \boldsymbol{\sigma}^{\mathrm{T}}(s,t)\boldsymbol{\Sigma}(s,t)\,\mathrm{d}s + \int_t^T \boldsymbol{\sigma}^{\mathrm{T}}(s,T)\,\mathrm{d}\mathbf{W}_s \\
&\quad + \int_0^t (\boldsymbol{\sigma}(s,T) - \boldsymbol{\sigma}(s,t))^T \,\mathrm{d}\mathbf{W}_s.
\end{aligned}
\qquad (5.41)
$$

The last term in the above equation cannot depend on $\{\mathrm{d}\mathbf{W}_s, s \in (t,T)\}$, so it can depend only on r_t. Because

$$\int_0^t \boldsymbol{\sigma}^{\mathrm{T}}(s,t)\,\mathrm{d}\mathbf{W}_s = r_t + \text{deterministic function}, \qquad (5.42)$$

we conclude that

$$\int_0^t \boldsymbol{\sigma}^{\mathrm{T}}(s,T)\,\mathrm{d}\mathbf{W}_s \qquad (5.43)$$

is also a deterministic function of r_t. Hence, there is

$$\text{Correlation}\left[\int_0^t \boldsymbol{\sigma}^{\mathrm{T}}(s,T)\,\mathrm{d}\mathbf{W}_s, \int_0^t \boldsymbol{\sigma}^{\mathrm{T}}(s,t)\,\mathrm{d}\mathbf{W}_s\right] = 1. \qquad (5.44)$$

The last equality can be rewritten into

$$
\begin{aligned}
&E^{\mathbb{Q}}\left[\left(\int_0^t \boldsymbol{\sigma}^{\mathrm{T}}(s,T)\,\mathrm{d}\mathbf{W}_s\right) \times \left(\int_0^t \boldsymbol{\sigma}^{\mathrm{T}}(s,t)\,\mathrm{d}\mathbf{W}_s\right)\right] \\
&= E^{\mathbb{Q}}\left[\left(\int_0^t \boldsymbol{\sigma}^{\mathrm{T}}(s,T)\,\mathrm{d}\mathbf{W}_s\right)^2\right]^{1/2} \times E^{\mathbb{Q}}\left[\left(\int_0^t \boldsymbol{\sigma}^{\mathrm{T}}(s,t)\,\mathrm{d}\mathbf{W}_s\right)^2\right]^{1/2}.
\end{aligned}
$$
$$(5.45)$$

By Ito's isometry, Equation 5.45 implies that

$$
\left| \int_0^t \boldsymbol{\sigma}^{\mathrm{T}}(s,T)\boldsymbol{\sigma}(s,t)\,\mathrm{d}s \right| = \left(\int_0^t \|\boldsymbol{\sigma}(s,T)\|^2\,\mathrm{d}s \right)^{1/2} \times \left(\int_0^t \|\boldsymbol{\sigma}(s,t)\|^2\,\mathrm{d}s \right)^{1/2},
$$
(5.46)

that is, the equality is achieved in the formula of the Cauchy–Schwartz inequality (Rudin, 1976), and the equality holds if and only if

$$
\boldsymbol{\sigma}(s,t) = \alpha(t,T)\boldsymbol{\sigma}(s,T), \quad 0 \le s \le t
$$
(5.47)

for some deterministic scalar function, α. Similarly, we also have

$$
\boldsymbol{\sigma}(s,t) = \alpha(t,T')\boldsymbol{\sigma}(s,T'), \quad 0 \le s \le t
$$
(5.48)

for any other T'. Assume that $\sigma_i(s,t) \ne 0$, we then have

$$
\frac{\sigma_i(s,T)}{\sigma_i(s,T')} = \frac{\alpha(t,T')}{\alpha(t,T)} = \frac{\alpha(0,T')}{\alpha(0,T)}, \quad i = 1,\ldots,n.
$$
(5.49)

Making $T' = s$, we have then proved that $\sigma_i(s,T)$ can be factorized:

$$
\sigma_i(s,T) = x_i(s)y_i(T).
$$
(5.50)

By differentiating the above equation with respect to T, we obtain

$$
\frac{\partial \sigma_i(s,T)}{\partial T} = x_i(s)y_i(T)\frac{\partial \ln y_i(T)}{\partial T}.
$$
(5.51)

Denote $\partial \ln y_i(T)/\partial T$ by $-\kappa_i(T)$; we arrive at Equation 5.40. $\square$

5.2.1 One-Factor Models

A one-factor model, for $n = 1$, can be cast as

$$
\begin{aligned}
\mathrm{d}r_t &= \left(\kappa(t)\left(f(0,t) - r_t\right) + f_t(0,t) + \varphi_t\right)\mathrm{d}t + \sigma(t,t)\,\mathrm{d}W_t,\\
\mathrm{d}\varphi_t &= \left(\sigma^2(t,t) - 2\kappa(t)\varphi_t\right)\mathrm{d}t,
\end{aligned}
$$
(5.52)

based on Equations 5.23 and 5.30. When the forward-rate volatility takes $\sigma(t,t) = \sigma_0$ and $\kappa > 0$, we have

$$
\varphi_t = \frac{\sigma_0^2}{2\kappa}\left(1 - e^{-2kt}\right).
$$
(5.53)

By substituting Equation 5.53 back to the short-rate equation in Equation 5.52, then, unsurprisingly, we reproduce the Hull–White model in terms of the short rate.

When the short-rate volatility takes the form

$$\sigma(t, t) = \sigma_0 \sqrt{r_t}, \tag{5.54}$$

the governing equations of the short rate become

$$dr_t = \left(\kappa\left(f(0, t) - r_t\right) + f_t(0, t) + \varphi(t)\right) dt + \sigma_0 \sqrt{r_t}\, dW_t,$$
$$d\varphi(t) = \left(\sigma_0^2 r_t - 2\kappa\varphi(t)\right) dt. \tag{5.55}$$

Solving for φ_t, we obtain

$$\varphi_t = \sigma_0^2 \int_0^t r_s e^{-2\kappa(t-s)}\, ds. \tag{5.56}$$

Eliminating φ_t in the equation for the short rate, we finally obtain

$$dr_t = \left(\kappa\left(f(0, t) - r_t\right) + f_t(0, t) + \sigma_0^2 \int_0^t r_s e^{-2\kappa(t-s)}\, ds\right) dt + \sigma_0 \sqrt{r_t}\, dW_t. \tag{5.57}$$

The insight here is that the CIR model of the form (Equation 5.10) that is consistent to the term structure has a path-dependent mean level, θ_t. We thus expect that, in general, one-factor short-rate models that have a state-dependent volatility,

$$dr_t = \kappa_t(\theta_t - r_t)\, dt + \sigma(r_t, t)\, dW_t, \tag{5.58}$$

may have a path-dependent drift term.

Finally, we present a result of Inui and Kijima (1998) on the positiveness of forward rates for the general model (Equation 5.58).

Theorem 5.2.3. *Suppose that the short-rate volatility, $\sigma(r, t)$, is Lipschitz continuous with $\sigma(0, t) = 0$. If $r_0 \geq 0$ and*

$$\kappa(t)f(0, t) + f_T(0, t) > 0, \quad 0 \leq t, \tag{5.59}$$

then the short rate and forward rates are positive almost surely.

Proof. Under the conditions of the theorem, the strong solution of r_t exists. Moreover, since zero is an unattainable boundary (Karlin and Taylor, 1981) and r_t has continuous sample paths, we know from Equation 5.52 that r_t is positive for all $t > 0$. On the other hand, from the bond price formula

$$P(t, T) = \frac{P(0, T)}{P(0, t)} \exp\left\{-\left(\beta(t, T)\chi(t) + \frac{1}{2}\beta^2(t, T)\varphi(t)\right)\right\}$$
$$= \frac{P(0, T)}{P(0, t)} \exp\left\{-\left(\beta(t, T)(r_t - f(0, t)) + \frac{1}{2}\beta^2(t, T)\varphi(t)\right)\right\}, \tag{5.60}$$

we have

$$f(t,T) = f(0,T) - e^{-\int_t^T \kappa(x)\,dx} f(0,t) + e^{-\int_t^T \kappa(x)\,dx}\left[r_t + \beta(t,T)\varphi(t)\right].$$
(5.61)

Under condition 5.59, the second term in the right-hand side of Equation 5.61 is monotonically increasing in t, implying

$$e^{-\int_t^T \kappa(x)\,dx} f(0,t) < e^{-\int_t^T \kappa(x)\,dx} f(0,t)\Big|_{t=T} = f(0,T).$$
(5.62)

According to Equation 5.61, there must be $f(t,T) > 0$. $\qquad\qquad\square$

5.2.2 Monte Carlo Simulations for Options Pricing

Owing to the Markovian property of short-rate models, path simulations by Monte Carlo methods can be carried out efficiently, which is important for pricing exotic and path-dependent options. Take the pricing of the option on a zero-coupon for example. The value can be expressed as

$$V_t = E_t^{\mathbb{Q}}\left[e^{-\int_t^T r_s\,ds}\left(P(T,\tau) - K\right)^+\right], \quad t < T < \tau,$$
(5.63)

where $\mathbb{Q}$ stands for the risk-neutral measure, r_t is given by Equation 5.20, and the bond price is given by Equation 5.39. Both variables are expressed in terms of $\chi_i(t)$ and $\varphi_i(t)$, $i = 1,\ldots,n$, which evolve according to Equation 5.23. The corresponding simulation scheme for $\chi_i(t)$ and $\varphi_i(t)$ is

$$\begin{aligned}
\varphi_i(t+\Delta t) &= \varphi_i(t) + \left(\sigma_i^2(t,t) - 2\kappa_i(t)\varphi_i(t)\right)\Delta t, \\
\chi_i(t+\Delta t) &= \chi_i(t) + \left(\varphi_i(t) - \kappa_i(t)\chi_i(t)\right)dt + \sigma_i(t,t)\Delta W_i(t),
\end{aligned}$$
(5.64)

which is simply the so-called Euler scheme. The bond option is priced by simulating many payoffs before taking an average.

In Inui and Kijima (1998), the following example is considered:

$$\boldsymbol{\sigma}(t,T) = \begin{pmatrix} c_1 r_t^\alpha \\ c_2 r_t^\beta e^{-\kappa(T-t)} \end{pmatrix},$$
(5.65)

where $c_i, i = 1, 2, \alpha, \beta$, and κ are non-negative constants. It can be verified that the components of the volatility vector satisfy

$$\frac{\partial \sigma_1(t,T)}{\partial T} = 0, \qquad \frac{\partial \sigma_2(t,T)}{\partial T} = -\kappa\sigma_2(t,T).$$
(5.66)

According to Theorem 5.2, the corresponding short-rate process is a Markovian process. For various strikes, we consider the pricing of two-year maturity European options on the seven-year maturity bond, that is, $t = 0, T = 2, \tau = 7$. Assume a flat initial term structure:

$$f(0,T) = 0.05, \quad \forall T,$$

TABLE 5.1: Simulated Call Values Versus Exact Values

Strike	Simulated Value	Exact Value	Difference
0.500	0.252346	0.252269	7.70E−05
0.525	0.229725	0.229649	7.60E−05
0.550	0.207105	0.207029	7.60E−05
0.575	0.184491	0.184414	7.70E−05
0.600	0.161906	0.161826	8.00E−05
0.625	0.139417	0.139333	8.40E−05
0.650	0.117207	0.117112	9.50E−05
0.675	0.095628	0.095517	1.11E−04
0.700	0.075271	0.075106	1.65E−04
0.725	0.056802	0.056581	2.21E−04
0.750	0.040866	0.040619	2.47E−04
0.775	0.027960	0.027675	2.85E−04
0.800	0.018086	0.017848	2.38E−04
0.825	0.011090	0.010881	2.09E−04
0.850	0.006422	0.006272	1.50E−04
0.875	0.003504	0.003420	8.40E−05

Reprinted from Inui K and Kijima M, 1998. *Journal of Financial and Quantitative Analysis* 33(3): 423–440. With permission.

and take the following model specifications:

$$\alpha = \beta = 0, \quad \kappa = 0.05, \quad c_1 = c_2 = 0.01.$$

The size for time stepping for Equation 5.64 is $\Delta t = 1/200$, and the number of paths is $N = 50,000$. The results by the simulation method, together with the exact results (Heath et al., 1992), are listed in Table 5.1. As can be seen from the table, the accuracy of simulation pricing is rather high. The simulation scheme performs just as robustly for $\alpha \neq 0$ or $\beta \neq 0$, but there is no exact solution for these cases.

5.3 Binomial Trees of Interest Rates

We have seen that interest-rate options can be priced by Monte Carlo simulations. When we are using a Markovian short-rate model, we actually have another choice of numerical method, namely, the lattice tree method. An interest-rate tree can also be regarded as a form of path simulations of interest rates, yet the paths are connected in certain ways, so that the number of states in each time step remains small, making computation efficient. Whenever possible, lattice tree methods (also called lattice or tree methods) are preferred over Monte Carlo simulation methods because the former offer higher efficiency at much less computational cost. Lattice methods are particularly powerful for pricing American options for which early exercises must be considered. In this section, we study the construction and calibration of trees for the short rate. Both binomial tree and trinomial tree methods will be considered.

An interest-rate tree can either be built directly or be derived by discretizing a continuous-time short-rate model. In either case, the tree must at first fit to the prices of underlying securities before being applied for derivatives pricing. This process is called calibration. For interest-rate derivatives, the underlying securities may include a spectrum of zero-coupon bonds or swaps.[1] Note that even if a tree is obtained from discretizing an already fitted continuous-time short-rate model, calibration may still be needed because the discretizing errors may have spoiled the original fitting. With a calibrated tree, we can price interest-rate options by calculating the expected payoff values through a procedure of backward induction, where the interest rates are also used for discounting.

5.3.1 A Binomial Tree for the Ho–Lee Model

The Ho–Lee model was first presented with a binomial tree. For a Gaussian short-rate model with mean and variance of change over $(t, t + \Delta t)$ given by

$$
\begin{aligned}
E^{\mathbb{Q}}[\Delta r_t] &= \theta_t \Delta t, \\
\text{VaR}(\Delta r_t) &= \sigma^2 \Delta t,
\end{aligned}
\tag{5.67}
$$

we consider a rather natural binomial tree approximation as illustrated in Figure 5.1, where, without loss of generality, the branching probabilities are uniformly one half.

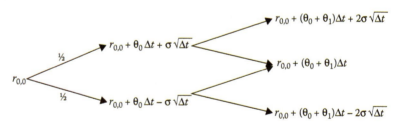

FIGURE 5.1: A binomial tree for the Ho–Lee model.

For notational efficiency, we let

$$
r_{i,n} = r_{0,0} + \Delta t \sum_{k=1}^{n-1} \theta_k + (2i - n)\,\sigma\sqrt{\Delta t}, \quad i = 0, 1, \dots, n.
\tag{5.68}
$$

Then we have a multi-period tree as shown in Figure 5.2.

---

[1] In either situation, we are fitting to either discount curve or the par-yield curve.

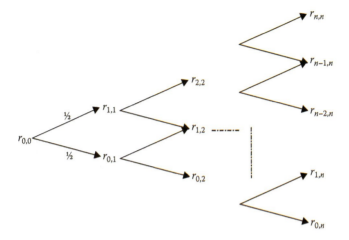

FIGURE 5.2: A general multi-period binomial interest-rate tree.

Before being applied to derivatives pricing, such a tree must first be calibrated to the current term structure of the interest rate. For the Ho–Lee model, we need to determine the drift, θ_t, by reproducing the prices of zero-coupon bonds of all maturities. This task can be efficiently achieved with the help of the so-called Arrow–Debreu prices.

5.3.2 Arrow–Debreu Prices

An Arrow–Debreu (1954) security is a canonical asset that has a cash flow of $1 if a particular state (of interest rate) is realized, or nothing otherwise. The pattern of payment is shown in Figure 5.3, where we let $Q_{i,j}$ denote the price of the security at time 0 that would pay $1 at time j if the state i is realized, or nothing if otherwise.

Note that a zero-coupon bond can be regarded as a portfolio of Arrow–Debreu securities. By linearity, the price of the zero-coupon bond maturing in time j is equal to

$$P(0, j) = \sum_{i=0}^{j} Q_{i,j}. \tag{5.69}$$

Given an interest-rate tree as in Figure 5.2, we can construct the Arrow–Debreu tree through a forward induction process. We begin with

$$Q_{0,0} = 1. \tag{5.70}$$

The calculations of $Q_{1,1}$ and $Q_{0,1}$ are done by "expectation pricing" using the trees in Figure 5.4, where $r_{0,0}$ is the discount rate at node $(0,0)$. Intuitively, the prices of the two Arrow–Debreu securities are given by

$$Q_{1,1} = Q_{0,1} = \frac{1}{2} e^{-r_{0,0} \Delta t}. \tag{5.71}$$

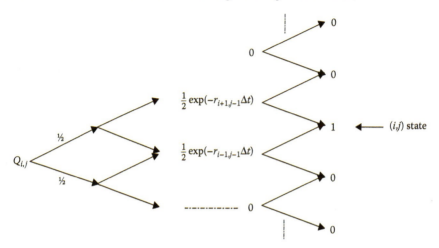

FIGURE 5.3: The payoff pattern and inductive tree for an Arrow–Debreu security.

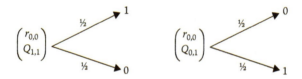

FIGURE 5.4: A one-period tree for Arrow–Debreu prices.

Suppose that we have already obtained

$$Q_{i,j}, \quad i = 0, 1, \ldots, j, \text{ and } j = 0, 1, \ldots, n - 1.$$

We then proceed to the calculation of $Q_{\cdot,n}$ using the formula

$$Q_{0,n} = \frac{e^{-r_{0,n-1}\Delta t}}{2} Q_{0,n-1},$$

$$Q_{i,n} = \frac{e^{-r_{i,n-1}\Delta t}}{2} Q_{i,n-1} + \frac{e^{-r_{i-1,n-1}\Delta t}}{2} Q_{i-1,n-1}, \quad i = 1 : (n-1),$$

$$Q_{n,n} = \frac{e^{-r_{n-1,n-1}\Delta t}}{2} Q_{n-1,n-1}. \tag{5.72}$$

The rationale of the above inductive scheme can be seen in Figure 5.5. After one step of backward induction from time n, there are only two non-zero nodal values, at nodes $(i - 1, n - 1)$ and $(i, n - 1)$, for which we have already determined the corresponding Arrow–Debreu prices. The Arrow–Debreu price for the cash flow at node (i, n), therefore, can be calculated by the second formula in Equation 5.72. The first and the third formula in Equation 5.72 are for calculating Arrow–Debreu prices corresponding to the first and the last node at time n.

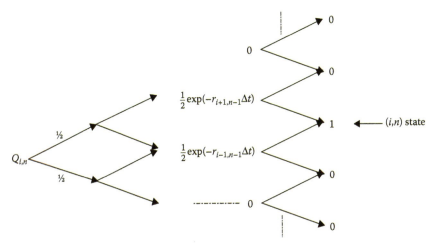

FIGURE 5.5: An inductive pricing tree for $Q_{i,n}$.

Eventually, the above process generates the Arrow–Debreu price tree shown in Figure 5.6.

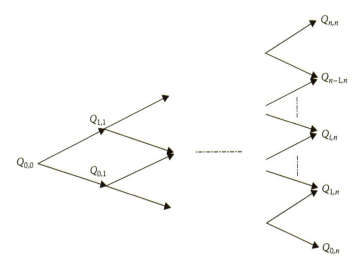

FIGURE 5.6: An Arrow–Debreu price tree.

5.3.3 A Calibrated Tree for the Ho–Lee Model

In reality, Arrow–Debreu prices must be determined as a part of the calibration procedure, when we must pin down θ_t using price information from discount bonds. The procedure is described below.

The first step is to determine θ_0 by using the price of $P(0, 2\Delta t)$, as

$$P(0, 2\Delta t) = Q_{2,2} + Q_{1,2} + Q_{0,2}$$
$$= e^{-\left(r_{0,0}+\theta_0\Delta t+\sigma\sqrt{\Delta t}\right)\Delta t}Q_{1,1} + e^{-\left(r_{0,0}+\theta_0\Delta t-\sigma\sqrt{\Delta t}\right)\Delta t}Q_{0,1} \quad (5.73)$$
$$= e^{-\theta_0(\Delta t)^2}\left(e^{-\left(r_{0,0}+\sigma\sqrt{\Delta t}\right)\Delta t}Q_{1,1} + e^{-\left(r_{0,0}-\sigma\sqrt{\Delta t}\right)\Delta t}Q_{0,1}\right),$$

where $Q_{\cdot,1}$ are independent of θ_0 and are already given in Equation 5.71. It follows that

$$\theta_0 = \frac{1}{(\Delta t)^2}\left[\ln\left(e^{-\left(r_{0,0}+\sigma\sqrt{\Delta t}\right)\Delta t}Q_{1,1} + e^{-\left(r_{0,0}-\sigma\sqrt{\Delta t}\right)\Delta t}Q_{0,1}\right)\right.$$
$$\left.- \ln P(0, 2\Delta t)\right]. \quad (5.74)$$

With the newly obtained θ_0, we can calculate the nodal values of the interest rates at time 1 by

$$r_{1,1} = r_{0,0} + \theta_0\Delta t + \sigma\sqrt{\Delta t},$$
$$r_{0,1} = r_{0,0} + \theta_0\Delta t - \sigma\sqrt{\Delta t}, \quad (5.75)$$

and, finally, calculate $Q_{i,2}$, $i = 0, 1, 2$ using Equation 5.72.

Assume now that we have found

$$\theta_j \quad \text{for } j = 0 : (n-2),$$
$$r_{i,j} \quad \text{for } i = 0 : j, \; j = 0 : (n-1),$$
$$Q_{i,j} \quad \text{for } i = 0 : j, \; j = 0 : n.$$

We can proceed to find θ_{n-1}, $r_{\cdot,n}$, and $Q_{\cdot,n+1}$ by matching to the price of $P(0, (n+1)\Delta t)$ as follows:

$$P\left(0, (n+1)\Delta t\right)$$
$$= \sum_{i=0}^{n-1} Q_{i,n}e^{-\left(r_{i,n-1}+\theta_{n-1}\Delta t-\sigma\sqrt{\Delta t}\right)\Delta t} + Q_{n,n}e^{-\left(r_{n-1,n-1}+\theta_{n-1}\Delta t+\sigma\sqrt{\Delta t}\right)\Delta t}$$
$$= e^{-\theta_{n-1}(\Delta t)^2}\left(\sum_{i=0}^{n-1} Q_{i,n}e^{-\left(r_{i,n-1}-\sigma\sqrt{\Delta t}\right)\Delta t} + Q_{n,n}e^{-\left(r_{n-1,n-1}+\sigma\sqrt{\Delta t}\right)\Delta t}\right).$$
$$(5.76)$$

We again have the explicit expression for θ_{n-1}:

$$\theta_{n-1} = \frac{1}{(\Delta t)^2}\ln\left(\frac{\sum_{i=0}^{n-1} Q_{i,n}e^{-\left(r_{i,n-1}-\sigma\sqrt{\Delta t}\right)\Delta t} + Q_{n,n}e^{-\left(r_{n-1,n-1}+\sigma\sqrt{\Delta t}\right)\Delta t}}{P\left(0, (n+1)\Delta t\right)}\right).$$
$$(5.77)$$

Next, we calculate $r_{.,n}$ according to

$$r_{i,n} = r_{i,n-1} + \theta_{n-1}\Delta t - \sigma\sqrt{\Delta t}, \quad i = 0 : (n-1)$$

$$r_{n,n} = r_{n-1,n-1} + \theta_{n-1}\Delta t + \sigma\sqrt{\Delta t}. \tag{5.78}$$

Finally, we calculate the Arrow–Debreu prices, $Q_{i,n+1}, i = 0, 1, \ldots, n+1$, using Equation 5.72.

TABLE 5.2: The Discount Curve on March 23, 2007

Year	Discount Factor	Year	Discount Factor	Year	Discount Factor
0.5	0.97584	10.5	0.61896	20.5	0.38537
1	0.95223	11	0.60497	21	0.37604
1.5	0.92914	11.5	0.59125	21.5	0.3669
2	0.90712	12	0.5778	22	0.35798
2.5	0.88629	12.5	0.5646	22.5	0.34925
3	0.86643	13	0.55165	23	0.34073
3.5	0.84724	13.5	0.53895	23.5	0.33241
4	0.82856	14	0.52649	24	0.32428
4.5	0.81032	14.5	0.51427	24.5	0.31635
5	0.7925	15	0.50229	25	0.30862
5.5	0.77506	15.5	0.49055	25.5	0.30107
6	0.75799	16	0.47903	26	0.29372
6.5	0.74127	16.5	0.46774	26.5	0.28655
7	0.72489	17	0.45668	27	0.27957
7.5	0.70884	17.5	0.44584	27.5	0.27277
8	0.69312	18	0.43523	28	0.26615
8.5	0.6777	18.5	0.42483	28.5	0.25971
9	0.66258	19	0.41464	29	0.25345
9.5	0.64776	19.5	0.40467	29.5	0.24736
10	0.63322	20	0.39492	30	0.24145

Note: Constructed using Treasury yield data from Reuters.

A succinct algorithm is described below.

1. Let $r_{0,0} = r_0$ and $Q_{0,0} = 1$.

2. For $j = 1 : n - 1$,

   (a) compute $Q_{i,j}$ using Equation 5.72 for $i = 0 : j$;

   (b) compute $r_{i,j} = r_{i,j-1} - \sigma\sqrt{\Delta t}, i = 0 : (j-1)$ and $r_{j,j} = r_{j-1,j-1} + \sigma\sqrt{\Delta t}$;

   (c) sum up $Q_{i,j}\exp(-r_{i,j}\Delta t)$ for $i = 0 : j$;

   (d) compute θ_{n-1} according to Equation 5.77; and

   (e) update $r_{i,j}$ according to $r_{i,j} = r_{i,j} + \theta_{j-1}\Delta t, i = 0 : j$.

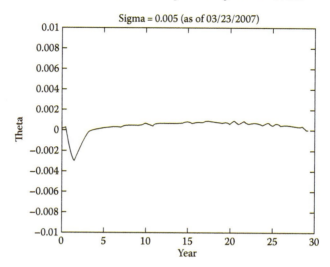

FIGURE 5.7: The Ho–Lee drift, $\theta(t)$, calibrated using the discount curve of March 23, 2007, and a short-rate volatility of $\sigma = 0.005$.

3. End

We now calibrate the Ho–Lee model to the term structure of March 23, 2007. The discount bond prices are first bootstrapped and listed in Table 5.2. We take the volatility of the short rate, the only exogenous parameter of the model, to be $\sigma = 0.005$, meaning a 50 basis points standard deviation of the short rate over one year. The time step is $\Delta t = 0.5$.

The θ_n calculated using the algorithm above is plotted in Figure 5.7. In this example, the drift term for the Ho–Lee model is rather small, and it can be negative. The surface plot for the Arrow–Debreu prices are shown in Figure 5.8 (note that the price only exists for the subdiagonal part of the rectangular area), and a plot for the nodal values of the interest-rate tree is given in Figure 5.9, where we can see that while the interest rates span from -15% to 30%, the Arrow–Debreu prices are very small for interest rates that are either too high or too low.

5.4   A General Tree-Building Procedure

The procedure for tree building presented in the last section can be generalized to other diffusion models for short rates. The major disadvantage of the resulting tree is that the interest-rate span can be too wide and fall too deep in the negative territory. In this section, we introduce a more sophisticated method of tree building for short-rate models with mean reversion. Although

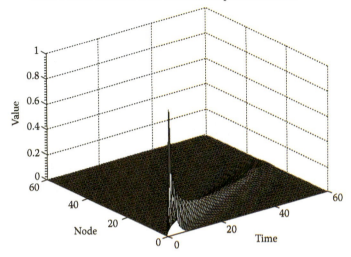

FIGURE 5.8: A surface plot of the Arrow–Debreu prices.

the resulting tree does not guarantee that negative interest rates are avoided, it does provide a tree with a much narrower rate span around the level of the mean interest rate.

5.4.1 A Truncated Tree for the Hull–White Model

Without loss of generality, we consider a trinomial tree approximation of a general short-rate model with a state-dependent drift term,

$$dr_t = \mu(r_t, t; \theta_t)\, dt + \sigma_t\, dW_t. \tag{5.79}$$

An immediate example of μ is the drift term for the Hull–White model, $\kappa(\theta_t - r_t)$, but the method can be applied to other models by replacing r_t with a function $f(r_t)$. Literally, the discrete version of Equation 5.79 is

$$\Delta r_t = \mu(r_t, t; \theta_t)\Delta t + \sigma_t \Delta W_t. \tag{5.80}$$

Its first two moments are

$$E^{\mathbb{Q}}\left[\Delta r_t \mid r_t\right] = \mu(r_t, t; \theta_t)\Delta t,$$
$$E^{\mathbb{Q}}\left[(\Delta r_t)^2 \mid r_t\right] = \sigma_t^2 \Delta t + \mu^2(r_t, t; \theta_t)\Delta t^2. \tag{5.81}$$

Consider an element with trinomial branching as depicted in Figure 5.10, where we choose the middle branch for the next time step, $r_{k,n+1}$, according to the principle that it is the one closest to the expected value of the short-rate conditional on $r_{j,n}$,

$$k = k(j) = \arg\min_i \left| E^{\mathbb{Q}}\left[r_{\cdot,n+1} \mid r_{j,n}\right] - (r_{0,n+1} + i\delta r)\right|, \tag{5.82}$$

$$r_{0,n+1} = r_{0,n} + \mu(r_{0,n}, t_n; \theta_n)\Delta t.$$

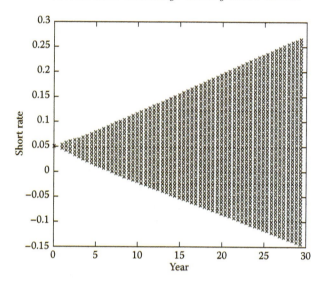

FIGURE 5.9: A surface plot of the interest-rate tree.

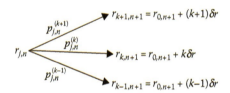

FIGURE 5.10: Trinomial tree branching of the short rate.

Through moment matching, we obtain the following governing equations for
the branching probabilities:

$$
\begin{aligned}
E_t^{\mathbb{Q}}\left[\Delta r_t - \xi_{j,n}\right] &= \mu(r_t, t; \theta_t)\Delta t - \xi_{j,n}, \\
E_t^{\mathbb{Q}}\left[(\Delta r_t - \xi_{j,n})^2\right] &= \sigma_t^2 \Delta t + (\mu(r_t, t; \theta_t)\Delta t - \xi_{j,n})^2,
\end{aligned}
\tag{5.83}
$$

where

$$
\xi_{j,n} = (k(j) - j)\delta r + \mu(r_{0,n}, t_n; \theta_n)\Delta t.
\tag{5.84}
$$

As a random variable, $\Delta r_t - \xi_{j,n}$ takes the following possible values with
corresponding probabilities,

$$
\Delta r_t - \xi_{j,n} =
\begin{cases}
\delta r, & p_{j,n}^{(k+1)}, \\
0, & p_{j,n}^{(k)}, \\
-\delta r, & p_{j,n}^{(k-1)}.
\end{cases}
\tag{5.85}
$$

Hence, Equation 5.83 becomes

$$p_{j,n}^{(k+1)} \delta r - p_{j,n}^{(k-1)} \delta r = \mu(r_{j,n}, t_n; \theta_n) \Delta t - \xi_{j,n},$$
$$p_{j,n}^{(k+1)} \delta r^2 + p_{j,n}^{(k-1)} \delta r^2 = \sigma_n^2 \Delta t + (\mu(r_{j,n}, t_n; \theta_n) \Delta t - \xi_{j,n})^2, \tag{5.86}$$

subject to the following natural conditions for probabilities,

$$p_{j,n}^{(k+1)} + p_{j,n}^{(k)} + p_{j,n}^{(k-1)} = 1,$$
$$p_{j,n}^{(k+1)}, \ p_{j,n}^{(k)}, \ p_{j,n}^{(k-1)} \geq 0. \tag{5.87}$$

Let

$$\eta_{j,n} = \mu(r_{j,n}, t_n; \theta_n) \Delta t - \xi_{j,n}. \tag{5.88}$$

Then the solutions to the probabilities can be expressed as

$$p_{j,n}^{(k+1)} = \frac{1}{2} \left(\frac{\sigma_n^2 \Delta t + \eta_{j,n}^2}{\delta r^2} + \frac{\eta_{j,n}}{\delta r} \right),$$

$$p_{j,n}^{(k-1)} = \frac{1}{2} \left(\frac{\sigma_n^2 \Delta t + \eta_{j,n}^2}{\delta r^2} - \frac{\eta_{j,n}}{\delta r} \right), \tag{5.89}$$

$$p_{j,n}^{(k)} = 1 - \frac{\sigma_n^2 \Delta t + \eta_{j,n}^2}{\delta r^2}.$$

A good choice for the ratio $\Delta t / \delta r^2$ is to set

$$\frac{\sigma_n^2 \Delta t}{\delta r^2} = \frac{1}{3}, \tag{5.90}$$

which will match the third moments of the continuous and discrete processes (and also implies a varying time step if σ_n is not a constant). When the short rate is mean-reverting, the above discretization method will automatically produce a truncated tree. To see that, we take as an example, $\mu(r, t) = \kappa(\theta_t - r)$. We can justify that the nodes that can be reached at time t_n are within the band

$$r_{0,n} - \left(\left\lfloor \frac{1}{2\kappa \Delta t} \right\rfloor + 1 \right) \delta r \leq r_{j,n} \leq r_{0,n} + \left(\left\lfloor \frac{1}{2\kappa \Delta t} \right\rfloor + 1 \right) \delta r, \tag{5.91}$$

where $\lfloor x \rfloor$ stands for the integer part of x. The maximum number of nodes is

$$n_{\max} = 2 \left\lfloor \frac{1}{2\kappa \Delta t} \right\rfloor + 3, \tag{5.92}$$

The resulting multi-period trinomial tree is shown in Figure 5.11.

Yet again, the trinomial tree [and $\theta(t_n)$ as well] must be calibrated to the current term structure of interest rates for the sake of arbitrage pricing. When

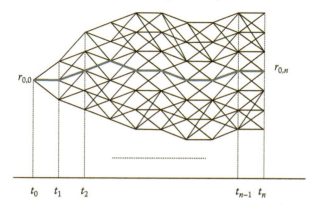

FIGURE 5.11: A truncated trinomial tree.

$\eta_{j,n}$ of Equation 5.88 is independent of θ_n, the calibration can be achieved very efficiently. This is the case for the Hull–White model, which has the drift term

$$\mu(r, t; \theta_t) = \kappa(\theta_t - r). \tag{5.93}$$

The corresponding $\eta_{j,n}$ is

$$\begin{aligned} \eta_{j,n} &= \mu(r_{j,n}, t_n; \theta_n)\Delta t - \xi_{j,n} \\ &= ((1 - \kappa\Delta t)\, j - k)\, \delta r, \end{aligned} \tag{5.94}$$

which contains no θ_n. Furthermore, from Equation 5.94, we can determine the range for the index from

$$|j| \leq \left[\frac{1}{2\kappa\Delta t}\right] + 1 \overset{\Delta}{=} J. \tag{5.95}$$

Note that $k = k(j)$ minimizes $|\eta_{j,n}|$ and $n_{\max} = 2J + 1$. In addition, the central node of branching from a node j is

$$k(j) = \begin{cases} -J + 1, & \text{if } j = -J, \\ j, & \text{if } |j| < J, \\ J - 1, & \text{if } j = J. \end{cases} \tag{5.96}$$

Starting from $Q_{0,0} = 1$, we determine $Q_{i,j}$ using the following procedure:

1. Taking $Q_{0,0} = 1$, $\theta_0 = r_{0,0} = r_0$.

2. For $j = 1 : n - 1$, repeat the following steps.

    (a) Compute $Q_{k,j}$ for according to

$$Q_{k,j} = \sum_{|i|\leq j-1} Q_{i,j-1} p^{(k)}_{i,j-1} \exp(-r_{i,j-1}\Delta t). \tag{5.97}$$

   Here $p^{(k)}_{i,j-1}$ are calculated using Equations 5.88 and 5.89.

(b) Define
$$r_{i,j} = (1 - \kappa\Delta t)\, r_{0,j-1} + i\delta r, \quad -j \le i \le j. \qquad (5.98)$$

(c) Calculate θ_{j-1} from the equation

$$P\left(0, (j+1)\Delta t\right) = e^{-\kappa\theta_{j-1}(\Delta t)^2} \sum_{i=-j}^{j} Q_{i,j} e^{-r_{i,j}\Delta t}. \qquad (5.99)$$

Then,

$$\theta_{j-1} = \frac{1}{\kappa(\Delta t)^2} \ln\left(\frac{\sum_{i=-j}^{j} Q_{i,j} e^{-r_{i,j}\Delta t}}{P(0, (j+1)\Delta t)}\right). \qquad (5.100)$$

(d) Update $r_{i,j}$ according to
$$r_{i,j} = r_{i,j} + \kappa\theta_{j-1}\Delta t, \quad i = -j : j. \qquad (5.101)$$

3. End of the loop.

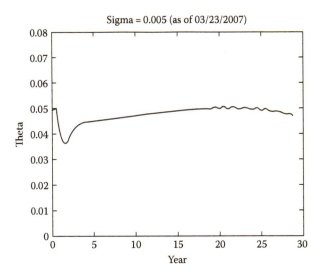

FIGURE 5.12: Theta for the Hull–White model.

We now take
$$\kappa = 0.25 \quad \text{and} \quad \sigma = 0.5\%, \qquad (5.102)$$

and calibrate the tree for the Hull–White model for the discount curve in Table 5.2. This set of parameters means that there is a half-life of mean reversion,

$$T = \frac{\ln 2}{\kappa} = 2.8 \text{ years,}$$

and a one-year standard deviation of 50 basis point for the short rate. The algorithm performs robustly, and the results are displayed in Figures 5.12 through 5.14. Figure 5.12 shows the θ_t obtained by calibration. Interestingly, the shape of the curve is very similar to that of θ_t for the Ho–Lee model, which is shown in Figure 5.7. As the level for mean reversion, θ_t appears in the very reasonable range of interest rates, from about 3.5 to 5%. The Arrow–Debreu prices also look very nice. What is perhaps most interesting is the interest-rate tree itself displayed in Figure 5.14, which has a small span, evolving around the mean level, represented by the "$*$" plots. Computationally, such a naturally truncated tree only renders more efficiency.

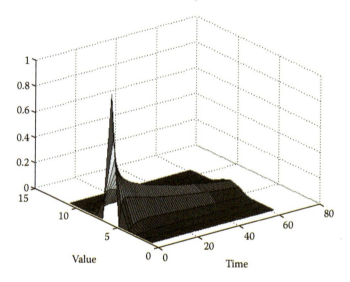

FIGURE 5.13: Arrow–Debreu prices for the Hull–White model.

5.4.2 Trinomial Trees with Adaptive Time Steps

In applications, we often need to value a portfolio of interest-rate derivatives. The cash-flow dates of these derivatives will not necessarily be Δt period apart, unless Δt is very small, which, however, would result in a dense tree and extensive calculations. The trick we introduce here is to use an adaptive tree whose node points are positioned to the cash-flow dates of the portfolio, whenever necessary. For the tree to be recombining, we need to fix δr, the difference between two adjacent nodes at the same time step. The size of time stepping becomes adaptive, yet it has to observe the following constraint,

$$\frac{\sigma_n^2 \Delta t}{\delta r^2} \leq 1, \tag{5.103}$$

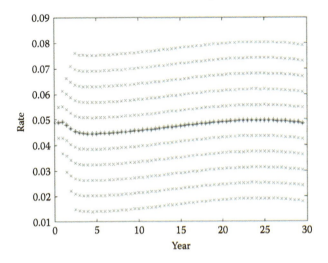

FIGURE 5.14: A short-rate tree for the Hull–White model.

to ensure that branching probabilities are positive. The changes to the algorithm above are very limited. Potentially, trees with adaptive time steps are very useful. In Figure 5.15, we draw one such tree with variable time steps.

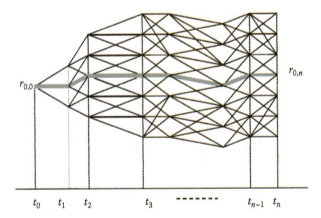

FIGURE 5.15: An interest-rate tree with adaptive time steps.

5.4.3 The Black–Karasinski Model

The discussions of tree methods for short-rate models have focused on the Ho–Lee and Hull–White models. In this section, we discuss a more general class of short-rate models and their lattice implementations. This class of models can be cast in the form

$$\mathrm{d}f(r_t) = \kappa(\theta_t - f(r_t))\,\mathrm{d}t + \sigma_t\,\mathrm{d}W_t, \tag{5.104}$$

for some monotonic function, $f(x)$, such that $f^{-1}(y)$ exists. For a given function of θ_t, we can apply the procedure from Equations 5.82 through 5.89 to build a tree for $R_t = f(r_t)$. The interest-rate tree for r_t then results from the one-to-one correspondence between R_t and r_t. For the calibration procedure, however, there is a small difference from the one for the Hull–White model: due to the non-linearity of $f(x)$, θ_n will be solved through a root-finding procedure, instead of being obtained explicitly. Specifically, with r_t taken over by R_t elsewhere in the algorithm, we need to replace Equations 5.99 and 5.100 with a single equation

$$P(0, (j+1)\Delta t) = \sum_{i=-j}^{j} Q_{i,j} e^{-f^{-1}(R_{i,j}+\kappa\theta_{j-1}\Delta t)\Delta t}, \tag{5.105}$$

and solve for θ_{j-1} iteratively.

As a major model in the class of Equation 5.104, we now introduce the Black and Karasinski (1991) model, which corresponds to $f(x) = \ln x$. Alternatively, we can write the short rate as an exponential function,

$$r_t = e^{X_t},$$

while X_t follows a Vasicek process,

$$\mathrm{d}X_t = \sigma_t\,\mathrm{d}W_t + \kappa_t(\theta_t - X_t)\,\mathrm{d}t.$$

Here σ_t and κ_t can be deterministic functions of time, and θ_t the subject to the term structure. The Black and Karasinski model is partly motivated by the objective of ensuring positive interest rates, and it also has the desired feature of mean reversion. It is not yet clear, however, how to cast the model as a special case under the HJM framework, although it has been a popular model for years.

The Black–Karasinski model has three parameters. In applications, users of the model have managed to gain extra freedom by turning these parameters into time-dependent functions. These time-dependent functions can be determined by fitting the model to the following three term structures:

- The zero-coupon yield curve.

- The volatility curve of the zero-coupon yields.

- The local volatility of $\ln r_t$, $\forall t$.

For the details of the algorithm, we refer to Black and Karasinski (1991). Note that the stylized algorithm introduced in the last section can also be modified for such a comprehensive calibration task.

We complete this section with a comment. Originally, the Black–Karasinski model did not start from a continuous time as in Equation 5.104. Instead, it started from simple random walks of the interest rate. Black, Derman, and Toy (1990) managed to fit a simple random walk to

- the zero-coupon yield curve; and

- the volatility curve of the zero-coupon yields.

It was realized later that under the limit $\Delta t \to 0$, the simple random walk converges to the continuous-time process,

$$\mathrm{d}X_t = \frac{\sigma_t'}{\sigma_t}\left(\theta_t + X_t\right)\mathrm{d}t + \sigma_t\,\mathrm{d}W_t, \tag{5.106}$$

which is called the Black–Derman–Toy model (1990). This model is not necessarily mean-reverting unless there is $\sigma_t < 0$, $\forall t$.

Exercises

1. Derive the drift term of a short-rate model whose volatility is proportional to the short rate. Take the approach of Section 5.2.1.

2. Let $\sigma(s,t)$ be a deterministic function. Given Equation 5.45, prove Equation 5.46.

3. Price a European option on a bond using a trinomial tree under the Hull–White model. The strike price and maturity of the option is $100 and two years, respectively. The underlying bond has a coupon rate of 5%. At the maturity of the option, the bond has a remaining life of two years. The parameters for the model are $\kappa = 0.1$, $\sigma = 0.08$, and the yield curve is flat at 5%. Take $\Delta t = 1/2$ year.

4. Price an American option on a bond using the trinomial tree under the Hull–White model, using the same input parameters as in Problem 3.

5. Redo the pricing of the European option in Problem 3 with the Black–Karasinski model. Take the same model input except $\sigma = 0.2$.

6. Use a binomial tree for the one-factor Ho–Lee model to price the following bond options with payoff at T_0 given by

$$\left(\sum_{i=1}^{N} \Delta T \cdot c \cdot P_{T_0}^{T_i} + P_{T_0}^{T_N} - 1\right)^+,$$

where $T_i = T_0 + i\Delta T$, $\Delta T = 0.25$ or 0.5, and c is the swap rate for the period $[T_0, T_N]$ seen at time $t = 0$, defined by

$$c = \frac{\left(P_0^{T_0} - P_0^{T_N}\right)}{\sum_{i=1}^{N} \Delta T P_0^{T_i}}.$$

The forward-rate volatility of the model takes the value $\sigma_0 = 0.008544$. You need to do the following:

(a) Take the discount factors given in Table 5.2.

(b) Fill in the column under Ho–Lee in Table 5.3, using both $\Delta t = 0.25$ and $\Delta t = 0.5$ in your calculations.

TABLE 5.3: Prices of Par-Bond Options in Basis Points

Maturity, T_0	Tenor, $T_n - T_0$	Ho–Lee
	Bond Options	
1	0.25	
2	0.25	
5	0.25	
1	5	
2	5	
5	5	

One basis point corresponds to one cent for the notional of $100.

Chapter 6

The LIBOR Market Model

LIBOR is an acronym for London Interbank Offer Rates. It is a set of reference interest rates at which banks lend unsecured loans to other banks in the London wholesale money market. The so-called LIBOR market is largely an over-the-counter (OTC) market for loans and interest-rate derivatives of various currencies based on LIBOR; it crosses the geographical boundaries of countries and is outside the jurisdiction of the banking authority of any single government. Because of that, the LIBOR market has enjoyed a high degree of flexibility in financial innovations and has evolved rapidly over the years. Today, for almost all major currencies, the turnover of interest-rate derivatives in the LIBOR market has well surpassed that of each domestic market. In this chapter, we first describe the major instruments of the LIBOR market, before introducing the LIBOR market model, namely, once the dominant model for LIBOR derivatives.

6.1 LIBOR Market Instruments

The LIBOR market began when USD were deposited into non-U.S. banks in Europe. After the Second World War, the amount of USD in Europe increased enormously, both as a result of trading with the United States and the Marshall Plan. During the Cold War period (1950–1989), the Eastern Bloc countries deposited most of their USD assets into British or other European banks, for fear of the possibility that the United States would freeze these assets if they were held in U.S. banks. The USD owned by non-U.S. entities and circulated in Europe formed the basis of the so-called Eurodollar market. Over the years, as a result of the United States's successive commercial deficits, dollar-denominated assets also came to be held in many countries around the world. Such a situation effectively turned the Eurodollar market into a global market.

Since the Eurodollar market does not fall under the banking regulations of the U.S. government, banks in the Eurodollar market can operate on narrower margins than can banks in the United States. Thus, the Eurodollar market has expanded largely as a means of avoiding the regulatory costs involved

TABLE 6.1: USD LIBOR Rates of November 27, 2018

Term	Rate (%)
1 month	2.35
3 months	2.71
6 months	2.88
12 months	3.13

Available at www.bankrate.com.

in dollar-denominated financial intermediation. Such financial intermediation has fostered various financial innovations, particularly derivatives. Many currently popular interest-rate derivative instruments originate in the Eurodollar market.

Gradually, derivatives on interest rates of other major currencies also became tradable in the LIBOR market, including those on the German Mark, the Japanese Yen, the Swiss Franc, and of course, the Pound Sterling. In 1999, the Euro became the official currency of the European Union. Nowadays, about 27 European countries have adopted the Euro, making it another major currency parallel to the USD. To some extent, the LIBOR market acted like an offshore market for these currencies. But today, thanks to globalization, the LIBOR market has crossed the geographical boundaries of most industrialized nations, and LIBOR instruments of various currencies can be traded in any major financial center in the world. In the next section, we introduce LIBOR instruments. We begin with standardized instruments and finish with exotic ones.

6.1.1 LIBOR Rates

LIBOR is a set of reference interest rates at which banks lend unsecured loans to other banks in the London wholesale money market. The LIBOR rates are benchmark rates for CDs. The terms of these CDs are typically 1 month, 3 months, 6 months, and 12 months (1 year). For example, the LIBOR rates of November 27, 2018, for USD are listed in Table 6.1.

The day-count convention for USD and Euro CDs is actual/360.

6.1.2 Forward-Rate Agreements

A forward-rate agreement (FRA) is a contract between two parties to exchange interest-rate payments, calculated using fixed and floating interest rates, respectively, over a certain period on a notional amount. The floating rate is usually a LIBOR rate. FRAs are OTC derivatives and are cash-settled, meaning that only the net payment is made, which, in particular, is made at the maturity of the FRA, not at the end of interest accrual period. The payer of the fixed interest rate is also known as the borrower or the buyer, while the receiver of the fixed interest rate is the lender or the seller. The features of a typical FRA are listed below.

1. Notional amount: 1 million dollars ($1 m)

2. Compounding convention: simple

3. Term of lending: one year from now for three months

4. Fixed rate: negotiated

5. Floating rate: three-month LIBOR

To express the net payment, we let the maturity of an FRA be T, the fixed rate be f, the interest accrual period be ΔT, and the floating rate for the term ΔT at the maturity T be f_T. Then the net payment to the seller at maturity T is

$$P(T, T + \Delta T) \times \text{Notional} \times \Delta T \times (f - f_T). \tag{6.1}$$

Here $P(T, T + \Delta T)$ is the discount factor from $T + \Delta T$ to T. Apparently, Equation 6.1 is equivalent to the payment of

$$\text{Notional} \times \Delta T \times (f - f_T)$$

at time $T + \Delta T$.

Conventionally, the value of an FRA at initiation is made zero, so that neither party needs to make an upfront payment. Like forward contracts on commodities, this is achieved by taking a special fixed rate (or strike rate). This special fixed rate can be determined through the following arbitrage arguments.[1] For notational simplicity, we let the notional amount be $1. Apparently, the value of the floating leg at time T is the notional amount, $1, while that value of the fixed leg at time $T + \Delta T$ is $(1 + \Delta T f)$ dollars. An FRA has no value if and only if the present values of the two legs offset each other. Hence, by equating the present values of the two legs, we have

$$P(0, T) = P(0, T + \Delta T)(1 + \Delta T f), \tag{6.2}$$

which yields the following unique fixed rate,

$$f = \frac{1}{\Delta T}\left(\frac{P(0, T)}{P(0, T + \Delta T)} - 1\right) \triangleq f_{\Delta T}(0, T), \tag{6.3}$$

This fixed rate nullifies the FRA. Based on the criterion of arbitrage, the fixed rate defined above is the only rate fair to both parties.

If the fixed rate is set differently, say, $f > f_{\Delta T}(0, T)$, then the seller can arbitrage with the following additional transactions:

---

[1]Negotiation occurs if there are counterparty risk or liquidity issues.

1. At time 0, short $P(0,T)/P(0,T+\Delta T)$ units of a $(T+\Delta T)$-maturity zero-coupon bond, and long one unit of a T-maturity zero-coupon bond.

2. At time T, cover the short position while the long position matures naturally.

The P&L of these transactions is

$$1 - P(T,T+\Delta T)\frac{P(0,T)}{P(0,T+\Delta T)}$$
$$= P(T,T+\Delta T) \times \Delta T \times (f_T - f_{\Delta T}(0,T)). \tag{6.4}$$

Adding Equation 6.4 to the net payment of the short position in the FRA, Equation 6.1, the seller secures a total profit of

$$P(T,T+\Delta T) \times \Delta T \times (f - f_{\Delta T}(0,T)) \tag{6.5}$$

with only zero-net transactions, which constitute an arbitrage.

In the marketplace, the typical maturities (T) for FRAs are 1 month, 3 months, 6 months, 9 months, and 12 months, and the typical terms (ΔT) are 1 month, 2 months, 3 months, and 6 months. According to the arguments above, a fair fixed rate for an FRA is also an arbitrage-free deposit rate for a CD with a corresponding term in the future. In contrast to the spot forward rates described in Section 6.1.1, the arbitrage-free deposit rate for a future CD defined by Equation 6.3 is called a forward LIBOR rate or simply a forward rate.

For later use, we make the remark that the forward rate can be treated as the value of the portfolio consisting of a long T-zero-coupon bond and a short $(T+\Delta T)$-zero-coupon bond, relative to the value of the latter.

6.1.3   Repurchasing Agreement

A repurchasing agreement (repo) is very similar to an FRA, except that it requires collateral for borrowing, which typically takes the form of U.S. Treasury bonds or municipal bonds. Owing to the use of collateral, a repo is considered free of credit risk, and thus the interest rate implied by the repo price is essentially a riskless interest rate. Note that repos were invented first in U.S. capital market as a means of short-term financing, and they are not necessarily considered to be LIBOR instruments. Playing a role similar to FRAs for short-term borrowing/lending, repos enjoy high liquidity and thus deserve mentioning here.

6.1.4   Eurodollar Futures

Eurodollar futures contracts are perhaps the most popular futures contracts in the global capital markets. These contracts are traded on the

International Monetary Market (IMM) of the Chicago Mercantile Exchange (CME) and the London International Financial Futures Exchanges (LIFFE). Underlying these futures contracts are the interest payments of three-month Eurodollar CDs of one million dollars notional. Let t be the current time, T be the maturity of a Eurodollar futures, and $\Delta T = 0.25$, then the simple interest for the CD (to be initiated at time T) will be

$$\$ \, 1{,}000{,}000 \times \Delta T \times f_{\Delta T}^{(\text{fut})}(t, T), \tag{6.6}$$

where $f_{\Delta T}^{(\text{fut})}(t, T)$ is the annualized interest rate for the CD seen at time t. At the maturity of the futures contract, $t = T$, the futures rate will settle into the three-month LIBOR rate: $f_{\Delta T}^{(\text{fut})}(T, T) = f_{\Delta T}(T, T)$.

Like most futures contracts, Eurodollar futures contracts are cash-settled and marked to market on a daily basis. There is no delivery of a cash instrument upon expiration because cash Eurodollar time deposits are not transferable. For an outstanding futures position, the marking-to-market value is

$$\$1{,}000{,}000 \times \Delta T \times \left(f_{\Delta T}^{(\text{fut})}(t + \Delta t, T) - f_{\Delta T}^{(\text{fut})}(t, T) \right), \tag{6.7}$$

where Δt represents one trading day. According to Equation 6.7, one basis point increment in the futures rate will generate a profit of \$25 to the holder, which is the usual price tick for the futures contract. Trading can also occur in the minimum ticks of 0.0025%, or 1/4 ticks, representing \$6.25 per contract and in 0.005%, or 1/2 ticks, representing \$12.50 per contract.

Let us consider an example of a Eurodollar futures investment for profit and loss. Assume that, at initiation in October 2018, the rate for six-month maturity Eurodollar futures (on three-month Eurodollar CDs) is 5.5%; and, at maturity in April 2019, the three-month LIBOR is set or fixed at 6%. Then the final P&L is

$$\$1{,}000{,}000 \times 0.25 \times (6\% - 5.5\%) = \$1250, \tag{6.8}$$

plus a small amount of accrued interest in the margin account due to marking to market. We emphasize here that, because of the daily marking to market, the implied futures rate is not equal to but larger than the corresponding LIBOR rate, as otherwise an arbitrage opportunity can occur. The issue regarding the difference of the two rates is addressed in Chapter 8 in depth.

Note that, in the marketplace, Eurodollar futures are not quoted by futures rates, but instead by the "index prices," similar to those for U.S. Treasury bill futures contracts. The index price, Z, corresponding to a futures rate is defined by

$$Z = 100 \left(1 - f_{\Delta T}^{(\text{fut})}(t, T) \right). \tag{6.9}$$

Hence, for the futures rates of 5.5% and 6%, the quote prices are 94.5 and 94, respectively.

In CME, the contracts have maturities in March, June, September, and December for up to 10 years into the future.

6.1.5 Floating-Rate Notes

A floating-rate note (FRN) is a coupon bond that pays floating-rate coupons indexed to some specific interest rate. The most basic FRN is indexed to LIBOR,

$$f_{\Delta T_j}(T_j, T_j) = \frac{1}{\Delta T_j}\left(\frac{P(T_j, T_j)}{P(T_j, T_{j+1})} - 1\right), \tag{6.10}$$

such that, at times T_{j+1}, $j = 0, 1, \ldots, n-1$, a bond holder receives a coupon payment in the dollar amount of

$$\text{Pr} \times \Delta T_j \times f_{\Delta T_j}(T_j, T_j). \tag{6.11}$$

Here Pr stands for the principal of the bond. At the maturity of the bond, T_n, the bondholder is also paid back the principal.

Let us consider the pricing of the FRN. As was remarked in the end of Section 6.1.2, the LIBOR in Equation 6.10 can be interpreted as the T_{j+1}-forward price of a portfolio consisting of a long T_j-maturity zero-coupon bond and a short T_{j+1}-maturity zero-coupon bond. According to arbitrage pricing theory (Harrison and Pliska, 1981), the T_{j+1}-forward price of the portfolios must be a martingale under the T_{j+1}-forward measure, $\mathbb{Q}_{T_{j+1}}$. Without loss of generality, we let the bond principal Pr $= \$1$. Assume that $T_k \le t \le T_{k+1}$. Then the next coupon is known for certain and its present value is

$$C_{k+1} = P(t, T_{k+1})\left(\frac{1}{P(T_k, T_{k+1})} - 1\right) = \frac{P(t, T_{k+1})}{P(T_k, T_{k+1})} - P(t, T_{k+1}). \tag{6.12}$$

The present value of the subsequent coupons, meanwhile, can be evaluated using cash flow measures, namely, the forward measures with delivery at respective cash flow dates:

$$\begin{aligned} C_{j+1} &= P(t, T_{j+1})\mathbf{E}^{\mathbb{Q}_{j+1}}\left[\left(\frac{P(T_j, T_j)}{P(T_j, T_{j+1})} - 1\right)\Big|\mathcal{F}_t\right] \\ &= P(t, T_{j+1})\left(\frac{P(t, T_j)}{P(t, T_{j+1})} - 1\right) = P(t, T_j) - P(t, T_{j+1}). \end{aligned} \tag{6.13}$$

Here $\mathbb{Q}_{j+1}$ stands for $\mathbb{Q}_{T_{j+1}}$ for notational simplicity. The value of the FRN is obtained by summing up the present value of the outstanding coupons and principal:

$$
\begin{aligned}
V_t &= \sum_{j=k}^{n-1} C_{j+1} + P(t, T_n) \\
&= \frac{P(t, T_{k+1})}{P(T_k, T_{k+1})} - P(t, T_{k+1}) \\
&\quad + \sum_{j=k+1}^{n-1} [P(t, T_j) - P(t, T_{j+1})] + P(t, T_n) \\
&= \frac{P(t, T_{k+1})}{P(T_k, T_{k+1})}.
\end{aligned}
\tag{6.14}
$$

The above formula for FRN is a rather clean one, and it states that the value of an FRN equals par at $t = T_k^+$, that is the moment immediately after a coupon payment. After that moment, the value will increase (as long as the short rate remains positive) and reach $1/P(T_k, T_{k+1})$ at $t = T_{k+1}^-$, immediately before the next coupon payment.

We finish this section with two remarks. First, Equation 6.13 implies an important property of LIBOR rates, which is stated as a lemma due to its importance in the pricing of various LIBOR derivatives.

Lemma 6.1.1. *The LIBOR rate, $f(t; T_j, \Delta T_j)$, is a martingale under the T_{j+1}-forward measure:*

$$
f_{\Delta T_j}(t, T_j) = E_t^{Q_{j+1}} \left[f_{\Delta T_j}(T_j, T_j) \right].
\tag{6.15}
$$

Second, an FRN can be replicated by rolling CDs. In fact, all we need to do is to invest in a CD with the term $\Delta T_j = T_{j+1} - T_j$ and roll it over until T_n.

6.1.6 Swaps

A swap is a contract to exchange interest payments out of a notional principal. Most interest-rate swap contracts exchange floating-rate payments for fixed-rate payments, and only the net payment is made. The party who swaps a fixed-rate payment for a floating-rate payment (to pay the fixed rate and to receive the floating rate) is said to hold a payer's swap, whereas the counter party is said to hold a receiver's swap. While an FRA or Eurodollar futures contract allows its holder to lock in a short rate in the future, a payer's swap allows its holder to lock in a long-term yield.

Let us take a look at the cash flow of a payer's swap as depicted in Figure 6.1. For generality, we let the period over which interest payments are to be exchanged be (T_m, T_n), which is called the term of the swap. The cash flows occur at the end of each accrual period. In Figure 6.1, the solid-line arrows stand for fixed cash flows, whereas the dotted-line arrows stand for floating and uncertain cash flows.

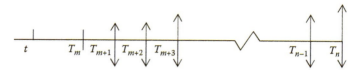

FIGURE 6.1: The cash flow pattern for a payer's swap.

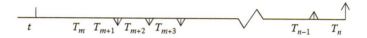

FIGURE 6.2: Net cash flows of a payer's swap.

As was indicated earlier, only the net payment is made. The net cash flow is depicted in Figure 6.2, which is uncertain at the initiation of the swap contract (except, when $m = 0$, for the first piece of the cash flow due at T_1).

Next, let us consider the pricing of the payer's swaps, which take LIBOR, given in Equation 6.10, as the reference rate for the floating leg. Let the fixed rate be K and the number of interest-rate payments be n. Then, according to Lemma 6.1, the value of the swap at $t \leq T_m$ can be derived as follows:

$$
\begin{aligned}
&\text{Swap}(t; T_m, T_n, K) \\
&= \sum_{j=m}^{n-1} \Delta T_j P(t, T_{j+1}) \, \mathbf{E}_t^{Q_{j+1}} \left[f_{\Delta T_j}(T_j, T_j) \right] - \sum_{j=0}^{n-1} \Delta T_j K P(t, T_{j+1}) \\
&= \sum_{j=m}^{n-1} \Delta T_j P(t, T_{j+1}) \, f_{\Delta T_j}(t, T_j) - \sum_{j=0}^{n-1} \Delta T_j K P(t, T_{j+1}) \\
&= \sum_{j=m}^{n-1} \left(P(t, T_j) - P(t, T_{j+1}) \right) + P(t, T_n) \\
&\quad - \left(\sum_{j=m}^{n-1} \Delta T_j K P(t, T_{j+1}) + P(t, T_n) \right) \\
&= P(t, T_m) - P(t, T_n) - \sum_{j=m}^{n-1} \Delta T_j K P(t, T_{j+1}).
\end{aligned}
\tag{6.16}
$$

As yet another market convention, when a swap is initiated, it takes zero value, meaning that there is no upfront payment by either party. The corresponding fixed rate of the contract is called the prevailing swap rate of the market. Denote the prevailing swap rate for the term (T_m, T_n) as $R_{m,n}(t)$. According to Equation 6.16, the prevailing swap rate is given by the formula

$$
R_{m,n}(t) = \frac{P(t, T_m) - P(t, T_n)}{\sum_{j=m}^{n-1} \Delta T_j P(t, T_{j+1})}.
\tag{6.17}
$$

When $t = T_m$, we say that the swap contract is "spot-starting." Otherwise, if $t < T_m$, we say that it is "forward-starting."

A swap rate, in fact, is a synonym of a par yield. In fact, according to the swap rate formula (Equation 6.17), there is

$$P(t, T_m) = \sum_{j=m}^{n-1} \Delta T_j R_{m,n}(t) P(t, T_{j+1}) + P(t, T_n), \qquad (6.18)$$

which implies that $R_{m,n}(t)$ is the coupon rate for a par bond (or a forward-starting par bond if $t < T_m$). Note that par bonds are also liquid LIBOR instruments.

For later use, we need to draw a direct connection between the prevailing swap rates and forward rates. By setting the second line of Equation 6.16 to zero, we can express the swap rate in terms of a set of forward rates:

$$R = \sum_{j=m}^{n-1} \alpha_j f_{\Delta T_j}(t, T_j), \qquad (6.19)$$

with

$$\alpha_j = \frac{\Delta T_j P(t, T_{j+1})}{\sum_{k=m}^{n-1} \Delta T_k P(t, T_{k+1})}. \qquad (6.20)$$

Equations 6.19 and 6.20 show that the swap rate is the weighted average of the forward rates over the cash flow period of the swap. We thus understand that, for modeling purposes, the swap-rate dynamics is determined by the forward-rate dynamics.

We finish this section with some remarks. The swap contract described in this section is called a vanilla swap. According to Equation 6.17, the prevailing swap rate can be treated as the value of the portfolio consisting of a long T_m-maturity zero-coupon bond and a short T_n-maturity zero-coupon bond, relative to the value of an annuity, $\{\Delta T_j P(t, T_{j+1})\}_{j=m}^{n-1}$.

6.1.7 Caps

A cap contract consists of a series of options on LIBOR, and, sequentially, these options matures at time T_{j+1}, with payoff

$$\Delta T_j \left(f_{\Delta T_j}(T_j, T_j) - K \right)^+, \quad j = 0, 1, \ldots, n-1. \qquad (6.21)$$

In the derivatives business, each option is called a caplet. When $t < T_0$, we call this cap forward-starting. Most caps traded in the market, however, are spot-starting with $t = T_0$. For a spot-starting cap, the payoff of the first caplet is known with certainty.

To price caplets, we can take zero-coupon bonds as state variables and treat forward rates as relative prices of bond portfolios. Using the forward measure, Q_{j+1}, we can price the jth caplet as follows:

$$C_{j+1} = \Delta T_j P(t, T_{j+1}) \mathbf{E}^{Q_{j+1}} \left[\left(f_{\Delta T_j}(T_j, T_j) - K \right)^+ \big| \mathcal{F}_t \right]$$

$$= P(t, T_{j+1}) \mathbf{E}^{Q_{j+1}} \left[\left(\frac{P(T_j, T_j)}{P(T_j, T_{j+1})} - (1 + \Delta T_j K) \right)^+ \big| \mathcal{F}_t \right]. \qquad (6.22)$$

By making use of the lognormal martingale property of the relative price under Q_{j+1}, we obtain

$$C_{j+1} = P(t, T_{j+1}) \left(\frac{P(t, T_j)}{P(t, T_{j+1})} \Phi(d_1^{(j)}) - (1 + \Delta T_j K) \Phi(d_2^{(j)}) \right)$$

$$= P(t, T_j) \Phi(d_1^{(j)}) - (1 + \Delta T_j K) P(t, T_{j+1}) \Phi(d_2^{(j)}), \qquad (6.23)$$

where

$$d_1^{(j)} = \frac{\ln \left(P(t, T_j) / (1 + \Delta T_j K) P(t, T_{j+1}) \right) + (1/2) \int_t^{T_j} \| \boldsymbol{\Sigma}(s, T_j)}{\sqrt{\int_t^{T_j} \| \boldsymbol{\Sigma}(s, T_j) - \boldsymbol{\Sigma}(s, T_{j+1}) \|^2 \, ds}},$$

$$d_2^{(j)} = d_1^{(j)} - \sqrt{\int_t^{T_j} \| \boldsymbol{\Sigma}(s, T_j) - \boldsymbol{\Sigma}(s, T_{j+1}) \|^2 \, ds}. \qquad (6.24)$$

Note that the range of integration for the variance of the relative price is (t, T_j) instead of (t, T_{j+1}), which reflects the fact that the forward rate, $f(t; T_j, \Delta T_j)$, is fixed (or "dead") at time T_j. Black's formula (Equation 6.23) is the price formula of caplets under the HJM model for interest rates. We point out here that this is merely a theoretical result and is not popular in practice.

6.1.8 Swaptions

A swaption entitles its holder to enter into a swap contract at a specific fixed rate in the future. Let the specific fixed rate be K and the value of the corresponding payer's swap with tenor (T_m, T_n) be $\text{swap}(t; T_m, T_n, K)$. Then the payoff of the swaption at its maturity, T, is

$$\text{swap}(T; T_m, T_n, K)^+ = \max(\text{swap}(T; T_m, T_n, K), 0). \qquad (6.25)$$

Swaptions allow investors to hedge interest-rate risk for longer terms. An investor who is concerned about higher interest rates over a period from T_m to T_n could opt to buy a swaption on the payer's swap with the prevailing swap rate for the period. If interest rates do go up, the holder will be compensated by the appreciation of the swaption.

The swaption payoff in Equation 6.25 can be expressed in terms of the prevailing swap rate at the maturity of the option. It is understood that the prevailing swap rate for the period (T_m, T_n) seen at time t is defined by

$$R_{m,n}(t) = \frac{P(t, T_m) - P(t, T_n)}{\sum_{j=m}^{n-1} \Delta T_j P(t, T_{j+1})}, \tag{6.26}$$

which implies that

$$P(T, T_m) = \sum_{j=m}^{n-1} \Delta T_j R_{m,n}(T) P(T, T_{j+1}) + P(T, T_n). \tag{6.27}$$

According to Equation 6.16, we have the following expression for the payoff of the swaption,

$$\text{swap}(T; T_m, T_n, K) = P(T, T_m) - P(T, T_n) - \sum_{j=m}^{n-1} \Delta T_j K P(T, T_{j+1})$$

$$= \left(\sum_{j=m}^{n-1} \Delta T_j P(T, T_{j+1}) \right) (R_{m,n}(T) - K). \tag{6.28}$$

The payoff for the swaption can also be expressed as

$$\text{swap}(T; T_m, T_n, K)^+ = \left(\sum_{j=m}^{n-1} \Delta T_j P(T, T_{j+1}) \right) (R_{m,n}(T) - K)^+. \tag{6.29}$$

This expression allows us to price the swaption as an option on the swap rate, using an appropriate pricing measure.

6.1.9 Bermudan Swaptions

A Bermudan swaption allows its holder to exercise the option at a set of pre-specified dates prior to the maturity of the contract. There are two possibilities for the term of the underlying swap, namely, variable term and fixed term. Note that the former is much more popular. To describe the term, we let T_i be the date of (early) exercise. The variable term corresponds to the swap term of $(T_{i \vee m}, T_n)$, whereas the fixed term corresponds to the swap term of (T_i, T_{i+n}). Here T_m is the earliest starting date of the swap. Let T_b and T_e be the first and the last dates for exercise, respectively, and τ be the stochastic time of optimal early exercise. Then the value of the variable-term Bermudan swaption can be expressed as

$$\sup_{T_b \leq \tau \leq T_e} \mathbf{E}^{\mathbb{Q}} \left[B_\tau^{-1} \left(\sum_{j=i \vee m}^{n-1} \Delta T P(\tau, T_{j+1}) \right) (R_{i \vee m, n}(\tau) - K)^+ \right]. \tag{6.30}$$

while the value of the fixed-term Bermudan swaption is

$$\sup_{T_b \leq \tau \leq T_e} \mathbf{E}^{\mathbb{Q}} \left[B_\tau^{-1} \left(\sum_{j=i}^{i+n-1} \Delta T P(\tau, T_{j+1}) \right) (R_{i,i+n}(\tau) - K)^+ \right], \qquad (6.31)$$

Here, $\mathbb{Q}$ stands for the risk-neutral measure.

The evaluation of Equations 6.30 and 6.31 involves the construction of an optimal strategy with early exercise, which is not trivial under a multi-factor model. Later in this chapter, we will introduce a Monte Carlo simulation method for Bermudan swaptions. Note that Bermudan swaptions are not considered to be vanilla derivative instruments like the ones we have described so far. But their popularity or liquidity warrants them a place in this list of standard LIBOR products.

6.1.10 LIBOR Exotics

The LIBOR market is known for its vivid financial innovations due to relatively few regulations and the high level of investor participation. Many interesting exotic derivative products were first invented in this market, and these products have gained reputations as "LIBOR exotics." In this section, we introduce some of them; our list is by no means exhaustive.

Flexible Cap

A flexible cap is an interest-rate cap with a predetermined number of possible exercises or uses. The cap is automatically used if the reference interest rate is above the strike level. If the reference interest rate fixes above the strike more times than the agreed number of exercises, then the flexible cap is terminated. If the predetermined number of possible exercises equals the total number of caplets, then the flexible cap reduces to a standard cap.

Chooser Cap

A chooser cap is similar to the flexible cap except that the exercises are chosen by the buyer, who will choose a strategy of the most profitable exercises.

Cancellable Swap

A cancellable swap is an interest-rate swap where one of the parties has the right to terminate the contract on a predetermined date at no cost. The right is equivalent to a swaption to enter a reverse swap. Owing to the swaption, the interest-rate payment of the counterparty will be reduced for compensation.

TABLE 6.2: Contractual Details of a Dollar/Euro Swap

	Dollar	Euro
Principal	1.25 m	1 m
Interest rate (%)	5	4.5

Resettable Swap

A resettable swap is a swap where the fixed rate paid depends on the market swap rate for the remaining maturity. The change in the swap rate is triggered by a barrier. With the right to reset, the payer can take advantage of a potential fall of the market swap rate.

Ratchet Swap

A ratchet swap is a swap where the floating leg is capped with a strike that is reset at the beginning of each interest-rate period. For example, a floating-rate payment at T_{j+1} can take the form

$$\min\big(f_{\Delta T_j}(T_j, T_j) + s_1, \ f_{\Delta T_{j-1}}(T_{j-1}, T_{j-1}) + s_2 \big),$$

where s_1 and s_1 are two pre-specified spreads.

Constant Maturity Swaps

A constant maturity swap (CMS) is a variant of the vanilla swap. One leg of the swap is either fixed or reset at LIBOR. The "constant maturity" leg is indexed to a long-term interest rate, typically a long-term swap rate. When the long-term rate is a yield of a government bond, for example, a 10-year Treasury rate, then the swap is called a constant maturity treasury (CMT) swap.

Cross-Currency Swaps

A cross-currency swap is a swap of cash flows between two currencies. For example, a currency swap between USD and Euros can be structured as follows: the term of the contract is five years, the frequency of interest-rate payment is two per year, and the principal and interest rate for each currency are listed in Table 6.2.

A major difference between a cross-currency swap and a usual vanilla swap is that, with the former, the principals are also exchanged at the maturity of the swap. Currency swaps are very sensitive to exchange-rate risk. They are perhaps the most popular OTC instruments for hedging against exchange-rate risk.

Dual Currency Basis Swap

A dual currency basis swap allows one party to switch from one currency to another on a specific date at a specified exchange rate. By giving such a right, the counterparty should be compensated with a lower interest rate.

Callable Range Accrual Swap

A callable range accrual swap is a swap where the payment on one of the legs depends on the daily fixing of an interest-rate index within a specific range. The interest payment on each payment date can be written as nK/N, where K is a fixed coupon rate, N the total number of days of the observation period, and n the number of days the index lies within the range.

Range Accruals Note

A range accrual note is an FRN with a cap and a floor on the floating rate.

Callable Range Accruals Note

A callable range accrual note is like a range accrual note, except that the issuer has the right to redeem the note after a lockout period. Owing to this call feature, the issuer needs to compensate the investor with higher interest rates.

Target Redemption Note

A target redemption note is an inverse floater structure that pays coupons equal to $\Delta T_j \max\big[K - \alpha f_{\Delta T_j}(T_j, T_j), K_{\text{floor}}\big]$, as long as the sum of all previous coupons is below a predetermined barrier. Here K_{floor} is the floor and α the gearing factor.

6.2   The LIBOR Market Model

The so-called LIBOR market model can be regarded as a model of practitioners. Before any rigorous justification was made, traders had already treated forward rates and swap rates as lognormally distributed variables with zero drift, and therefore had priced caplets, floorlets, and swaptions with Black's formula followed by discounting from settlement dates.

These market practices had been a serious concern to academicians. The pricing was not done under any term structure model, so it was not clear whether arbitrage opportunities could be generated. Moreover, it had been well-known that forward rates (as well as swap rates) are related to one another, and they could not all be lognormal under a single pricing measure. The popularity of this practice motivated a lot of studies aimed at reconciling the practice with an established arbitrage-free framework like Heath et al. (1992). A breakthrough came in 1997, when a rigorous foundation to the market's approach to pricing those LIBOR derivatives was established. The underlying term structure model is called the LIBOR market model (or the market model for short).

In this section, we present a systematic derivation of the LIBOR market model where forward rates are indeed treated as lognormal martingales under their respective forward measures. We show how this model is embedded into the general no-arbitrage framework of Heath et al. (1992), which implies that the model is arbitrage-free. Under the LIBOR model, caplets and floorlets are naturally priced with Black's formula. As for European swaptions, we will demonstrate that a swap rate is approximately a lognormal martingale under its own swap measure. This approximation paves the way for the application of Black's formula to swaptions, and thus also justifies the market practice of swaption pricing.

The market model caused a shift of paradigm: the state variables of the model are LIBOR instead of the instantaneous forward rates as in the HJM model. The derivation starts with the identification of martingale measures for forward term rates and swap rates. As shown in the last section, a LIBOR or a swap rate can be regarded as the value of a tradable portfolio (with a long and a short position in two zero-coupon bonds) relative to the numeraire asset of either a zero-coupon bond or an annuity, respectively. Hence, the LIBOR and the swap rates should be martingales under the martingale measures corresponding to those numeraire assets. In subsequent analysis, we make the above statement precise for forward rates in the next lemma.

Lemma 6.2.1. *The process of a forward term rate, $f_{\Delta T}(t, T)$, is a martingale under the forward measure $\mathbb{Q}_{T+\Delta T}$.*

Proof. The ΔT-term forward LIBOR is defined as

$$f_{\Delta T}(t, T) = \frac{1}{\Delta T} \left(\frac{P(t, T)}{P(t, T + \Delta T)} - 1 \right). \tag{6.32}$$

By Ito's lemma, there is

$$\begin{aligned} d\left(\frac{P(t, T)}{P(t, T + \Delta T)} \right) = \frac{P(t, T)}{P(t, T + \Delta T)} \left(\mathbf{\Sigma}(t, T) - \mathbf{\Sigma}(t, T + \Delta T) \right)^{\mathrm{T}} \\ \times \left(d\mathbf{W}_t - \mathbf{\Sigma}(t, T + \Delta T) dt \right), \end{aligned} \tag{6.33}$$

where $\mathbf{W}_t$ is a Brownian motion under the risk-neutral measure, $\mathbb{Q}$. It follows that

$$\begin{aligned} df_{\Delta T}(t, T) = \frac{1 + \Delta T f_{\Delta T}(t, T)}{\Delta T} \left(\mathbf{\Sigma}(t, T) - \mathbf{\Sigma}(t, T + \Delta T) \right)^{\mathrm{T}} \\ \times \left(d\mathbf{W}_t - \mathbf{\Sigma}(t, T + \Delta T) dt \right). \end{aligned} \tag{6.34}$$

As is already known,

$$\hat{\mathbf{W}}_t = \mathbf{W}_t - \int_0^t \mathbf{\Sigma}(s, T + \Delta T) \mathrm{d}s \tag{6.35}$$

is a $\mathbb{Q}_{T+\Delta T}$-Brownian motion, so $f_{\Delta T}(t, T)$ is a $\mathbb{Q}_{T+\Delta T}$-martingale. $\square$

Equation 6.34 suggests that the forward rate is a Gaussian $\mathbb{Q}_{T+\Delta T}$-martingale under conventional HJM specifications. This however does not stop us from turning the forward rate into a lognormal $\mathbb{Q}_{T+\Delta T}$-martingale. In formalism, we can change the process of the forward term rate into a lognormal one by casting Equation 6.34 into

$$\frac{\mathrm{d}f_{\Delta T}(t, T)}{f_{\Delta T}(t, T)} = \boldsymbol{\gamma}^{\mathrm{T}}(t, T) \left(\mathrm{d}\mathbf{W}_t - \mathbf{\Sigma}(t, T + \Delta T) \mathrm{d}t \right), \tag{6.36}$$

where $\boldsymbol{\gamma}(t, T)$, considered the "volatility" of the forward term rate, is given by

$$\boldsymbol{\gamma}(t, T) = \frac{1 + \Delta T f_{\Delta T}(t, T)}{\Delta T f_{\Delta T}(t, T)} \left(\mathbf{\Sigma}(t, T) - \mathbf{\Sigma}(t, T + \Delta T) \right). \tag{6.37}$$

As it appears, $\boldsymbol{\gamma}(t, T)$ is rather complex: it is a function of zero-coupon bond volatilities as well as the forward rate. Because Equations 6.36 and 6.37 are merely the consequences of the HJM model, they do not seem to offer any added value to derivatives pricing.

What follows is a change in the paradigm: we take the forward term rate as the state variable and prescribe its volatility directly. By doing so, we produce a lognormal model for forward rates under their corresponding forward measures. For the new model of the form (Equation 6.36) with a deterministic function, $\boldsymbol{\gamma}(t, T)$, Equation 6.37 in fact allows us to determine $\mathbf{\Sigma}(t, T + \Delta T)$ in terms of $\boldsymbol{\gamma}$ and the forward rates. If $\mathbf{\Sigma}(t, T)$ can be defined for any $T \geq t$ and can satisfy certain regularity conditions, we can define the volatility of the instantaneous forward rate as $\boldsymbol{\sigma}(t, T) = -\mathbf{\Sigma}_T(t, T)$, thus identifying the new model as a special case of the HJM model.

Next, we proceed to define $\mathbf{\Sigma}(t, T + \Delta T)$ and investigate its properties as a function. In view of Equation 6.37, we have the following expression of $\mathbf{\Sigma}(t, T + \Delta T)$ in terms of forward rates and their volatilities by a recursive procedure:

$$\boldsymbol{\Sigma}(t, T + \Delta T) = \boldsymbol{\Sigma}(t, T) - \frac{\Delta T f_{\Delta T}(t, T)}{1 + \Delta T f_{\Delta T}(t, T)} \boldsymbol{\gamma}(t, T)$$

$$= \boldsymbol{\Sigma}(t, T - \Delta T) - \frac{\Delta T f_{\Delta T}(t, T - \Delta T)}{1 + \Delta T f_{\Delta T}(t, T - \Delta T)}$$

$$\times \boldsymbol{\gamma}(t, T - \Delta T) - \frac{\Delta T f_{\Delta T}(t, T)}{1 + \Delta T f_{\Delta T}(t, T)} \boldsymbol{\gamma}(t, T)$$

$$= \cdots$$

$$= \boldsymbol{\Sigma}(t, t + \varepsilon) - \sum_{k=0}^{[(T-t)/\Delta T]} \frac{\Delta T f_{\Delta T}(t, T - k\Delta T)}{1 + \Delta T f_{\Delta T}(t, T - k\Delta T)}$$

$$\times \boldsymbol{\gamma}(t, T - k\Delta T), \tag{6.38}$$

where $[x]$ means the integer part of a positive number x, and

$$0 \le \varepsilon = (T - t) - \left[\frac{(T - t)}{\Delta T} \right] \Delta T < \Delta T.$$

Apparently, $\boldsymbol{\Sigma}(t, t + \varepsilon)$ is the volatility of a zero-coupon bond with a very short maturity, and thus should be close to zero. According to Musiela and Rutkowski (1995), we can set $\boldsymbol{\Sigma}(t, t + \varepsilon) = 0$ for any $\varepsilon \in [0, \Delta T)$, without causing any problem. Substituting Equation 6.38 back to Equation 6.36, we obtain a concise expression for the dynamics of the T-maturity forward rate:

$$\frac{df_{\Delta T}(t, T)}{f_{\Delta T}(t, T)} = \boldsymbol{\gamma}^{\mathrm{T}}(t, T) \cdot \left[d\mathbf{W}_t + \left(\sum_{k=0}^{[(T-t)/\Delta T]} \frac{\Delta T f_{\Delta T}(t, T - k\Delta T)}{1 + \Delta T f_{\Delta T}(t, T - k\Delta T)} \right. \right.$$

$$\left. \left. \times \boldsymbol{\gamma}(t, T - k\Delta T) \right) dt \right]. \tag{6.39}$$

Note that, in the LIBOR market, the forward rate, $f_{\Delta T}(t, T)$, is fixed at time $t = T$. Beyond that, the forward rate is fixed or "dead" and, accordingly, there is $\boldsymbol{\gamma}(t, T) = 0$ for $t \ge T$. With Equation 6.39, we have created the formalism of the new model. But the model will not be completely established until we have proved that the SDE (Equation 6.39) admits a meaningful solution.

The existence of a global solution to Equation 6.39 was established by Brace, Gatarek, and Musiela (1997). Under very general conditions, they proved that a unique and strictly positive solution exists for all t. In addition, the solution is found to have a number of interesting properties. First, it is mean-reverting: the forward rates tend to drop when they are too high and they tend to rise when they are too low. Second, the solution is bounded from above and below by two lognormal processes. Third, if both $\boldsymbol{\gamma}(t, T)$ and $f_{\Delta T}(0, T)$ are smooth in T, then the solution $f_{\Delta T}(t, T)$ has the same degree of smoothness. For details of the results, we refer readers to Brace et al.'s (1997) paper. Other papers that make important contributions to an understanding of the LIBOR model include Jamshidian (1997) and Milternsen, Sandmann, and Sonderman (1997).

Because of the positiveness of $f_{\Delta T}(t, T)$, $\boldsymbol{\Sigma}(t, T)$ is well defined and bounded. In addition, if both $\boldsymbol{\gamma}(t, T)$ and $f_{\Delta T}(0, T)$ are differentiable in T, $\boldsymbol{\Sigma}(t, T)$ is also differentiable in T. Therefore, we can identify the market model as a special case of HJM's no-arbitrage framework with, especially, instantaneous forward-rate volatility given by $\boldsymbol{\sigma}(t, T) = -\boldsymbol{\Sigma}_T(t, T)$.

In applications, the market model takes a stream of "spanning" forward rates as state variables:

$$f_j(t) \triangleq f_{\Delta T_j}(t, T_j), \quad j = 1, 2, \ldots, N, \tag{6.40}$$

and the forward rates evolve according to

$$\frac{df_j(t)}{f_j(t)} = \boldsymbol{\gamma}_j^{\mathrm{T}}(t) \left[d\mathbf{W}_t + \left(\sum_{k=\eta(t)}^{j} \frac{\Delta T f_k(t)}{1 + \Delta T f_k(t)} \boldsymbol{\gamma}_k(t) \right) dt \right], \quad j = 1, 2, \ldots, N, \tag{6.41}$$

with $\eta(t) = \min\{j \mid T_j > t\}$, is the smallest index of the first forward rates that are alive. The number of forward rates depends on the horizon of time over which we want to model the interest-rate dynamics. Clearly, Equation 6.41 can be a term structure model for the entire forward-rate curve. Typically, $f_j(t)$ represents three- or six-month LIBORs. As a result, Equation 6.41 is often called the LIBOR market model.

We remark here that under this market model, money market accounts accrue by simple compounding with LIBOR:

$$B(t) = \left(\prod_{j=0}^{\eta(t)-2} (1 + f_j(T_j)\Delta T_j) \right) (1 + f_{\eta(t)-1}(T_{\eta(t)-1})(t - T_{\eta(t)-1})), \tag{6.42}$$

which depends only on forward rates that are "dead" by time t.

We want to highlight the martingale property of the forward rates. Recall that

$$\boldsymbol{\Sigma}(t, T_{j+1}) = -\sum_{k=\eta(t)}^{j} \frac{\Delta T f_k(t)}{1 + \Delta T f_k(t)} \boldsymbol{\gamma}_k(t) \tag{6.43}$$

is the volatility of the T_j-maturity zero-coupon bond. Under the T_{j+1}-forward measure, $f_j(t)$ is indeed a lognormal martingale:

$$\frac{df_j(t)}{f_j(t)} = \boldsymbol{\gamma}_j^{\mathrm{T}}(t) d\mathbf{W}_t^{j+1}, \tag{6.44}$$

where $\mathbf{W}_t^{j+1}$ is a $\mathbb{Q}_{j+1}$-Brownian motion. It is this martingale property that readily justifies Black's formula for caplets and floorlets.

The LIBOR model (Equation 6.41) takes the initial term structure, $\{f_j(0)\}_{j=1}^N$, as its initial conditions. Hence, the construction of the initial term structure is part of the preparatory work for the model. In the marketplace, forward rates can be directly observed only for short maturities, up to a year, in CDs, FRAs, or repos. Forward rates of maturities beyond a year are implied by Eurodollar futures rates and, particularly, swap rates, and they need to be extracted through bootstrapping and/or other techniques, including convexity adjustment, which is discussed in Chapter 8.

6.3 Pricing of Caps and Floors

Black's formula has been a market standard for caplet pricing since the early 1990s. Yet, a rigorous mathematical foundation did not come to exist until 1997, when the market model was established. Under such an arbitrage-free term structure model, Black's formula is readily justified.

Cap pricing is a vivid example of the appropriate use of measures in derivatives pricing. Having noticed that $f_j(t)$ is a lognormal martingale under its own forward measure, $\mathbb{Q}_{j+1}$, we naturally adopt this measure to price caplets on $f_j(t)$. Hence, caplets of different maturities are priced using different forward measures. A cap's price is obtained as the sum of those of caplets:

$$\text{Cap}(n) = \sum_{j=0}^{n-1} P_0^{T_{j+1}} \Delta T_j E^{\mathbb{Q}_{j+1}} \left[(f_j(T_j) - K)^+ \right]$$

$$= \sum_{j=0}^{n-1} P_0^{T_{j+1}} \Delta T_j \left[f_j(0) N \left(d_1^{(j)} \right) - K N \left(d_2^{(j)} \right) \right], \qquad (6.45)$$

where

$$d_1^{(j)} = \frac{\ln\left(f_j(0)/K\right) + (1/2)\sigma_j^2 T_j}{\sigma_j \sqrt{T_j}}, \quad d_2^{(j)} = d_1^{(j)} - \sigma_j \sqrt{T_j},$$

$$\sigma_j^2 = \frac{1}{T_j} \int_0^{T_j} \|\gamma_j(t)\|^2 \, dt, \qquad (6.46)$$

for $j = 0, \ldots, n-1$. Note that when $T_0 = 0$, the cash flow of the first caplet is known for certain and is equal to $(f_0(T_0) - K)^+$, which is also the result of Black's formula for $\sigma_0 = 0$.

The price formula for floors can be derived by using the call–put parity: for the same strike, a long cap and a short floor are equivalent to a swap, meaning

$$\text{Cap}(n) - \text{Floor}(n) = \text{Swap}(n). \qquad (6.47)$$

The price formula for a floor is thus

$$\text{Floor}(n) = \text{Cap}(n) - \text{Swap}(n)$$

$$= \sum_{j=0}^{n-1} P_0^{T_{j+1}} \Delta T_j \left[K\Phi(-d_2^{(j)}) - f_j(0)\Phi(-d_1^{(j)}) \right]. \qquad (6.48)$$

Note that, when the strike rate equals the equilibrium swap rate, Swap$(n) = 0$, yielding

$$\text{Floor}(n) = \text{Cap}(n).$$

6.4 Pricing of Swaptions

Swaption pricing could be made as convenient as caplet pricing by assuming the lognormal dynamics for a swap rate under its corresponding martingale measure. The problem in so doing, however, is the apparent fact that a swap rate and its related forward rates cannot simultaneously be lognormal, even under their own martingale measures. Nonetheless, we will show that, under the LIBOR market model, a swap-rate process can be so accurately approximated by a lognormal process that pricing errors for swaptions due to the approximation are generally small and negligible. Specifically, for modest swaption maturities, the errors can be smaller than the typical bid-ask spreads. In this regard, this section serves as a justification for the simultaneous use of lognormal dynamics for forward rates and swap rates.

A crucial treatment in swaption pricing is to convert an option on a swap into an option on a prevailing swap rate. Recall that a swaption is an option on a swap with payoff given by Equation 6.25. Typically, the maturity of the option is T_m, the fixing date of the prevailing swap rate. According to Equation 6.29, the payoff of the swaption on the payer's swap is

$$V_{T_m} = \left(\sum_{j=m}^{n-1} \Delta T_j P(T_m, T_{j+1}) \right) (R_{m,n}(T_m) - K)^+, \qquad (6.49)$$

where $R_{m,n}(t)$ is the prevailing swap rate at time t, given by

$$R_{m,n}(t) = \frac{P(t, T_m) - P(t, T_n)}{\sum_{j=m}^{n-1} \Delta T_j P(t, T_{j+1})}. \qquad (6.50)$$

Thus, the payoff is an option on the prevailing swap rate multiplied by the *annuity*:

$$A_{m,n}(t) \triangleq \sum_{j=m}^{n-1} \Delta T_j P(t, T_{j+1}), \quad t \leq T_m. \qquad (6.51)$$

In view of Equation 6.50, we can treat the swap rate as the price of a portfolio of zero-coupon bonds (a long T_m-zero-coupon bond and a short T_n-zero-coupon bond) relative to the annuity, $A_{m,n}(t)$, yet another tradable asset. Therefore, if we choose $A_{m,n}(t)$ as a new numeraire and let $\mathbb{Q}_S$ be the corresponding martingale measure, then, according to arbitrage pricing theory (Harrison and Pliska, 1981), the price of any tradable asset relative to the annuity is a $\mathbb{Q}_S$-martingale. When the asset is the swaption, its value at time t satisfies

$$\frac{V_t}{A_{m,n}(t)} = E^{\mathbb{Q}_S}\left[\frac{V_{T_m}}{A_{m,n}(T_m)}\ \middle|\ \mathcal{F}_t\right]$$

$$= E^{\mathbb{Q}_S}\left[(R_{m,n}(T_m) - K)^+\ \middle|\ \mathcal{F}_t\right]. \tag{6.52}$$

By the same token, we understand that $R_{m,n}(t)$ is also a $\mathbb{Q}_S$-martingale. The focus now shifts to evaluating the expectation in Equation 6.52. Note that, in the literature, $\mathbb{Q}_S$ is called a *forward swap measure* or simply a *swap measure*.

To evaluate the expectation in Equation 6.52, we must (1) characterize the forward swap measure, $\mathbb{Q}_S$; and (2) specify the $\mathbb{Q}_S$-dynamics of the swap rate. According to the general theory of change of measures (in Chapter 4), the Radon–Nikodym derivative of $\mathbb{Q}_S$ with respect to the risk-neutral measure, $\mathbb{Q}$, is given by

$$\left.\frac{d\mathbb{Q}_S}{d\mathbb{Q}}\right|_{\mathcal{F}_t} = \frac{A_{m,n}(t)/A_{m,n}(0)}{B(t)/B(0)}$$

$$= \frac{1}{A_{m,n}(0)} \sum_{j=m}^{n-1} \Delta T_j \frac{P(t, T_{j+1})}{B(t)} \triangleq \zeta(t), \quad t \le T_m. \tag{6.53}$$

Using Equation 4.84, the price process of zero-coupon bonds discounted by the money market account, we obtain the following equation for the Radon–Nikodym process, $\zeta(t)$:

$$d\zeta(t) = \frac{1}{A_{m,n}(0)} \sum_{j=m}^{n-1} \Delta T_j\, d\left(\frac{P(t, T_{j+1})}{B(t)}\right)$$

$$= \frac{1}{A_{m,n}(0)} \sum_{j=m}^{n-1} \Delta T_j \left(\frac{P(t, T_{j+1})}{B(t)}\right) \Sigma^{\mathrm{T}}(t, T_{j+1})\, d\mathbf{W}_t$$

$$= \frac{A_{m,n}(t)}{A_{m,n}(0)B(t)} \sum_{j=m}^{n-1} \frac{\Delta T_j P(t, T_{j+1})}{A_{m,n}(t)} \Sigma^{\mathrm{T}}(t, T_{j+1})\, d\mathbf{W}_t$$

$$= \zeta(t) \sum_{j=m}^{n-1} \alpha_j(t) \Sigma^{\mathrm{T}}(t, T_{j+1})\, d\mathbf{W}_t, \tag{6.54}$$

where

$$\alpha_j(t) = \frac{\Delta T_j P(t, T_{j+1})}{A_{m,n}(t)}. \tag{6.55}$$

According to the CMG theorem, the process $\mathbf{W}_t^S$, defined as

$$
\begin{aligned}
d\mathbf{W}_t^S &= d\mathbf{W}_t - \left\langle d\mathbf{W}_t, \frac{d\zeta(t)}{\zeta(t)} \right\rangle \\
&= d\mathbf{W}_t - \sum_{j=m}^{n-1} \alpha_j(t) \mathbf{\Sigma}(t, T_{j+1}) \, dt \\
&= d\mathbf{W}_t - \mathbf{\Sigma}_A(t) \, dt,
\end{aligned}
\tag{6.56}
$$

is a $\mathbb{Q}_S$-Brownian motion. Here, as a notation,

$$
\mathbf{\Sigma}_A(t) = \sum_{j=m}^{n-1} \alpha_j(t) \mathbf{\Sigma}(t, T_{j+1}),
\tag{6.57}
$$

which is the "weighted average" of forward-rate volatilities, since

$$
0 < \alpha_j(t) \leq 1 \quad \text{and} \quad \sum_{j=m}^{n-1} \alpha_j(t) = 1.
$$

The dynamics of the swap rate under the forward swap measure can be derived by brute force. We have already learned that $R_{m,n}(t)$ is the weighted average of forward rates,

$$
R_{m,n}(t) = \sum_{j=m}^{n-1} \alpha_j f_j(t),
\tag{6.58}
$$

where the α_js, defined in Equation 6.55, depend on $f_j(t)$, $j = m, \ldots, n-1$ only. To see that, we plug

$$
P(t, T_{j+1}) = \frac{P(t, T_m)}{\prod_{k=m}^{j} (1 + \Delta T f_k)}
\tag{6.59}
$$

into Equation 6.55, and then cancel $P(t, T_m)$, thus rendering $R_{m,n}(t)$ an explicit function of $f_j(t)$, $j = m, \ldots, n-1$ only. Using Ito's lemma and by brute force, we can derive the following process of $R_{m,n}(t)$ under the risk-neutral measure, $\mathbb{Q}$:

$$
dR_{m,n}(t) = \sum_{j=m}^{n-1} \frac{\partial R_{m,n}(t)}{\partial f_j(t)} f_j(t) \boldsymbol{\gamma}_j^{\mathrm{T}}(t) \left(d\mathbf{W}_t - \mathbf{\Sigma}_A(t) dt \right).
\tag{6.60}
$$

In terms of the $\mathbb{Q}_S$-Brownian motion, $\mathbf{W}_t^S$, defined in Equation 6.56, we obtain

$$
dR_{m,n}(t) = \sum_{j=m}^{n-1} \frac{\partial R_{m,n}(t)}{\partial f_j(t)} f_j(t) \boldsymbol{\gamma}_j^{\mathrm{T}}(t) \, d\mathbf{W}_t^S,
\tag{6.61}
$$

which reconfirms that $R_{m,n}(t)$ is a $\mathbb{Q}_S$-martingale. The explicit expression of the partial derivative in Equation 6.60 or 6.61 is detailed in the next lemma, which is useful in the derivation of Equation 6.60.

Lemma 6.4.1. *The partial derivative of $R_{m,n}(t)$ with respect to $f_j(t)$ is given by*

$$\frac{\partial R_{m,n}(t)}{\partial f_j(t)} = \alpha_j + \frac{\Delta T_j}{1 + \Delta T_j f_j(t)} \sum_{k=m}^{j-1} \alpha_k \left(f_k(t) - R_{m,n}(t) \right). \qquad (6.62)$$

Proof. By differentiating Equation 6.58 with respect to $f_j(t)$, we have

$$\frac{\partial R_{m,n}(t)}{\partial f_j(t)} = \alpha_j + \sum_{k=m}^{n-1} \frac{\partial \alpha_k}{\partial f_j} f_k. \qquad (6.63)$$

According to Equation 6.55, there is

$$\frac{\partial \alpha_k}{\partial f_j} = \Delta T_k \frac{\left[(\partial P(t, T_{k+1})/\partial f_j) A_{m,n}(t) - P(t, T_{k+1}) (\partial A_{m,n}(t)/\partial f_j) \right]}{(A_{m,n}(t))^2}. \qquad (6.64)$$

The key to the proof is to calculate the partial derivative of the zero-coupon bonds with respect to $f_j(t)$. Based on Equation 6.59, we have

$$\frac{\partial P(t, T_{k+1})}{\partial f_j} = \begin{cases} 0, & j > k, \\ \dfrac{-\Delta T_j}{1 + \Delta T_j f_j} \dfrac{P(t, T_m)}{\prod_{l=m}^{k} (1 + \Delta T_l f_l)}, & j \leq k, \end{cases}$$

$$= \frac{-\Delta T_j}{1 + \Delta T_j f_j} P(t, T_{k+1}) \mathbf{1}_{k \geq j}. \qquad (6.65)$$

It follows that

$$\frac{\partial \alpha_k}{\partial f_j} = \Delta T_k \frac{[(-\Delta T_j/1 + \Delta T_j f_j) P(t, T_{k+1}) \mathbf{1}_{k \geq j} A_{m,n}(t) - P(t, T_{k+1}) \sum_{l=m}^{n-1} \Delta T_l (-\Delta T_j/1 + \Delta T_j f_j) P(t, T_{l+1}) \mathbf{1}_{l \geq j}]}{(A_{m,n}(t))^2}$$

$$= \frac{-\Delta T_j}{1 + \Delta T_j f_j} \alpha_k \left\{ \mathbf{1}_{k \geq j} - \sum_{l=j}^{n-1} \alpha_l \right\}. \qquad (6.66)$$

Substituting the above expression back into Equation 6.63, we finally arrive at

$$\frac{\partial R_{m,n}(t)}{\partial f_j} = \alpha_j - \frac{\Delta T_j}{1 + \Delta T_j f_j} \sum_{k=m}^{n-1} \alpha_k \left\{ \mathbf{1}_{k \geq j} - \sum_{l=j}^{n-1} \alpha_l \right\} f_k$$

$$= \alpha_j - \frac{\Delta T_j}{1 + \Delta T_j f_j} \left(\sum_{k=j}^{n-1} \alpha_k f_k - \left(\sum_{l=j}^{n-1} \alpha_l \right) \left(\sum_{k=m}^{n-1} \alpha_k f_k \right) \right)$$

$$= \alpha_j - \frac{\Delta T_j}{1 + \Delta T_j f_j} \sum_{k=j}^{n-1} \alpha_k \left(f_k(t) - R_{m,n}(t) \right)$$

$$= \alpha_j + \frac{\Delta T_j}{1 + \Delta T_j f_j} \sum_{k=m}^{j-1} \alpha_k \left(f_k(t) - R_{m,n}(t) \right). \qquad (6.67)$$

This completes the proof. □

As it appears, the swap-rate process, Equation 6.61, has no analytical tractability and thus has little value for swaption pricing. To reconcile the market's practice of swaption pricing, we consider the following lognormal approximation to Equation 6.61: for $u \geq t$ and conditional on $\mathcal{F}_t$,

$$dR_{m,n}(u) = \sum_{j=m}^{n-1} \frac{\partial R_{m,n}(u)}{\partial f_j} f_j(u) \gamma_j^{\mathrm{T}}(u) d\mathbf{W}_u^S$$

$$= R_{m,n}(u) \sum_{j=m}^{n-1} \frac{\partial R_{m,n}(u)}{\partial f_j} \frac{f_j(u)}{R_{m,n}(u)} \gamma_j^{\mathrm{T}}(u) d\mathbf{W}_u^S$$

$$\approx R_{m,n}(u) \sum_{j=m}^{n-1} \frac{\partial R_{m,n}(t)}{\partial f_j} \frac{f_j(t)}{R_{m,n}(t)} \gamma_j^{\mathrm{T}}(u) d\mathbf{W}_u^S$$

$$= R_{m,n}(u) \left(\sum_{j=m}^{n-1} \omega_j \gamma_j(u) \right)^{\mathrm{T}} d\mathbf{W}_u^S$$

$$\triangleq R_{m,n}(u) \gamma_{m,n}^{\mathrm{T}}(u) d\mathbf{W}_u^S, \tag{6.68}$$

where

$$\gamma_{m,n}(u) \triangleq \sum_{j=m}^{n-1} \omega_j \gamma_j(u), \tag{6.69}$$

with

$$\omega_j = \frac{\partial R_{m,n}(t)}{\partial f_j} \frac{f_j(t)}{R_{m,n}(t)}, \quad \text{for } j = m, \ldots, n-1. \tag{6.70}$$

The approximate swap-rate process (Equation 6.68) is now a lognormal one. The approximation made in Equation 6.68 is an example of the "frozen coefficient" technique, which is often used in the industry, whenever appropriate, to get rid of the dependence of model coefficients on state variables. The approximation we made here is based on two important observations: (1) $R_{m,n}(u)$ is a positive process; and (2) the stochastic coefficient,

$$\frac{\partial R_{m,n}(u)}{\partial f_j} \frac{f_j(u)}{R_{m,n}(u)},$$

indeed has a very low variability. Note that, after the approximation, the percentage volatility of the swap rate is simply a weighted average of the forward-rate volatilities. When $n = m + 1$, Equation 6.68 reduces to the lognormal process for $f_m(t)$ under its forward measure, $\mathbb{Q}_{m+1}$.

With the lognormal swap-rate process, from Equations 6.68 through 6.70, we easily derive the following Black's formula for the swaption:

$$V_t = A_{m,n}(t) E^{\mathbb{Q}_S} \left[(R_{m,n}(T_m) - K)^+ \mid \mathcal{F}_t \right]$$

$$= A_{m,n}(t) (R_{m,n}(t) \Phi(d_1) - K \Phi(d_2)), \tag{6.71}$$

where

$$d_1 = \frac{\ln\left(R_{m,n}(t)/K\right) + (1/2)\sigma_{m,n}^2 \left(T_m - t\right)}{\sigma_{m,n}\sqrt{T_m - t}}, \quad d_2 = d_1 - \sigma_{m,n}\sqrt{T_m - t},$$

$$(6.72)$$

and

$$\sigma_{m,n}^2 = \frac{1}{T_m - t}\int_t^{T_m} \|\gamma_{m,n}(s)\|^2 ds$$

$$= \frac{1}{T_m - t}\int_t^{T_m} \sum_{j,k=m}^{n-1} \omega_j\omega_k\gamma_j^{\mathrm{T}}(s)\gamma_k(s)\, ds. \qquad (6.73)$$

Note that $\sigma_{m,n}^2(T_m - t)$ is the variance of $\ln\left(R_{m,n}(T_m)\right)$, and the swaption price is quoted using $\sigma_{m,n}$, the so-called Black's volatility. When $n = m + 1$, the swaption reduces to a caplet and formulae 6.71 through 6.73 reduce to formulae 6.45 and 6.46 for caplets.

Black's formula, Equation 6.71, also implies a replicating (or hedging) strategy for the swaption using an ATM swap and a money market account:

1. long $\Phi(d_1(t))$ units of the ATM swap; and

2. long $A_{m,n}(t)\left(R_{m,n}(t)\Phi(d_1(t)) - K\Phi(d_2(t))\right)$ units of cash in the money market account, or $R_{m,n}(t)\Phi(d_1(t)) - K\Phi(d_2(t))$ units of the numeraire annuity.

The above hedging strategy is not unique. In fact, we can rewrite Black's formula as

$$V_t = [P(t, T_m) - P(t, T_n)]\, \Phi(d_1(t)) - KA_{m,n}(t)\Phi(d_2(t)), \qquad (6.74)$$

which suggests an alternative hedging strategy:

1. long $\Phi(d_1(t))$ unit of T_m-maturity zero-coupon bonds and short $\Phi(d_1(t))$ unit of T_n-maturity zero-coupon bonds, respectively; and

2. short $K\Phi(d_2(t))$ units of the annuity, or short $KA_{m,n}(t)\Phi(d_2(t))$ units of the money market account.

Note that the first hedging strategy is more practical because it uses the ATM swap, a very liquid security at no cost.

The swaption on a receiver's swap can be treated as a put option on the swap rate. Using the call–put parity, we readily obtain its formula:

$$V_t = A_{m,n}(t)\left(K\Phi(-d_2(t)) - R_{m,n}(t)\Phi(-d_1(t))\right). \qquad (6.75)$$

Black's formula offers only approximate prices for European swaptions, yet they are found to be very accurate and the formula has already become an industrial standard. For typical industrial applications, empirical studies (e.g., Rebonato, 1999; Sidenius, 2000) show that the pricing errors are well

TABLE 6.3: Monte Carlo Simulation and Black's Formula Prices for a Three-Factor Model

Maturity	Tenor	Strike (%)	Black	MC	STE
1	0.25	4.15	17.51	17.55	0.18
5	0.25	4.75	29.92	29.99	0.32
10	0.25	5.50	33.49	33.61	0.36
1	1	4.18	33.63	33.70	0.34
5	1	4.78	57.94	58.05	0.62
10	1	5.53	65.30	65.49	0.68
1	5	4.47	129.07	128.96	1.26
5	5	5.07	237.61	237.40	2.35
10	5	5.81	278.63	278.26	2.65
1	10	4.80	209.59	208.99	1.98
5	10	5.39	406.51	404.82	3.78
10	10	6.13	483.25	479.87	4.24

within one *kappa*, the usual market bid-ask spread defined as the change in the present value with 1% change in volatility. To demonstrate the accuracy by Black's formula, we compare the prices obtained by Black's formula and by a Monte Carlo simulation method using a two-factor model in Example 6.4.1.

Example 6.4.1. *We consider swaption pricing under a two-factor model, which takes the following forward-rate volatility:*

$$\gamma_j(T_k) = \left(0.08 + 0.1e^{-0.05(j-k)}, \ 0.1 - 0.25e^{-0.1(j-k)}\right), \quad k \leq j.$$

The initial term structure of forward rates is

$$f_j(0) = 0.04 + 0.00075j, \quad \text{for all } j.$$

Here we take $\Delta T_j = 0.5$. The results for ATM swaptions are listed in Table 6.3, where the first and second column are the maturity of the options and the tenor of the underlying swaps, respectively; the third column contains the prevailing swap rates, which are taken as the strike rates of the swaptions; the fourth column lists swaption prices, in basis points, obtained by Black's formula, the fifth column is for the prices obtained by the Monte Carlo simulations, and the last column is for the standard errors of the simulation results. The simulations are made with the standard LIBOR market model under the risk-neutral measure. One can see that the prices are very close across various maturities and tenors.

We finish the section with more comments on the accuracy of the lognormal approximation of the swap-rate process, Equation 6.68. Brigo and Liinev (2003) manage to calculate the Kullback–Leibler entropy distance between the original swap-rate distribution and the approximate lognormal distribution, and they find that the distance is encouragingly small. Precisely, for an implied Black's volatility of approximately 20%, the Kullback–Leibler entropy distance will be translated to a difference of about 0.1% in implied volatility, which is much smaller than 1%, the size of the bid-ask spread in swaption transactions.

6.5 Specifications of the LIBOR Market Model

The risk-neutral processes of the LIBOR rates are fully specified by their volatility vectors, $\gamma_j(t)$, or, equivalently, by the information on covariance between pairs of all forward rates:

$$\xi_{j,k}(t) = \gamma_j^{\mathrm{T}}(t)\gamma_k(t), \quad \forall j, k. \tag{6.76}$$

In fact, given the covariance matrix, $\mathbf{G}(t) = (\xi_{j,k}(t))$, which is a symmetric and non-negative definite matrix, we can perform a so-called Cholesky decomposition on $\mathbf{G}(t)$:

$$\mathbf{G}(t) = \mathbf{A}^{\mathrm{T}}(t)\mathbf{A}(t), \tag{6.77}$$

such that $\mathbf{A}(t)$ is an upper triangular matrix. We then set the volatility vector to

$$\gamma_j(t) = \mathbf{A}(t)\mathbf{e}_j, \quad j = 1, 2, \ldots, \tag{6.78}$$

where $\mathbf{e}_j$ is the jth column of the identity matrix of the same dimensions as that of $\mathbf{G}(t)$.

The information on the covariance of the forward rates comes from two sources. The first is the time series data on the forward rates. The second is the covariance implied by market prices of swaptions or other liquid derivatives that are sensitive to the covariance. For the purpose of pricing derivatives, the implied covariance should be more relevant than the historical covariance. Unfortunately, it is very difficult, if not impossible, to estimate the implied covariance in a reliable way. Because of that, we must rely, to varying degrees, on the covariance estimated from the time series data on the forward rates. Next, we describe the patterns of variances and correlations observed in the time series data of forward rates. Jointly, they provide information on the covariance. Such understanding will enable us to make reasonable parameterizations of the forward-rate volatilities.

Empirical studies on forward rates with various times to maturity have found that the norm of the volatility vector, $\|\gamma_j(t)\|$, as a function of $T_j - t$ is hump-shaped. Figure 6.3 shows such a hump-shaped function for USD, which is estimated using the daily yield data over the 10-year period, from 1996 to 2006. Similar humped-shaped volatility functions also appear in other currency.[2] Market participants generally agree that the forward-rate volatilities

---

[2]See, for example, Dodds (1998) on GBP.

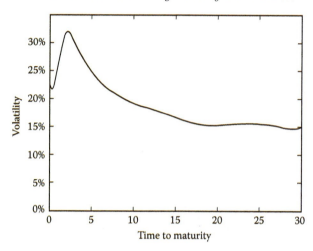

FIGURE 6.3: Historical volatilities for USD forward rates. (Estimated from USD yield data, from Reuters, from June 1996 to May 2006.)

should have some sort of time homogeneity, meaning that the norm of volatility vectors should depend on the time to maturity only. As such, after a series of trials, Rebonato (2002) suggested adopting the following parameterization:

$$\left\|\gamma_j(t)\right\| = \delta_j \left[(a + b(T_j - t)) \, e^{-c(T_j - t)} + d\right]. \tag{6.79}$$

Here a, b, c, and d are all constants, and δ_j the another constant that depends on j. This δ_j provides extra freedom to calibrate the model to option prices. The parameterization (Equation 6.79) is in agreement with empirical observations and is widely appreciated.

The shape for correlations is more variable. An example of the historical correlation among the forward rate of Pound Sterling (GBP) is displayed in Figure 6.4. Note that the correlations are mostly positive, but, away from the diagonal line, the correlations drop rather quickly. This is called the *decorrelation effect* of the LIBOR rates.

The parameterization of the correlations depends on the number of factors in the model. There is not yet a general consensus on how to parameterize the correlations. Partly, this is because the role of the correlations may not be as important as that of the volatilities, as the prices of caps do not depend on the correlations, while swaption prices are fairly insensitive to the correlations. Nonetheless, there are instruments that are sensitive to the correlations. Furthermore, the correlations are necessary to have the model fully specified. The rule of thumb is to utilize the information offered by historical correlations. For a parameterization of the correlation function, Rebonato (1999) makes

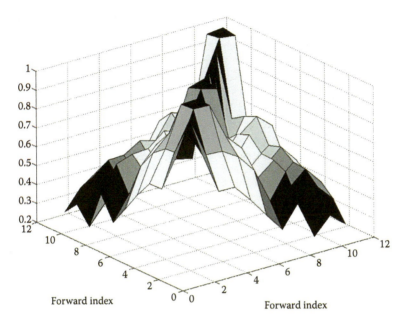

FIGURE 6.4: Historical correlations of GBP. (Estimated using one-year data from 1994.) (Data from Brace A, Gatarek D, and Musiela M, 1997. *Mathematical Finance* 7(2): 127–147. With permission.)

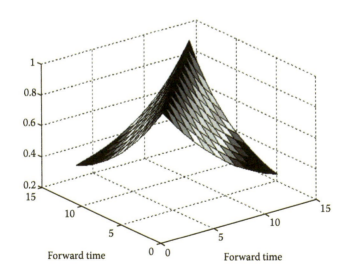

FIGURE 6.5: An artificial correlation matrix.

some suggestions, one of which is

$$\rho_{i,j}(t) = \text{LongCorr} + (1 - \text{LongCorr}) \exp\left[-\beta(T_i - T_j)\right],$$
$$\beta = d_1 - d_2 \max(T_i, T_j). \tag{6.80}$$

The above specification has the feature that the correlation first decays from one for $i = j$, but the decay slows down as $|i - j|$ increases, which mimics the decorrelation effect depicted in Figure 6.4. A surface plot for the correlations is given in Figure 6.5, where we have used LongCorr $= 0.3$, $d_1 = -0.12$, $d_2 = 0.005$, and $\Delta T = 0.5$. We call an exogenously specified correlation function, like the one in Equation 6.80, a model correlation.

6.6 Monte Carlo Simulation Method

6.6.1 The Log–Euler Scheme

Except for the few vanilla derivatives, namely, caps, floors, and swaptions, there is no closed-form formula for general LIBOR derivatives. It is also not feasible to implement the LIBOR market model, a multi-factor model, via a lattice tree. Hence, to price general LIBOR derivatives, we resort to Monte Carlo methods. In a Monte Carlo simulation method, we need to advance the spanning LIBOR rates in discrete time steps. The most natural scheme for the model is the log-Euler scheme, which is the result of applying the so-called Euler scheme to the stochastic equation of the logs of the forward rates under the risk-neutral measure:

$$d \ln f_j(t) = -\left(\gamma_j^{\mathrm{T}}(t)\Sigma_{j+1}(t) + \frac{1}{2}\|\gamma_j(t)\|^2\right) dt + \gamma_j^{\mathrm{T}}(t)d\mathbf{W}_t, \tag{6.81}$$

where $\Sigma_{j+1}(t) = \Sigma(t, T_{j+1})$. The log-Euler scheme is

$$\ln f_j(t + \Delta t) = \ln f_j(t) - \left(\gamma_j^{\mathrm{T}}(t)\Sigma_{j+1}(t) + \frac{1}{2}\|\gamma_j(t)\|^2\right) \Delta t + \gamma_j^{\mathrm{T}}(t)\Delta\mathbf{W}_t, \tag{6.82}$$

or

$$f_j(t + \Delta t) = f_j(t)e^{-(\gamma_j^{\mathrm{T}}(t)\Sigma_{j+1}(t)+(1/2)\|\gamma_j(t)\|^2)\Delta t+\gamma_j^{\mathrm{T}}(t)\Delta\mathbf{W}_t}. \tag{6.83}$$

The volatility of a zero-coupon bond can also be calculated from a recursive scheme,

$$\Sigma_{\eta(t)} = 0,$$
$$\Sigma_{j+1} = \Sigma_j - \frac{\Delta T_j f_j(t)}{1 + \Delta T_j f_j(t)}\gamma_j, \quad j \geq \eta(t), \tag{6.84}$$

where $\eta(t) = \min\{j \mid T_j > t\}$. Let N denote the number of forward rates at time $t = 0$, and let n_f denote the number of factors. Then, for the first step in the iteration of Equation 6.82, we need to perform $2(N+1)n_f + 3$ multiplications. For each path, which typically has N steps, the total number of multiplications is

$$\sum_{j=1}^{N} (2(j+1)n_f + 3) = N^2 + (2n_f + 4)N,$$

which is quite manageable.

The log-Euler scheme is a first-order scheme in Δt and it is appreciated for its simplicity and intuitiveness. More sophisticated or accurate schemes for the market model have also been developed. Hunter, Jackel, and Joshi (2001) consider a predictor–corrector scheme, aiming at raising the order of accuracy to two. Pietersz, Pelsser, and Regenmortel (2004) introduce a Brownian bridge discretization for the drift. To preclude possible arbitrage arising from discretization, Glasserman and Zhao (2000) propose to simulate zero-coupon bonds first before computing the forward rates. Most of these methods achieve some improvements on pricing accuracy over the simple log-Euler method.

6.6.2 Calculation of the Greeks

Next, we address the calculation of the sensitivity parameters of an option, the so-called Greeks, with respect to various risk factors. For simplicity, we restrict ourselves to the calculation of the sensitivities with respect to the spot forward-rate curve, which will be useful for hedging with FRAs or Eurodollar futures contracts. The calculation proceeds as follows. Suppose the maturity of the option is T_m. We consider the pricing under the risk-neutral measure,

$$V_0 = E^Q \left[B_{T_m}^{-1} V_{T_m} \right].$$

Differentiating with respect to $f_i(0)$, we obtain

$$\frac{\partial V_0}{\partial f_i(0)} = E^Q \left[\frac{\partial B_{T_m}^{-1}}{\partial f_i(0)} V_{T_m} + B_{T_m}^{-1} \frac{\partial V_{T_m}}{\partial f_i(0)} \right], \quad i = 1, \ldots, m-1.$$

According to Equation 6.42,

$$\frac{\partial B_{T_m}^{-1}}{\partial f_i(0)} = B_{T_m}^{-1} \sum_{j=i}^{m-1} \frac{\Delta T_j}{1 + \Delta T_j f_j(T_j)} \frac{\partial f_j(T_j)}{\partial f_i(0)},$$

while, by the chain rule of differentiation,

$$\frac{\partial V_{T_m}}{\partial f_i(0)} = \sum_{j=i}^{N} \frac{\partial V_{T_m}}{\partial f_j(T_m)} \frac{\partial f_j(T_m)}{\partial f_i(0)}.$$

For notational simplicity, we denote

$$f_{j,i}(t) = \frac{\partial f_j(t)}{\partial f_i(0)} \quad \text{and} \quad \Sigma_{j,i}(t) = \frac{\partial \Sigma_j(t)}{\partial f_i(0)}.$$

Differentiating Equation 6.83 with respect to $f_i(0)$, we have

$$f_{j,i}(t + \Delta t) = f_j(t + \Delta t) \left(\frac{f_{j,i}(t)}{f_j(t)} - \Delta t \gamma_j^{\mathrm{T}} \Sigma_{j+1,i}(t) \right), \tag{6.85}$$

where

$$\Sigma_{j+1,i}(t) = \Sigma_{j,i}(t) - \frac{\Delta T_j f_{j,i}(t)}{(1 + \Delta T_j f_j(t))^2} \gamma_j. \tag{6.86}$$

Equations 6.85 and 6.86 can be implemented alongside the Monte Carlo simulation for the derivatives value. This example demonstrates the spirit of the Greek calculations in terms of the Monte Carlo simulation method.

6.6.3   Early Exercise

The payoff of an exotic option may depend on the minimum, maximum, or average values of the state variables over the life of the option, or its cash flows can also be triggered by pre-specified barriers or a provision of early exercise. With Monte Carlo simulations, options on the minimum, maximum, or average value can be evaluated straightforwardly. But this is not the case for options with the provision of early exercise; such options are called American options. In fact, pricing the feature of early exercise in the context of Monte Carlo simulations had long been expensive and slow, due to the need for making decisions on early exercises throughout the usual backward induction procedure. This difficulty was overcome by a regression-based method developed between 1996 and 2000. In this section, we introduce this method with the pricing of Bermudan swaptions, which is also one of the most important applications of the method.

Consider a Bermudan swaption of either the first or the second kind that allows the holder to enter into a swap with a fixed rate at a pre-specified sequence of observation dates. At any of these observation dates, the holder must decide whether to exercise the option or hold on to it. These two actions result in two different values to the option. To the holder, the principle of action is simple: take the one that generates higher value for the option. As is well-known, numerical option pricing is a backward induction procedure. Let $h(\mathbf{f}_i)$ denote the payoff function of a Bermudan swaption at time T_i, where $\mathbf{f}_i = \{ f_j(T_i) \}_{j=i}^N$ denotes the forward rates that are still alive. The pricing of the Bermudan swaption may be done through the following dynamic programming procedure:

$$V_m(\mathbf{f}_m) = h(\mathbf{f}_m),$$
$$V_i(\mathbf{f}_i) = \max\{ h(\mathbf{f}_i), P(T_i, T_{i+1}) E^{\mathbb{Q}} [V_{i+1}(\mathbf{f}_{i+1}) \mid \mathbf{f}_i] \}, \tag{6.87}$$
$$i = m - 1, m - 2, \ldots, 0.$$

The first term in the second equation of Equation 6.87 is called the intrinsic value, the second term is called the "holding value" or "continuation value," and $P(T_i, T_{i+1})$ is the discount factor, given by

$$P(T_i, T_{i+1}) = \frac{1}{1 + \Delta T_j f_j(T_j)}.$$

Equation 6.87 is a mathematical interpretation of the holder's action: if the exercise value is higher, then exercise; otherwise, hold. The key in the implementation of the backward induction scheme lies in the valuation of the continuation value, which, however, had not been easy until a regression-based solution was first proposed by Carriere (1996) and later improved by Longstaff and Schwartz (2001). Carriere proposed to approximate the conditional expectation in Equation 6.87 by a function of the state variables:

$$E^{\mathbb{Q}}\left[V_{i+1}(\mathbf{f}_{i+1}) \mid \mathbf{f}_i = \mathbf{f}\right] = \sum_{j=1}^{M} \beta_{i,j} \psi_j(\mathbf{f}) \triangleq C_i(\mathbf{f}), \tag{6.88}$$

for some basis function, ψ_j, and constants, $\beta_{i,j}, j = 1, \dots, M$. Let

$$\begin{aligned} \Psi(\mathbf{f}) &= (\psi_1(\mathbf{f}), \dots, \psi_M(\mathbf{f}))^{\mathrm{T}}, \\ \boldsymbol{\beta}_i &= (\beta_{i,1}, \dots, \beta_{i,M})^{\mathrm{T}}. \end{aligned} \tag{6.89}$$

Then, by linear regression, we have the following formula for $\boldsymbol{\beta}_i$:

$$\begin{aligned} \boldsymbol{\beta}_i &= \left(E^{\mathbb{Q}}\left[\Psi(\mathbf{f}_i)\Psi(\mathbf{f}_i)^{\mathrm{T}}\right]\right)^{-1} E^{\mathbb{Q}}\left[\Psi(\mathbf{f}_i)V_{i+1}(\mathbf{f}_{i+1}) \mid \mathbf{f}\right] \\ &\triangleq B_{\Psi}^{-1} B_{\Psi V}, \end{aligned} \tag{6.90}$$

where B_Ψ is an M-by-M matrix, and $B_{\Psi V}$ an M-by-1 matrix. These two matrices can be approximated by using sample data as follows:

$$\left(\hat{B}_\Psi\right)_{q,r} = \frac{1}{N} \sum_{j=1}^{N} \psi_q\left(\mathbf{f}_i^{(j)}\right) \psi_r\left(\mathbf{f}_i^{(j)}\right), \tag{6.91}$$

and

$$\left(\hat{B}_{\Psi V}\right)_r = \frac{1}{N} \sum_{j=1}^{N} \psi_r\left(\mathbf{f}_i^{(j)}\right) V_{i+1}\left(\mathbf{f}_{i+1}^{(j)}\right). \tag{6.92}$$

The estimate of the continuation value is then given by

$$\hat{C}_i(\mathbf{f}_i) = P(T_i, T_{i+1})\hat{\boldsymbol{\beta}}_i^{\mathrm{T}} \Psi(\mathbf{f}_i) \triangleq P(T_i, T_{i+1})\hat{B}_\Psi^{-1} \hat{B}_{\Psi V} \Psi(\mathbf{f}_i). \tag{6.93}$$

With the continuation values, the decision about the early exercise can be made. Here is the regression-based Monte Carlo simulation algorithm.

Algorithm 6.6.1.

1. *Simulate N independent paths, $\{\mathbf{f}_i^{(j)}, \dots, \mathbf{f}_m^{(j)}\}, j = 1, \dots, N$, of the Markov chain.*

2. At the terminal nodes, calculate $\hat{V}_m^{(j)} = h(\mathbf{f}_m^{(j)}), j = 1, \ldots, N$.

3. Apply backward induction: for $i = m - 1, \ldots, 1, 0$, do the following:

    a. Given estimated value $\hat{V}_{i+1}^{(j)}, j = 1, \ldots, N$, use regression formulae 6.91 through 6.93 to calculate $\hat{C}_i \left(\mathbf{f}_i^{(j)} \right)$.

    b. Set

$$\hat{V}_i^{(j)} = \max\left\{ h \left(\mathbf{f}_i^{(j)} \right), \hat{C}_i \left(\mathbf{f}_i^{(j)} \right) \right\}, \quad j = 1, \ldots, N. \tag{6.94}$$

4. Take

$$\hat{V}_0 = \max\left\{ h(\mathbf{f}_0), \frac{P(0, T_1)}{N} \sum_{j=1}^{N} \hat{V}_1^{(j)} \right\} \tag{6.95}$$

Although easy to comprehend, the above estimator is, however, biased upward. Here is the explanation (Glasserman, 2003). Conditional to $\mathbf{f}_i$, there is

$$
\begin{aligned}
E^{\mathbb{Q}} \left[\hat{V}_i(\mathbf{f}_i) \right] &= E^{\mathbb{Q}} \left[\max\{h(\mathbf{f}_i), P(T_i, T_{i+1})\hat{C}_i(\mathbf{f}_i)\} \right] \\
&\geq \max \left\{ h(\mathbf{f}_i), P(T_i, T_{i+1})E^{\mathbb{Q}}[\hat{C}_i(\mathbf{f}_i)] \right\} \\
&= \max\{h(\mathbf{f}_i), P(T_i, T_{i+1})E^{\mathbb{Q}}[\hat{V}_{i+1}(\mathbf{f}_{i+1})|\mathbf{f}_i]\}.
\end{aligned}
$$

due to the Jensen's inequality. If, conditional on $\mathbf{f}_{i+1}$, we already have

$$E^{\mathbb{Q}} \left[\hat{V}_{i+1}(\mathbf{f}_{i+1}) \right] \geq V_{i+1}(\mathbf{f}_{i+1}),$$

which is at least true for $i = m - 1$, then, by the tower law,

$$
\begin{aligned}
E^{\mathbb{Q}} \left[\hat{V}_{i+1}(\mathbf{f}_{i+1}) \, | \, \mathbf{f}_i \right] &= E^{\mathbb{Q}} \left[E^{\mathbb{Q}} \left[\hat{V}_{i+1}(\mathbf{f}_{i+1}) \right] \Big| \, \mathbf{f}_i \right] \\
&\geq E^{\mathbb{Q}} \left[V_{i+1}(\mathbf{f}_{i+1}) \, | \, \mathbf{f}_i \right].
\end{aligned}
$$

It then follows that

$$E^{\mathbb{Q}} \left[\hat{V}_i(\mathbf{f}_i) \right] \geq \max\{h(\mathbf{f}_i), P(T_i, T_{i+1})E^{\mathbb{Q}} \left[V_{i+1}(\mathbf{f}_{i+1}) \, | \, \mathbf{f}_i \right] \} = V_i(\mathbf{f}_i).$$

The upward bias of the estimator may be attributed to the use of the same information in deciding whether to exercise and the estimation of the continuation value.

To remove the bias, we must separate the exercise decision making from the valuation of continuation value. This is in fact the key to removing the bias in all Monte Carlo methods. To show the idea, we consider the problem of estimating

$$\max\{a, E\left[y\right]\}$$

from iid replicates $Y_1, \ldots, Y_N$. The estimator,

$$\max\{a, \bar{Y}\},$$

is upward biased since

$$E\left[\max\{a, \bar{Y}\}\right] \geq \max\{a, E\left[\bar{Y}\right]\}$$
$$= \max\{a, E\left[Y\right]\}.$$

To fix such a bias is nonetheless simple: separate $\{Y_i\}$ into two disjointed subsets; calculate sample means, $\bar{Y}_1$ and $\bar{Y}_2$; and set

$$\hat{V} = \begin{cases} a, & \text{if } \bar{Y}_1 \leq a, \\ \bar{Y}_2, & \text{if } \bar{Y}_1 > a. \end{cases} \tag{6.96}$$

Then,

$$E\left[\hat{V}\right] = P(\bar{Y}_1 \leq a)a + (1 - P(\bar{Y}_1 \leq a))E\left[\bar{Y}_2\right]$$
$$\leq \max\{a, E\left[Y\right]\}.$$

Hence, the new estimator is biased downward.

In the regression-based method, valuations like the one in Equation 6.96 ought to be made at every time step for all paths. While $\bar{Y}_1$ is obtained through regression, $\bar{Y}_2$ should be obtained by pricing a similar Bermudan option, for which we face the same problem of decision making on early exercise. It seems that the problem is not solved unless we can find an inexpensive estimation of $\bar{Y}_2$. The simplest way (Glasserman, 2003) is to simulate a single additional path starting from $\tilde{\mathbf{f}}_i = \mathbf{f}_i$, which is independent of the existing paths used to calculate the continuation value; calculate the continuation values, $\tilde{C}_l^{(j)}, i \leq l \leq m - 1$, along the path; and define the stopping time:

$$\hat{\tau} = \min_{i \leq l \leq m} \left\{ l : h(\tilde{\mathbf{f}}_l) \geq \tilde{C}_l(\tilde{\mathbf{f}}_l) \right\}. \tag{6.97}$$

Here $\{\tilde{\mathbf{f}}_i, \tilde{\mathbf{f}}_{i+1}, \ldots, \tilde{\mathbf{f}}_m\}$ is the additional path. With this stopping time, $\hat{\tau}$, we define an estimator to be

$$\hat{V}_i(\mathbf{f}_i) = P(T_i, T_{\hat{\tau}})h_{\hat{\tau}}(\tilde{\mathbf{f}}_{\hat{\tau}}). \tag{6.98}$$

Hence, the estimator for $\bar{Y}_2$ is defined as the exercise value of the option as soon as it is equal to or higher than the continuation value. Since any policy is prone to be suboptimal, Equation 6.98 is thus a low estimator. One can imagine that this estimator based on a single path is not accurate enough, but the resulted value functions have the right distribution for the American option in all time steps.

Longstaff and Schwartz (2001) present an alternative solution to the upward-bias problem of Algorithm 6.1 by looking at the cash flow date along all paths. At time step T_i, they first estimate the continuation value for all paths by the regression method of Carriere. Then, if intrinsic values are larger than the continuation values, they exercise the option and update the cash flow dates. At the end of the backward induction, they obtain a matrix of cash flow dates. The continuation value at time T_0 is defined as the average value of discounted cash flows, and the value of the American option is taken as the maximum one between the intrinsic value and the continuation value. To make the estimation of continuation values more accurate, Longstaff and Schwartz use only the paths that are in the money in the regression. Moreover, discounted values of future cash flows are used in the regression whenever the paths are in the money. The convergence of the Longstaff–Schwartz algorithm is proved by Clement, Lamberton, and Protter (2002) for $j \to \infty$. More precisely, the limit obtained yields the true price if the functional representation (Equation 6.88) holds exactly; otherwise, it is smaller than the true price.

Through a careful comparison, we understand that the approach of Longstaff and Schwartz is equivalent to taking the existing path, $\{\mathbf{f}_i, \mathbf{f}_{i+1}, \ldots, \mathbf{f}_m\}$, for the additional path, $\{\tilde{\mathbf{f}}_i, \tilde{\mathbf{f}}_{i+1}, \ldots, \tilde{\mathbf{f}}_m\}$, in Glasserman's algorithm described earlier. We thus can present the algorithm as follows:

$$\hat{V}_i^{(j)} = \begin{cases} h(\mathbf{f}_i^{(j)}), & h(\mathbf{f}_i^{(j)}) > C_i(\mathbf{f}_i^{(j)}), \\ P(T_i, T_{i+1})\hat{V}_{i+1}^{(j)}, & h(\mathbf{f}_i^{(j)}) \le C_i(\mathbf{f}_i^{(j)}). \end{cases} \tag{6.99}$$

By backward induction with Equation 6.99, we no longer need the cash flow matrix. By replacing Equation 6.94 with 6.99, we obtain a robust algorithm that is currently very popular in the industry.

For completeness, we comment on the functional form, Ψ, in the estimation of the conditional expectation in Equation 6.88. The simplest example of functional approximation to the conditional expectation is the univariate quadratic polynomial with a certain state variable. Take the pricing of a Bermudan swaption on swaps with fixed tenor for example. One may take

$$\Psi(\mathbf{f}) = \left(1, R, R^2\right)^{\mathrm{T}}, \tag{6.100}$$

where $R = R_{i,i+n}$ is for the underlying swap rate. Bivariate functional approximations of the conditional expectation can also be considered (Pedersen, 1999). One example is the bivariate quadratic polynomial in a money market account, B_{T_i}, and the swap rate, $R_{i,i+n}$,

$$\Psi(\mathbf{f}) = \left(1, R, B, R^2, RB, B^2\right)^{\mathrm{T}}. \tag{6.101}$$

In Pedersen's test, regressions using the univariate polynomial approximation and various bivariate quadratic polynomial approximations perform similarly and all produce rather accurate results.

6.7 Notes

There are two important issues not addressed in this chapter. The first is the construction of initial term structure of LIBOR (curve), and the second is the phenomenon of volatility smiles in the option markets.

After the 2008 financial crisis, both the U.S. Federal Reserve and European Central Bank adopted quantitative easing (QE) policy. The direct effect is the ultra low interest rates for all maturities which have never been seen before. When a conventional method is adopted for the construction of the LIBOR curve, often part(s) of the curve is below the level of zero, meaning negative forward rates for some maturities, then arbitrage occurs. The problem was solved by Hagan and West (2006). They gave up some smoothness or regularity property in exchange for the overall positiveness of the LIBOR curve.

The phenomenon of volatility smiles has existed since the Black Monday of 1987, which means that, for a fixed maturity, the implied Black's volatilities across strikes form a smile shape. This phenomenon exists in equity, commodity, forex and interest rate derivatives. Since the LIBOR market model cannot generate a smile-shaped volatility curve, it is somewhat invalidated by the markets.

Many competitive models have been developed to cope with the implied volatility smiles. These models either add additional risk factors, like jumps or stochastic volatility, to the otherwise lognormal model, or turn the local volatility state dependent function. In Chapter 10 and 11, we will introduce the best known smile models. The Black's formula, meanwhile, has become a device for quoting the call or put option prices.

Exercises

1. Price the two-year maturity caplet under a one-factor Ho–Lee model. Let the current forward-rate term structure be

$$f(0,T) = 0.02 + 0.001T,$$

the forward-rate volatility be $\sigma = 0.015$, the cash flow interval be $\Delta T = 0.25$, and the strike price be the *spot forward term rate*, $K = f_{0.25}(0,2)$.

2. Use the risk-neutral measure and the tower law to show that, at time $t \leq T_0$, the price of an FRN with interest rates $f_{\Delta T_{i-1}}(T_{i-1}, T_{i-1}), T_i = T_0 + i\Delta T, i = 1, \ldots, n$, is equal to $P(0, T_0)$, the price of a T_0-maturity zero-coupon bond.

3. It is understood that a forward rate satisfies

$$f_{\Delta T}(0,T) = E^{\mathbb{Q}_{T+\Delta T}}\left[f_{\Delta T}(T,T)\right],$$

whereas the corresponding futures implied rate satisfies

$$f_{\Delta T}^{(fut)}(0,T) = E^{\mathbb{Q}}\left[f_{\Delta T}(T,T)\right].$$

Prove that

$$f_{\Delta T}(0,T) - f_{\Delta T}^{(fut)}(0,T)$$
$$= \frac{\mathrm{Cov}^{\mathbb{Q}}(f_{\Delta T}(T,T), B^{-1}(T+\Delta T))}{E^{\mathbb{Q}}[B^{-1}(T+\Delta T)]}.$$

4. Prove the *call–put parity* between a cap and a floor with the same strike rate and tenor:
$$\mathrm{Cap} - \mathrm{Floor} = \mathrm{Swap},$$
where the swap has the same strike rate and tenor as those of the cap and floor.

5. Suggest a strategy to hedge a swap with a sequence of FRAs, and use the strategy to argue that

$$R_{m,n}(t) = \sum_{j=m}^{n-1} \alpha_j f_j(t), \quad \alpha_j = \frac{\Delta T P(t, T_{j+1})}{A_{m,n}(t)}.$$

6. Describe the swaption hedging strategy using the ATM swap and the annuity. Prove it if you think that the strategy is a self-financing one.

7. Prove the alternative version of the CMG theorem: define a new measure, $\tilde{\mathbb{Q}}$, as

$$\left.\frac{d\tilde{\mathbb{Q}}}{d\mathbb{Q}}\right|_{\mathcal{F}_t} = m(t),$$

where $m(t)$ is a $\mathbb{Q}$-martingale with $m(0) = 1$ and $m(t) > 0$. Let $\mathbf{W}_t$ be a vector of independent $\mathbb{Q}$-Brownian motions. Then, $\tilde{\mathbf{W}}_t$ defined as

$$d\tilde{\mathbf{W}}_t = d\mathbf{W}_t - \left\langle d\mathbf{W}_t, \frac{dm(t)}{m(t)} \right\rangle,$$

is a vector of independent $\tilde{\mathbb{Q}}$-Brownian motions (here $\langle \cdot \rangle$ means covariance).

8. For all problems below, use the spot term structure

$$f(0,T) = 0.02 + 0.001T,$$

for the instantaneous forward rates. Pricing is done under the market model (which assumes lognormal forward rates or swap rates). The payment frequency is a quarter year for caps and half a year for swaps.

(a) Price the 10-year cap with the strike rate to be the 10-year par yield (such a cap is called *ATM-forward cap*). Take the implied cap volatility to be 20%.

(b) Price the in-5-to-10 ATM swaption (maturity of the option: 5; tenor of the underlying swap: 10). Assume a 20% swap-rate volatility.

9. Redo the pricing of swaption in Problem 8b under the LIBOR market model using the method of Monte Carlo simulation. Use the dynamics of LIBOR under the forward measure with delivery at the maturity day (also called terminal measure) of the swaption.

10. This problem deals with call options on coupon bonds with par strike.

(a) Show that any coupon-bond options with par strike can be treated as swaptions.

(b) Based on (a), develop a closed-form formula under the market model for coupon bond options with par strike,

(c) Based on (b) suggest a hedging strategy for these options using ATM swaps.

11. Provide a closed-form price formula under the market model for a LIBOR corridor, which consists of a series of cash flow, $\Delta T_j c_j$, at time $T_{j+1}, j = 0, 1, \ldots, N - 1$, with

$$
c_j = \begin{cases} K_1, & \text{if} \quad f_j(T_j) < K_1, \\ f_j(T_j), & \text{if} \quad K_1 \leq f_j(T_j) \leq K_2, \\ K_2, & \text{if} \quad K_2 \leq f_j(T_j). \end{cases}
$$

Explain how to hedge the corridor using forward-rate agreements.

12. Consider the pricing of a cancelable swap (that exchanges LIBOR for a fixed rate). The maturity of the swap is ten years, and the payer has the right to cancel the swap in five years. How to solve for the fair swap rate so that the value of the cancelable swap is zero?

Chapter 7

Calibration of LIBOR Market Model

The LIBOR market model is a term structure model driven by multiple random factors. Owing to the multiple factors, the model has a rich capacity to accommodate rather general covariance structures for the forward rates. In other words, this model can be calibrated, or fitted, to various structures of covariance and various price inputs of benchmark derivatives, which is very desirable for a production model. Nonetheless, these advantages have long been hard to exploit. For a long time, the nonparametric calibration of the model to fit input correlations and input prices simultaneously had been an unresolved challenge to model users, and it had in fact become a bottleneck for the model's applications. The situation was changed essentially in 2003, when a satisfying solution to calibration was introduced by Wu (2003). This chapter is devoted to the methodologies of Wu's solution.

A comprehensive, nonparametric calibration of the market model consists of two parts: fitting the model to input correlations and fitting the model to input prices. These two objectives of fitting are, luckily, essentially decoupled. The fitting of input correlations becomes a problem because the input correlation matrix usually has a full rank, equal to the number of forward rates, while the model correlation matrix can only have a low rank, no more than the number of driving factors. Note that based on PCA of yield curves, the number of factors used in applications is much smaller than the number of state variables. The challenge here is to find a series of low-rank approximations to the full-rank input correlation matrix. The fitting of input prices is about finding the norm of the local volatility vector of forward rates. One of the most interesting treatments in Wu's approach is to fit the implied Black's volatilities rather than to fit the prices themselves, as fitting the implied Black's volatilities leads to a kind of quadratic programming problem, which is unconventional but can still be solved efficiently. Once the norms are obtained (usually as time-dependent functions), we can determine the components of the volatility vector based on model correlations, as was already explained in Section 6.5. Through the calibration of the LIBOR market model, we demonstrate that model calibrations in finance can be very delicate in mathematics as well as in computations.

7.1   Implied Cap and Caplet Volatilities

As a preparation for subsequent sections, in this section, we consider calibrating the LIBOR market model based on cap prices only. Cap prices imply prices of caplets of various maturities, and the Black's volatilities corresponding to the caplet prices will be taken to be the norms of the local volatility vectors. The volatility vectors of forward rates can be determined based on the norms and, in addition, information on forward-rate correlations.

Let $\text{Cap}(n)$ denote the market price of a cap that has n caplets, and let K_n denote the strike rate. In the marketplace, a cap is quoted using the implied cap volatility, $\bar{\sigma}_n$, defined as the single non-negative number that, when plugged in into Black's formula for all caplets, reproduces the market price of the cap. That is, $\bar{\sigma}_n$ satisfies

$$\text{Cap}(n) = \sum_{j=0}^{n-1} P_0^{T_{j+1}} \Delta T_j \cdot BC(f_j(0), K_n, T_j, \bar{\sigma}_n), \qquad (7.1)$$

where $BC(f, K, T, \sigma)$ stands for Black's call formula,

$$BC(f, K, T, \sigma) = f\, \Phi(d_1) - K\Phi(d_2), \qquad (7.2)$$

with

$$d_1 = \frac{\ln(f/K) + (1/2)\sigma^2 T}{\sigma\sqrt{T}}, \quad d_2 = d_1 - \sigma\sqrt{T}. \qquad (7.3)$$

Given a sequence of implied cap volatilities, $\bar{\sigma}_n$, $n = 1, 2, \ldots, N$, for increasing cap maturities, we can bootstrap the values of all caplets of maturities up to the longest maturity of the caps. The caplet prices can be quoted using their Black's volatilities, $\{\sigma_j\}$. In terms of the Black's volatilities of caplets, we can recast the cap price as

$$\begin{aligned}
\text{Cap}(n) &= \sum_{j=0}^{n-1} P_0^{T_{j+1}} \Delta T_j \cdot BC(f_j(0), K_n, T_j, \sigma_j) \\
&= \sum_{j=0}^{n-1} \text{Caplet}(j+1, n).
\end{aligned} \qquad (7.4)$$

Here $\text{Caplet}(j, n)$ stands for the jth caplet of the nth cap. Equation 7.4 can be used to bootstrap the value of the caplets, which proceeds as follows. First, we identify the value of the first caplet of all caps as their intrinsic values:

$$\text{Caplet}(1, n) = P_0^{T_1} \Delta T_0 \left(f_0(T_0) - K_n \right)^+, \quad \forall n. \qquad (7.5)$$

This is based on the fact that $f_0(t)$, the first forward rate, has already been fixed at time 0. Next, we remove the known value of the first caplet, $\text{Caplet}(1, n)$, from the nth cap for $n \geq 2$,

$$
\begin{aligned}
\text{Cap}(n) - \text{Caplet}(1, n) &= \sum_{j=1}^{n-1} \text{Caplet}(j+1, n) \\
&= \sum_{j=1}^{n-1} P_0^{T_{j+1}} \Delta T_j \cdot BC(f_j(0), K_n, T_j, \sigma_j).
\end{aligned}
\tag{7.6}
$$

When $n = 2$, we have, in particular, the dollar price of the second caplet of the second cap:

$$
\text{Caplet}(2, 2) = P_0^{T_2} \Delta T_1 \cdot BC(f_1(0), K_2, T_1, \sigma_1),
\tag{7.7}
$$

which allows us to solve for the implied Black's volatility for all caplets on $f_1(t)$. We write

$$
\sigma_1 = \text{implied vol}(\text{Caplet}(2, 2)).
\tag{7.8}
$$

With this σ_1, we define the price of the second caplet of all caps according to

$$
\text{Caplet}(2, n) = P_0^{T_2} \Delta T_1 \cdot BC(f_1(0), K_n, T_1, \sigma_1), \quad \text{for } n \geq 2.
\tag{7.9}
$$

Then, we strip the value of the second caplet from caps for $n \geq 3$, single out the value of $\text{Caplet}(3, 3)$, and calculate the implied volatility, σ_2, for $f_2(t)$. Suppose that we have carried out this procedure and already obtained σ_j, $j \leq k - 1$. We then proceed to strip the kth caplet from the remaining caps:

$$
\text{Cap}(n) - \sum_{j=0}^{k-1} \text{Caplet}(j+1, n) = \sum_{j=k}^{n-1} \text{Caplet}(j+1, n), \quad \text{for } n \geq k+1.
\tag{7.10}
$$

Yet again, for $n = k + 1$, there is only one term, $\text{Caplet}(k+1, k+1)$, on the right-hand side of Equation 7.10, from which we obtain

$$
\sigma_k = \text{implied vol}(\text{Caplet}(k+1, k+1)).
\tag{7.11}
$$

Continuing with this procedure, we obtain all Black's volatilities of caplets, σ_j, $\forall j \leq N - 1$, which are taken as the implied volatility of $f_{k-1}(t), k \leq N$. The bootstrapping procedure can be visualized as in Figure 7.1, where we proceed from left to right by columns.

We use the yield curve and the cap data for February 3, 1995, for illustration. The discount curve and the cap prices in USD are listed in Tables 7.1 and 7.2, respectively. The column under "ATM strike" contains the prevailing swap rate for various maturities, which are also called the ATM swap rates. The implied caplet volatilities are calculated and listed in Table 7.3, which are in the same magnitude of the implied volatility of the caps. Note that, in the calculations, we have used the built-in root-finding function, *fzero*, in Matlab.

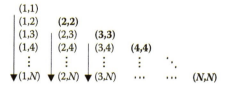

FIGURE 7.1: The order for bootstrapping the diagonal caplets.

TABLE 7.1: Discount Curve

Maturity	Price
0.00	1.00000000
0.25	0.98317518
0.50	0.96533801
1.00	0.92713249
1.50	0.88814477
2.00	0.84964678
2.50	0.81226987
3.00	0.77629645
4.00	0.71122696
5.00	0.64912053
7.00	0.54020582
9.00	0.45339458
10.00	0.41531609
11.00	0.37873810

Reprinted from Wu L, 2003. *Journal of Computational Finance* 6(2): 39–77. With permission.

7.2   Calibrating the LIBOR Market Model to Caps

Calibration here means to specify $\{\gamma_j(t)\}$, the local volatility vector of forward rates, so that the market prices of all caplets are reproduced. According to Black's formula, the implied Black's volatility, σ_j, relates to the local volatility vector of the jth forward rate via

$$\sigma_j^2 = \frac{1}{T_j} \int_0^{T_j} \left\| \gamma_j(t) \right\|^2 \mathrm{d}t. \tag{7.12}$$

The right-hand side is the mean variance of the jth forward rate. Equation 7.12 allows us to determine $\|\gamma_j(t)\|$. For a parametric calibration, we may take the functional specification of Equation 6.79 and solve for the parameters through an optimization procedure that minimizes the aggregated squared errors between the two sides of Equation 7.12. The procedure is a rather typical one.

The focus of this section, however, is on nonparametric calibration.

TABLE 7.2: Input Prices of Caps

Contract Type	Length	ATM Strike (%)	Black's Volatilities (%)	Market Prices (bps)
Cap	1	7.88	15.5	27
Cap	2	8.39	17.75	100
Cap	3	8.64	18	185
Cap	4	8.69	17.75	267
Cap	5	8.79	17.75	360
Cap	7	8.9	16.5	511
Cap	10	8.89	15.5	703

Reprinted from Brace A, Gatarek D, and Musiela M, 1997. *Mathematical Finance* 7: 127–154. With permission.

TABLE 7.3: Stripped Caplet Prices and Their Implied Volatilities

Contract Type	Length	ATM Strike %	Black's Volatilities %	Market Prices (bps)
Caplet	0.25	7.88	15.23	1.57
Caplet	0.5	7.88	15.23	8.94
Caplet	1	8.39	18.53	16.71
Caplet	2	8.64	18.17	23.55
Caplet	3	8.69	17.21	22.33
Caplet	4	8.79	17.89	24.34
Caplet	5	8.90	14.05	22.25
Caplet	7	8.89	13.44	17.70
Caplet	9	8.89	13.44	15.08

Reprinted from Wu L, 2003. *Journal of Computational Finance* 6(2): 39–77. With permission.

Suppose that $\gamma_j(t)$ is constant in t, we can simply take

$$\|\gamma_j(t)\| = \sigma_j, \quad 0 \le t \le T_j, \quad j = 1, \dots, N. \tag{7.13}$$

The plot of $\{\|\gamma_j(t)\|\}$ is called the local volatility surface of the LIBOR market model. Figure 7.2 shows the local volatility surface for February 3, 1995, generated using the implied Black's volatility of the caplets in Table 7.3. Note that the part of the surface over the area of $t \le T$ is the only relevant part.

To specify the LIBOR model, we also need the pair-wise instantaneous correlations among forward rates:

$$\frac{\langle \mathrm{d}f_j, \mathrm{d}f_k \rangle}{\sqrt{\langle \mathrm{d}f_j, \mathrm{d}f_j \rangle}\sqrt{\langle \mathrm{d}f_k, \mathrm{d}f_k \rangle}} = \rho_{jk}(t). \tag{7.14}$$

Recall that, in the LIBOR model, a forward rate follows

$$\mathrm{d}f_j(t) = f_j(t)\gamma_j^{\mathrm{T}}(t)\left[\mathrm{d}\mathbf{W}_t - \mathbf{\Sigma}(t, T_{j+1})\,\mathrm{d}t\right]. \tag{7.15}$$

The instantaneous covariance between $\mathrm{d}f_j(t)$ and $\mathrm{d}f_k(t)$ is

$$\mathrm{Cov}_{jk}(t) = f_j f_k \gamma_j^{\mathrm{T}} \gamma_k \,\mathrm{d}t. \tag{7.16}$$

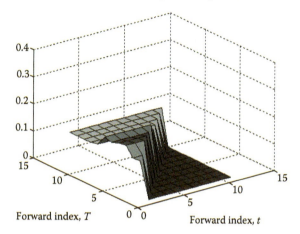

FIGURE 7.2: The local volatility surface for a calibrated LIBOR model.

In terms of the volatility vectors, the model correlation between df_j and df_k is

$$\text{Corr}_{jk}(t) = \frac{\text{Cov}_{jk}(t)}{\sqrt{\text{Cov}_{jj}(t)}\sqrt{\text{Cov}_{kk}(t)}} = \frac{\gamma_j^{\text{T}}(t)\gamma_k(t)}{\|\gamma_j(t)\|\|\gamma_k(t)\|}. \tag{7.17}$$

Let us assume that the full set of correlations, $\{\rho_{jk}(t)\}$, is available. Then, we try to determine $\{\gamma_j(t)\}$ by matching the model correlations to the input correlations,

$$\frac{\gamma_j^{\text{T}}(t)\gamma_k(t)}{\|\gamma_j(t)\|\|\gamma_k(t)\|} = \rho_{jk}(t), \quad \forall j,k \geq \eta(t). \tag{7.18}$$

A solution of $\{\gamma_j(t)\}$ can be easily defined by factorizing a corresponding correlation matrix. To see this, we introduce

$$A(t) = \left(\mathbf{a}_{\eta(t)}\ \mathbf{a}_{\eta(t)+1}\cdots\mathbf{a}_N\right), \tag{7.19}$$

where $\eta(t) = \min\{j|T_j > t\}$ is the index of the first live forward rate, and

$$\mathbf{a}_j = \frac{\gamma_j}{\|\gamma_j\|}, \quad j \geq \eta(t), \tag{7.20}$$

which is a vector of length $N - \eta(t) + 1$. In terms of the matrix $A(t)$ equations, Equation 7.18 becomes

$$A^{\text{T}}(t)A(t) = (\rho_{jk}(t)), \quad \text{for } j,k \geq \eta(t). \tag{7.21}$$

Given the matrix of the input correlations, $(\rho_{jk}(t))$, we can define an $A(t)$ by either a Cholesky decomposition or an eigenvalue decomposition. Although possible factorizations are not unique, they make no difference from the viewpoint of distribution. Once we have obtained $A(t)$, we set γ_j to be $\|\gamma_j\|$ times the jth column of A, that is,

$$\gamma_j(t) = \|\gamma_j(t)\|A\mathbf{e}_j, \quad j = \eta(t),\ldots,N, \tag{7.22}$$

TABLE 7.4: A Historical Correlation Matrix for GBP Forward Rates

	0.25	0.5	1	1.5	2	2.5	3	4	5	7	9
0.25	1.0000	0.8415	0.6246	0.6231	0.5330	0.4287	0.3274	0.4463	0.2439	0.3326	0.2625
0.5	0.8415	1.0000	0.7903	0.7844	0.7320	0.6346	0.4521	0.5812	0.3439	0.4533	0.3661
1	0.6246	0.7903	1.0000	0.9967	0.8108	0.7239	0.5429	0.6121	0.4426	0.5189	0.4251
1.5	0.6231	0.7844	0.9967	1.0000	0.8149	0.7286	0.5384	0.6169	0.4464	0.5233	0.4299
2	0.5330	0.7320	0.8108	0.8149	1.0000	0.9756	0.5676	0.6860	0.4969	0.5734	0.4771
2.5	0.4287	0.6346	0.7239	0.7286	0.9756	1.0000	0.5457	0.6583	0.4921	0.5510	0.4581
3	0.3274	0.4521	0.5429	0.5384	0.5676	0.5457	1.0000	0.5942	0.6078	0.6751	0.6017
4	0.4463	0.5812	0.6121	0.6169	0.6860	0.6583	0.5942	1.0000	0.4845	0.6452	0.5673
5	0.2439	0.3439	0.4426	0.4464	0.4969	0.4921	0.6078	0.4845	1.0000	0.6015	0.5200
7	0.3326	0.4533	0.5189	0.5233	0.5734	0.5510	0.6751	0.6452	0.6015	1.0000	0.9889
9	0.2625	0.3661	0.4251	0.4299	0.4771	0.4581	0.6017	0.5673	0.5200	0.9889	1.0000

Reprinted from Brace A, Gatarek D, and Musiela M, 1997. *Mathematical Finance* 7: 127–154. With permission.

where e_j is the jth unit vector such that its jth element is 1 and the rest of the elements are 0. At this point, the model is fully specified. For actual implementation of the LIBOR model, we prefer to take the solution from a Cholesky decomposition of the correlation matrix, which renders an upper triangular matrix for $A(t)$.

As an example, we list a set of modified historical correlations of GBP forward rates, $(\rho_{jk}(0))$, in Table 7.4. The original estimated correlation matrix (Brace, Gatarek, and Musiela, 1997) is found to have a negative eigenvalue, so the matrix is an indefinite and Cholesky decomposition breaks down. The correlation matrix shown in Table 7.4 is an approximation to the original one and it is obtained through the rank reduction algorithm described in Section 7.4.1 with rank 10. When we visualize the correlation matrix through a surface plot, we see a surface that is almost identical to the one in Figure 6.4. The matrix A out of the Cholesky decomposition of $(\rho_{jk}(0))$ is presented in Table 7.5.

7.3 Calibration to Caps, Swaptions, and Input Correlations

In the fixed-income market, caps, floors, and ATM swaptions dominate liquidity, and hence they are considered to be benchmark derivatives. For hedging purposes, swaptions are used more often than are caps and floors. Hence, it is highly desirable to have a LIBOR model calibrated to swaptions, in addition to caps and floors. If we take a parametric approach for the calibration, we will have a trivial numerical problem for which we do not have much to say. All we need to do is to adopt the parameterization for $\|\gamma_j(t)\|$ like Equation 6.79, and solve for the parameters through a minimization procedure. In this section, we instead will take a nonparametric approach and avoid putting a structure on the solution, in hopes of gaining insights into the "objective" local volatility function.

Unlike conventional approaches of option calibrations that match prices, we instead take an alternative approach to match Black's volatilities. The calibration to ATM swaptions is described as follows. Let $\{\sigma_{m,n}\}$ be a set of implied Black's volatilities of benchmark swaptions. We look for $\{\gamma_j(t)\}$, which makes the Black's volatilities of the model equal to the implied Black's volatilities:

$$\sigma_{m,n}^2 = \frac{1}{T_m - t} \int_t^{T_m} \sum_{j,k=m+1}^{n} \omega_j \omega_k \gamma_j^{\mathrm{T}}(s) \gamma_k(s) \, \mathrm{d}s, \quad \text{for } (m,n) \in \Gamma.$$

(7.23)

Here, we use Γ to denote the index set of the input swaptions. Note that when $n = m + 1$, the swaption reduces to a caplet and Equation 7.23 reduces to Equation 7.12, so Equation 7.23 applies to both caplets and swaptions. When there are only caps (or equivalently, caplets) in the input set, the local volatility can be time-independent, corresponding to a volatility function that is flat in the direction of time, as was shown in Figure 7.2. If there are, in addition, swaptions in the input set, we will need to bend the volatility function or, in other words, make the local volatility function time-dependent as well.

Including the matching of input correlations, a comprehensive calibration problem for production use can be stated as follows: given the implied Black's volatilities, $\{\sigma_{m,n}\}$, of the ATM caplets and swaptions and a time-homogeneous correlation matrix, $\mathbf{C}(t) = (\rho_{j,k}(t))$, of the LIBOR rates, solve for $\{\gamma_j(t)\}$ from the joint equations

$$\begin{cases} \sigma_{m,n}^2 = \dfrac{1}{T_m - t} \displaystyle\int_t^{T_m} \sum_{j,k=m+1}^{n} \omega_j \omega_k \gamma_j^{\mathrm{T}}(s) \gamma_k(s) \, \mathrm{d}s, & \text{for } (m,n) \in \Gamma, \\[2ex] \rho_{j,k}(t) = \dfrac{\gamma_j^{\mathrm{T}}(t) \gamma_k(t)}{\left\|\gamma_j(t)\right\| \left\|\gamma_k(t)\right\|}, & \forall j, k. \end{cases}$$

(7.24)

We consider a nonparametric solution to Equation 7.24 and look for the volatilities in the form of a piece-wise constant function in t:

$$\gamma_j(t) = \gamma_j^i = s_{i,j}\left(a_{1,j}^{(i)}, a_{2,j}^{(i)}, \ldots, a_{n_i,j}^{(i)}\right)^{\mathrm{T}} = s_{i,j} \mathbf{a}_j^i, \quad \text{for } T_{i-1} \le t \le T_i, \ i \le j,$$

(7.25)

TABLE 7.5: Cholesky Decomposition of the Correlation Matrix in Table 7.4

1.0000	0.8415	0.6246	0.6231	0.5330	0.4287	0.3274	0.4463	0.2439	0.3326	0.2625
	0.5403	0.4900	0.4814	0.5247	0.5069	0.3269	0.3806	0.2567	0.3208	0.2689
		0.6081	0.6112	0.3631	0.3417	0.2931	0.2415	0.2705	0.2531	0.2129
			0.0807	0.1029	0.1072	−0.0259	0.0991	0.0685	0.0868	0.0837
				0.5461	0.6337	0.2158	0.2757	0.2325	0.2326	0.2017
					0.1716	0.0323	0.0362	0.0603	0.0147	−0.0012
						0.8073	0.2413	0.3895	0.4195	0.4016
							0.6725	0.0678	0.2066	0.2019
								0.7635	0.2366	0.1404
									0.6243	0.7378
										0.0000

where $n_i = N - i + 1$, and

$$s_{i,j} = \left\|\boldsymbol{\gamma}_j^i\right\| \quad \text{and} \quad \left\|\mathbf{a}_j^i\right\| = 1.$$

By casting $\boldsymbol{\gamma}_j(t)$ in the form of Equation 7.25, we will take full advantage of the decoupling between the fitting of prices and the fitting of correlations. Note that $\{s_{i,j}\}$ are subject only to input prices, whereas $\{\mathbf{a}_j^i\}$ are subject only to input correlations, and they can be solved separately. We call both $\{\|\boldsymbol{\gamma}_j(t)\|\}$ and $\{s_{i,j}\}$ the local volatility surface of the LIBOR market model.

The determination of $\mathbf{a}_j^i$ depends only on part of the input correlations

$$\mathbf{C}^{(i)} \triangleq \mathbf{C}(t) = (\rho_{j,k}(t)), \quad t \in (T_{i-1}, T_i), \quad j, k \geq \eta(t).$$

The procedure described in Section 7.2 also applies to the solution of $\mathbf{a}_j^i$. In practice, however, we can estimate $\mathbf{C}^{(1)}$ only, the spot correlations of the forward rates, from the time series data of forward rates. The inputs of $\mathbf{C}^{(i)}$, for $i > 1$, must be postulated. Based on financial and economical arguments, we can assume time homogeneity for future correlations, that is,

$$\rho_{j,k}(t) = \rho(T_j - t, T_k - t), \tag{7.26}$$

which is equivalent to taking

$$\mathbf{C}^{(i)} = \mathbf{C}_{N-i+1,N-i+1}, \tag{7.27}$$

where $\mathbf{C}_{n,n}$ stands for the first n-by-n principal submatrix of $\mathbf{C}$. As a matter of fact, the assumption of time homogeneity here fills the information gap.

There is, however, a major complication in the calibration of correlations. According to the procedure described in Section 7.2, $\mathbf{a}_j^i$ will have a length equal to the rank of $\mathbf{C}^{(i)}$. For a correlation matrix obtained from time series data of forward rates, $\mathbf{C}^{(i)}$ usually has full rank that is equal to the number of live forward rates, n_i. Yet, in many applications, we often take only a few driving factors, as was suggested by the PCA on the time series data of forward rates, which can be much smaller than n_i. To utilize the historical information, we should replace $\mathbf{C}^{(i)}$ by a lower rank approximation, $\hat{\mathbf{C}}^{(i)}$. Mathematically, $\hat{\mathbf{C}}^{(i)}$ can be chosen to be the solution that solves the following constrained minimization problem:

$$\min_{\hat{\mathbf{C}}^{(i)}} \left\|\mathbf{C}^{(i)} - \hat{\mathbf{C}}^{(i)}\right\|_F,$$
$$\text{s.t. } \hat{\mathbf{C}}^{(i)} \geq \mathbf{0}, \ \text{rank}(\hat{\mathbf{C}}^{(i)}) \leq \hat{n}_i, \ \hat{C}_{j,j}^{(i)} = 1, \quad \forall j, \tag{7.28}$$

where the subindex F represents Frobeniu's norm of matrices, $\hat{\mathbf{C}}^{(i)} \geq \mathbf{0}$ means that $\hat{\mathbf{C}}^{(i)}$ is a non-negative matrix, and $\hat{n}_i \leq n_i$. We will consider the solution of problem 7.28 in Section 7.4.1. Note that in the determination of $\{s_{i,j}\}$, $\{\hat{\mathbf{C}}^{(i)}\}$ will be part of the input.

We now formulate a mathematical problem for the local volatility surface, $\{s_{i,j}\}$, in terms of which the first equation of Equation 7.24 becomes

$$
\begin{aligned}
\sigma^2_{m,n} &= \frac{1}{T_m - t} \sum_{j,k=m,n-1} w_j w_k \sum_{i=1}^{m} s_{i,j} s_{i,k} \hat{C}^i_{j,k} \Delta T_{i-1} \\
&= \frac{1}{T_m - t} \sum_{i=1}^{m} \Delta T_{i-1} \sum_{j,k=m,n-1} s_{i,j} s_{i,k} \left(w_j w_k \hat{C}^i_{j,k} \right), \quad \text{for } (m,n) \in \Gamma.
\end{aligned}
$$

$$(7.29)$$

Note that, in Equation 7.29, $\hat{C}^{(i)}_{j,k}$ is an element of $\hat{\mathbf{C}}^{(i)}$, the low-rank approximation to correlation matrix $\mathbf{C}^{(i)}$. The total number of unknowns, $\{s_{i,j}\}$, is $N(N+1)/2$, which is typically much higher than the number of input prices. Hence, Equation 7.29 alone is underdetermined and solutions are not unique. For uniqueness and, in addition, smoothness of the solution as a function of i and j, we adopt the following regularization to the solutions of Equation 7.29,

$$
\min_s \left\| \nabla s \right\|^2 + \varepsilon \left\| s - s^{(0)} \right\|^2 = \min_s - (s, (\nabla \cdot \nabla)s) + \varepsilon \left\| s - s^{(0)} \right\|^2
$$
for some $\varepsilon > 0$, $$(7.30)$$

where $(\nabla \cdot \nabla)$ stands for the so-called discrete Laplacian operator,

$$
(\nabla \cdot \nabla)s_{i,j} = -s_{i,j-1} - s_{i,j+1} - s_{i-1,j} - s_{i+1,j} + 4s_{i,j}, \quad (7.31)
$$

and $s^{(0)}$ is a prior volatility function that may be, for instance, the local volatility function of the previous day. The objective function regulates both uniqueness and smoothness of the solution. In detail, the right-hand side of Equation 7.30 reads

$$
\begin{aligned}
&- \sum_{i=1}^{N} \sum_{j=i}^{N} s_{i,j} \left(s_{i,j-1} + s_{i,j+1} + s_{i-1,j} + s_{i+1,j} - 4s_{i,j} \right) \\
&+ \varepsilon \sum_{i=1}^{N} \sum_{j=i}^{N} \left(s_{i,j} - s^{(0)}_{i,j} \right)^2 .
\end{aligned}
$$

$$(7.32)$$

Notice that, in Equation 7.32, there are some "ghost" variables whose sub- or sup-indices go beyond the designated range, from 1 to N. These "ghost" variables can be eliminated by using "Neumann's boundary condition"[1]:

$$
\begin{aligned}
s_{0,j} &= s_{1,j}, \quad j = 1, \ldots, N, \\
s_{i,N+1} &= s_{i,N}, \\
s_{i+1,i} &= s_{i,i}, \quad i = 1, \ldots, N, \\
s_{i,i-1} &= s_{i,i}.
\end{aligned}
$$

---

[1]The Neumann's boundary condition corresponds to zero normal derivatives at the boundary. The use of the Neumann's boundary condition here may cause multiple values for the "ghost" variables. This is, however, harmless.

By minimizing the function in Equation 7.30 subject to the equality constraint of Equation 7.29, we expect to obtain a solution that is both unique and smooth. There is a distinct feature in this constrained minimization problem: both the objective function and constraints are quadratic in $\{s_{i,j}\}$. We take advantage of such a feature in Section 7.4.2 when we study the solution of Equations 7.29 and 7.30.

For an efficient solution of Equations 7.29 and 7.30, we will use matrix operations. For this purpose, we line up the volatilities in a one-dimensional array,

$$X = (\mathbf{s}_1, \mathbf{s}_2, \ldots, \mathbf{s}_N)^{\mathrm{T}},$$

where

$$\mathbf{s}_i = (s_{i,i}, s_{i,j}, \ldots, s_{i,N}).$$

We then define the matrix corresponding to the discrete Laplacian operator,[2] Equation 7.31, by

$$B = \mathrm{diag}(-1, -1, 4, -1, -1).$$

Finally, we associate each instrument with a "weight matrix,"

$$W_{m,n} = \mathrm{diag}(0, \ldots, 0, w_m, \ldots, w_{n-1}, 0, \ldots, 0),$$

and a "covariance matrix,"

$$G_{m,n} = (T_m - t)^{-1}\mathrm{diag}(\Delta T_{\eta(t)}W_{m,n}\hat{C}^{(1)}W_{m,n}, \Delta T_{\eta(t)+1}W_{m,n}\hat{C}^{(2)}$$
$$\times W_{m,n}, \ldots, \Delta T_{m-1}W_{m,n}\hat{C}^{(m)}W_{m,n}, 0, \ldots, 0).$$

Then, in terms of these matrix notations, we can cast the solution of $\{s_{i,j}\}$ in a rather neat form:

$$\min_x X^{\mathrm{T}}BX + \varepsilon(X - X_0)^{\mathrm{T}}(X - X_0), \tag{7.33}$$
$$\text{s.t. } X^{\mathrm{T}}G_{m,n}X = \sigma_{m,n}^2, \quad \text{for } (m, n) \in \Gamma.$$

The objective function in problem 7.33 can be simplified even further. Expanding the objective function, there will be

$$X^{\mathrm{T}}BX + \varepsilon(X - X_0)^{\mathrm{T}}(X - X_0)$$
$$= (X - \varepsilon(B + \varepsilon I)^{-1}X_0)^{\mathrm{T}}(B + \varepsilon I)(X - \varepsilon(B + \varepsilon I)^{-1}X_0)$$
$$+ \varepsilon X_0^{\mathrm{T}}(I - \varepsilon(B + \varepsilon I)^{-1})X_0,$$

where the last term is a constant and thus can be ignored during optimization. Denote

$$A = B + \varepsilon I,$$

---

[2]The discrete Laplacian operator is a difference operator, also called a five-point stencil, that approximates the second-order differential operator.

which is a positive-definite matrix, and

$$\tilde{X}_0 = \varepsilon(B + \varepsilon I)^{-1} X_0. \tag{7.34}$$

We end up with the following formulation for price calibration:

$$\min_X \left(X - \tilde{X}_0\right)^{\mathrm{T}} A \left(X - \tilde{X}_0\right), \tag{7.35}$$
$$\text{s.t. } X^{\mathrm{T}} G_{m,n} X = \sigma_{m,n}^2, \quad \text{for } (m, n) \in \Lambda.$$

We reiterate here that both the objective function and constraints are quadratic functions, and we consider problem 7.35 an unconventional quadratic programming problem.[3] We discuss an innovative methodology in Section 7.4.2 to solve this problem.

7.4   Calibration Methodologies

In this section, we develop numerical methods to solve the constrained minimization problems 7.28 and 7.35 separately. The methodology to be used is a combination of the Lagrange multiplier and the steepest descent methods. In developing the numerical methods, we have taken full advantage of the quadratic form of the objective function and constraints. We rigorously justify the usefulness of the Lagrange multiplier problems and demonstrate the performance of the methods with market data.

7.4.1   Rank-Reduction Algorithm

To fit a LIBOR market model optimally to input correlations, we must first find the low-rank approximations to the input correlation matrices, as was explained in the last section. This is formulated as a constrained minimization problem 7.28, which is a special case of the so-called rank reduction problem for matrices. A somewhat more general rank-reduction problem can be formulated as follows. For a symmetric matrix, $C \in R^{N \times N}$, where $R^{N \times N}$ represents the collection of N-by-N real matrices, the rank-n approximation is defined as the solution to the following problem:

$$\min_X \|C - X\|_F, \tag{7.36}$$
$$\text{s.t. } \mathrm{rank}(X) \leq n < N, \quad \mathrm{diag}(X) = \mathrm{diag}(C).$$

We denote any solution to problem 7.36 as C^*. Note that problem 7.36 does not impose the explicit condition of the non-negativity of a solution. This,

---

[3]A conventional quadratic programming problem has linear constraints.

however, is not a concern, because we can prove that C^* will be automatically a non-negative matrix provided that C is one. The proof is beyond the scope of this book and we refer to Zhang and Wu (2003) for details. For later use, we denote the feasible set of problem 7.36 as

$$\mathcal{H} = \left\{ X \in R^{N \times N} \,\middle|\, \text{rank}(X) \leq n, \quad \text{diag}(X) = \text{diag}(C) \right\}.$$

Following the general approach of Lagrange methods, we transform the above constrained minimization problem into an equivalently min-max problem. Let $\mathcal{R}_n$ be the subset of $R^{N \times N}$ for matrices with ranks less or equal to n. The Lagrange multiplier problem corresponding to problem 7.36 is

$$\min_{d} \max_{X \in \mathcal{R}_n} L(X, d), \tag{7.37}$$

where $L(X, d)$ is called the Lagrange function, defined by

$$L(X, d) = -\|C - X\|_F^2 - 2d^{\mathrm{T}} \text{diag}(C - X), \tag{7.38}$$

and d is the vector of the multipliers. For later use, we emphasize here that $L(X, d)$ is linear in d in the following sense:

$$L(X, td + (1 - t)\hat{d}) = tL(X, d) + (1 - t)L(X, \hat{d}). \tag{7.39}$$

We justify later that the min-max problem 7.37 is equivalent to the original problem 7.36.

In numerical implementations, the min-max problem 7.37 is solved as a minimization problem of the form

$$\min_{d} V(d), \tag{7.40}$$

with the objective function defined by

$$V(d) = \max_{X \in \mathcal{R}_n} L(X, d). \tag{7.41}$$

Hence, it is a matter of finding efficient methods for the maximization problem 7.41 and minimization problem 7.40, respectively.

For solving the maximization problem 7.41, it is crucial to notice that the Lagrange function can be rewritten as

$$L(X, d) = -\|C + D - X\|_F^2 + \|d\|^2, \tag{7.42}$$

where D is the diagonalized matrix for the vector d : $D = \text{diag}(d)$. For a given d, the maximizer to problem 7.41 can be obtained by an eigenvalue decomposition of $C + D$, a symmetric matrix. Let

$$C + D = U \Lambda U^{\mathrm{T}} \tag{7.43}$$

be the eigenvalue decomposition with orthogonormal matrix U and eigenvalue matrix

$$\Lambda = \text{diag}(\lambda_1, \lambda_2, \ldots, \lambda_N).$$

The eigenvalues are, in particular, put in decreasing order in absolute values,

$$|\lambda_1| \geq |\lambda_2| \geq \cdots \geq |\lambda_N|.$$

Then, the solutions to problem 7.41, the best rank-n approximations of $C+D$, are obviously given by

$$C(d) \overset{\Delta}{=} C_n(d) = U_n \Lambda_n U_n^{\mathrm{T}}, \tag{7.44}$$

where U_n is the matrix consisting of the first n columns of U, and $\Lambda_n = \text{diag}(\lambda_1, \ldots, \lambda_n)$ the principal submatrix of Λ of degree n. Consequently, we have the following solution to problem 7.40:

$$V(d) = - \sum_{j=n+1}^{N} \lambda_j^2 + \|d\|^2. \tag{7.45}$$

When $|\lambda_n| > |\lambda_{n+1}|$, the solution to problem 7.40 is unique. In the case of $|\lambda_n| = |\lambda_{n+1}|$, the solutions become non-unique. The complications for the case of $|\lambda_n| = |\lambda_{n+1}|$ were studied by Zhang and Wu (2003), and the analysis is quite demanding in matrix algebra. Hence, we restrict ourselves to the case of $|\lambda_n| > |\lambda_{n+1}|$ here. Note that both U and Λ depend on the vector of the multiplier, d; thus, whenever necessary, we will also write them as $U(d)$ and $\Lambda(d)$.

Before addressing the solutions of the minimization problem 7.40, we perform some analysis on the min-max problem, problems 7.40 and 7.41, which will shed light on the numerical solution of problem 7.40. We start from the existence of the solutions.

Theorem 7.4.1. *There exists at least one solution to problems 7.40 and 7.41, and any local minimum to problems 7.40 and 7.41 is a global minimum.*

Proof. We use the method of contradiction to prove the existence of the solution. Suppose that there was no solution to problems 7.40 and 7.41, then there would exist a sequence of $d^{(j)} = \{d_i^{(j)}\}$, such that $\{d_i^{(j)}\} \to \infty$ while $V(d^{(j)})$ decreases. Write $D^{(j)} = \text{diag}(d^{(j)})$ as a direct sum of two diagonal matrices, $D^{(j)} = D_1^{(j)} + D_2^{(j)}$, with $\text{rank}(D_1^{(j)}) \leq n$ and $\left\|D_1^{(j)}\right\|_\infty = \left\|D^{(j)}\right\|_\infty \to +\infty$, where the infinity norm stands for the maximum absolute value of a matrix. Since

$$V(d^{(j)}) = \max_{X \in \mathcal{R}_n} L(X, d^{(j)}) \geq L(D_1^{(j)}, d^{(j)}),$$

we have, using form Equation 7.42 for $L(X, d)$,

$$V(d^{(j)}) \geq - \left\| C + D_2^{(j)} \right\|_F^2 + \left\| D^{(j)} \right\|_F^2$$

$$= - \|C\|_F^2 - 2tr(CD_2^{(j)}) + \left\| D_1^{(j)} \right\|_F^2$$

$$\geq - \|C\|_F^2 + \left(\left\| D^{(j)} \right\|_\infty - 2tr(|C|) \right) \left\| D^{(j)} \right\|_\infty \rightarrow +\infty.$$

Here $|C|$ denotes the matrix whose entries are the absolute values of those of C, and $tr(|C|)$ means the summation of the diagonal entries of $|C|$. The divergence of $V(d^{(j)})$ contradicts the assumption of a decreasing $V(d^{(j)})$. Hence, the existence of the solution(s) follows.

The property that any local minimum must be at the same time a global minimum is due to the convexity property of $V(d)$. To see this, we consider any two points, $d^{(1)}$ and $d^{(2)}$. By the linearity of the Lagrange function, we have, for any $t \in (0, 1)$,

$$V(td^{(1)} + (1 - t)d^{(2)})$$
$$= \max_X - \|C - X\|_F^2 - 2(td^{(1)} + (1 - t)d^{(2)})^T \operatorname{diag}(C - X)$$
$$= \max_X t \left(- \|C - X\|_F^2 - 2(d^{(1)})^T \operatorname{diag}(C - X) \right)$$
$$+ (1 - t) \left(- \|C - X\|_F^2 - 2(d^{(2)})^T \operatorname{diag}(C - X) \right)$$
$$\leq t \max_X \left(- \|C - X\|_F^2 - 2(d^{(1)})^T \operatorname{diag}(C - X) \right)$$
$$+ (1 - t) \max_X \left(- \|C - X\|_F^2 - 2(d^{(2)})^T \operatorname{diag}(C - X) \right)$$
$$= tV(d^{(1)}) + (1 - t)V(d^{(2)}).$$

Suppose that there are two local minimums, d^* and d^{**}, such that $V(d^*) > V(d^{**})$. Let $d(t) = td^* + (1 - t)d^{**}$ for any $t \in (0, 1)$. Then, by the convexity of V, there is

$$V(d(t)) = V(td^* + (1 - t)d^{**})$$
$$\leq tV(d^*) + (1 - t)V(d^{**}) \qquad (7.46)$$
$$< tV(d^*) + (1 - t)V(d^*) = V(d^*).$$

When t approaches 1, $d(t)$ approaches d^*, while $V(d(t))$ approaches $V(d^*)$ from below. This is in contradiction to the assumption that d^* is a local minimum. Hence, we arrive at the second conclusion of the theorem. $\square$

For the analytical properties of various functions involved in the solution of problem 7.41, we have

Theorem 7.4.2. *When $|\lambda_n(C + D)| > |\lambda_{n+1}(C + D)|$, we have the following conclusions.*

1. *The optimal solution $C(d)$ to problem 7.41 is unique and is differentiable in d.*

2. *$V(d)$ is second-order continuously differentiable in a neighborhood of d.*

3. *If $\nabla V(d) = 0$, then d must be a global minimizer of $V(d)$.*

Proof. The uniqueness is obtained by construction, as all solutions must be in the form of Equation 7.44, where Λ_n is unique when $|\lambda_n(C+D)| > |\lambda_{n+1}(C+D)|$, as is $C(d)$. Since C is symmetric and D is only a diagonal matrix, Gerschgorin's theorem in linear algebra (see e.g., Stewart and Sun, 1990) implies that all eigenvalues (which are not necessarily distinct from each other) and eigenvectors are differentiable in a neighborhood of d. The differentiability of $V(d)$ and $C(d)$ then follows and their partial derivatives are related by the chain rule:

$$\frac{\partial V(d)}{\partial d_k} = \sum_{i,j} \frac{\partial L(C(d), d)}{\partial X_{ij}} \frac{\partial X_{ij}}{\partial d_k} + \frac{\partial L(C(d), d)}{\partial d_k}, \quad 1 \le k \le N.$$

For an optimal and fixed d, the optimality implies that

$$\frac{\partial L(C(d), d)}{\partial X_{ij}} = 0, \quad \text{for all } i \text{ and } j. \tag{7.47}$$

Consequently, we have

$$\frac{\partial V(d)}{\partial d_k} = \frac{\partial L(C(d), d)}{\partial d_k} = -2(C_{kk} - C_{kk}(d)). \tag{7.48}$$

Differentiating Equation 7.48 yet again with respect to d_l yields

$$\frac{\partial^2 V(d)}{\partial d_k \partial d_l} = 2\frac{\partial C_{kk}(d)}{\partial d_l}, \quad \forall k \text{ and } l,$$

whose continuity follows from that of $\partial C_{kk}(d)/\partial d_l$.

Next, we show that any critical points must be the global minimum. In fact, if $\nabla V(d) = 0$ for some d that is not a global minimum, then we must have another point, $\hat{d}$, such that $V(\hat{d}) < V(d)$. From the convexity property of V, we have

$$V(t\hat{d} + (1-t)d) \le tV(\hat{d}) + (1-t)V(d), \quad \text{for any } t \in (0, 1).$$

The above equation can be rewritten as

$$\frac{V(t\hat{d} + (1-t)d) - V(d)}{t} \le V(\hat{d}) - V(d) < 0.$$

Letting $t \to 0$, we would then arrive at

$$\frac{(\hat{d} - d)}{\|\hat{d} - d\|} \cdot \nabla_d V(d) \le V(\hat{d}) - V(d) < 0, \tag{7.49}$$

which contradicts the condition of $\nabla_d V(d) = 0$ for a critical point d. The lemma is thus proved. $\square$

Based on the existence of a minimizer and the differentiability of the value function, we can establish the equivalence between the constrained minimization problem, 7.36, and the Lagrange multiplier problem, 7.37.

Theorem 7.4.3. *Let d^* be any minimizer of the Lagrange multiplier problem, 7.37. If $|\lambda_n(C+D^*)| > |\lambda_{n+1}(C+D^*)|$ and $\mathrm{diag}(C) > 0$, then d^* is the unique minimizer and $C(d^*)$ solves the constrained minimization problem, 7.36.*

Proof. Assume to the contrary that there exists $d^{**} \neq d^*$ such that $V(d^{**}) = V(d^*)$. Denote $d(t) = d^* + t(d^{**} - d^*)$. The convexity property of $V(d)$ yields $V(d(t)) = V(d^*)$ for all $t \in [0,1]$. Denote $C(t)$ as $C(d(t))$ for simplicity. Owing to the linearity of $L(X,d)$ in d, we have, for any $t \in [0,1]$,

$$
\begin{aligned}
V(d) = V(d(t)) &= L(C(t), d(t)) \\
&= (1-t)L(C(t), d^*) + tL(C(t), d^{**}) \\
&\leq (1-t)\max_{X \in \mathcal{H}} L(X, d^*) + t\max_{X \in \mathcal{H}} L(X, d^{**}) \\
&= (1-t)V(d^*) + tV(d^{**}) = V(d^*).
\end{aligned}
$$

From the above equalities, we obtain $L(C(t), d^*) = V(d^*)$, or

$$
\|C + D^* - C(t)\|_F = \min_{X \in \mathcal{R}_n} \|C + D^* - X\|_F .
$$

It follows that, by the uniqueness of $C(d^*)$,

$$
C(t) = C(0)
$$

is the optimal solution corresponding to $d(t)$, $t \in [0,1]$. Let $D(t) = \mathrm{diag}(d(t))$. The eigen-decomposition of matrix, $C + D(t)$, is

$$
C + D(t) = U(t)\Lambda(t)U(t)^{\mathrm{T}} = C(0) + E(t),
$$

where

$$
\begin{aligned}
C(0) &= U_n(0)\Lambda_n(0)U_n^{\mathrm{T}}(0), \\
E(t) &= G_n(t)\Theta_n(t)G_n^{\mathrm{T}}(t),
\end{aligned}
\tag{7.50}
$$

and $G_n(t)$ consists of the last $N - n$ columns of $U(t)$. Since the columns of $G_n(0)$ and $G_n(t)$ form two orthogonal bases of the null space of $C(0)$, there must exist an $(N - n)$-by-$(N - n)$ orthogonal matrix $W(t)$ such that $G_n(t) = G_n(0)W(t)$. Therefore,

$$
E(t) = G_n(0)\left(W(t)\Theta_n(t)W^{\mathrm{T}}(t)\right)G_n^{\mathrm{T}}(0).
$$

By substituting the above expression into the equality

$$
E(t) - E(0) = t(D - D^*) \quad \text{for } t \geq 0,
$$

we then have

$$D - D^* = t^{-1} G_n(0)(W(t)\Theta_n(t)W^{\mathrm{T}}(t) - \Theta_n(0))G_n^{\mathrm{T}}(0).$$

After post-multiplying $U_n(0)$ to the above equation and recalling the orthogonality between $G_n(0)$ and $U_n(0)$, we arrive at

$$(D - D^*)U_n(0) = 0.$$

If any of the diagonal elements of $D - D^*$ is not zero, then the corresponding rows $U_n(0)$ must be a zero row, which, according to the first equation of Equation 7.50, implies that the corresponding row of $C(0)$ is zero as well. But this is impossible because $\mathrm{diag}(C(0)) = \mathrm{diag}(C) > 0$. Hence, there cannot be more than one minimizer.

If d^* solves the min-max problem 7.37, then, due to the differentiability of $V(d)$, d^* must be a critical point of $V(d)$ and its gradient, according to Theorem 7.2, must vanish:

$$0 = \frac{\partial V(d)}{\partial d_k} = -2(C_{kk} - C_{kk}(d)), \quad 1 \le k \le N.$$

Hence, for any other matrix, $\tilde{C} \in \mathcal{H}$, we have

$$\begin{aligned} V(d^*) &= -\|C - C(d^*)\|_F^2 \\ &= \max_{X \in \mathcal{H}} -\|C - X\|_F^2 \\ &\ge -\left\|C - \tilde{C}\right\|_F^2, \end{aligned}$$

meaning that $C(d^*)$ is a solution to the constrained minimization problem 7.36. □

We have shown that the inner maximization problem 7.41 can be solved nicely by a single eigenvalue decomposition. The outer minimization problem 7.40, instead, will be solved by the method of steepest descent, which usually works well for convex objective functions. The algorithm is described below.

Algorithm. Take $D^{(0)}$ to be a null matrix and repeat the following steps:

1. compute the decomposition: $C + D^{(k)} = U^{(k)}\Lambda^{(k)}\left(U^{(k)}\right)^{\mathrm{T}}$; set $\alpha^{(k)} = 1$ and $\nabla V(d^{(k)}) = -2\,\mathrm{diag}\left(C - U_n^{(k)}\Lambda_n^{(k)}(U_n^{(k)})^{\mathrm{T}}\right)$;

2. define $d^{(k+1)} = d^{(k)} - \alpha^{(k)}\nabla V(d^{(k)})$;

3. if $V(d^{(k+1)}) > V(d^{(k)}) - \left(\alpha^{(k)}/2\right)\left\|\nabla V(d^{(k)})\right\|^2$, take $\alpha^{(k)} \triangleq \alpha^{(k)}/2$, go back to step 2;

4. if $\left\|d^{(k+1)} - d^{(k)}\right\|_2 > \mathrm{tol.}$, go back to step 1; and

5. take $d^* = d^{(k+1)}$ and $C^* = U_n^{(k)} \Lambda_n^{(k)} (U_n^{(k)})^{\mathrm{T}}$.

We have the following results on the convergence of the above descending search algorithm.

Theorem 7.4.4. *The sequence $\{d^{(k)}\}$ is bounded and hence has accumulation points. Let d^* be an accumulation point such that $|\lambda_n(C + D^*)| > |\lambda_{n+1}(C + D^*)|$. Then d^* is the unique global minimizer, and $\{d^{(k)}\}$ converges to d^*.*

Proof. The boundedness of $\{d^{(k)}\}$ follows from the monotonic decreasing of function $V(d^{(k)})$. If the boundedness is not true, then there must be a subset K_0 of the index such that $\|d^{(k)}\| \to +\infty$ for $k \in K_0$. Repeating the relevant arguments in Theorem 7.3, we would again end up in a contradiction that is against the decreasing property of $V(d^{(k)})$. Hence, $\{d^{(k)}\}$ must be bounded. The existence of the accumulation points follows.

Let d^* be an accumulation point such that $d^{(k)} \to d^*$ for ks in some subset of K_0. Under the condition $|\lambda_n(C + D^*)| > |\lambda_{n+1}(C + D^*)|$, we claim that d^* must be a critical point of $V(d)$ such that $\nabla_d V(d^*) = 0$. Suppose that this is not the case. Then, according to the line search and the continuity of $\nabla V(d)$ around d^*, there must be

$$\lim_{k \in K_0, k \to \infty} \alpha^{(k)} = 0,$$

and

$$V(d^{(k)} - \hat{\alpha}^{(k)} \nabla V(d^{(k)})) > V(d^{(k)}) - \frac{\hat{\alpha}^{(k)}}{2} \left\| \nabla V(d^{(k)}) \right\|^2. \tag{7.51}$$

Here $\hat{\alpha}^{(k)} = 2\alpha^{(k)}$. Note that there is

$$\lim_{\hat{\alpha}^{(k)} \to 0} \frac{V(d^* - \hat{\alpha}^{(k)} \nabla_d V(d^*)) - V(d^*)}{\hat{\alpha}^{(k)}} = -\|\nabla_d V(d^*)\|.$$

Based on the continuity of the second-order derivatives, we have, for sufficiently large $k \in K_0$,

$$\frac{V(d^{(k)} - \hat{\alpha}^{(k)} \nabla_d V(d^{(k)})) - V(d^{(k)})}{\hat{\alpha}^{(k)}} = -\left\| \nabla_d V(d^{(k)}) \right\|^2 + O((\hat{\alpha}^{(k)})^2)$$
$$\leq -\frac{1}{2} \left\| \nabla_d V(d^{(k)}) \right\|^2. \tag{7.52}$$

Clearly, Equation 7.52 is contradictory to Equation 7.51. Hence, there must be $\nabla V(d^*) = 0$. According to Theorem 7.3, d^* is a global minimizer. $\square$

We now try our algorithm with the historical correlation matrix of GBP, shown in Figure 6.4 as well as in Table 7.4. We calculate (a) rank-one, (b) -two, (c) -three, (d) -four, (e) -six, and (f) -ten approximations, and visualize the approximate correlation matrices, also with surface plots, in Figure 7.3. The surface plots begin, without surprise, with a flat surface for rank-one

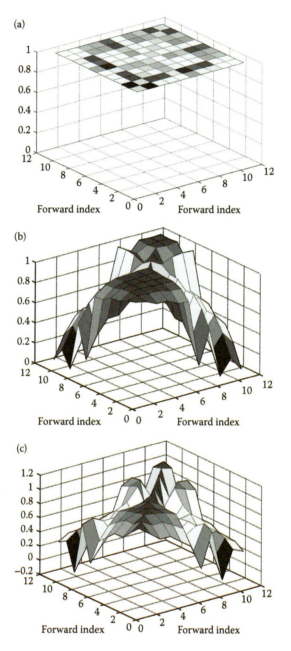

FIGURE 7.3: Lower-rank approximations of the correlation matrix. (Reprinted from Wu L, 2003. *Journal of Computational Finance* 6(2): 39–77. With permission.)

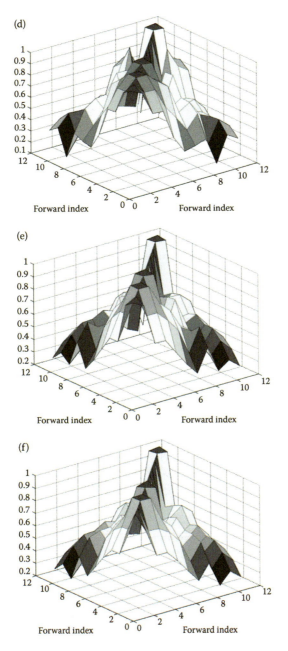

FIGURE 7.3: Continued.

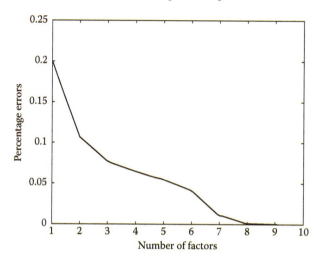

FIGURE 7.4: Convergence trend with increasing rank. (Reprinted from Wu L, 2003. *Journal of Computational Finance* 6(2): 39–77. With permission.)

approximation. The quality of the approximation improves with the increasing rank. When the rank reaches six, we see hardly any difference between the approximate correlation surface and the historical correlation surface, in Figure 6.4, which suggests that the forward-rate curve is driven by no more than six factors. The convergence is confirmed by the decreasing percentage error in the Frobenius norm, as is shown in Figure 7.4. For each number of factors, the calibration requires less than seven functional valuations (of $V(d)$). This is a very small-scale application for the powerful algorithm developed in this section, which can be applied to large-scale problems.

Next, we examine the principal components of the approximate correlation matrices. We take, for example, the rank-four approximation and compare its first three principal components (marked by Δ) with those (marked by $\circ$) of the original correlation matrix. As we can see in Figure 7.5, the first three principal components are very close to their counterparts of the original matrix. When the rank increases, we witness convergence.

We conclude this section with a remark. The above method can be easily extended to Frobenius norms with "weights." In some applications, the correlations between some forward rates are considered to be more important than the rest of the correlations, so we may want to place proper emphasis on these correlations in a calibration procedure. One way to achieve this is to adopt a Frobenius norm with "weights":

$$\|A\|_{W,F}^2 \triangleq \left\|\sqrt{W}A\sqrt{W}\right\|_F^2,$$

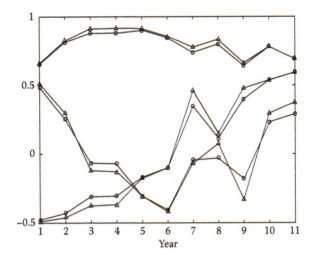

FIGURE 7.5: The first three principal components of the original correlation matrix and its rank-four approximation. (Reprinted from Wu L, 2003. *Journal of Computational Finance* 6(2): 39–77. With permission.)

where W is a diagonal matrix,

$$W = \text{diag}(w_1, \ldots, w_N),$$

with positive diagonal entries. For example, if we think the correlations among the first i_0 forward rates are more important than the rest of the correlations, we can take $w_i = 1$ for $i \le i_0$ and $w_i < 1$ for $i > i_0$. In computations, we only need to substitute C in the aforementioned algorithm by $\sqrt{W}C\sqrt{W}$.

7.4.2 The Eigenvalue Problem for Calibrating to Input Prices

The problem of fitting to the implied Black's volatilities is preliminarily formulated in the form of problem 7.35, where both the objective function and the constraints are in quadratic forms. If we proceed with the method of Lagrange multipliers to solve problem 7.35, we will see that the Lagrange function may not have a finite minimum or maximum, so that the algorithm can break down. The remedy, however, is easy. All we need to alter is to make the objective function dominate the constraints. This can be achieved by superimposing a convex function on the original objective function:

$$\min_X F\left((X - \tilde{X}_0)^{\mathrm{T}} A (X - \tilde{X}_0)\right),$$
$$\text{s.t. } X^{\mathrm{T}} G_{m,n} X = \sigma_{m,n}^2, \quad \text{for } (m, n) \in \Gamma, \tag{7.53}$$

where $F(x)$ must be a superlinear and monotonically increasing function for $x \geq 0$, which is also a convex function. Owing to the convexity of $F(x)$, problem 7.53 shares the same constrained minimum(s) with problem 7.35. For efficient numerical solutions, we also require $F(x)$ to be differentiable up to the second order. For such an $F(x)$, problem 7.53 can be solved efficiently by using Lagrange multipliers. Examples of $F(x)$ are handy, including $F(x) = x^2$, $x \ln x$ or e^x. It is not difficult to argue that the problems of problem 7.53 with any of these $F(x)$ share the same solution(s). Judging on computational efficiency, we will take $F(x) = x^2$.

The corresponding Lagrange multiplier problem is

$$\min_{d} \max_{X} L(X, d), \tag{7.54}$$

where

$$L(X, d) = -\left((X - X_0)^{\mathrm{T}} A (X - X_0)\right)^2 + 2 \sum_{i=1}^{N} d_i (X^{\mathrm{T}} G_i X - h_i). \tag{7.55}$$

Note that we have used $\{G_i, h_i\}$ in place of $\{G_{m,n}, \sigma^2_{m,n}\}$ and have dropped "~" over X_0 for notational simplicity. For numerical solutions of problem 7.54, we must be able to evaluate the inner maximization problem before solving the outer minimization problem. Yet again, we denote the value function for the outer minimization problem by

$$V(d) = \max_{X} L(X, d), \tag{7.56}$$

and its feasible set of solutions by

$$\mathcal{H} = \{X \mid X^{\mathrm{T}} G_i X = h_i, \quad i = 1, \dots, N\}.$$

By virtue of the positive-definiteness of matrix A, we understand that the maximizer of $L(X, d)$ is finite for any fixed d, so $V(d)$ exists for all d.

We now consider the solution of problem 7.56. For a fixed d, a necessary condition for X to be a maximizer is being a critical point of the function where the gradient either vanishes or does not exist. Because the Lagrange function, $L(X, d)$, is smooth everywhere, at any critical point, the gradient must vanish, yielding the so-called first-order condition

$$\left((X - X_0)^{\mathrm{T}} A (X - X_0)\right) A (X - X_0) = \left(\sum d_i G_i\right) X. \tag{7.57}$$

Denote, for simplicity, $B_d = \left(\sum d_i G_i\right)$ and $Y = X - X_0$. Equation 7.57 then becomes

$$\left[(Y^{\mathrm{T}} A Y) A - B_d\right] Y = B_d X_0. \tag{7.58}$$

The solutions to Equation 7.58, Y, may yield the maximizers to problem 7.56 through $X = Y + X_0$. The nonlinear Equation 7.58 can be solved through an eigenvalue decomposition. For a given d, we can solve the following general eigenvalue problem,

$$B_d \mathbf{u}_i = \lambda_i A \mathbf{u}_i, \qquad i = 1, \ldots, N. \tag{7.59}$$
$$\mathbf{u}_i^T A \mathbf{u}_i = 1,$$

We call $(\lambda_i, \mathbf{u}_i)$ an eigen-pair of (B_d, A). Because both B_d and A are symmetric, and A is positive definite, all eigenvalues and eigenvectors are real. We put λ_is in descending order, $\lambda_i \geq \lambda_{i+1}$, $\forall i$, and denote by U the A-orthogonormal eigenvector matrix:

$$U = (\mathbf{u}_1, \mathbf{u}_2, \ldots, \mathbf{u}_N).$$

To solve Equation 7.58 for Y, we let $\alpha = Y^T A Y$ and pre-multiply Equation 7.58 by U^T, thus transforming the equation to

$$[\alpha I - \Lambda] U^{-1} Y = U^T B_d X_0, \tag{7.60}$$

where

$$\Lambda = \text{diag}(\lambda_1, \ldots, \lambda_n)$$

is the matrix of the eigenvalues. From Equation 7.60, we can write the solution to Equation 7.58 as

$$Y = Y(\alpha) = U [\alpha I - \Lambda]^{-1} U^T B_d X_0,$$

where the scalar α is subject to the nonlinear equation

$$Y(\alpha) A Y^T(\alpha) = \alpha.$$

The last equation for α can be solved in a few iterative steps.

The above procedure for the solution of problem 7.56, however, does not apply to the case of $X_0 = 0$. For this particular case, Equation 7.58 reduces to an eigenvalue problem and thus admits multiple solutions, which are

$$Y_i = \sqrt{\max(\lambda_i, 0)} \, \mathbf{u}_i, \qquad i = 1, \ldots, N. \tag{7.61}$$

We will focus on the particular case, $X_0 = 0$, for the following reasons. $X_0 = 0$ can be associated with the situation that we have input no prior volatility surface. The solution to problem 7.54 will thus offer an "objective" volatility surface, which can be compared to the volatility surface estimated using time series data. The analysis done for the case of $X_0 = 0$ can also be carried over to the case $X_0 \neq 0$ accordingly.

Now consider the case $X_0 = 0$. Suppose that there is at least $\lambda_1 > 0$. In view of Equations 7.55, 7.58, and 7.61, the value function of problem 7.56 is

$$\begin{aligned}
V(d) &= \max_{j, \lambda_j \geq 0} - \left(Y_j^T A Y_j\right)^2 + 2 \sum_{i=1}^{N} d_i (Y_j^T G_i Y_j - h_i) \\
&= \max_{j, \lambda_j \geq 0} \lambda_j^2 - 2 \sum_{i=1}^{N} d_i h_i \\
&= \lambda_1^2 - 2 \sum_{i=1}^{N} d_i h_i.
\end{aligned} \tag{7.62}$$

It is equivalent to say, for a fixed d, that function $L(X, d)$ achieves its maximum at $Y = Y_1$. Here, in Equation 7.62, we have used the fact that the lambdas are put in descending order. The corresponding Y_1 for the optimal d that minimizes $V(d)$ will give us the calibrated volatility surface. Let us make a comment here. It is well-known in matrix theory that the eigenvector corresponding to the largest eigenvalue of a positive semidefinite matrix is often the smoothest one among all eigenvectors. Hence, the result of Equation 7.62 establishes the connection between the smoothest fit of a volatility surface and the smoothest eigenvector of a generalized eigenvalue problem.

To develop a numerical method for the minimization of $V(d)$, we need to derive its gradient and the Hessian matrix. The next theorem summarizes the analytical properties of $\lambda_1(d)$, $Y_1(d)$, and $V(d)$.

Theorem 7.4.5. *If $\lambda_1(d) > \max\{\lambda_2(d), 0\}$, then*

1. *$(\lambda_1(d), Y_1(d))$ and $V(d)$ are differentiable with respect to d;*

2. *the gradient of $V(d)$ is*

$$
\nabla V(d) = 2 \begin{pmatrix} Y_1^{\mathrm{T}} G_1 Y_1 - h_1 \\ Y_1^{\mathrm{T}} G_2 Y_1 - h_2 \\ \vdots \\ Y_1^{\mathrm{T}} G_N Y_1 - h_N \end{pmatrix} ; \quad and
$$

3. *the elements of the Hessian matrix are given by*

$$
H_{ij}(d) = \frac{\partial^2 V}{\partial d_i \partial d_j} = 2 Y_1^{\mathrm{T}} G_i U \Phi^{-1} U^{\mathrm{T}} G_j Y_1, \tag{7.63}
$$

where

$$
\Phi = \lambda_1 I + 2 \begin{pmatrix} \lambda_1 & & \\ & 0 & \\ & & \ddots & \\ & & & 0 \end{pmatrix} - \Lambda. \tag{7.64}
$$

Moreover, the Hessian matrix is a non-negative definite matrix.

Proof. The proof for statements 1 and 2 is similar to that of Theorem 7.2, and thus is omitted for brevity. We prove the last statement only.

The solution $Y_1 = Y_1(d)$ satisfies Equation 7.58 for $X_0 = 0$, and, in addition, $Y_1^{\mathrm{T}} A Y_1 = \lambda_1$. By differentiating both sides of this equation with respect to d_i, we obtain

$$
(Y_1^{\mathrm{T}} A Y_1) A \frac{\partial Y_1}{\partial d_i} + 2 A Y_1 Y_1^{\mathrm{T}} A \frac{\partial Y_1}{\partial d_i} = G_i Y_1 + B_d \frac{\partial Y_1}{\partial d_i}.
$$

We thus have the following equation for $\partial Y_1 / \partial d_i$:

$$
\left[\lambda_1 A + 2 A Y_1 Y_1^{\mathrm{T}} A - B_d \right] \frac{\partial Y_1}{\partial d_i} = G_i Y_1.
$$

We pre-multiply both sides by U^{T} and then have

$$\left[\lambda_1 U^{\mathrm{T}} A U + 2U^{\mathrm{T}} A Y_1 Y_1^{\mathrm{T}} A U - U^{\mathrm{T}} B_d U\right] U^{-1} \frac{\partial Y_1}{\partial d_i}$$

$$= \left[\lambda_1 I + 2 \begin{pmatrix} \lambda_1 & & 0 \\ & \ddots & \\ 0 & & 0 \end{pmatrix} - \Lambda\right] U^{-1} \frac{\partial Y_1}{\partial d_i}$$

$$= U^{\mathrm{T}} G_i Y_1,$$

i.e.,

$$\Phi U^{-1} \frac{\partial Y_1}{\partial d_i} = U^{\mathrm{T}} G_i Y_1, \tag{7.65}$$

where matrix Φ is defined in Equation 7.64. When $\lambda_1 > \lambda_2$, Φ is guaranteed to be positive definite, and it follows that

$$\frac{\partial Y_1}{\partial d_i} = U \Phi^{-1} U^{\mathrm{T}} G_i Y_1. \tag{7.66}$$

According to statement 2, the expression for the gradient is, component-wise,

$$\frac{\partial V}{\partial d_i} = 2(Y_1^{\mathrm{T}} G_i Y_1 - h_i), \quad i = 1, \ldots, N.$$

Differentiating this equation with respect to d_j produces the elements of a Hessian matrix:

$$\begin{aligned} \frac{\partial^2 V}{\partial d_i \partial d_j} &= 4Y_1^{\mathrm{T}} G_i \frac{\partial Y_1}{\partial d_i} \\ &= 4Y_1^{\mathrm{T}} G_i U \Phi^{-1} U^{\mathrm{T}} G_j Y_1. \end{aligned} \tag{7.67}$$

For a concise expression of the Hessian matrix, we denote

$$G = [G_1, G_2, \ldots, G_N],$$

and define the matrix as

$$G \otimes Y = [G_1 Y, G_2 Y, \ldots, G_N Y].$$

The non-negative definiteness of the Hessian matrix follows from the following expression:

$$H(d) = 4 \left(G \otimes Y_1\right)^{\mathrm{T}} U \Phi^{-1} U^{\mathrm{T}} \left(G \otimes Y_1\right). \tag{7.68}$$

This completes the proof. □

We make the following remarks for future reference. For a calibrated model, there are

$$0 = \frac{\partial V(d^*)}{\partial d_i} = 2(Y_1^{\mathrm{T}} G_i Y_1 - h_i), \quad i = 1, \ldots, N.$$

This means that the implied Black's volatilities are matched. The above equations also allow us to treat the inputs of h_i as functions of d^*:

$$h_i = h_i(d^*) = Y_1^T G_i Y_1 \big|_{d=d^*}, \quad i = 1, \ldots, N.$$

Differentiating h_i with respect to d^*, we obtain

$$
\begin{aligned}
\frac{\partial h_i}{\partial d_j^*} &= 2Y_1^T G_i \frac{\partial Y_1}{\partial d_j} \bigg|_{d=d^*} \\
&= \frac{1}{2} \frac{\partial^2 V(d^*)}{\partial d_i \partial d_j} = \frac{1}{2} H_{ij}(d^*).
\end{aligned}
\tag{7.69}
$$

This relation indicates that, for a calibrated model, the Hessian matrix defines the sensitivities of the input prices with respect to the Lagrange multipliers. It will be more useful, however, to know the opposite, that is, the sensitivities of the Lagrange multipliers with respect to the input prices. Evaluating such sensitivities is a matter of calculating the inverse matrix, since (Rockafellar, 1970)

$$\frac{\partial d_j^*}{\partial h_i} = \left(\left(\frac{\partial h_i}{\partial d_i^*} \right)^{-1} \right)_{ij} = 2 \left(H^{-1}(d^*) \right)_{ij}. \tag{7.70}$$

The above result can be used to calculate the sensitivities of a derivative instrument with respect to the benchmark instruments.

To ensure that the min-max problem has at least one solution, we introduce the concept of *nonarbitrage implied volatilities*.

Definition 7.4.1. *We call $h = \{h_i\}$ a set of nonarbitrage implied volatilities if there are $\varepsilon_h > 0$ and $\delta_h > 0$, such that, for any $\{\varepsilon_i\}$ satisfying $|\varepsilon_i| \le \varepsilon_h$, $\forall i$, there exists at least one solution, bounded in the sense of $\|X\| \le \delta_h$, to the equations*

$$X^T G_i X = h_i + \varepsilon_i, \quad i = 1, 2, \ldots, N.$$

We should understand the above concept from the viewpoint of the price-volatility correspondence. We anticipate that, for a set of realistic prices of market instruments, a market model with a reasonable number of driving factors should be able to "rationalize" the prices through generating a reasonable volatility surface. Furthermore, we want to see that small changes in prices will be accommodated by a small variation of the volatility surface. If such accommodation cannot be achieved, then either the model suggests the existence of an arbitrage opportunity, or the model simply lacks enough risk factors to describe reality.

For the existence of the global minimizer, we have

Theorem 7.4.6. *If $h = \{h_i\}$ is a set of no-arbitrage implied volatilities, then there is at least one solution to problem $V(d)$. Moreover, any local minimum is a global minimum.*

Proof. If $h = \{h_i\}$ is a set of no-arbitrage implied volatilities, then the feasible set, $\mathcal{H}$, is not empty. There is thus a finite lower bound for the value function:

$$V(d) \geq \max_{X \in \mathcal{H}} - \left(X^{\mathrm{T}} A X\right)^2 = V^* > -\infty.$$

Assume on the contrary that $V(d)$ has no minimizer. Then, there must exist a sequence $d^{(j)} \to \infty$, such that $V(d^{(j)}) \to V^*$ monotonically from above. Since $h = \{h_i\}$ is a set of no-arbitrage implied volatilities, we can choose a bounded sequence of $X^{(j)}$, such that

$$(X^{(j)})^T G_i X^{(j)} - h_i = \varepsilon^* sign(d^{(j)})$$

for some fixed $\varepsilon^* > 0$. We then have, in view of Equation 7.55,

$$V(d^{(j)}) \to +\infty.$$

This contradicts the assumption of the monotonic decreasing of $V(d^{(j)})$. Hence, any sequence $\{d^{(j)}\}$ such that $V(d^{(j)}) \to V^*$ monotonically must be a bounded sequence and thus have an accumulation point, and the accumulation point is a global minimum.

The conclusion that any local minimum is also a global minimum follows from the convexity of $V(d)$. $\qquad\square$

For the uniqueness of the solution, we have

Theorem 7.4.7. *Let d^* be a minimizer of $V(d)$. If $\lambda_1(d^*) > \max\{\lambda_2(d^*), 0\}$ and the Hessian $H(d^*)$ is positive definite, then d^* is the unique minimizer of $V(d)$, and $Y_1(d^*)$ solves the constrained minimization problem 7.54.*

Proof. If $\lambda_1(d^*) > \max\{\lambda_2(d^*), 0\}$, then, according to Theorem 7.5, $V(d)$ is differentiable near d^* and there is $\nabla V(d^*) = 0$. The positive-definiteness of $H(d^*)$ implies that d^* must be the only local minimizer in its immediate neighborhood. Assume to the contrary that there is another minimizer, say, d^{**}. Then, by the linearity of $L(X, d)$ in d and the convexity of $V(d)$, we have $V(d(t)) = V(d^*)$ for all $d(t) = td^* + (1 - t)d^{**}$, $t \in [0, 1]$. This implies that d^* is not the only minimum in its immediate neighborhood, which is a contradiction.

The conclusion that the solution to the Lagrange multiplier problem solves the constrained minimization problem can be argued in a way similar to the last part of the proof of Theorem 7.3. $\qquad\square$

Two comments are in order. First, the convergence of the gradient-based algorithm can be proved in a way similar to the convergence proof in the last section for the lower-rank approximations algorithm, and thus is omitted for brevity. Second, the availability of the Hessian matrix in closed form potentially enables very fast calibration to swaption prices. It is a common sense

TABLE 7.6: Input Prices of Swaptions

Contract Type	Maturity × Tenor	ATM Strike (%)	Black's Volatilities (%)	Market Prices (bps)
Swaption	0.25 × 2	8.57	16.75	50
Swaption	0.25 × 3	8.75	16.5	73
Swaption	1 × 4	9.1	15.5	172
Swaption	0.25 × 5	8.9	15	103
Swaption	0.25 × 7	9	13.75	123
Swaption	0.25 × 10	8.99	13.25	151
Swaption	1 × 9	9.12	13.25	271
Swaption	2 × 8	9.16	12.75	312

Reprinted from Brace A, Gatarek D, and Musiela M, 1997. *Mathematical Finance* 7:127–154. With permission. Maturity and tenor mean the option's maturity and the life of the underlying swap, respectively.

that, whenever the Hessian matrix is available, we should use a Hessian-based algorithm for numerical optimizations. Our numerical tests and other people's tests on swaption calibration have confirmed that Hessian-based algorithms significantly outperform gradient-based algorithms: the former can calibrate the LIBOR model in real time, whereas the latter take much longer time to get the job done.

Finally, we use the cap/swaption data from Brace et al. (1997) to demonstrate the performance of the Hessian-based calibration method. Specifically, we calibrate the market model to the implied volatilities of a set of caplets and swaptions as well as historical correlations of the pound Sterling, which are listed in Tables 7.3, 7.6, and 7.4, respectively. Table 7.4 is for spot correlations, out of which we can infer *forward correlations* (the correlation seen at a future date) under the assumption of time homogeneity. We consider LIBOR market model for various numbers of driving factors, which require solving many low-rank approximations of correlation matrices.

Solutions to the local volatility surface, $\{s_{i,j}\}$, are obtained by solving the min-max problem 7.54. The inner maximization problem is solved by eigen-decompositions, whereas the outer minimization problem is solved using a built-in function of Matlab, *fminunc*, which utilizes both gradient and Hessian matrices in descending searches. The stopping criterion is

$$\frac{\left\|Y^{\mathrm{T}}G_{m,n}Y - \sigma_{m,n}^2\right\|}{\left\|\sigma_{m,n}^2\right\|} \leq 10^{-4}.$$

Let us explain what this criterion means to the calibration error in terms of Black's volatilities. Suppose that an input-implied Black's volatility is under 50%. Once the iteration of the algorithm stops, we will achieve a percentage error under 0.25% for the Black's volatility of the model, which is much smaller than the usual 1% bid-ask spread for swaptions. We calculate the volatility surface for models with: (a) one factor, (b) two factors, (c) three factors, (d) four factors, (e) six factors, and (f) ten factors. Each calibration run takes about seven functional valuations. The entire calibration (to both input correlations and input implied Black's volatilities) is completed within 20 s. The results are presented in various subplots of Figure 7.6, which will be explained shortly. Note that the gradient-based minimization also converges to the same solution, but it takes a much longer time to do so.

The calibration algorithm performed very robustly and produced nice results, as shown in Figure 7.6. We make a cautionary note that it is possible that the model cannot be calibrated to all inputs. In fact, the volatility surfaces shown in Figure 7.6 are the results obtained by calibrating to all input prices but those of the last two swaptions. When we add the second-to-the-last swaption volatility into the pool of calibration prices, the calibration error increases to 4%. When we add both volatilities to the pool, the iteration of the outer minimization does not converge even with the 10-factor model. The cause is not yet fully understood, but there are at least two possibilities. The first is the failure of the algorithm. This could happen if the Hessian matrix is not positive definite. The second is price inconsistency under the LIBOR market model with less than 10 factors. In fact, suppose that the model has captured the forward-curve dynamics properly. Then, being unable to calibrate the input prices probably implies the existence of arbitrage. The failure to calibrate all input prices reminds us of the possible limitations of both the algorithm and the model.

Now let us take a close look at the figures. We can see that the volatility surfaces look incredibly close, although, as Figure 7.7 confirms, they are different. Different numbers of factors correspond to different input correlations, which can differ significantly from one another, as shown in Figure 7.3. The closeness of the volatility surfaces only suggests that they are insensitive to the input correlations. This viewpoint is also echoed in the study by Choy, Dun, and Schlogl (2004), where they have compared swaption prices under various correlation inputs, only to see that the results are very close. One of the implications of such insensitivity is that we can hardly estimate implied correlations based on vanilla swaptions. In fact, Brace and Womersley (2000) make such an effort using a semidefinite programming technique, and they recognize that the problem is highly underdetermined.

Before finishing this section, we want to draw the reader's attention to the shape of the local volatility functions of the forward rate obtained by the non-parametric calibration, which is considered to be a function of calendar time, t. The volatility function is demonstrated in Figure 7.8 using $\|\gamma(t,5)\|$, $t \leq 5$. It has a humped shape, increasing first and then decreasing after the peak is reached. The forward-rate volatility of other maturities has a similar shape, as can be seen from the plots of Figure 7.6. The calibrated volatilities therefore have similar shapes to the volatilities estimated from time series data, as demonstrated in Figure 6.3. This is a rather plausible agreement, which to some extent suggests that the LIBOR derivative market is efficient.

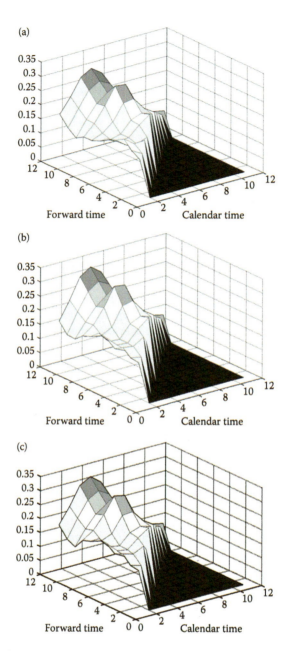

FIGURE 7.6: Local volatility surfaces of models with various numbers of factors. (Reprinted from Wu L, 2003. *Journal of Computational Finance* 6(2): 39–77. With permission.)

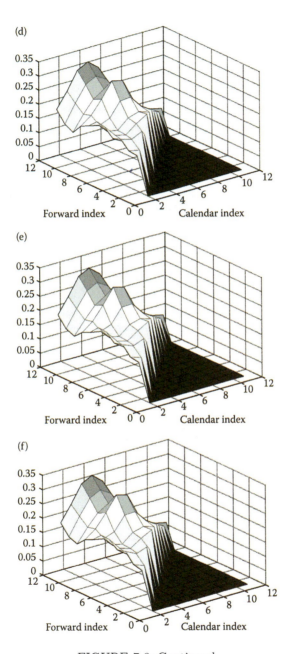

FIGURE 7.6: Continued.

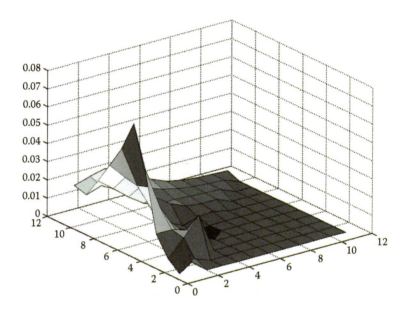

FIGURE 7.7: The difference between the volatility surface of 1- and 10-factor models. (Reprinted from Wu L, 2003. *Journal of Computational Finance* 6(2): 39–77. With permission.)

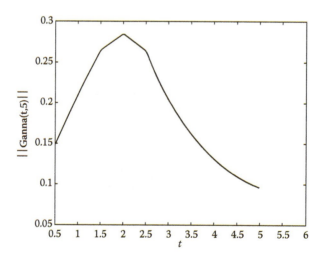

FIGURE 7.8: The humped shape of forward-rate volatility.

7.5 Sensitivity with Respect to the Input Prices

In this section, we discuss the pricing and hedging of a LIBOR derivative with a model calibrated to the prices of benchmark instruments and correlation matrices. The theory we develop is in the spirit of Avellaneda et al. (1998) for calibrating an equity derivatives model via relative-entropy minimization. Without loss of generality, we suppose that a derivative is sold at time $t = 0$, which promises to receive a sequence of cash flows, $\{F_i\}$, at time T_i, $i = 1, \ldots, N$, and the cash flows are contingent on the forward rates, $\{f_j\}_{j=1}^N$, in the future. By using cash flow measures, we can express the value of the derivative as

$$\varphi(0, \mathbf{f}(0)) = \sum_{i=1}^{N} P(0, T_i) E^{Q_j} [F_i]. \tag{7.71}$$

The expectation in Equation 7.71 can be calculated by Monte Carlo simulations.

Consider hedging the short position using benchmark instruments, for which we need to calculate the *hedge ratios*. Let the prices of the benchmarks be $\{C_j\}$. By the chain rule, we have

$$\frac{\partial \varphi}{\partial C_j} = \sum_{i=1}^{I} \frac{\partial \varphi}{\partial d_i} \frac{\partial d_i}{\partial h_j} \frac{\partial h_j}{\partial C_j}. \tag{7.72}$$

Hence, we must calculate the three derivative chains as follows:

$$\left\{ \frac{\partial h_j}{\partial C_j} \right\}, \quad \left\{ \frac{\partial d_i}{\partial h_j} \right\}, \quad \text{and} \quad \left\{ \frac{\partial \varphi}{\partial d_i} \right\}.$$

The first two derivative chains are not difficult to obtain. The partial derivative of volatilities with respect to prices can be calculated by making use Black's formula (which includes a caplet as a special case). In fact,

$$\frac{\partial h_j}{\partial C_j} = \frac{1}{\partial C_j / \partial h_j}.$$

By direct differentiation of Black's formula with respect to Black's volatility, we obtain

$$\frac{\partial C_j}{\partial h_j} = A_{m,n} \left[R_{m,n}(t) n(g_+) g_+'(h_j) - K n(g_-) g_-'(h_j) \right],$$

where

$$g_+(h) = \frac{\ln\left(R_{m,n}(t)/K\right) + (1/2)h\left(T_m - t\right)}{\sqrt{h(T_m - t)}},$$

$$g_-(h) = g_+(h) - \sqrt{h(T_m - t)},$$

$$g'_\pm(h) = -\frac{1}{2h\sqrt{h(T_m - t)}}\ln\frac{R_{m,n}(t)}{K} \pm \frac{1}{4}h^{-(1/2)}\sqrt{T_m - t},$$

$$n(x) = \frac{1}{\sqrt{2\pi}}e^{-(x^2/2)}.$$

For $\partial d_i / \partial h_j$, we recall Equation 7.70,

$$\frac{\partial d_i}{\partial h_j} = 2\left(H^{-1}\right)_{i,j},$$

where H is the Hessian matrix whose elements are given by Equation 7.67. Finally, to calculate $\partial \varphi / \partial d_i$, we differentiate Equation 7.71 with respect to d_i:

$$\frac{\partial \varphi}{\partial d_i}(0, \mathbf{f}(0)) = \sum_{j=1}^{N} P(0, T_j)E^{Q_j}\left[\frac{\partial F_j}{\partial d_i}\right].$$

Here,

$$\frac{\partial F_j}{\partial d_i} = \sum_{k=1}^{K} \frac{\partial F_j}{\partial f_k}\frac{\partial f_k}{\partial d_i}.$$

Denote $\partial f_k / \partial d_i$ by $f_{k,i}$, which evolves according to

$$\begin{aligned}
\mathrm{d}f_{k,i}(t) = {} & \left(f_{k,i}(t)s_k(t) + f_k(t)\frac{\partial s_k(t)}{\partial d_i}\right)\mathrm{d}W_t^j \\
& + \left(f_{k,i}(t)s_k(t)\sigma_{k+1}(t) + f_k(t)\frac{\partial s_k(t)}{\partial d_i}\sigma_{k+1}(t)\right. \\
& \left. + f_k(t)s_k(t)\frac{\partial \sigma_{k+1}(t)}{\partial d_i}\right)\mathrm{d}t,
\end{aligned} \qquad (7.73)$$

where

$$\frac{\partial \sigma_{k+1}(t)}{\partial d_i} = \sum_{l=j}^{k} \frac{\Delta T_l}{1 + \Delta T_l f_l(t)}\left(\frac{\Delta T_l f_{l,i}}{1 + \Delta T_l f_l(t)}s_l(t) + f_l\frac{\partial s_l(t)}{\partial d_i}\right)\rho_{l,j},$$

and

$$\frac{\partial s_l(t)}{\partial d_i} = \left(\frac{\partial Y}{\partial d_i}\right)_l = \left(U\Phi^{-1}U^T G_i Y\right)_l.$$

Note that Equation 7.73 is linear in $f_{i,k}$ and thus the solution does exist. By putting all calculated partial derivatives back into Equation 7.72, we obtain the hedging ratio with respect to the ith benchmark instrument.

Chapter 8

Volatility and Correlation Adjustments

In Chapter 6, we introduced a series of LIBOR derivatives. In that series, Eurodollar futures and swaps of various kinds are particularly important to interest-rate derivatives markets. Unlike other derivatives, futures and swaps cost nothing to enter (except for deposits as guarantee funds), making them popular tools for either hedging or speculation. Probably because of that, Eurodollar futures and swaps have long dominated the liquidity of the interest-rate derivatives markets. Nonetheless, they are not necessarily easy to price. In this chapter, we consider the pricing of futures and non-vanilla swaps relative to forward contracts and vanilla swaps. In concrete terms, we derive formulae for futures prices and non-vanilla swap rates in terms of available forward prices and vanilla swap rates, respectively.

The so-called correlation adjustment is a way to calculate futures prices based on forward prices. Due to marking to market, a futures contract cannot be replicated statically, so its pricing is model dependent. In the next section, we derive a formula for futures prices under general asset dynamics. According to the formula, a positive asset–short rate correlation will make a futures price higher than the forward price, whereas a negative correlation will make it lower. Based on this result and, in addition, the convexity property of bond prices as functions of yields, we further develop a formula for futures rates under the HJM model. Correlation adjustment is also essential for pricing the so-called quanto derivatives, where a payoff is measured in one currency but paid in another without currency conversion. In the end of this chapter, we address the pricing of quanto futures and quanto options.

The so-called volatility adjustment essentially is about calculating the expectation of a bond yield under a forward measure. Note that the bond yield is in general not a martingale under the forward measure, but the forward price of the bond is. By utilizing the martingale property of the forward bond price, we produce an approximation formula for the expected bond yield in terms of the forward bond yield, that is, the yield of the forward bond price. Due to the convex feature of the bond price as a function of its yield, the expected bond yield is bigger than the forward bond yield with certainty. Hence, volatility adjustment is also called convexity adjustment.

From a mathematical viewpoint, what we do in this chapter can be summarized as follows: given the expectation of a random variable under one measure, compute the expectation of the variable under another equivalent measure. For an asset, it is a matter of changing the measures. But for a yield, a change in measures does not help much because the dynamics of the yield can become too complicated. For an expected yield, we produce an approximation formula by taking advantage of the martingale property of the forward bond price.

8.1 Adjustment due to Correlations

We begin with the convexity adjustment for tradable assets. The need for convexity adjustment arises from futures trading when we need to calculate futures prices, which, as we will show, are equal to the expected terminal asset prices under the risk-neutral measure. In this section, we give a detailed description of the pricing of futures contracts for general tradable assets. Part of the results will then be applied to pricing interest-rate futures contracts.

8.1.1 Futures Price versus Forward Price

The value of a forward contract with maturity, T, and strike, K, on a tradable asset is

$$V_0 = E^{\mathbb{Q}}\big[B_T^{-1}(S_T - K)\big|\mathcal{F}_0\big], \tag{8.1}$$

where $\mathbb{Q}$ stands for the risk-neutral measure, S_T the asset price at maturity, and B_T the balance of the money market account at T:

$$B_T = \exp\left(\int_0^T r_s\,\mathrm{d}s\right). \tag{8.2}$$

The forward price is defined as the strike price that nullifies the value of the forward contract. In view of Equation 8.1, we know that the forward price for the contract satisfies

$$F_0^T = \frac{E_0^{\mathbb{Q}}\big[B_T^{-1}S_T\big]}{E_0^{\mathbb{Q}}\big[B_T^{-1}\big]} = E_0^{\mathbb{Q}^T}[S_T], \tag{8.3}$$

that is, it is the expectation of the terminal asset price under the forward measure, $\mathbb{Q}_T$. We have already learned in Chapter 4 that the above expectation equals

$$F_0^T = \frac{\hat{S}_0}{P(0,T)}, \tag{8.4}$$

where $\hat{S}_t$ is the stripped-dividend price of the asset at time t.

With everything else the same, a futures contract differs from a forward contract by "marking to market," meaning that the P&L from holding the contract is credited to or debited from the holder's margin account on a daily basis. The margin account, meanwhile, is accrued using risk-free interest rates. A futures price parallels the forward price and nullifies the value of a futures contract. Let $\tilde{F}_t^T$ be the futures price observed at time $t \leq T$. At the maturity of the futures contract, the futures price is fixed or set to the price of the underlying security, that is, $\tilde{F}_T^T = S_T$.

Our next goal is to derive $\tilde{F}_t^T$ for $t < T$. Let us consider the price dynamics of the futures contract. At any moment, $t < T$, the value of the futures contract is reflected in the balance in the margin account, V_t. The change to the futures price comes from two sources: the accrual of the margin account and the change in the futures price. Hence, in differential form, the change in the value of the futures contract is

$$dV_t = r_t V_t \, dt + d\tilde{F}_t^T. \tag{8.5}$$

From Equation 8.5, we have the dynamics of the futures price,

$$d\tilde{F}_t^T = B_t d\left(B_t^{-1} V_t\right). \tag{8.6}$$

It follows that

$$\tilde{F}_t^T - \tilde{F}_0^T = \int_0^t B_s d(B_s^{-1} V_s). \tag{8.7}$$

Being the value of the futures contract, a tradable security, the balance of the margin account after discounting must be a martingale under the risk-neutral measure. This gives rise to the following general formula for the futures price:

$$\tilde{F}_0^T = E^{\mathbb{Q}}\left[\tilde{F}_T^T\right] = E^{\mathbb{Q}}[S_T]. \tag{8.8}$$

We have a more intuitive approach to derive the futures price formula (Equation 8.8) that is based on the so-called dividend-yield analogy.[1] At any time, $t \leq T$, we will maintain B_{t+dt} units of the futures contract. Here,

$$B_{t+dt} = (1 + r_t \, dt)B_t,$$

which is known at time t. This is a dynamical strategy. We start with $B_{dt} > 1$ units of the futures contract at time 0, which generates a P&L in the amount of $B_{dt}\left(\tilde{F}_{dt}^T - \tilde{F}_0^T\right)$ at time dt. Suppose, at time t, that we have already accumulated B_t units of the futures contract and $B_t\left(\tilde{F}_t^T - \tilde{F}_0^T\right)$ dollars in the margin account, and we add $r_t B_t dt$ units of the futures contract at no cost.

---

[1] This means the following: by receiving continuous share dividends or converting cash dividends into shares, we will increase the number of shares from one at time 0 to $\exp\left(\int_0^t q_s ds\right)$ at any later time, t, where q_s is the dividend yield.

At time $t + dt$, due to interest accrual and marking to market, the balance of the margin account will become

$$(1 + r_t dt) B_t \left(\tilde{F}_t^T - \tilde{F}_0^T \right) + (1 + r_t dt) B_t \left(\tilde{F}_{t+dt}^T - \tilde{F}_t^T \right)$$
$$= B_{t+dt} \left(\tilde{F}_{t+dt}^T - \tilde{F}_0^T \right). \tag{8.9}$$

Continuing with the above strategy until $T - dt$, then, at time T, when the futures contract expires, we will have a balance in the amount of

$$V_T = B_T (S_T - \tilde{F}_0^T) \tag{8.10}$$

in the margin account. The expected present value of the terminal balance is thus

$$V_0 = E^Q \left[B_T^{-1} B_T \left(S_T - \tilde{F}_0^T \right) \right] = E^Q[S_T] - \tilde{F}_0^T. \tag{8.11}$$

By making $V_0 = 0$, we rederive Equation 8.8.

Our next issue of interest is on the difference between the forward price and the futures price. When the interest rate is deterministic, the two prices are, apparently, equal. When the interest rate is stochastic, they are generally not the same. Let us see what causes the difference. Straightforwardly, we have

$$E^Q[S_T] = E^Q \left[B_T^{-1} B_T S_T \right]$$
$$= P(0, T) E^{Q_T} [B_T S_T]. \tag{8.12}$$

When $S_T = 1$, Equation 8.12 yields the expectation of the money market account under the forward measure:

$$E^{Q_T}[B_T] = \frac{1}{P(0, T)}. \tag{8.13}$$

It then follows that

$$\begin{aligned} \tilde{F}_0^T - F_0^T &= E^Q[S_T] - E^{Q_T}[S_T] \\ &= P(0, T) E^{Q_T}[B_T S_T] - E^{Q_T}[S_T] \\ &= P(0, T) \left(E^{Q_T}[B_T S_T] - E^{Q_T}[B_T] E^{Q_T}[S_T] \right) \\ &= P(0, T) \times \text{Cov}^{Q_T}(B_T, S_T) \\ &= P(0, T) \times \text{Cov}^{Q_T}(B_T, F_T^T). \end{aligned} \tag{8.14}$$

Hence, if the asset is positively correlated to the money market account, then there will be $\tilde{F}_0^T > F_0^T$, or, otherwise, $\tilde{F}_0^T \leq F_0^T$. Such orders can well be explained by intuition: given a positive correlation between the asset and the short rate, a gain in the margin account will likely be accrued with a higher interest rate, whereas a loss in the account will likely be accrued with a lower interest rate. This is to the advantage of the party who longs the futures contract. To make the futures a fair game, the futures price should be set somewhat higher than the forward price.

To figure out the exact difference between the two prices, we need an asset price model. For generality, we take the lognormal risk-neutral dynamics as the price of the stripped-dividend asset and zero-coupon bonds,

$$dS_t = S_t(r_t dt + \boldsymbol{\Sigma}_S^T d\mathbf{W}_t),$$
$$dP(t,T) = P(t,T)(r_t dt + \boldsymbol{\Sigma}_P^T d\mathbf{W}_t). \tag{8.15}$$

In Equation 8.15, we have replaced $\boldsymbol{\Sigma}(t,T)$ by $\boldsymbol{\Sigma}_P$ and omitted the hat over S_t for simplicity. In terms of the volatility vectors, we have

$$\tilde{F}_0^T = E^Q\left[S_0 e^{\int_0^T (r_s - \frac{1}{2}\|\boldsymbol{\Sigma}_S\|^2)ds + \boldsymbol{\Sigma}_S^T d\mathbf{W}_t}\right]$$
$$= S_0 E^Q\left[B_T e^{\int_0^T -\frac{1}{2}\|\boldsymbol{\Sigma}_S\|^2 ds + \boldsymbol{\Sigma}_S^T d\mathbf{W}_t}\right]. \tag{8.16}$$

Substituting

$$B_T = \frac{1}{P(0,T)} \exp\left\{\int_0^T \frac{1}{2}\|\boldsymbol{\Sigma}_P\|^2 ds - \boldsymbol{\Sigma}_P^T d\mathbf{W}_t\right\} \tag{8.17}$$

into Equation 8.16, we then obtain

$$\tilde{F}_0^T = S_0 E^Q\left[\frac{1}{P(0,T)} e^{\int_0^T -\frac{1}{2}(\|\boldsymbol{\Sigma}_S\|^2 - \|\boldsymbol{\Sigma}_P\|^2)dt + (\boldsymbol{\Sigma}_S - \boldsymbol{\Sigma}_P)^T d\mathbf{W}_t}\right]$$
$$= F_0^T E^Q\left[e^{\int_0^T (\|\boldsymbol{\Sigma}_P\|^2 - \boldsymbol{\Sigma}_S^T \boldsymbol{\Sigma}_P)dt - \frac{1}{2}\|\boldsymbol{\Sigma}_S - \boldsymbol{\Sigma}_P\|^2 dt + (\boldsymbol{\Sigma}_S - \boldsymbol{\Sigma}_P)^T d\mathbf{W}_t}\right]. \tag{8.18}$$

Suppose, for simplicity, that the volatility vector of both S_t and $P(t,T)$ are non-stochastic. It then follows that

$$\tilde{F}_0^T = F_0^T e^{\int_0^T (\|\boldsymbol{\Sigma}_P\|^2 - \boldsymbol{\Sigma}_S^T \boldsymbol{\Sigma}_P)dt} = F_0^T e^{-\int_0^T \boldsymbol{\Sigma}_F^T \boldsymbol{\Sigma}_P dt}, \tag{8.19}$$

where $\boldsymbol{\Sigma}_F = \boldsymbol{\Sigma}_S - \boldsymbol{\Sigma}_P$ is the volatility of the forward price. The futures-forward difference is thus

$$\tilde{F}_0^T - F_0^T = F_0^T\left(e^{-\int_0^T \boldsymbol{\Sigma}_F^T \boldsymbol{\Sigma}_P dt} - 1\right). \tag{8.20}$$

Note that Equation 8.20 is a variant of Equation 8.14. In fact, by making use of Equation 8.17, we can directly prove that

$$\text{Cov}^{Q_T}(B_T, S_T) = \frac{F_0^T}{P(0,T)}\left(e^{-\int_0^T \boldsymbol{\Sigma}_F^T \boldsymbol{\Sigma}_P dt} - 1\right). \tag{8.21}$$

Hence, we have yet again shown that $\tilde{F}_0^T > F_0^T$ provides a positive correlation between the asset and the money market account.

We now comment on the hedging of the futures contract. Futures are often used for hedging due to high liquidity and low cost (for deposits in the margin account). But when a futures price is too much off the line, futures can be arbitraged against using, for instance, the underlying security. Although the futures cannot be hedged statically like a forward contract, it can still be dynamically hedged, using the delta

$$\Delta = \frac{\partial \tilde{F}_t^T}{\partial S_t} = \frac{1}{P(t,T)} e^{\int_t^T (\|\Sigma_P\|^2 - \Sigma_S^T \Sigma_P) ds}, \tag{8.22}$$

which requires information on the volatilities.

8.1.2   Convexity Adjustment for LIBOR Rates

A LIBOR rate, $f_{\Delta T}(t, T)$, and its corresponding futures rate, $E^{\mathbb{Q}}[f_{\Delta T}(T, T) \mid \mathcal{F}_0]$, are tradable quantities through an FRA or a Eurodollar futures, respectively. Here $\mathbb{Q}$ stands for the risk-neutral measure. In this section, we look at the difference between these two interest rates. Because $f_{\Delta T}(t, T)$ does not follow the general asset price process as in Equation 8.15, the result of Equation 8.20 does not apply to the forward–futures difference. This difference, instead, will be figured out under the general framework of HJM for interest rates. We begin, however, with establishing a definitive order between the two rates that is model independent.

Owing to the negative correlation between the money market account and zero-coupon bonds, we can see from Equation 8.14 that there will generally be

$$E^{\mathbb{Q}}[P(T, T + \Delta T)] < E^{\mathbb{Q}_T}[P(T, T + \Delta T)] = \frac{P(0, T + \Delta T)}{P(0, T)}. \tag{8.23}$$

Owing to the convex relationship between a bond price and its yield, we obtain, using Jensen's inequality,

$$E^{\mathbb{Q}}[f_{\Delta T}(T, T)] \geq \frac{1}{\Delta T} \left(\frac{1}{E^{\mathbb{Q}}[P(T, T + \Delta T)]} - 1 \right)$$

$$> \frac{1}{\Delta T} \left(\frac{P(0, T)}{P(0, T + \Delta T)} - 1 \right) = f_{\Delta T}(0, T). \tag{8.24}$$

In fact, a similar order between a pair of futures and forward yields can also be established based on the convex feature of bond prices as functions of their yields. For financial transactions, we need to know not only the order but also the actual difference. In Section 8.2, we present a general methodology for estimating the difference between pairs of forward and futures yields. In the finance literature, such differences are called convexity adjustments.

We now consider calculating the actual futures–forward difference under the HJM model, the general no-arbitrage framework. Recall that

$$E^{\mathbb{Q}}[f_{\Delta T}(T, T)] = \frac{1}{\Delta T} \left(E^{\mathbb{Q}} \left[\frac{P(T, T)}{P(T, T + \Delta T)} \right] - 1 \right). \tag{8.25}$$

Under the general HJM framework,

$$
E^{\mathbb{Q}}\left[\frac{P(T,T)}{P(T,T+\Delta T)}\right]
$$

$$
= \frac{P(0,T)}{P(0,T+\Delta T)} \exp\left\{\int_0^T -\frac{1}{2}\left(\|\boldsymbol{\Sigma}(t,T)\|^2 - \|\boldsymbol{\Sigma}(t,T+\Delta T)\|^2\right) dt\right\}
$$

$$
\times\ E^{\mathbb{Q}}\left[\exp\left\{\int_0^T \left(\boldsymbol{\Sigma}(t,T) - \boldsymbol{\Sigma}(t,T+\Delta T)\right)^{\mathrm{T}} d\hat{\mathbf{W}}_t\right\}\right]
$$

$$
= \frac{P(0,T)}{P(0,T+\Delta T)} \exp\left\{\int_0^T \boldsymbol{\Sigma}^{\mathrm{T}}(t,T+\Delta T)\left(\boldsymbol{\Sigma}(t,T+\Delta T)\right.\right.
$$

$$
\left.\left. -\boldsymbol{\Sigma}(t,T)\right) dt\right\}. \tag{8.26}
$$

Here, we have implicitly assumed that the volatilities of discount bonds are deterministic functions. It follows that

$$
E^{\mathbb{Q}}[f_{\Delta T}(T,T)] - f_{\Delta T}(0,T) = (1 + \Delta T f_{\Delta T}(0,T))
$$

$$
\times\ \frac{\exp\left\{\int_0^T \boldsymbol{\Sigma}^{\mathrm{T}}(t,T+\Delta T)\left(\boldsymbol{\Sigma}(t,T+\Delta T) - \boldsymbol{\Sigma}(t,T)\right) dt\right\} - 1}{\Delta T}. \tag{8.27}
$$

Generally, the variance in the discount price of $P(t,T+\Delta T)$ is bigger than that of $P(t,T)$, that is,

$$
\int_0^T \|\boldsymbol{\Sigma}(t,T+\Delta T)\|^2 dt \geq \int_0^T \|\boldsymbol{\Sigma}(t,T)\|^2 dt. \tag{8.28}
$$

By Schwartz's inequality (Rudin, 1976),

$$
\int_0^T \|\boldsymbol{\Sigma}(t,T+\Delta T)\|^2 dt \geq \int_0^T \boldsymbol{\Sigma}^{\mathrm{T}}(t,T+\Delta T)\boldsymbol{\Sigma}(t,T) dt. \tag{8.29}
$$

Thus, the exponent term in Equation 8.27 is positive, yielding again the order

$$
E^{\mathbb{Q}}[f_{\Delta T}(T,T)] > f_{\Delta T}(0,T). \tag{8.30}
$$

Let us take a look at the limiting case of Equation 8.27 for $\Delta T \to 0$. Letting $\Delta T \to 0$ in Equation 8.27, we end up with

$$
E^{\mathbb{Q}}[f(T;T)] = f(0;T) + \int_0^T \boldsymbol{\Sigma}^{\mathrm{T}}(t,T)\boldsymbol{\Sigma}_T(t,T) dt
$$

$$
= f(0;T) + \int_0^T \boldsymbol{\sigma}^{\mathrm{T}}(t,T)\left(\int_t^T \boldsymbol{\sigma}(t,s) ds\right) dt. \tag{8.31}
$$

The last equation can also be derived directly from the HJM equation. Hence, the drift term of the HJM equation is for "convexity adjustment." When forward rates of different maturities are positively correlated, that is, $\boldsymbol{\sigma}^{\mathrm{T}}(t,T)\boldsymbol{\sigma}(t,s) \geq 0$, we have a positive value for the convexity adjustment with instantaneous forward rates.

We finish this section with more comments. As we have seen, the futures–forward difference for an interest rate is due to two mechanisms. The first is the negative correlation between the money market account and the yield, and the second is the convex price–yield relationship. In Section 8.2, we consider calculating the difference between the expectations of a yield under a martingale measure and a cash flow measure, which will be entirely due to the second mechanism.

8.1.3 Convexity Adjustment under the Ho–Lee Model

In the fixed-income market, it had once been popular to evaluate the convexity adjustment based on the Ho–Lee model (1986). In the Ho–Lee model, we have the zero-coupon bond volatility given by $\boldsymbol{\Sigma}(t,T) = -\sigma(T-t)$. Thus, the exponential term in Equation 8.27 becomes

$$\int_0^T \boldsymbol{\Sigma}^{\mathrm{T}}(t,T+\Delta T)\big(\boldsymbol{\Sigma}(t,T+\Delta T) - \boldsymbol{\Sigma}(t,T)\big)\mathrm{d}t$$

$$= \int_0^T \sigma^2\,(T+\Delta T - t)\,\Delta T\,\mathrm{d}t$$

$$= \frac{1}{2}\sigma^2 T^2 \Delta T \left(1 + 2\frac{\Delta T}{T}\right). \tag{8.32}$$

It follows that

$$f_{\Delta T}^{(\mathrm{fut})}(0,T) - f_{\Delta T}(0,T)$$

$$= (1 + \Delta T f_{\Delta T}(0,T))\,\frac{e^{\frac{1}{2}\sigma^2 T^2 \Delta T(1+2\Delta T/T)} - 1}{\Delta T}$$

$$\approx (1 + \Delta T f_{\Delta T}(0,T)) \times \frac{1}{2}\sigma^2 T^2 \left(1 + 2\Delta T/T\right). \tag{8.33}$$

The last approximation is rather accurate because $\sigma^2 T^2$ is, typically, small.

8.1.4 An Example of Arbitrage

What happens if a forward rate is taken to be a futures rate in a market? The answer is arbitrage. In the following example, we make arbitrage transactions using a futures contract and forward contracts on zero-coupon bonds.

Assume that the one-year rate tree evolves according to a binomial tree, like the one in Figure 8.1, under the risk-neutral measure.

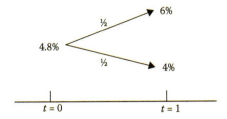

FIGURE 8.1: Evolution of a one-year interest rate.

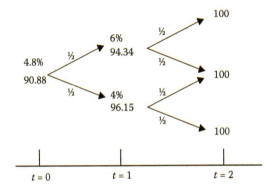

FIGURE 8.2: Price tree of a two-year bond.

Then, the price of a two-year maturity zero-coupon bond can be calculated through backward induction as demonstrated in Figure 8.2.

Meanwhile, the price of the one-year maturity zero-coupon bond is

$$P(0,1) = \frac{100}{1 + 4.8\%} = 95.42. \tag{8.34}$$

The forward rate for the second year is thus

$$f_{\Delta T}(0,1) = \frac{P(0,1)}{P(0,2)} - 1 = 4.9905\%, \tag{8.35}$$

with $\Delta T = 1$. Note that $f_{\Delta T}(0,1)$ has nothing to do with the spot rate, 4.8%. The expected futures rate for the second year is

$$E^{\mathbb{Q}}[f_{\Delta T}(1,1))] = 0.5 \times 6\% + 0.5 \times 4\% = 5\% > f_{\Delta T}(0,1). \tag{8.36}$$

If we long a futures contract at rate $E^{\mathbb{Q}}[f_{\Delta T}(1,1)] = 5\%$, we have the pattern of contingent cash flows as shown in Figure 8.3, which is clearly a fair game.

If we long an FRA at time $t = 0$, and close it out at time $t = 1$, the cash flow pattern is shown in Figure 8.4, which is also a fair game.

If, at $t = 0$, the futures contract is traded at $f_{\Delta T}(0,1)$ instead of $E^{\mathbb{Q}}[f_{\Delta T}(1,1)]$, then the futures contract is underpriced, and we can arbitrage as follows:

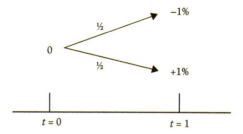

FIGURE 8.3: Cash flow pattern of the interest-rate futures on $f_{\Delta T}(1,1)$.

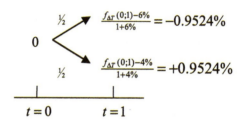

FIGURE 8.4: Cash flow pattern of the FRA on $f_{\Delta T}(1,1)$.

1. long one futures contract, and

2. short 1.04 units of the FRA.

The cash flow pattern of the transactions is shown in Figure 8.5.

According to the pattern of cash flow at time $t = 0$ and $t = 1$, we know that the above long/short transactions constitute an arbitrage. Hence, to avoid arbitrage, the futures rate must be set somewhat higher than the corresponding forward rate.

8.2   Adjustment due to Convexity

In some FRNs and swaps, the cash flows depend on bond yields or swap rates of a fixed maturity. The pricing of such cash flows requires taking expectations under a series of forward measures with delivery at the cash flow days (the so-called cash flow measures). At the same time, forward bond yields and swap rates are observable (directly or indirectly) in the marketplace, which are known to be close to the expected yields or swap rates. This section presents a fine-tuning procedure that generates a good approximation of an expected yield based on its known forward yield. As we shall see, the mechanism of the adjustment (from a forward yield to an expected forward yield) lies in the convex relationship between the bond price and its yield, which leads to the

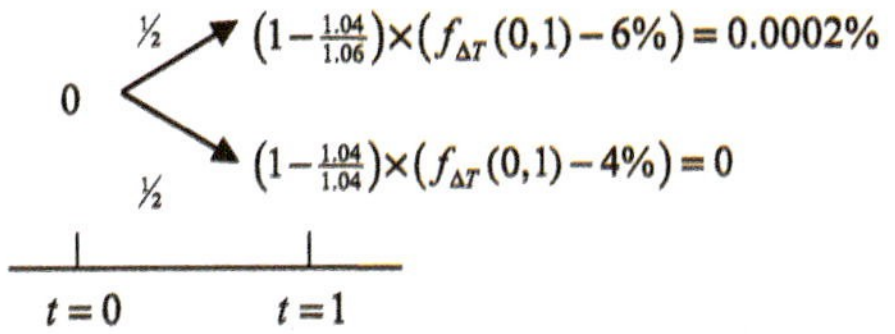

$$\frac{1}{2} \nearrow \left(1-\frac{1.04}{1.06}\right)\times\left(f_{\Delta T}(0,1)-6\%\right) = 0.0002\%$$

$$0$$

$$\frac{1}{2} \searrow \left(1-\frac{1.04}{1.04}\right)\times\left(f_{\Delta T}(0,1)-4\%\right) = 0$$

$$t=0 \qquad t=1$$

FIGURE 8.5: The payoff pattern for the arbitrage strategy.

name of the technique. Convexity adjustments of this kind are often needed in pricing various contingent claims on yields.

8.2.1 Payment in Arrears versus Payment in Advance

We motivate the convexity adjustments with the following example.

Example 8.2.1. *"Paying in arrears vs. paying in advance." Consider a derivative that pays $f_3(3)$, the one-year simple rate three years forward, according to two payment schedules.*

1. *"Paying in arrears": The rate is fixed in the end of the third year but the payment is made a year later. The price of the derivative is given by the expression*

$$V_0 = P(0,4)E^{\mathbb{Q}_4}[f_3(3)]$$
$$= P(0,4)f_3(0). \tag{8.37}$$

   Here, we have made use of the martingale property of the forward rate, $f_j(t) = f_{\Delta T}(t;T_j)$, under the forward measure, $\mathbb{Q}_{j+1}$, $\forall j$. Note that $f_j(0)$ is observable and the pricing is done rather conveniently.

2. *"Paying in advance": The rate is fixed in the end of the third year and the payment is made at the same time. Accordingly, we have*

$$V_0 = P(0,3)E^{\mathbb{Q}_3}[f_3(3)]. \tag{8.38}$$

   Note that $f_3(t)$ is not a $\mathbb{Q}_3$-martingale, and it is certain that $E^{\mathbb{Q}_3}[f_3(3)] \neq f_3(0)$. Additional efforts are needed to evaluate the expectation.

Next, we show how to evaluate $E^{\mathbb{Q}_j}[f_j(T_j)]$ under the LIBOR model. The forward rate follows

$$\frac{\mathrm{d}f_j(t)}{f_j(t)} = \gamma_j^{\mathrm{T}}(t)[\mathrm{d}\mathbf{W}_t - \mathbf{\Sigma}(t,T_{j+1})\,\mathrm{d}t]$$

$$= \gamma_j^{\mathrm{T}}(t)\left[\mathrm{d}\mathbf{W}_t^j + \frac{\Delta T f_j(t)}{1+\Delta T f_j(t)}\gamma_j(t)\,\mathrm{d}t\right], \tag{8.39}$$

where $\mathbf{W}_t^j = \mathbf{W}_t - \int_0^t \mathbf{\Sigma}(t, T_j)\,dt$ is a $\mathbb{Q}_j$-Brownian motion. Solving for $f_j(T_j)$ and then taking the $\mathbb{Q}_j$-expectation, we have

$$E^{\mathbb{Q}_j}[f_j(T_j)] = f_j(0)E^{\mathbb{Q}_j}\left[\exp\left(\int_0^{T_j} \frac{\Delta T f_j(t)}{1 + \Delta T f_j(t)}\|\gamma_j(t)\|^2\,dt\right)\right.$$

$$\times \left. \exp\left(\int_0^{T_j} -\frac{1}{2}\|\gamma_j(t)\|^2\,dt + \gamma_j^T(t)\,d\mathbf{W}_t^j\right)\right]. \qquad (8.40)$$

Without loss of generality, we assume constant volatility, $\|\gamma_j(t)\| = \|\gamma_j(0)\|$. We also freeze the state variable at time 0,

$$\frac{\Delta T f_j(t)}{1 + \Delta T f_j(t)} \approx \frac{\Delta T f_j(0)}{1 + \Delta T f_j(0)}. \qquad (8.41)$$

Then we obtain the following approximation to the $\mathbb{Q}_j$-expectation of the forward rate,

$$E^{\mathbb{Q}_j}[f_j(T_j)] \approx f_j(0)\exp\left(\int_0^{T_j} \frac{\Delta T f_j(0)}{1 + \Delta T f_j(0)}\|\gamma_j(t)\|^2\,dt\right)$$

$$\approx f_j(0) + \frac{\Delta T (f_j(0))^2}{1 + \Delta T f_j(0)}\|\gamma_j(0)\|^2\,T_j. \qquad (8.42)$$

Although the treatment of freezing the state variable looks crude, the resulting approximation, in Equation 8.42, is actually a rather good one.

8.2.2 Geometric Explanation for Convexity Adjustment

We now offer some geometrical insights into convexity adjustment, using a simple example. Suppose that, in one year, the one-year zero-coupon bond has the following two possible prices with equal probabilities:

$$P_{-1} = 94 \quad \text{and} \quad P_{+1} = 96, \qquad (8.43)$$

which correspond to two forward rates,

$$f_{-1} = 6.3830\% \quad \text{and} \quad f_{+1} = 4.1667\%, \qquad (8.44)$$

respectively. The forward measure of interest is $\mathbb{Q}_1 = \{\frac{1}{2}, \frac{1}{2}\}$. The expected value of the bond price under $\mathbb{Q}_1$ is

$$P_0 = \frac{1}{2}P_{-1} + \frac{1}{2}P_{+1} = 95. \qquad (8.45)$$

The forward rate, as the yield of P_0, is

$$f_0 = \frac{100}{95} - 1 = 5.2632\%. \qquad (8.46)$$

Price

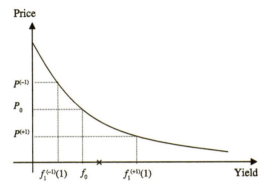

FIGURE 8.6: Relative locations of the expected futures rate and forward rate.

Let $f = f_1(1)$ represent the one-year rate for one-year forward. The expected value of this future interest rate under $\mathbb{Q}_1$ is[2]

$$E^{\mathbb{Q}_1}[f] = \frac{1}{2}(4.1667\% + 6.3830\%) = 5.2748\% > f_0. \qquad (8.47)$$

The geometric explanation for the order, $E^{\mathbb{Q}_1}[f] > f_0$, lies in the convex feature of the forward bond price curve as a function of the forward yield. Such a feature is demonstrated in Figure 8.6, where "×" represents the value of the expected future rate, which always lies to the right of the forward rate due to the convexity.

8.2.3   General Theory of Convexity Adjustment

In fixed-income derivative markets, convexity adjustment is needed in many circumstances and for various interest rates or yields. Here is another example.

Example 8.2.2. *Consider a bond with coupon rate c, to be paid at times $T_i = T_0 + i\Delta T, i = 1, \ldots, N$. Let y_T denote the YTM of the bond at time T. We consider the pricing of the following option on the yield,*

$$V_T = 100\,(y_T - K)^+, \qquad (8.48)$$

which matures at $T \le T_N$. Here is the market practice. We assume that the bond yield, y_t, follows a lognormal process under the T-forward measure, $\mathbb{Q}_T$, with a constant volatility, σ_y, and we price the option using Black's formula:

$$V_0 = 100P(0,T)\left(E^{\mathbb{Q}_T}[y_T]\,\Phi(d_1) - K\Phi(d_2)\right), \qquad (8.49)$$

---

[2]It is known that f is a $\mathbb{Q}_2$-martingale.

where

$$d_1 = \frac{\ln\left(E^{Q_T}[y_T]/K\right) + (1/2)\sigma_y^2 T}{\sigma_y \sqrt{T}},$$

(8.50)

$$d_2 = d_1 - \sigma_y \sqrt{T}.$$

To price the option, we must value $E^{Q_T}[y_T]$. In general, the arbitrage-free drift of y_t is state dependent and can be so complicated that a direct evaluation of the expectation becomes impossible.

The general technique for convexity adjustment is attributed to Brotherton-Ratcliffe and Iben (1993). The key of this technique is to take advantage of the martingale property of the forward bond price under the corresponding forward measure. According to the definition of the "stripped-dividend price" in Chapter 4, we define the T-forward price of the bond seen at time $t \leq T$ as

$$F_t^T \triangleq \frac{B^c(t) - \sum_{T_i \leq T} \Delta T c P(t, T_i)}{P(t, T)}.$$

(8.51)

The forward bond yield, y_t^T, is defined to be the YTM of the forward price, that is, y_t^T solves the equation

$$\sum_{T_i > T}^{T_N} \frac{\Delta T c}{\left(1 + \Delta T y_t^T\right)^{(T_i - T)/\Delta T}} + \frac{1}{\left(1 + \Delta T y_t^T\right)^{(T_N - T)/\Delta T}} = F_t^T.$$

(8.52)

Note that, at time T, the forward bond yield reduces to the spot bond yield, $y_T^T = y_T$. It follows that $E^{Q_T}[y_T] = E^{Q_T}[y_T^T]$.

Before its evaluation, we show that the convexity adjustment is positive in general. Let $B_F(y_t^T)$ denote the left-hand side of Equation 8.52. The martingale property of the forward bond price means

$$B_F(y_0^T) = E^{Q_T}[B_F(y_T^T)].$$

(8.53)

Since $B_F(y)$ is a monotonically decreasing and convex function of y, $y_t^T = B_F^{-1}(F_t^T)$ exists and it is also a convex function of F_t^T. By Jensen's inequality,

$$E^{Q_T}[y_T^T] = E^{Q_T}[B_F^{-1}(F_T^T)]$$
$$\geq B_F^{-1}(E^{Q_T}[F_T^T]) = B_F^{-1}(F_0^T) = y_0^T.$$

(8.54)

To evaluate $E^{Q_T}[y_T^T]$, we consider the Taylor expansion of $B_F(y_T^T)$ around y_0^T and retain up to the second-order terms:

$$B_F\left(y_T^T\right) = B_F\left(y_0^T\right) + B_F'\left(y_0^T\right)\left(y_T^T - y_0^T\right) + \frac{1}{2}B_F''\left(y_0^T\right)\left(y_T^T - y_0^T\right)^2.$$

(8.55)

Then, we take the $\mathbb{Q}_T$-expectation with the above equation, obtaining

$$
\begin{aligned}
B_F\left(y_0^T\right) &= E^{\mathbb{Q}_T}\left[B_F\left(y_T^T\right)\right] \\
&= B_F\left(y_0^T\right) + B_F'\left(y_0^T\right) E^{\mathbb{Q}_T}\left[y_T^T - y_0^T\right] \\
&\quad + \frac{1}{2}B_F''\left(y_0^T\right) E^{\mathbb{Q}_T}\left[\left(y_T^T - y_0^T\right)^2\right].
\end{aligned}
\tag{8.56}
$$

This gives an approximation to the expectation of the yield:

$$
E^{\mathbb{Q}_T}\left[y_T^T\right] = y_0^T - \frac{1}{2}E^{\mathbb{Q}_T}\left[\left(y_T^T - y_0^T\right)^2\right]\frac{B_F''\left(y_0^T\right)}{B_F'\left(y_0^T\right)}.
\tag{8.57}
$$

Because $B_F(y)$ is monotonically decreasing and convex, we have $B_F'(y) < 0$ and $B_F''(y) > 0$, implying that the order $E^{\mathbb{Q}_T}\left[y_T^T\right] \geq y_0^T$ holds with certainty.

The only problem left now is to estimate $E^{\mathbb{Q}_T}\left[\left(y_T^T - y_0^T\right)^2\right]$. Standard market practice is to assume that y_t^T follows lognormal dynamics under $\mathbb{Q}_T$,

$$
dy_t^T = y_t^T\left(\mu\,dt + \sigma_y\,dW_t\right),
\tag{8.58}
$$

where σ_y is constant while μ is a small deterministic function. The formal solution to Equation 8.58 is

$$
y_T^T = y_0^T\, e^{\left(\bar{\mu} - \frac{1}{2}\sigma_y^2\right)T + \sigma_y W_T},
\tag{8.59}
$$

where $\bar{\mu} = \int_0^T \mu\,dt/T$. Furthermore, by assuming $\bar{\mu}T \ll 1$, we have

$$
\begin{aligned}
E^{\mathbb{Q}_T}\left[\left(y_T^T - y_0^T\right)^2\right] &= \left(y_0^T\right)^2 E^{\mathbb{Q}_T}\left[\left(e^{\left(\bar{\mu} - \frac{1}{2}\sigma_y^2\right)T + \sigma_y W_T} - 1\right)^2\right] \\
&= \left(y_0^T\right)^2\left[e^{\left(2\bar{\mu} + \sigma_y^2\right)T} - 2e^{\bar{\mu}T} + 1\right] \\
&\approx \left(y_0^T\right)^2 \sigma_y^2 T.
\end{aligned}
\tag{8.60}
$$

Substituting the approximation (Equation 8.60) back to Equation 8.57, we finally arrive at a general formula for convexity adjustment:

$$
E^{\mathbb{Q}_T}\left[y_T^T\right] = y_0^T - \frac{1}{2}\left(y_0^T\right)^2 \sigma_y^2 T\frac{B_F''\left(y_0^T\right)}{B_F'\left(y_0^T\right)}.
\tag{8.61}
$$

As an example of the application of Equation 8.61, we take $f_j(T_j)$ for y_T^T and reinvestigate the forward–futures difference studied in Section 8.2.1. The price–yield relationship for the ΔT_j-maturity zero-coupon bond is

$$
B_F\left(y_t^T\right) = \frac{1}{1 + \Delta T_j y_t^T}.
\tag{8.62}
$$

Differentiating $B_F\left(y_t^T\right)$ repeatedly, we have

$$B_F'\left(y_t^T\right) = -\frac{\Delta T_j}{\left(1 + \Delta T_j y_t^T\right)^2},$$

$$B_F''\left(y_t^T\right) = \frac{2\left(\Delta T_j\right)^2}{\left(1 + \Delta T_j y_t^T\right)^3}. \tag{8.63}$$

Substituting Equation 8.63 into 8.61, we obtain the correction formula

$$E^{Q_T}\left[y_T^T\right] = y_0^T + \left(y_0^T\right)^2 \sigma_y^2 T_j \frac{\Delta T_j}{1 + \Delta T_j y_0^T}, \tag{8.64}$$

where $y_0^T = f_j(0)$. Note that the right-hand side of Equation 8.64 is nothing else but Equation 8.42. The above result suggests that the general formula (Equation 8.61) for convexity adjustment is consistent with the LIBOR market model.

We finish this section with the pricing of the contingent claim in Example 8.1.

Example 8.2.3. *Consider the valuation of Equation 8.38 using the following input parameters and term structure.*

1. *The volatility of the $f_3(t)$ is 20%.*

2. *The current yield curve is flat at 5% (with annual compounding).*

   Solution: The value of the instrument is

$$V = (1.05)^{-3} \times E^{Q_3}[y], \tag{8.65}$$

where y is the yield of the one-year bond three years forward. The relation between the yield and price of a one-year zero-coupon bond is

$$B_F(y) = \frac{1}{1+y}. \tag{8.66}$$

There are

$$B_F'(y) = \frac{-1}{(1+y)^2} \quad and \quad B_F''(y) = \frac{2}{(1+y)^3}. \tag{8.67}$$

Taking $y = 0.05$, we have

$$B_F'(0.05) = \frac{-1}{(1+0.05)^2} = -0.9070,$$

$$B_F''(0.05) = \frac{2}{(1+0.05)^3} = 1.7277. \tag{8.68}$$

The convexity adjustment is

$$0.5 \times 0.05^2 \times 0.2^2 \times 3 \times \frac{1.7277}{0.9070} = 0.00057, \tag{8.69}$$

or 5.7 basis points, so that the value of the instrument is $0.05057/(1.05)^3 = 0.043684$. Note that the price error would be 1.1% if there were no adjustment.

8.2.4 Convexity Adjustment for CMS and CMT Swaps

In the marketplace, convexity adjustment is most often applied to par yields or, equivalently, to swap rates. We now explain how the general formula (Equation 8.61) works for a swap rate. It is understood that a swap rate is the yield of a par bond. Suppose that the tenor of a forward starting swap is (T_m, T_n). Then its T-forward price of the par bond seen at time t is given by

$$F_t^T = \frac{P(t, T_m)}{P(t, T)}, \qquad (8.70)$$

which is a $\mathbb{Q}_T$-martingale. Let $y_t = R_{m,n}(t)$ be the prevailing swap rate seen at $t \le T_m$. Then y_t is the uniform discount rate for coupon bonds with tenor (T_m, T_n). Suppose that a swap contract is entered at time $t = 0$ with the swap rate y_0. Then, at any later time, t, the T-forward price of the fixed-rate leg of the swap initiated at $t = 0$ is given by

$$B_F(y_t^T) = \sum_{T_i > T, i > m} \frac{\Delta T y_0^T}{\left(1 + \Delta T y_t^T\right)^{(T_i - T)/\Delta T}} + \frac{1}{\left(1 + \Delta T y_t^T\right)^{(T_N - T)/\Delta T}} = F_t^T. \qquad (8.71)$$

The convexity adjustment formula for the swap rate follows as

$$E^{\mathbb{Q}_T}\left[y_T^T\right] = y_0^T - \frac{1}{2}(y_0^T)^2 \sigma_y^2 T \frac{B_F''(y_0^T)}{B_F'(y_0^T)},$$

Because $y_0^T = y_0$ and $y_T^T = y_T$, the formula can be recast as

$$E^{\mathbb{Q}_T}[y_T] = y_0 - \frac{1}{2}(y_0)^2 \sigma_y^2 T \frac{B_F''(y_0)}{B_F'(y_0)}. \qquad (8.72)$$

Let us apply the results above to an example.

Example 8.2.4. *Evaluate the derivative on the in-3-to-3 swap rate with payoff*

$$V_3 = 100 \times (R_{3,6}(3) - R_{3,6}(0))^+. \qquad (8.73)$$

Assume that the current yield curve is flat at 5%, and the volatility of the swap rate, $R_{3,6}(t)$, is 20%.

Solution: According to Equations 8.49 and 8.50, the value of the option is given by

$$V_0 = 100 P(0, 3) \left(E^{\mathbb{Q}_3}[R_{3,6}(3)] \, \Phi(d_1) - K \Phi(d_2)\right), \qquad (8.74)$$

where

$$P(0, 3) = (1.05)^{-3},$$

$$d_1 = \frac{\ln\left(E^{\mathbb{Q}_3}[R_{3,6}(3)]/R_{3,6}(0)\right) + (1/2)\,(0.2)^2 \times 3}{0.2\sqrt{3}}, \qquad (8.75)$$

$$d_2 = d_1 - 0.2\sqrt{3}.$$

The only input we do not yet know in Black's formula is $E^{\mathbb{Q}_3}[R_{3,6}(3)]$. For simplicity, we let $y_0 = R_{3,6}(0)$ and $y = R_{3,6}(3)$. For the par bond over the period $(3,6)$, the forward price–forward yield relation for the three-year bond is

$$B_F(y) = \frac{y_0}{1+y} + \frac{y_0}{(1+y)^2} + \frac{1+y_0}{(1+y)^3}. \tag{8.76}$$

Its first and second derivatives are

$$
\begin{aligned}
B_F'(y) &= -\frac{y_0}{(1+y)^2} - \frac{2y_0}{(1+y)^3} - \frac{3(1+y_0)}{(1+y)^4}, \\
B_F''(y) &= \frac{2y_0}{(1+y)^3} + \frac{6y_0}{(1+y)^4} + \frac{12(1+y_0)}{(1+y)^5}.
\end{aligned}
\tag{8.77}
$$

Taking $y = y_0 = 5\%$ in Equation 8.77, we obtain

$$B_F'(y_0) = -2.7232, \quad B_F''(y_0) = 10.2056. \tag{8.78}$$

Hence, the adjustment is

$$\frac{1}{2} \times 0.05^2 \times 0.2^2 \times 3 \times \frac{10.2056}{2.7232} = 0.00225, \tag{8.79}$$

and thus

$$E^{\mathbb{Q}_3}[R_{3,6}(3)] = 0.05 + 0.00225 = 0.05225. \tag{8.80}$$

Inserting Equation 8.80 into Equations 8.74 and 8.75, we finally obtain $V_0 = 0.7093$.

A very important application of convexity adjustment is in the evaluation of the CMSs and CMT swaps, which are agreements to exchange a particular par yield or Treasury yield for a fixed rate or LIBOR. For example, the most popular CMT exchanges 10-year Treasury yields for LIBOR or a fixed rate. Suppose the CMT swap is spot-starting, then the price of the CMT leg is given by

$$V_0 = \sum_{j=m+1}^{n} \Delta T P(0, T_j) E^{\mathbb{Q}_j}[y_N(T_j, T_j)] + P(0, T_n), \tag{8.81}$$

where $y_N(t, T_j)$ is the N-year Treasury yield beyond T_j seen at time t. The relationship between $y_N(t, T_j)$ and its corresponding T_j-forward bond price is

$$B_F(y_N) = \sum_{i=1}^{N/\Delta T} \frac{\Delta T y_N(0, T_j)}{(1 + \Delta T y_N)^i} + \frac{1}{(1 + \Delta T y_N)^{N/\Delta T}}, \tag{8.82}$$

where, for simplicity, we let y_N stand for $y_N(t, T_j)$. Then,

$$
\begin{aligned}
B_F'(y_N) &= \sum_{i=1}^{N/\Delta T} \frac{-i\,(\Delta T)^2 y_N(0, T_j)}{(1 + \Delta T y_N)^{i+1}} + \frac{-N}{(1 + \Delta T y_N)^{(N/\Delta T)+1}}, \\
B_F''(y_N) &= \sum_{i=1}^{N/\Delta T} \frac{i(i+1)\,(\Delta T)^3 y_N(0, T_j)}{(1 + \Delta T y_N)^{i+2}} + \frac{N(N + \Delta T)}{(1 + \Delta T y_N)^{(N/\Delta T)+2}}.
\end{aligned}
\tag{8.83}
$$

The convexity adjustment for the CMT swap rate is

$$E^{\mathbb{Q}_j}[y_N(T_j, T_j)] = y_N(0, T_j) - \frac{1}{2}\left(y_N(0, T_j)\right)^2 \sigma_{j,N}^2 T_j \frac{B_F''(y_N(0, T_j))}{B_F'(y_N(0, T_j))},$$

(8.84)

for $j = m+1, 2, \ldots, n$, where $\sigma_{j,N}$ is the percentage volatility of the swap rate, $y_N(t, T_j)$.

8.3 Timing Adjustment

In the previous section, we determined the difference between a T-forward yield and the $\mathbb{Q}_T$-expectation of the yield. The general formula for convexity adjustment, Equation 8.72, can be applied to pricing many derivatives on the yields that are fixed at payment dates. Yet, in the interest-rate derivatives market, it is also quite common to have payments made "in arrears," that is, to pay sometime later after fixing. A representative example is a CMS swap whose CMS leg is fixed at time T_j but the payment is made at time $T_{j+1}, j = 0, 1, \ldots, N - 1$. To price products with the feature of "paying in arrears," we need to develop a general formula for the valuation of $E^{\mathbb{Q}_{T+\Delta T}}[y_T^T]$, where y_t^T is the forward yield defined in Equation 8.71.

The expectation $E^{\mathbb{Q}_{T+\Delta T}}[y_T^T]$ can be derived from $E^{\mathbb{Q}_T}[y_T^T]$ by a change of measure. In fact, it can be easily seen that

$$\left.\frac{d\mathbb{Q}_{T+\Delta T}}{\mathbb{Q}_T}\right|_{\mathcal{F}_t} = \exp\left(\int_0^t -\frac{1}{2}\left\|\hat{\boldsymbol{\Sigma}}(s)\right\|^2 ds + \hat{\boldsymbol{\Sigma}}^T(s)\, d\mathbf{W}_s^T\right) = \zeta_t,$$

where $\mathbf{W}_s^T$ is a $\mathbb{Q}_T$-Brownian motion, and

$$\hat{\boldsymbol{\Sigma}}(t) = \boldsymbol{\Sigma}(t, T+\Delta T) - \boldsymbol{\Sigma}(t, T).$$

Also, we understand that

$$y_T^T = E^{\mathbb{Q}_T}[y_T^T] \exp\left(\int_0^T -\frac{1}{2}\left\|\boldsymbol{\sigma}_y\right\|^2 dt + \boldsymbol{\sigma}_y^T\, d\mathbf{W}_t^T\right).$$

(8.85)

Hence,

$$
\begin{aligned}
E^{Q_{T+\Delta T}} \left[y_T^T \right] &= E^{Q_T} \left[y_T^T \zeta_T \right] \\
&= E^{Q_T} \left[E^{Q_T} \left[y_T^T \right] \exp \left(\int_0^T -\frac{1}{2} \left(\| \boldsymbol{\sigma}_y \|^2 + \left\| \hat{\boldsymbol{\Sigma}} \right\|^2 \right) dt \right. \right. \\
&\qquad \left. \left. + \left(\boldsymbol{\sigma}_y + \hat{\boldsymbol{\Sigma}} \right)^T dW_t^T \right) \right] \\
&= E^{Q_T} \left[y_T^T \right] \exp \left(\int_0^T \boldsymbol{\sigma}_y^T \hat{\boldsymbol{\Sigma}} \, dt \right).
\end{aligned}
\tag{8.86}
$$

Under the market model,

$$
\hat{\boldsymbol{\Sigma}} = -\frac{\Delta T f_{\Delta T}(t, T)}{1 + \Delta T f_{\Delta T}(t, T)} \boldsymbol{\gamma}(t, T) \approx -\frac{\Delta T f_{\Delta T}(0, T)}{1 + \Delta T f_{\Delta T}(0, T)} \boldsymbol{\gamma}(t, T).
\tag{8.87}
$$

Assume that both $\boldsymbol{\sigma}_y$ and $\boldsymbol{\gamma}(t, T)$ are constants. We then end up with a general formula for timing adjustment:

$$
E^{Q_{T+\Delta T}} \left[y_T^T \right] = E^{Q_T} \left[y_T^T \right] \exp \left(-\rho \frac{\Delta T f_{\Delta T}(0, T)}{1 + \Delta T f_{\Delta T}(0, T)} \| \boldsymbol{\sigma}_y \| \cdot \| \boldsymbol{\gamma}(0, T) \| T \right),
\tag{8.88}
$$

where ρ is the correlation between y_t^T and $f(t; T, \Delta T)$.

Example 8.3.1. *Based on the set-up and the results from Example 8.3, we consider calculating $E^{Q_4}[f_3(3)]$ using Equation 8.88. Here, we have $T = 3$, $\Delta T = 1$ and $E^{Q_3}[f_3(3)] = 0.05057$. Suppose, in addition, that the yield curve moves in parallel. Then, there must be*

$$
\| \boldsymbol{\sigma}_y \| = \| \boldsymbol{\gamma}(0, 3) \| = 20\%, \quad \text{and} \quad \rho = 1.
\tag{8.89}
$$

Inserting all relevant numbers into Equation 8.88, we obtain

$$
E^{Q_4}[f_3(3)] = 0.05028.
$$

Note that the exact value is 0.05.

8.4 Quanto Derivatives

Quanto derivatives are cross-currency derivatives where two or more currencies are involved. They can be categorized into two types. The first type is a derivative that is measured and paid in a foreign currency, before being converted to the domestic currency; and the second type is a derivative that is measured in the foreign currency but paid in the domestic currency without conversion. An example of quanto derivatives of the second type is the CME futures contract on the Nikkei futures index, for which the underlying variable is the Nikkei index but the payoff is settled directly in USD. The pricing approaches of these two types of derivatives are very different, and the pricing of quanto derivatives of the first type is a lot easier.

We let

- $\mathbb{Q}_d$ be the martingale measure corresponding to the numeraire asset of a domestic money market account, $B_d(t)$;

- $\mathbb{Q}_f$ be the martingale measure corresponding to the numeraire asset of a foreign money market account, $B_f(t)$;

- x_t be the exchange rate: the value in domestic currency of one unit of foreign currency; and

- $y_t = 1/x_t$: the value in foreign currency of one unit of domestic currency.

Consider an option of the first type with a payoff function, $f(S_T)$, on a foreign asset, S_t. In domestic currency, the payoff is

$$V_T = x_T f(S_T). \tag{8.90}$$

To price this option, we take the foreign money market account as the numeraire and make use of the corresponding martingale measure, $\mathbb{Q}_f$. Note that the foreign money market account is a tradable asset with price $x_t B_f(t)$ in the domestic currency. Under $\mathbb{Q}_f$, we know that the price of the option is given by

$$V_0 = x_0 E^{\mathbb{Q}_f} \left[\frac{V_T}{x_T B_f(T)} \right]. \tag{8.91}$$

As an example, we consider a call option such that $f(S_T) = (S_T - K)^+$. Then,

$$V_0 = x_0 \left(S_0 \Phi(d_1) - P_f(0, T) K \Phi(d_2) \right), \tag{8.92}$$

where d_1 and d_2 take the usual values for a call option under the stochastic foreign interest rate. By examining Equations 8.91 and 8.92, we understand that general quanto options of the first type can be priced by taking expectations of the payoffs discounted in the foreign currency under the foreign risk-neutral measure, before converting the value to the domestic currency.

Quanto options of the second type are intriguing, and their pricing is less straightforward. We focus first on the pricing of quanto futures before applying relevant results to pricing quanto options.

Let S_t be a domestic asset whose price is, however, indexed to a foreign asset. As discussed in the first section of this chapter, the futures price of the asset is given by

$$\tilde{F}_t^d = E^{\mathbb{Q}_d}[S_T \mid \mathcal{F}_t] = E_t^{\mathbb{Q}_d}[S_T]. \tag{8.93}$$

We will try to calculate $\tilde{F}_t^{\mathrm{d}}$ based on the futures price in the foreign currency,

$$\tilde{F}_t^{\mathrm{f}} = E^{\mathbb{Q}_{\mathrm{f}}}[S_T \mid \mathcal{F}_t] = E_t^{\mathbb{Q}_{\mathrm{f}}}[S_T], \tag{8.94}$$

which is available in the foreign market. On the relationship between $\tilde{F}_t^{\mathrm{d}}$ and $\tilde{F}_t^{\mathrm{f}}$, we have

Theorem 8.4.1. *Suppose that both $\tilde{F}_t^{\mathrm{f}}$ and x_t are lognormally distributed with volatilities $\boldsymbol{\Sigma}_F$ and $\boldsymbol{\Sigma}_x$, respectively. Then, there is*

$$\tilde{F}_t^{\mathrm{d}} = \tilde{F}_t^{\mathrm{f}} e^{-\int_t^T \boldsymbol{\Sigma}_F^{\mathrm{T}} \boldsymbol{\Sigma}_x \mathrm{d}t}. \tag{8.95}$$

Proof. By a change of measure, we have

$$E_t^{\mathbb{Q}_{\mathrm{d}}}[S_T] = E_t^{\mathbb{Q}_{\mathrm{f}}}\left[S_T \frac{\zeta_T}{\zeta_t}\right], \tag{8.96}$$

where

$$\zeta_t = \left.\frac{\mathrm{d}\mathbb{Q}_{\mathrm{d}}}{\mathrm{d}\mathbb{Q}_{\mathrm{f}}}\right|_{\mathcal{F}_t} = \frac{y_t B_{\mathrm{d}}(t)}{y_0 B_{\mathrm{f}}(t)}, \tag{8.97}$$

and

$$B_{\mathrm{d}}(t) = e^{\int_0^t r_{\mathrm{d}}(s)\mathrm{d}s} \quad \text{and} \quad B_{\mathrm{f}}(t) = e^{\int_0^t r_{\mathrm{f}}(s)\mathrm{d}s}.$$

Here $r_{\mathrm{d}}(t)$ and $r_{\mathrm{f}}(t)$ are domestic and foreign short rates, respectively. Based on Equation 8.19, the futures–forward relationship, we can derive the following expression of the terminal value of asset prices under measure $\mathbb{Q}_{\mathrm{f}}$:

$$S_T = E_t^{\mathbb{Q}_{\mathrm{f}}}[S_T] e^{\int_t^T -\frac{1}{2}\|\boldsymbol{\Sigma}_F\|^2 \mathrm{d}s + \boldsymbol{\Sigma}_F^{\mathrm{T}}\mathrm{d}\mathbf{W}_{\mathrm{f}}(s)}, \tag{8.98}$$

where $\mathbf{W}_{\mathrm{f}}(t)$ is a $\mathbb{Q}_{\mathrm{f}}$-Brownian motion and $\boldsymbol{\Sigma}_F$ is the volatility of $\tilde{F}_t^{\mathrm{f}}$. Meanwhile, the process of y_t under $\mathbb{Q}_{\mathrm{f}}$ is

$$\mathrm{d}y_t = y_t\left((r_{\mathrm{f}} - r_{\mathrm{d}})\,\mathrm{d}t + \boldsymbol{\Sigma}_y^{\mathrm{T}}\mathrm{d}\mathbf{W}_{\mathrm{f}}(t)\right). \tag{8.99}$$

It follows that

$$\frac{\zeta_T}{\zeta_t} = \frac{y_T B_{\mathrm{d}}(T)/y_0 B_{\mathrm{f}}(T)}{y_t B_{\mathrm{d}}(t)/y_0 B_{\mathrm{f}}(t)} = e^{\int_t^T -\frac{1}{2}\|\boldsymbol{\Sigma}_y\|^2 \mathrm{d}s + \boldsymbol{\Sigma}_y^{\mathrm{T}}\mathrm{d}\mathbf{W}_{\mathrm{f}}(s)}. \tag{8.100}$$

By substituting Equations 8.98 and 8.100 into Equation 8.96, we then obtain

$$E_t^{\mathbb{Q}_{\mathrm{d}}}[S_T] = E_t^{\mathbb{Q}_{\mathrm{f}}}\left[E_t^{\mathbb{Q}_{\mathrm{f}}}[S_T] e^{\int_t^T -\frac{1}{2}\left(\|\boldsymbol{\Sigma}_F\|^2 + \|\boldsymbol{\Sigma}_y\|^2\right)\mathrm{d}s + (\boldsymbol{\Sigma}_F + \boldsymbol{\Sigma}_y)^{\mathrm{T}}\mathrm{d}\mathbf{W}_{\mathrm{f}}(s)}\right]$$

$$= E_t^{\mathbb{Q}_{\mathrm{f}}}[S_T] e^{\int_t^T \boldsymbol{\Sigma}_F^{\mathrm{T}}\boldsymbol{\Sigma}_y \mathrm{d}s}. \tag{8.101}$$

Finally, Equation 8.95 follows from the fact that $\boldsymbol{\Sigma}_x = -\boldsymbol{\Sigma}_y$. $\qquad\square$

Note that Equation 8.101 can be recast into the following form, which is better known:

$$E_t^{Q_d}[S_T] = E_t^{Q_f}[S_T]\, \mathrm{e}^{-\int_t^T \rho\sigma_F\sigma_x \mathrm{d}x}, \tag{8.102}$$

where $\sigma_F = \|\mathbf{\Sigma}_F\|$, $\sigma_x = \|\mathbf{\Sigma}_x\|$ and ρ is the correlation between F_t^T and x_t.

The proof for the theorem above carries useful insights on pricing quanto derivatives. Based on Equation 8.99, we can show, using the quotient rule of differentiation, that the Q_f-dynamics of x_t is

$$\mathrm{d}x_t = x_t\left[\left(r_d - r_f + \|\mathbf{\Sigma}_y\|^2\right)\mathrm{d}t - \mathbf{\Sigma}_y^{\mathrm{T}}\mathrm{d}\mathbf{W}_f(t)\right]. \tag{8.103}$$

On the other hand, it is well-known that the Q_d-dynamics of x_t is

$$\mathrm{d}x_t = x_t\left((r_d - r_f)\,\mathrm{d}t - \mathbf{\Sigma}_y^{\mathrm{T}}\mathrm{d}\mathbf{W}_d(t)\right), \tag{8.104}$$

where $\mathbf{W}_d(t)$ is a Q_d-Brownian motion. The coexistence of Equations 8.103 and 8.104 had once puzzled some people of the finance community. This is the so-called Siegel's paradox. Let us now dismiss this paradox. We can recast Equation 8.103 into

$$\mathrm{d}x_t = x_t\left[(r_d - r_f)\,\mathrm{d}t - \mathbf{\Sigma}_y^{\mathrm{T}}(\mathrm{d}\mathbf{W}_f(t) - \mathbf{\Sigma}_y\,\mathrm{d}t)\right]. \tag{8.105}$$

According to Equation 8.100, ζ_t satisfies

$$\mathrm{d}\zeta_t = \zeta_t\mathbf{\Sigma}_y^{\mathrm{T}}\mathrm{d}\mathbf{W}_f(t). \tag{8.106}$$

Therefore, we can write

$$\mathrm{d}\mathbf{W}_f(t) - \mathbf{\Sigma}_y\mathrm{d}t = \mathrm{d}\mathbf{W}_f(t) - \langle\mathrm{d}\mathbf{W}_f(t), \frac{\mathrm{d}\zeta_t}{\zeta_t}\rangle. \tag{8.107}$$

Because ζ_t is the Radon–Nikodym derivative of Q_d with respect to Q_f, according to the CMG theorem, $\mathbf{W}_d(t) = \mathbf{W}_f(t) - \int_0^t \mathbf{\Sigma}_y\,\mathrm{d}s$ is a Q_d-Brownian motion. Hence, Equation 8.104 is reconciled with Equation 8.103.

Next, let us try to figure out the intuition behind the adjustment formula (Equation 8.102). Let ρ be the correlation between x_t and $\tilde{F}_t^f$. To replicate the quanto futures contract, we should maintain y_t units of the foreign futures contract on the foreign asset at any time, $t < T$. This is a dynamic strategy. The P&L over a small interval of time $(t, t + \Delta t)$, is

$$\Delta\tilde{F}_t^f y_t \times x_{t+\Delta t} = \Delta\tilde{F}_t^f y_t(x_t + \Delta x_t)$$

$$= \Delta\tilde{F}_t^f + \Delta\tilde{F}_t^f\frac{\Delta x_t}{x_t}$$

$$\approx \Delta\tilde{F}_t^f + \rho\sigma_F\sigma_x\tilde{F}_t^f\Delta t. \tag{8.108}$$

In Equation 8.108, we can see that the P&L comes from two sources. The first is of course the change in the futures price, and the second is the change in the exchange rate. The first term of Equation 8.108 is symmetric to upward or downward moves of the futures price, yet the second term is not. Suppose, for example, that there is a positive correlation between x_t and $\tilde{F}_t$. Then, the second term is positive regardless of the direction of movement of the futures price. If the futures price moves up, the hedger makes a profit in the contract, which is likely to be accompanied by a higher exchange rate. If, on the other hand, the futures price moves down, then the hedger suffers a loss, but it is likely to be subject to a lower exchange rate. Thus, if the price for the quanto futures is set at $\tilde{F}_t^{\mathrm{d}} = \tilde{F}_t^{\mathrm{f}}$, smart investors will short the futures and hedge this position with the foreign futures, until the $\tilde{F}_t^{\mathrm{d}}$ is driven down to the level stated in Equation 8.102.

As a general result, we now derive the price process of the foreign asset under the domestic risk-neutral measure, $\mathbb{Q}_{\mathrm{d}}$. It is well-known that, under the foreign risk-neutral measure, $\mathbb{Q}_{\mathrm{f}}$, the price of a foreign asset follows

$$dS_t = S_t\left(r_{\mathrm{f}}\,dt + \boldsymbol{\Sigma}_S^{\mathrm{T}}d\mathbf{W}_{\mathrm{f}}(t)\right). \tag{8.109}$$

Define a $\mathbb{Q}_{\mathrm{d}}$-Brownian motion, $\mathbf{W}_{\mathrm{d}}(t)$, as

$$d\mathbf{W}_{\mathrm{d}}(t) = d\mathbf{W}_{\mathrm{f}}(t) - \langle d\mathbf{W}_{\mathrm{f}}(t), \frac{d\zeta_t}{\zeta_t}\rangle$$

$$= d\mathbf{W}_{\mathrm{f}}(t) + \boldsymbol{\Sigma}_x dt. \tag{8.110}$$

Then, in terms of $\mathbf{W}_{\mathrm{d}}(t)$, we have the price process of S_t under $\mathbb{Q}_{\mathrm{d}}$ to be

$$dS_t = S_t\left(\left(r_{\mathrm{f}} - \boldsymbol{\Sigma}_S^{\mathrm{T}}\boldsymbol{\Sigma}_x\right)dt + \boldsymbol{\Sigma}_S^{\mathrm{T}}d\mathbf{W}_{\mathrm{d}}(t)\right). \tag{8.111}$$

When interest rates in both currencies are deterministic, the above "risk-neutralized" process can be applied directly to pricing quanto options. Our interest, of course, is in the pricing of quanto options under stochastic interest rates.

Now, let us consider the pricing of a quanto call option of the second type. The payoff of the option,

$$V_T = (S_T - K)^+, \tag{8.112}$$

is indexed to the foreign asset in the foreign currency, but paid in the domestic currency without conversion. Because $\tilde{F}_T^{\mathrm{d}} = S_T$ by the definition of a futures price, we can rewrite the payoff as

$$V_T = \left(\tilde{F}_T^{\mathrm{d}} - K\right)^+ \text{ dollars.} \tag{8.113}$$

By the martingale property of the futures price, we have the following process of $\tilde{F}_t^{\mathrm{d}}$ under the domestic risk-neutral measure,[3] $\mathbb{Q}_{\mathrm{d}}$:

$$d\tilde{F}_t^{\mathrm{d}} = \tilde{F}_t^{\mathrm{d}}\boldsymbol{\Sigma}_F^{\mathrm{T}}\,d\mathbf{W}_{\mathrm{d}}(t). \tag{8.114}$$

---

[3]Equation 8.114 can also be derived directly from Equations 8.19 and 8.96.

Furthermore, under the forward measure, $\mathbb{Q}_d^T$, Equation 8.114 becomes

$$\mathrm{d}\tilde{F}_t^d = \tilde{F}_t^d \mathbf{\Sigma}_F^T \left(\mathrm{d}\mathbf{W}_d^T(t) + \mathbf{\Sigma}_{P_d} \mathrm{d}t \right), \tag{8.115}$$

where $\mathbf{W}_d^T(t)$ is a $\mathbb{Q}_d^T$-Brownian motion and $P_d = P(t,T)$ stands for the domestic zero-coupon bond. It follows that

$$V_0 = E^{\mathbb{Q}_d} \left[\frac{\left(\tilde{F}_T^d - K \right)^+}{B_d(T)} \right]$$

$$= P_d(0,T) E^{\mathbb{Q}_d^T} \left[\left(\tilde{F}_T^d - K \right)^+ \right]$$

$$= P_d(0,T) \left(E^{\mathbb{Q}_d^T} \left[\tilde{F}_T^d \right] \Phi(d_1) - K\Phi(d_2) \right), \tag{8.116}$$

with

$$d_1 = \frac{\ln \left(E^{\mathbb{Q}_d^T} \left[\tilde{F}_T^d \right] / K \right) + (1/2)\sigma_F^2 T}{\sigma_F \sqrt{T}}, \quad d_2 = d_1 - \sigma_F \sqrt{T},$$

$$\sigma_F^2 = \frac{1}{T} \int_0^T \| \mathbf{\Sigma}_F \|^2 \, \mathrm{d}t, \tag{8.117}$$

and, based on Equation 8.115,

$$E^{\mathbb{Q}_d^T} \left[\tilde{F}_T^d \right] = \tilde{F}_0^d \exp \left(\int_0^T \mathbf{\Sigma}_F^T \mathbf{\Sigma}_{P_d} \, \mathrm{d}t \right)$$

$$= \tilde{F}_0^d \exp \left(\int_0^T \rho_{F,P_d} \sigma_F \sigma_{P_d} \, \mathrm{d}t \right). \tag{8.118}$$

When $\rho_{F,P_d} = 0$, there is apparently $E^{\mathbb{Q}_d^T} \left[\tilde{F}_T^d \right] = \tilde{F}_0^d = E^{\mathbb{Q}_d} \left[\tilde{F}_T^d \right]$. Furthermore, when there is no correlation between x_t and $\tilde{F}_t^f$, Equation 8.116 reduces to Black's formula for call options on domestic futures.

8.5 Notes

In Equation 8.72, the convexity adjustment formula for a swap rate, we need to input σ_y, the volatility of the swap rate. The natural candidate of σ_y is the implied volatility of the ATM swaption of the same maturity. By doing so, we have the understanding that, if necessary, each piece of cash flows of a CMS swap (a so-called swaplet) will be hedged with the swaption on the corresponding ATM swap. Such a hedge ought to be replaced once the ATM swap rate has changed.

It was found by Hagan (2003) that a swaplet can be approximately replicated with swaptions of the same maturity across various strikes. In the marketplace, there are quotes for implied volatilities of at most a dozen strikes. To infer the implied volatilities for many other strikes needed for the replication, Hagan adopted the so-called SABR model (Hagan et al., 2002), a volatility smile model created by him and his colleagues, for swap rates. In this approach one will get an alternative formula for convexity adjustment, which is expressed as an integral of swaption values and is said to be "smile consistent." We refer readers to Mercurio and Pallavicini (2006) for an improved replication strategy and its corresponding adjustment formula.

Exercises

1. Prove the equality

$$\tilde{F}_0^T - F_0^T = \frac{-1}{P(0,T)} \times \mathrm{Cov}^{\mathbb{Q}}\big(B_T^{-1}, S_T\big),$$

where $\mathbb{Q}$ stands for the risk-neutral measure.

2. Price a cash flow in three years that is equal to \$100 times the three-year swap rate at the time. Let the forward-rate curve be

$$f_j(0) = 0.03 + 0.0006 \times j, \quad \Delta T_j = 0.5, \quad \forall j,$$

and the volatility of the swap rate be 30%. What will be the price if the payment is made in three and a half years?

3. (Continued from Problem 2) If the cash flow is instead from a call option on the swap rate with a strike rate of 4%, what should be the price of the option with and without payment in arrears?

4. For all problems below, we use the spot term structure

$$f_j(0) = 0.03 + 0.0003 \times j, \quad \Delta T_j = 0.25, \quad \forall j.$$

Pricing is done under the market model (which assumes lognormal forward rates or swap rates). The payment frequency is a quarter year for caps, and half a year for swaps.

   a. Price the ATM caplet (the strike rate equals to forward rate $f(0; 1, 0.25)$) that matures in one-year on three-month LIBOR rate. Suppose the payoff is made in one year instead of in 1.25 year (so convexity adjustment is needed). Assume a 20% forward-rate volatility.

b. Price the five-year maturity ATM option on five-year swap rate (i.e., $R_{10,20}(5)$; the strike rate is $R_{10,20}(0)$, and $\Delta T = 0.5$). Assume a 20% swap-rate volatility. (Note that this is not a swaption and convexity adjustment will be needed.)

5. Price the Nikkei quanto futures that pay in USD. The maturity of the futures contract is three months, and the correlation between the index and the Yen-dollar exchange rate (i.e., value in USD per Yen) is -20%, and the current Nikkei index is 12,117. Describe the hedging strategy using the standard Nikkei index futures.

6. (Continued from Problem 5) Suppose that the volatility of Nikkei index is 25%, price a quanto call option on the index that matures in three months with a strike price 12,100.

Chapter 9

Affine Term Structure Models

The class of ATSMs (Duffie and Kan, 1996; Duffie and Singleton, 1999; Dai and Singleton, 2000; etc.) holds an important position in the literature of interest-rate derivatives models. After the establishment of the HJM framework for arbitrage-free interest-rate models, research on interest-rate models has moved forward largely in two directions—represented by the market model and the ATSMs. The basic feature of ATSMs is that the short rate is an affine function of some Markov state variables; the latter can follow diffusion, jump-diffusion, or Lévy dynamics (Sato, 1999). There are several reasons that make the ATSMs attractive. First, an ATSM is usually parsimonious in its formulary. One only needs to specify a few parameters of the model, which still has a rich capacity to describe the dynamics of the term structure of interest rates. Second, the class of ATSMs has a relatively high degree of analytical tractability, so that vanilla options on bonds or interest rates can often be priced semi-analytically (through Laplace transforms). Third, the class of ATSMs serves as a rather natural framework for models that keep interest rates positive.[1] Last but not least, owing to their Markov property, affine models are well suited for pricing path-dependent options, through either Monte Carlo simulation methods, or when the number of factors is small, lattice, or finite difference methods. In the community of academic finance, ATSMs are often the choice of models for interest rates. There is also a huge literature on empirical studies with ATSMs.

In this chapter, we limit our discussions to affine models that are driven by Brownian diffusions only. We go over a cycle of model construction and model applications, from fitting a model to current term structures of interest rates, to pricing bond options and swaptions under either one-factor or multi-factor ATSMs. Moreover, we address the numerical implementation of the models, which makes use of fast Fourier transforms (FFTs) and thus represents another dimension of the numerical methodology for option pricing. For ATSMs based on more general dynamics, we refer readers to Duffie, Pan, and Singleton (2000), Filipovic (2001), and Duffie, Filipovic, and Schachermayer (2002), among others.

---

[1] In the context of credit risk modeling, affine models allow users to generate models for positive credit spreads.

9.1 An Exposition with One-Factor Models

Instead of starting with an affine function for the short rate, we start, equivalently, with an "exponential affine" function for zero-coupon bonds. Under the risk-neutral measure, the price of zero-coupon bonds can be written as an exponential affine function of an Ito's process, X_t, such that

$$P(t, T) = e^{\alpha(t, T) + \beta(t, T) X_t}, \tag{9.1}$$

where $\alpha(t, T)$ and $\beta(t, T)$ are deterministic functions satisfying

$$\alpha(t, t) = \beta(t, t) = 0, \quad \forall t,$$

and X_t follows

$$dX_t = \mu_X(X_t, t)\, dt + \sigma_X(X_t, t)\, dW_t. \tag{9.2}$$

Here W_t is a Brownian motion under the risk-neutral measure.

In an exponential affine model with zero-coupon bonds, yields, forward rates, and the short rate are all affine functions. In fact, from the price–yield relationship of zero-coupon bonds, we readily have

$$y(t, T) = -\frac{1}{T - t} \ln P(t, T) = -\frac{1}{T - t} (\alpha(t, T) + \beta(t, T) X_t). \tag{9.3}$$

For forward rates, $f(t, T)$ of any T, we have

$$\int_t^T f(t, s)\, ds = -\ln P(t, T) = -\alpha(t, T) - \beta(t, T) X_t. \tag{9.4}$$

Thus,

$$f(t, T) = -\alpha_T(t, T) - \beta_T(t, T) X_t, \tag{9.5}$$

where the subindex, "T," stands for the partial derivative with respect to T. Putting $T = t$, we obtain a linear functional for the short rate as well:

$$r_t = f(t, t) = -\alpha_T(t, t) - \beta_T(t, t) X_t. \tag{9.6}$$

For notational simplicity, we denote $\delta_0(t) = -\alpha_T(t, t)$, $\delta(t) = -\beta_T(t, t)$ hereafter. If we start from an affine function, (Equation 9.6), for the short rate, we can also show, through Lemma 9.1, that the zero-coupon bond price is an exponential affine function.

Next, we derive conditions on $\alpha(t, T)$ and $\beta(t, T)$ so that the model is arbitrage-free. The risk-neutral process of the zero-coupon bond is

$$dP(t, T) = P(t, T) (r_t\, dt + \Sigma(t, T)\, dW_t). \tag{9.7}$$

By Ito's lemma, we can obtain the following expressions for the short rate and the volatility of the zero-coupon bond in terms of $P(t, T)$:

$$r_t = \frac{1}{P(t,T)} \left[\frac{\partial P}{\partial t} + \mu_X \frac{\partial P}{\partial X} + \frac{1}{2} \sigma_X^2 \frac{\partial^2 P}{\partial X^2} \right],$$

$$\Sigma(t, T) = \frac{1}{P(t,T)} \frac{\partial P}{\partial X} \sigma_X. \tag{9.8}$$

Using the functional form of $P(t, T)$ in Equation 9.1, we can make Equation 9.8 more specific:

$$r_t = \alpha_t + \beta_t X_t + \mu_X \beta + \frac{1}{2} \sigma_X^2 \beta^2,$$

$$\Sigma(t, T) = \beta(t, T) \sigma_X (X_t, t). \tag{9.9}$$

By comparing the first equation of Equation 9.9 with Equation 9.6, we obtain the equality

$$\alpha_t + \beta_t X_t + \mu_X \beta + \frac{1}{2} \sigma_X^2 \beta^2 = \delta_0(t) + \delta(t) X_t, \tag{9.10}$$

or

$$\frac{1}{2} \sigma_X^2 \beta^2 + \mu_X \beta = (\delta(t) - \beta_t) X_t + (\delta_0(t) - \alpha_t). \tag{9.11}$$

The above equality suggests that σ_X^2 and μ_X should be affine functions as well. In fact, this is indeed the case. To see that, we take two distinct maturities, T_1 and T_2, for Equation 9.11, resulting in two joint equations for (μ_X, σ_X^2):

$$\begin{pmatrix} \beta(t, T_1) & (1/2)\beta^2(t, T_1) \\ \beta(t, T_2) & (1/2)\beta^2(t, T_2) \end{pmatrix} \begin{pmatrix} \mu_X \\ \sigma_X^2 \end{pmatrix}$$

$$= \begin{pmatrix} \delta(t) - \beta_t(t, T_1) \\ \delta(t) - \beta_t(t, T_2) \end{pmatrix} X_t + \begin{pmatrix} \delta_0(t) - \alpha_t(t, T_1) \\ \delta_0(t) - \alpha_t(t, T_2) \end{pmatrix}. \tag{9.12}$$

If the coefficient matrix is non-singular, we can solve for (μ_X, σ_X^2) and obtain a solution also in affine form. It is obvious that the coefficient matrix is singular only for the case when $\beta(t, T_1) = \beta(t, T_2)$ or the case when one of the betas equals zero. Suppose that the dependence of $\beta(t, T)$ on T is genuine. Then, the matrix is in general non-singular, so (μ_X, σ_X^2) must be affine functions. Hence, we can write

$$\mu_X(x, t) = K_0(t) + K_1(t)x,$$

$$\sigma_X^2(x, t) = H_0(t) + H_1(t)x. \tag{9.13}$$

Plugging in the above functional forms into Equation 9.11, we then obtain the equation

$$\left(\frac{1}{2} H_1(t)\beta^2 + K_1(t)\beta \right) X_t + \frac{1}{2} H_0(t)\beta^2 + K_0(t)\beta$$

$$= (\delta(t) - \beta_t) X_t + (\delta_0(t) - \alpha_t). \tag{9.14}$$

By equating the corresponding coefficients of the terms,[2] $\{1, X_t\}$, in both sides of Equation 9.14, we obtain the following equations for α and β:

$$
\begin{aligned}
\beta_t &= \delta(t) - K_1(t)\beta - \frac{1}{2}H_1(t)\beta^2, \quad \beta(T, T) = 0, \\
\alpha_t &= \delta_0(t) - K_0(t)\beta - \frac{1}{2}H_0(t)\beta^2, \quad \alpha(T, T) = 0.
\end{aligned}
\tag{9.15}
$$

The first equation in Equation 9.15 is a Riccarti equation. Its solution is finite under certain technical conditions on H_1 and K_1. The solution of α follows from that of β:

$$
\alpha(t, T) = -\int_t^T \delta_0(s)\,\mathrm{d}s + \bar{\alpha}(t, T),
\tag{9.16}
$$

where

$$
\bar{\alpha}(t, T) = \int_t^T \left[K_0(s)\beta(s, T) + \frac{1}{2}H_0(s)\beta^2(s, T) \right] \mathrm{d}s.
$$

Note that we have decomposed $\alpha(t, T)$ in such a way that the first term will depend on the term structure of interest rates, whereas the second term will not.

In applications, δ is often fixed as a positive constant. Because of that, both $\beta(t, T)$ and $\bar{\alpha}(t, T)$ do not depend on the term structure of interest rates. Function $\delta_0(t)$, meanwhile, is subject to the term structure of interest rates. In fact, given the term structure of $P(0, T), \forall T, \delta_0(t)$ satisfies the equation

$$
P(0, T) = \exp\left(-\int_0^T \delta_0(s)\,\mathrm{d}s + \bar{\alpha}(0, T) + \beta(0, T)X_0 \right),
\tag{9.17}
$$

or

$$
\int_0^T \delta_0(s)\,\mathrm{d}s = -\ln P(0, T) + \bar{\alpha}(0, T) + \beta(0, T)X_0.
\tag{9.18}
$$

Differentiating Equation 9.18 with respect to T, we obtain

$$
\delta_0(T) = f(0, T) + \bar{\alpha}_T(0, T) + \beta_T(0, T)X_0.
\tag{9.19}
$$

Hence, by taking $\delta_0(T)$ as in Equation 9.19, we ensure that the affine model is consistent with the term structure of interest rates, which is a premise for arbitrage-free pricing of derivatives.

We make another comment that, when the function of δs, Ks, and Hs are constants, we can argue that α and β are functions of time to maturity, that is,

$$
\begin{aligned}
\alpha(t, T) &= \alpha_0(T - t), \quad \alpha_0(0) = 0, \\
\beta(t, T) &= \beta_0(T - t), \quad \beta_0(0) = 0,
\end{aligned}
$$

---

[2] This is called the "matching principle" in the finance literature.

for some functions, α_0 and β_0.

Next, we consider option pricing under the ATSM. We define the moment-generating function with discounting as

$$\varphi(u; X_t, t, T) = E^{\mathbb{Q}}\left[\exp\left(-\int_t^T r(X_s, s)\,ds\right) e^{u X_T} \mid \mathcal{F}_t\right], \qquad (9.20)$$

where $\mathbb{Q}$ stands for the risk-neutral measure. For the function φ, we have:

Lemma 9.1.1. *Suppose that both H_0 and H_1 are bounded and there is*

$$E^{\mathbb{Q}}\left[\left(\int_0^T |X_t|\,dt\right)\right] < \infty. \qquad (9.21)$$

Then, we have

$$\varphi(u; X_t, t, T) = e^{\alpha_u(t,T)+\beta_u(t,T)X_t}, \qquad (9.22)$$

where α_u and β_u are bounded solutions to the following ordinary differential equations (ODE):

$$\begin{aligned}(\beta_u)_t &= \delta - K_1\beta_u - \frac{1}{2}H_1\,\beta_u^2, \quad \beta_u(T,T) = u,\\[4pt](\alpha_u)_t &= \delta_0 - K_0\beta_u - \frac{1}{2}H_0\,\beta_u^2, \quad \alpha_u(T,T) = 0.\end{aligned} \qquad (9.23)$$

Proof. Define an auxiliary function

$$\tilde{\varphi}_t(u) = \exp\left(-\int_0^t r(X_s, s)\,ds\right) e^{\alpha_u(t,T)+\beta_u(t,T)X_t}. \qquad (9.24)$$

It suffices to show that $\tilde{\varphi}_t$ is a martingale. Then, we have

$$\begin{aligned}E_t^{\mathbb{Q}}[\tilde{\varphi}_T(u)] &= E_t^{\mathbb{Q}}\left[\exp\left(-\int_0^T r(X_s, s)\,ds\right) e^{u X_T}\right]\\[4pt]&= \tilde{\varphi}_t(u)\\[4pt]&= \exp\left(-\int_0^t r_s\,ds\right) \varphi(u; X_t, t, T),\end{aligned} \qquad (9.25)$$

and Equation 9.22 then follows. To prove that $\tilde{\varphi}_t$ is a martingale, we look at the differential for $\tilde{\varphi}_t$:

$$d\tilde{\varphi}_t = \mu_\varphi(t)\,dt + \eta_\varphi(t)\,dW_t, \qquad (9.26)$$

where, by Ito's Lemma,

$$
\begin{aligned}
\mu_\varphi(t) &= \frac{\partial \tilde\varphi_t}{\partial t} + \mu_X \frac{\partial \tilde\varphi_t}{\partial X} + \frac{1}{2}\sigma_X^2 \frac{\partial^2 \tilde\varphi_t}{\partial X^2} \\
&= e^{-\int_0^t r_s ds} \left[(\alpha_u)_t + (\beta_u)_t X_t - r_t + \mu_X \beta_u + \frac{1}{2}\sigma_X^2 \beta_u^2 \right] \\
&= e^{-\int_0^t r_s ds} \left[(\alpha_u)_t + (\beta_u)_t X_t - (\delta_0 + \delta X_t) \right. \\
&\qquad \left. + (K_0 + K_1 X_t)\beta_u + \frac{1}{2}(H_0 + H_1 X_t)\beta_u^2 \right] \\
&= e^{-\int_0^t r_s ds} \left[\left((\alpha_u)_t - \delta_0 + K_0\beta_u + \frac{1}{2}H_0\beta_u^2 \right) \right. \\
&\qquad \left. + \left((\beta_u)_t - \delta + K_1\beta_u + \frac{1}{2}H_1\beta_u^2 \right) X_t \right] \\
&= 0. \tag{9.27}
\end{aligned}
$$

Meanwhile,

$$
\eta_\varphi = \sigma_X \frac{\partial \tilde\varphi_t}{\partial X} = \sigma_X \beta_u(t, T)\tilde\varphi_t. \tag{9.28}
$$

Owing to Equation 9.21 and the boundedness of H_0 and H_1, we also have

$$
E^{\mathbb{Q}} \left[\int_0^T (\sigma_X \beta_u(t, T))^2 \, dt \right] < \infty. \tag{9.29}
$$

Hence, $\tilde\varphi_t$ is a lognormal martingale.   □

Remark 9.1.1. When $u = 0$, $\varphi(u; X_t, t, T)$ *gives the price of the T-maturity zero-coupon bond. By comparing Equation 9.15 with 9.23, we know that* $\beta_0 = \beta$ *and* $\alpha_0 = \alpha$.

Next, we consider general option pricing under the affine model following an approach pioneered by Carr and Madan (1998). In this approach, the value of an option is regarded, whenever possible, as a convolution between the density function of the state variable(s) and the payoff function. By the Fourier convolution theorem (Rudin, 1976), the Laplace transform of the option is the product of the Laplace transforms of the density function and the payoff function. If we have the Laplace transforms of both the moment-generating function and the option payoff in closed form, we also have the Laplace transform of the option in closed form. The option's value, then, can be obtained by doing an inverse Laplace transform, which, in addition, can be implemented using the technology of FFT, thus making the method very fast.

Without loss of generality, we demonstrate Carr–Madan's approach with the pricing of a call option on a zero-coupon bond. The maturity of the bond is T', while the maturity of the option is $T < T'$. The payoff of the option is

$$V_T = \left(e^{\alpha + \beta X_T} - K\right)^+. \tag{9.30}$$

Here, $\alpha = \alpha_0$ and $\beta = \beta_0$, and we have suppressed the indices of α and β for notational simplicity. Let $k = \ln K$. Under the risk-neutral measure, $\mathbb{Q}$, we have

$$\begin{aligned}
V_t &= E_t^{\mathbb{Q}}\left[e^{-\int_t^T r_s\, ds}\left(e^{\alpha + \beta X_T} - K\right)^+\right]\\
&= E_t^{\mathbb{Q}}\left[e^{-\int_t^T r_s\, ds}\left(e^{\alpha + \beta X_T} - e^k\right)^+\right]\\
&\triangleq G(k). \tag{9.31}
\end{aligned}$$

Noticing that when $k \to -\infty$, $G(k)$ does not tend to zero, and thus the Fourier transform does not exist. For this reason, we pick a positive number, $a > 0$, and consider the damped option price, $g(k) = e^{ak}G(k)$. It can be shown (Lee, 2004) that $g(k)$ is bounded and decays exponentially for $|k| \to \infty$, its Fourier transform then exists and, moreover, we can apply the Fourier transform across the expectation in Equation 9.31:

$$\begin{aligned}
&\int_{-\infty}^{\infty} e^{(a+iu)k} G(k)dk\\
&= \int_{-\infty}^{\infty} e^{iuk} e^{ak} E_t^{\mathbb{Q}}\left[e^{-\int_t^T r_s ds}\left(e^{\alpha+\beta X_T} - e^k\right)^+\right]dk\\
&= E_t^{\mathbb{Q}}\left[e^{-\int_t^T r_s ds}\int_{-\infty}^{\infty} e^{(iu+a)k}\left(e^{\alpha+\beta X_T} - e^k\right)^+ dk\right]\\
&= E_t^{\mathbb{Q}}\left[e^{-\int_t^T r_s ds}\int_{-\infty}^{\alpha+\beta X_T} e^{(iu+a)k}\left(e^{\alpha+\beta X_T} - e^k\right) dk\right]\\
&= E_t^{\mathbb{Q}}\left[e^{-\int_t^T r_s ds}\left(\frac{e^{(iu+a+1)(\alpha+\beta X_T)}}{iu+a} - \frac{e^{(iu+a+1)(\alpha+\beta X_T)}}{iu+a+1}\right)\right]\\
&= \frac{e^{(iu+a+1)\alpha}}{(iu+a)(iu+a+1)}\varphi\left((1+a+iu)\beta; X_t, t, T\right)\\
&\triangleq \psi(u). \tag{9.32}
\end{aligned}$$

Note that the Fourier transform on the damped function is identical to the Laplace transform on the original function. With Equation 9.32, we show that the Laplace transform of $G(k)$ can be obtained explicitly in terms of the (discounted) moment-generating function of X_T. Once we have $\psi(u)$, we can calculate $G(k)$ through an inverse Laplace transform:

$$G(k) = \frac{e^{-ak}}{\pi}\int_0^{\infty} e^{-iuk}\psi(u)\, du. \tag{9.33}$$

We remark here that the above approach works as long as one has the Laplace transforms of both the moment-generating function and the payoff function. The fifth line of Equation 9.32 contains the Laplace transform of the payoff of a call option. For the Laplace transforms of other types of option payoffs, we refer readers to Lee (2004), where rigorous analysis is made on the feasibility to exchange the order of the Laplace transform and the expectation.

The inverse Laplace transform will be evaluated numerically. For that purpose, we need to truncate the infinite domain at a finite number. To choose this number, say, A, let us estimate the error of the numerical integration caused by the truncation. Let $z = (1+a)\beta$ and assume $X_t = 0$. According to the definition of $\varphi(\cdot)$, we have

$$|\varphi\left((1+a+iu)\,\beta; X_t, t, T\right)| \le |\varphi\left((1+a)\,\beta; X_t, t, T\right)| = e^{\alpha_z(t,T)}.$$

It follows that

$$|\psi(u)| \le \left|\frac{e^{(a+1)\alpha + \alpha_z(t,T)}}{(iu+a)(iu+a+1)}\right| \le \frac{\left|e^{(a+1)\alpha + \alpha_z(t,T)}\right|}{u^2 + a^2},$$

and

$$\left|\int_A^\infty e^{-iuk}\psi(u)\,du\right| \le \int_A^\infty \frac{\left|e^{(a+1)\alpha + \alpha_z(t,T)}\right|}{u^2 + a^2}\,du \le \frac{\left|e^{(a+1)\alpha + \alpha_z(t,T)}\right|}{A}.$$

Hence, to ensure accuracy on the order of one basis point, we may truncate the integral at $A = 10^4$. Our computational experiences, however, suggest that such a truncation bound is excessively large.

A direct numerical evaluation of the truncated integral of Equation 9.33 is easy but not efficient. To make the evaluation fast, Carr and Madan (1998) introduce a very delicate technique that converts the numerical integration into a discrete FFT. The technique is described below. After a truncation is adopted, we consider the *composite trapezoidal rule* for the numerical integration:

$$H(k) = \frac{1}{\pi}\left(\frac{\psi(0)}{2} + \sum_{m=1}^{N-1} e^{-iu_m k}\psi(u_m) + \frac{e^{-iu_N k}\psi(u_N)}{2}\right)\Delta u, \qquad (9.34)$$

where $u_m = m\Delta u$ and $\Delta u = A/N$. The composite trapezoidal rule has an order of accuracy of $O(\Delta u^2)$. Since we are interested mainly in the around-the-money options, we take k around zero:

$$k_n = -b + n\Delta k, \quad \text{for some } b > 0 \text{ and } n = 0, 1, \ldots, N-1,$$

with

$$\Delta k = \frac{2b}{N}.$$

Hence, for $n = 0, 1, \ldots, N-1$, we have

$$H(k_n) = \frac{1}{\pi} \left(\frac{\psi(0)}{2} + \sum_{m=1}^{N-1} e^{-i\Delta u \Delta k m n} \left[e^{i b u_m} \psi(u_m) \right] \right.$$
$$\left. + \frac{e^{-i\Delta u \Delta k N n} \psi(u_N)}{2} \right) \Delta u.$$

We now choose, in particular

$$\Delta u \Delta k = \frac{2\pi}{N}, \quad \text{or} \quad b = \frac{\pi N}{A},$$

which will then result in

$$H(k_n) = \frac{1}{\pi} \left(\frac{\psi(0)}{2} + \sum_{m=1}^{N-1} e^{-i\Delta u \Delta k m n} \left[e^{i b u_m} \psi(u_m) \right] \right.$$
$$\left. + \frac{e^{-i\Delta u \Delta k N n} \psi(u_N)}{2} \right) \Delta u, \quad n = 0, 1, \ldots, N-1.$$

The expression of $H(k_n)$ fits the definition of discrete Fourier transform, and it can be valued via FFT (see e.g., Press et al., 1992). For later references, we call the Fourier option pricing method the FFT method.

One can, of course, consider a more accurate numerical integration scheme. But our experiences suggest that the composite trapezoidal rule is accurate enough for most applications.

9.2 Analytical Solution of Riccarti Equations

We now focus on the best-known special case of the ATSM that has the following specification of the drift and volatility functions:

$$\mu_X(x, t) = \kappa(\theta - x),$$
$$\sigma_X^2(x, t) = \sigma_0^2 x, \tag{9.35}$$

where $\kappa, \theta \geq 0$. This results in the so-called square-root process

$$\begin{cases} dX_t = \kappa(\theta - X_t)\, dt + \sigma_0 \sqrt{X_t}\, dW_t, \\ X_0 \geq 0. \end{cases} \tag{9.36}$$

In Equation 9.36, θ and κ are called the mean level and the strength of the mean reversion, respectively, and X_t remains non-negative and is expected to evolve up and down around the mean level. When we take the following special affine specification of the short rate, $r_t = X_t$, we reproduce the famous CIR model for interest rates. One of the most important motivations of the CIR model is to keep the interest-rate positive. Using the square-root processes as building blocks, we can generate a variety of models that ensure interest rates to stay positive.

For the purpose of options pricing, we need the moment-generating function of X_t. According to Lemma 9.1, the moment-generating function is available in explicit form:

$$\psi(u; X_t, t, T) = e^{\alpha_u(t,T) + \beta_u(t,T)X_t}, \tag{9.37}$$

where α_u and β_u satisfy

$$\begin{cases} (\beta_u)_t = \delta + \kappa\beta_u - \frac{1}{2}\sigma_0^2\beta_u^2, & \beta_u(T,T) = u, \\ (\alpha_u)_t = \delta_0 - \kappa\theta\beta_u, & \alpha_u(T,T) = 0. \end{cases} \tag{9.38}$$

Note that δ is a constant while δ_0 is subject to the term structure of interest rates at time t. Having non-negative interest rates in the future requires $\beta_u(t,T)$ to be a decreasing function in T. The solution for α_u follows from that of β_u.

Next, let us focus on the solution of the Riccarti equation. We have

Lemma 9.2.1. *For constant coefficients, the solution to*

$$(\beta_u)_t = b_2\beta_u^2 + b_1\beta_u + b_0, \quad \beta_u(T,T) = u. \tag{9.39}$$

is

$$\beta_u(t,T) = -\frac{(b_1 + c)}{2b_2} \frac{\left(e^{c(T-t)} - 1\right)}{\left(ge^{c(T-t)} - 1\right)},$$

where

$$c = \sqrt{b_1^2 - 4b_0 b_2}, \quad g = \frac{2b_2 u + b_1 + c}{2b_2 u + b_1 - c}. \tag{9.40}$$

Proof. The roots for

$$0 = b_2\beta^2 + b_1\beta + b_0 \tag{9.41}$$

are

$$\beta_{\pm} = \frac{-b_1 \pm \sqrt{b_1^2 - 4b_0 b_1}}{2b_2} \triangleq \frac{-b_1 \pm c}{2b_2}. \tag{9.42}$$

Consider the difference, $Y(t,T) = \beta_u(t,T) - \beta_+$, which satisfies

$$\begin{aligned} Y_t &= b_2(Y + \beta_+)^2 + b_1(Y + \beta_+) + b_0 \\ &= b_2 Y^2 + (2b_2\beta_+ + b_1)Y \\ &= b_2 Y^2 + cY, \end{aligned} \tag{9.43}$$

with a final condition, $Y(T,T) = u - \beta_+$. The above ODE for Y can be rewritten into

$$\left(\frac{1}{Y}\right)_t = -b_2 - c\left(\frac{1}{Y}\right). \tag{9.44}$$

Let $z = 1/Y$. Then, we have

$$z_t = -b_2 - cz, \quad z(T) = \frac{1}{u - \beta_+} = \frac{2b_2}{2b_2u + b_1 - c}. \tag{9.45}$$

Solving Equation 9.45, we obtain

$$
\begin{aligned}
z(t) &= e^{c(T-t)}z(T) + \frac{b_2}{c}\left[e^{c(T-t)} - 1\right] \\
&= \frac{b_2}{c}\left[\left(\frac{c}{b_2}z(T) + 1\right)e^{c(T-t)} - 1\right] \\
&= \frac{b_2}{c}\left[\left(\frac{c}{b_2}\frac{2b_2}{2b_2u + b_1 - c} + 1\right)e^{c(T-t)} - 1\right] \\
&= \frac{b_2}{c}\left[\left(\frac{2b_2u + b_1 + c}{2b_2u + b_1 - c}\right)e^{c(T-t)} - 1\right] \\
&= \frac{b_2}{c}\left[ge^{c(T-t)} - 1\right],
\end{aligned}
\tag{9.46}
$$

where g is defined in Equation 9.40. The solution of $z(t)$ gives rise to

$$Y(t, T) = \frac{1}{z(t)} = \frac{c}{b_2}\frac{1}{\left(ge^{c(T-t)} - 1\right)}. \tag{9.47}$$

We thus obtain the solution to Equation 9.39:

$$
\begin{aligned}
\beta_u(t, T) &= Y(t, T) + \beta_+ \\
&= \frac{c}{b_2}\frac{1}{\left(ge^{c(T-t)} - 1\right)} - \frac{b_1 - c}{2b_2} \\
&= \frac{2c + (c - b_1)\left(ge^{c(T-t)} - 1\right)}{2b_2\left(ge^{c(T-t)} - 1\right)} \\
&= \frac{(b_1 + c) - (b_1 + c)\,e^{c(T-t)}}{2b_2\left(ge^{c(T-t)} - 1\right)} \\
&= -\frac{(b_1 + c)}{2b_2}\frac{\left(e^{c(T-t)} - 1\right)}{\left(ge^{c(T-t)} - 1\right)}.
\end{aligned}
\tag{9.48}
$$

$\square$

Once β_u is obtained, α_u is calculated by integrations:

$$\alpha_u(t, T) = -\int_t^T \delta_0 \, ds + \kappa\theta \int_t^T \beta_u \, ds, \tag{9.49}$$

where

$$\int_t^T \beta_u \, ds = -\frac{b_1 + c}{2b_2} \int_t^T \frac{e^{c(T-s)} - 1}{ge^{c(T-s)} - 1} ds$$

$$= -\frac{b_1 + c}{2b_2} \int_0^{T-t} \frac{e^{c\tau} - 1}{ge^{c\tau} - 1} d\tau$$

$$= -\frac{b_1 + c}{2b_2} \left[(T - t) - \int_0^{T-t} \frac{(1 - g)e^{c\tau}}{1 - ge^{c\tau}} d\tau \right], \quad (9.50)$$

while

$$\int_0^{T-t} \frac{(1 - g)\,e^{c\tau}}{1 - ge^{c\tau}} d\tau = \frac{1}{c} \int_1^{e^{c(T-t)}} \frac{1 - g}{1 - gu} du$$

$$= \frac{1}{c} \left(\frac{g - 1}{g} \right) \ln \left(\frac{1 - ge^{c(T-t)}}{1 - g} \right)$$

$$= \frac{2}{b_1 + c} \ln \left(\frac{1 - ge^{c(T-t)}}{1 - g} \right). \quad (9.51)$$

Putting the terms in place, we have

$$\alpha_u(t, T) = -\int_t^T \delta_0 \, ds - \kappa\theta \left(\frac{b_1 + c}{2b_2} \right) \left[(T - t) - \frac{2}{b_1 + c} \right.$$

$$\left. \times \ln \left(\frac{1 - ge^{c(T-t)}}{1 - g} \right) \right]$$

$$= -\int_t^T \delta_0 \, ds - \kappa\theta \left[\left(\frac{b_1 + c}{2b_2} \right) (T - t) - \frac{1}{b_2} \right.$$

$$\left. \times \ln \left(\frac{1 - ge^{c(T-t)}}{1 - g} \right) \right]. \quad (9.52)$$

Recall that

$$b_2 = -\frac{1}{2}\sigma_0^2, \quad b_1 = \kappa, \quad b_0 = \delta. \quad (9.53)$$

When $u = 0$, $\alpha_0(t, T)$, and $\beta_0(t, T)$ give the prices of zero-coupon bonds, $P(t, T) = e^{\alpha_0 + \beta_0 X_t}$. Since $b_0 b_2 < 0$, we have

$$c = \sqrt{b_1^2 - 4b_0 b_2} > b_1 > 0, \quad (9.54)$$

which results in

$$g = \frac{b_1 + c}{b_1 - c} < 0.$$

By examining Equation 9.48, we conclude that $\beta_0 < 0$, meaning that the zero-coupon bond is indeed a monotonically decreasing function of X_t under the square-root process for the state variable.

9.3   Pricing Options on Coupon Bonds

The payoff of an option on a coupon bond is

$$V_T = \left(\sum_{j \geq 1} \Delta T c P(T, T_j) + P(T, T_n) - K \right)^+, \tag{9.55}$$

where c is the coupon rate and K the strike price. Under a one-factor ATSM,

$$V_T = \left(\sum_{j \geq 1} \Delta T c e^{\alpha(T,T_i)+\beta(T,T_j)X_T} + e^{\alpha(T,T_n)+\beta(T,T_n)X_T} - K \right)^+. \tag{9.56}$$

Jamshidian (1989) suggests the following approach that decomposes the option on the coupon bond into a portfolio of options on zero-coupon bonds. In view of $P(T, T_j)$ being a monotonically decreasing function of X_T, one can solve for X^* such that

$$\sum_{j \geq 1} \Delta T c e^{\alpha(T,T_j)+\beta(T,T_j)X^*} + e^{\alpha(T,T_n)+\beta(T,T_n)X^*} - K = 0. \tag{9.57}$$

Then, we define

$$K_j = e^{\alpha(T,T_j)+\beta(T,T_j)X^*}. \tag{9.58}$$

Note that there is

$$K = \sum_{j \geq 1} \Delta T c K_j + K_n, \tag{9.59}$$

and we substitute the right-hand side of Equation 9.59 for K in Equation 9.56. Owing to the monotonicity of $P(T, T_j)$, we have

$$V_T = \sum_{j=1}^{n-1} \Delta T c \left(e^{\alpha(T,T_j)+\beta(T,T_j)X_T} - K_i \right)^+ + (1 + \Delta T c)$$

$$\times \left(e^{\alpha(T,T_n)+\beta(T,T_n)X_T} - K_n \right)^+. \tag{9.60}$$

The price of the option is given by

$$V_t = \sum_{j=1}^{n-1} \Delta T c E_t^{\mathbb{Q}} \left[e^{-\int_t^T r_s \, ds} \left(e^{\alpha(T,T_j)+\beta(T,T_j)X_T} - K_i \right)^+ \right]$$

$$+ (1 + \Delta T c) E_t^{\mathbb{Q}} \left[e^{-\int_t^T r_s ds} \left(e^{\alpha(T,T_n)+\beta(T,T_n)X_T} - K_n \right)^+ \right], \tag{9.61}$$

where $\mathbb{Q}$ stands for the risk-neutral measure. The terms under expectation are the discounted payoff of zero-coupon options, for which we already have a closed-form solution using the transform inversion formula. This approach, however, does not apply to multi-factor ATSMs.

9.4 Distributional Properties of Square-Root Processes

The square-root process, in Equation 9.36, is often used as the basic building block for one-factor or multi-factor ATSMs. It helps us to understand the distributional properties of the process, so that we can make proper parameterization in applications.

Lemma 9.4.1. *Suppose that $X_0 > 0, \kappa \geq 0, \theta \geq 0$.*

1. *If*

$$\kappa\theta \geq \frac{1}{2}\sigma_0^2, \tag{9.62}$$

   then the SDE (Equation 9.36) admits a solution, X_t, that is strictly positive for all $t > 0$.

2. *If*

$$0 < \kappa\theta < \frac{1}{2}\sigma_0^2, \tag{9.63}$$

   then the SDE (Equation 9.36) admits a unique solution, which is nonnegative but occasionally hits $X = 0$.

3. *If $\kappa\theta = 0$, the process, X_t, vanishes at a finite time and remains equal to zero thereafter.*

4. *If $\kappa\theta > 0$, then as $t \to \infty$, X_t has an asymptotic Gamma distribution with density function*

$$f(x) = \frac{\left(2\kappa/\sigma_0^2\right)^{\left(2\kappa\theta/\sigma_0^2\right)}}{\Gamma(2\kappa\theta/\sigma_0^2)} x^{\left(2\kappa\theta/\sigma_0^2\right)-1} \exp\left(-\frac{2\kappa}{\sigma_0^2}x\right), \tag{9.64}$$

   where

$$\Gamma(p) = \int_0^\infty x^{p-1} e^{-x}\, dx \tag{9.65}$$

   is the Gamma function.

For the proof of the above proposition, we refer to Feller (1971) or Ikeda and Watanabe (1989).

9.5 Multi-Factor Models

Under a multi-factor model, the short rate, $r(t)$, is an affine function of a vector of the unobserved state variable, $\mathbf{X}_t$, such that

$$r(t) = \delta_0 + \boldsymbol{\delta}^{\mathrm{T}}\mathbf{X}_t, \tag{9.66}$$

and $\mathbf{X}_t = \left(X_1(t), X_2(t), \ldots, X_N(t)\right)^{\mathrm{T}}$ follows an "affine diffusion,"

$$\mathrm{d}\mathbf{X}_t = \boldsymbol{\mu}(\mathbf{X}_t, t)\,\mathrm{d}t + \boldsymbol{\sigma}\left(\mathbf{X}_t, t\right)\mathrm{d}\mathbf{W}_t, \tag{9.67}$$

with

$$\boldsymbol{\mu}(\mathbf{X}_t, t) = K_0(t) + K_1(t)\mathbf{X}_t,$$
$$\boldsymbol{\sigma}\left(\mathbf{X}_t, t\right)\boldsymbol{\sigma}^{\mathrm{T}}(\mathbf{X}_t, t) = H_0(t) + \sum_{i=1}^{N} X_i(t)H_i(t), \tag{9.68}$$

where in Equation 9.68, K_0 is an $N \times 1$ matrix, $K_1(t)$ and $H_i(t), i = 0, \ldots, N$ the $N \times N$ matrices, and $\mathbf{W}_t$ an N-dimensional Brownian motion under the risk-neutral measure, $\mathbb{Q}$.

Suppose that $\boldsymbol{\sigma}\left(\mathbf{X}_t, t\right)$ is well defined. It is shown that (Duffie and Kan, 1996), similar to the one-factor model, zero-coupon bond prices are exponential affine functions of the form,

$$P(t, T) = e^{\alpha(t,T) + \boldsymbol{\beta}^{\mathrm{T}}(t,T)\mathbf{X}_t}, \tag{9.69}$$

where $\alpha(t, T)$ is a scalar, $\boldsymbol{\beta}(t, T)$ a vector,

$$\boldsymbol{\beta}(t, T) = (\beta_1(t, T), \beta_2(t, T), \ldots, \beta_N(t, T))^{\mathrm{T}}, \tag{9.70}$$

and they satisfy the following Riccarti equations

$$\boldsymbol{\beta}_t = \boldsymbol{\delta} - K_1^{\mathrm{T}}\boldsymbol{\beta} - \frac{1}{2}\boldsymbol{\beta}^T H\boldsymbol{\beta}, \quad \boldsymbol{\beta}(T, T) = 0,$$
$$\alpha_t = \delta_0 - K_0^{\mathrm{T}}\boldsymbol{\beta} - \frac{1}{2}\boldsymbol{\beta}^T H_0\boldsymbol{\beta}, \quad \alpha(T, T) = 0. \tag{9.71}$$

In Equation 9.71,

$$\boldsymbol{\beta}^{\mathrm{T}} H\boldsymbol{\beta} = \left(\boldsymbol{\beta}^{\mathrm{T}} H_1\boldsymbol{\beta}, \boldsymbol{\beta}^{\mathrm{T}} H_2\boldsymbol{\beta}, \ldots, \boldsymbol{\beta}^{\mathrm{T}} H_N\boldsymbol{\beta}\right)^{\mathrm{T}}.$$

Equation 9.71 can at least be solved numerically by the Runge–Kutta method (Press et al., 1992). Once we have obtained $\boldsymbol{\beta}$, we calculate α through an integral:

$$\alpha(t, T) = \int_t^T \left[-\delta_0 + K_0^{\mathrm{T}}\boldsymbol{\beta}(s, T) + \frac{1}{2}\boldsymbol{\beta}^T H_0\boldsymbol{\beta}(s, T)\right]\mathrm{d}s. \tag{9.72}$$

Under a multi-factor ATSM, the volatilities of forward rates can be state-dependent, and the forward rates can be guaranteed to be positive. In both academic literature and applications, a lot of attention has been given to the following subclass of the affine models (Duffie and Kan, 1996; Duffie and Singleton, 1999; Dai and Singleton, 2000; etc.):

$$\mathrm{d}\mathbf{X}_t = \mathcal{K}(\boldsymbol{\theta} - \mathbf{X}_t)\,\mathrm{d}t + \boldsymbol{\Sigma}\sqrt{V(t)}\,\mathrm{d}\mathbf{W}_t. \tag{9.73}$$

In Equation 9.73, $\boldsymbol{\theta}$ is an $N \times 1$ matrix, $\mathcal{K}$ and $\boldsymbol{\Sigma}$ the $N \times N$ matrices, and $V(t)$ a diagonal matrix, such that

$$V(t) = \text{diag}(V_i(t)) = \text{diag}\left(a_i + \mathbf{b}_i^{\text{T}} \mathbf{X}_t\right). \tag{9.74}$$

In formalism, Equation 9.73 is a multi-dimensional version of the square-root process (1979), which carries the important features of mean reversion and state-dependent volatility. For the generalized CIR process, Equation 9.73, the coefficient matrices for the Equation 9.71 take the form

$$K_0 = \mathcal{K}\boldsymbol{\theta}, \quad K_1 = \mathcal{K}, \quad H_0 = \boldsymbol{\Sigma}\text{diag}(a_j)\boldsymbol{\Sigma}^{\text{T}},$$
$$H_j = \boldsymbol{\Sigma}\text{diag}(b_{j,\cdot})\boldsymbol{\Sigma}^{\text{T}}, \quad j = 1, \ldots, N, \tag{9.75}$$

where $\text{diag}(a_i)$ stands for the diagonal matrix with diagonal elements a_j, $j = 1, \ldots, N$, and $\text{diag}(b_{j,\cdot})$ the diagonal matrix with diagonal elements $b_{j,i}, i = 1, \ldots, N$. For later reference, we denote

$$\mathbf{a} = (a_1, a_2, \ldots, a_N)^{\text{T}}, \quad \mathcal{B} = (\mathbf{b}_1, \mathbf{b}_2, \ldots, \mathbf{b}_N).$$

9.5.1 Admissible ATSMs

For a prescription of the multi-factor affine model, Equation 9.73, to be valid, one must guarantee that $V(t)$ stays non-negative all the time. Any affine model that guarantees $V(t) \geq 0$ is called an admissible model. For an admissible model, coefficient matrices of $\mathbf{X}_t$ should satisfy certain conditions. Denote the class of admissible ATSM of N state variables such that rank $(\mathcal{B}) = m$ by $\mathbb{A}_m(N)$. It is efficient to state these conditions only with some kind of standardized formalism of the affine models, which is called the canonical representation and is defined below.

Definition 9.5.1. *(Canonical Representation of $\mathbb{A}_m(N)$). For each m, we partition $\boldsymbol{X}_t^{\text{T}}$ as $\boldsymbol{X}_t^{\text{T}} = \left(\left(\boldsymbol{X}_t^{\text{B}}\right)^{\text{T}}, \left(\boldsymbol{X}_t^{\text{D}}\right)^{\text{T}}\right)$, where $\boldsymbol{X}_t^{\text{B}}$ is $m \times 1$ and $\boldsymbol{X}_t^{\text{D}}$ is $(N - m) \times 1$, and we define the canonical representation of $\mathbb{A}_m(N)$ as the special case of Equation 9.73 with*

$$\mathcal{K} = \begin{bmatrix} \mathcal{K}_{m \times m}^{BB} & 0_{m \times (N-m)} \\ \mathcal{K}_{(N-m) \times m}^{DB} & \mathcal{K}_{(N-m) \times (N-m)}^{BB} \end{bmatrix}, \tag{9.76}$$

for $m > 0$, and $\mathcal{K}$ is a lower triangular matrix for $m = 0$,

$$\theta = \begin{pmatrix} \theta^B_{m \times 1} \\ 0_{(N-m) \times 1} \end{pmatrix},$$

$$\Sigma = I,$$

$$a = \begin{pmatrix} 0_{m \times 1} \\ 1_{(N-m) \times 1} \end{pmatrix}, \tag{9.77}$$

$$\mathcal{B} = \begin{pmatrix} I_{m \times m} & B^{BD}_{m \times (N-m)} \\ 0_{(N-m) \times m} & 0_{(N-m) \times (N-m)} \end{pmatrix},$$

with the following parametric restrictions imposed:

$$\delta_i \geq 0, \qquad m + 1 \leq i \leq N,$$

$$\mathcal{K}_i \theta = \sum_{j=1}^{m} \mathcal{K}_{ij} \theta_j > 0, \quad 1 \leq i \leq m,$$

$$\mathcal{K}_{ij} \leq 0, \qquad 1 \leq j \leq m, j \neq i, \tag{9.78}$$

$$\theta_i \geq 0, \qquad 1 \leq i \leq m,$$

$$\mathcal{B}_{ij} \geq 0, \qquad 1 \leq i \leq m, \ m + 1 \leq j \leq N.$$

It is shown that any ATSM in the class of $\mathbb{A}_m(N)$ can be transformed into the canonical form using the so-called invariant transformation (Dai and Singleton, 2000), which preserves admissibility and leaves the short rate unchanged. The canonical representation also allows us to produce specific models for production use. The justification for the canonical representation is found in Dai and Singleton (2000).

We remark here that the conditions for the canonical representation to be admissible are sufficient rather than necessary. But they are known to be the set of minimal conditions so far. Furthermore, the canonical representation is not unique. The representation from Equations 9.76 through 9.78 was chosen because, with it, it is relatively easier to identify some existing cases and verify their admissibility.

9.5.2 Three-Factor ATSMs

According to Litterman and Scheinkman (1991), the term structure of the interest rates is essentially driven by three factors. Because of that, three-factor models have received more attention from both researchers and practitioners. A number of three-factor models have been proposed and subjected to empirical studies. In this section, we introduce four of these three-factor ATSMs, and identify their connections with the canonical representation introduced earlier.

$$\mathbb{A}_0(3)$$

If $m = 0$, then none of the $\mathbf{X}_t$s affect the volatility of $\mathbf{X}_t$, and $\mathbf{X}_t$ follows a three-dimensional Gaussian diffusion. The coefficient sets of the canonical representation of $\mathbb{A}_0(3)$ are given by

$$K = \begin{bmatrix} \kappa_{11} & & \\ \kappa_{21} & \kappa_{22} & \\ \kappa_{31} & \kappa_{32} & \kappa_{33} \end{bmatrix}, \quad \Sigma = \begin{bmatrix} 1 & & \\ & 1 & \\ & & 1 \end{bmatrix}, \quad \boldsymbol{\theta} = \begin{bmatrix} 0 \\ 0 \\ 0 \end{bmatrix},$$

$$\mathbf{a} = \begin{pmatrix} 1 \\ 1 \\ 1 \end{pmatrix}, \quad \mathbf{b}_i = \mathbf{0}, \quad i = 1, 2, 3,$$

where $\kappa_{11} > 0, \kappa_{22} > 0$, and $\kappa_{33} > 0$.

$$\mathbb{A}_1(3)$$

One member of $\mathbb{A}_1(3)$ is the model by Balduzzi, Das, Foresi, and Sundaram (BDFS) (1996):

$$\begin{aligned} du(t) &= \mu(\bar{u} - u(t)) \, dt + \eta\sqrt{u(t)} \, dW_u(t), \\ d\theta(t) &= \nu(\bar{\theta} - \theta(t)) \, dt + \zeta \, dW_\theta(t), \\ dr(t) &= \kappa(\theta(t) - r(t)) \, dt + \sqrt{u(t)} \, dW_r(t), \end{aligned} \tag{9.79}$$

with the only a non-zero diffusion correlation between W_u and W_r. It can be verified that Equation 9.79 is equivalent to the following affine representation:

$$r(t) = \delta_0 + X_2(t) + X_3(t), \tag{9.80}$$

where $X_2(t)$ and $X_3(t)$ satisfy

$$d\begin{pmatrix} X_1(t) \\ X_2(t) \\ X_3(t) \end{pmatrix} = \begin{bmatrix} \kappa_{11} & & \\ & \kappa_{22} & \\ \kappa_{31} & & \kappa_{33} \end{bmatrix}\left[\begin{pmatrix} \theta_1 \\ 0 \\ 0 \end{pmatrix} - \begin{pmatrix} X_1(t) \\ X_2(t) \\ X_3(t) \end{pmatrix}\right] dt$$

$$+ \begin{bmatrix} 1 & & \\ & 1 & \\ & & 1 \end{bmatrix}\begin{bmatrix} \sqrt{V_1(t)} & & \\ & \sqrt{V_2(t)} & \\ & & \sqrt{V_3(t)} \end{bmatrix} d\mathbf{W}_t, \tag{9.81}$$

with

$$\begin{aligned} V_1(t) &= X_1(t), \\ V_2(t) &= b_{1,2}X_1(t), \\ V_3(t) &= a_3. \end{aligned} \tag{9.82}$$

The model in Equation 9.79 is called the BDFS model for short. It is a short-rate model where the short rate is correlated with its stochastic volatility.

$$\mathbb{A}_2(3)$$

The $\mathbb{A}_2(3)$ family is characterized by the assumption that volatilities of $\mathbf{X}_t$ are determined by affine functions of two of the three Xs. A member of this subfamily is the model proposed by Chen (1996):

$$
\begin{aligned}
dv(t) &= \mu(v - v(t))\,dt + \eta\sqrt{v(t)}\,dW_1(t),\\
d\theta(t) &= v(\theta - \theta(t))\,dt + \zeta\sqrt{\theta(t)}\,dW_2(t),\\
dr(t) &= \kappa(\theta(t) - r(t))\,dt + \sqrt{v(t)}\,dW_3(t),
\end{aligned}
\tag{9.83}
$$

where the Brownian motions are assumed to be mutually independent. As in the BDFS model, v and θ are interpreted as the stochastic volatility and central tendency of $r(t)$, respectively. The equivalent affine representation of the model is given below:

$$
r(t) = X_2(t) + X_3(t),
\tag{9.84}
$$

where $X_i(t)$, $i = 1, 2, 3$ evolve according to

$$
d\begin{pmatrix} X_1(t) \\ X_2(t) \\ X_3(t) \end{pmatrix} = \begin{bmatrix} \kappa_{11} & & \\ & \kappa_{22} & \\ & & \kappa_{33} \end{bmatrix} \left[\begin{pmatrix} \theta_1 \\ \theta_2 \\ 0 \end{pmatrix} - \begin{pmatrix} X_1(t) \\ X_2(t) \\ X_3(t) \end{pmatrix} \right] dt
$$
$$
+ \begin{bmatrix} 1 & & \\ & 1 & \\ & & 1 \end{bmatrix} \begin{bmatrix} \sqrt{V_1(t)} & & \\ & \sqrt{V_2(t)} & \\ & & \sqrt{V_3(t)} \end{bmatrix} d\mathbf{W}_t,
\tag{9.85}
$$

with $\kappa_{11} > 0, \kappa_{22} > 0$ and

$$
\begin{aligned}
V_1(t) &= X_1(t),\\
V_2(t) &= X_2(t),\\
V_3(t) &= b_{1,3}X_1(t).
\end{aligned}
\tag{9.86}
$$

$$\mathbb{A}_3(3)$$

The final subfamily of the three-factor models has $m = 3$ so that all three Xs determine the volatility structure. The canonical representation of $\mathbb{A}_3(3)$ has parameters

$$
\mathcal{K} = \begin{bmatrix} \kappa_{11} & & \\ \kappa_{21} & \kappa_{22} & \\ \kappa_{21} & \kappa_{21} & \kappa_{33} \end{bmatrix}, \quad \boldsymbol{\Sigma} = \begin{bmatrix} 1 & 0 & 0 \\ 0 & 1 & 0 \\ 0 & 0 & 1 \end{bmatrix}, \quad \boldsymbol{\theta} = \begin{pmatrix} \theta_1 \\ \theta_2 \\ \theta_3 \end{pmatrix},
$$
$$
a = \begin{pmatrix} 0 \\ 0 \\ 0 \end{pmatrix}, \quad \mathbf{b}_i = \mathbf{e}_i, \quad i = 1, 2, 3,
\tag{9.87}
$$

where $\kappa_{ii} > 0$ and $\kappa_{ij} \le 0$ for $i \ne j$.

With both $\boldsymbol{\Sigma}$ and $\mathcal{B}$ equal to identity matrices, the diffusion term of this model is identical to that of the three-factor CIR processes:

$$dX_t = K(\theta - X_t) \, dt + \Sigma \sqrt{V(t)} \, dW_t,$$
$$X_0 \geq 0,$$

$$(9.88)$$

where $\theta = (\theta_i)_{3 \times 1}$, and

$$K = \text{diag}(k_i), \quad \Sigma \sqrt{V(t)} = \text{diag}(\sigma_i \sqrt{X_i}), \quad i = 1, 2, 3. \qquad (9.89)$$

We finish this section with some remarks. Frequently, the general multifactor affine model can be viewed as a blending of the Vasicek and CIR forms. The conditions from Equations 9.76 through 9.78 are also restrictions in multifactor term structure modeling with ATSMs. As a matter of fact, the CIR form perhaps offers the greatest flexibility in specifying the volatility dynamics of bond prices. However, this flexibility comes at a cost. The parameter restrictions for ensuring that Equation 9.66 provides a valid description of factor variances impose substantial restrictions on the permissible correlations among the factors. As is stated in Equations 9.88 and 9.89, in the extreme case of the pure multi-factor CIR model, the factors must be uncorrelated to ensure an admissible volatility specification.

9.6   Swaption Pricing under ATSMs

In Sections 9.1 and 9.2, we have studied the pricing of options on bonds under the ATSMs. In this section, we study the pricing of options on interest rates. Without loss of generality, let us focus on the pricing of swaptions, or options on swap rates, under ATSMs, as similar approaches apply to the pricing of options on other interest rates.

As was already determined in Chapter 6, the payoff of a swaption can be cast into

$$V_{T_m} = A_{m,n}(t) \left(R_{m,n}(T_m) - k \right)^+, \qquad (9.90)$$

where $A_{m,n}(t)$ is the annuity,

$$A_{m,n}(t) = \sum_{j=m}^{n-1} \Delta T_j P(t, T_{j+1}), \quad t \leq T_m, \qquad (9.91)$$

and $R_{m,n}(t)$ is the prevailing swap rate at time t, given by

$$R_{m,n}(t) = \frac{P(t, T_m) - P(t, T_n)}{\sum_{j=m}^{n-1} \Delta T_j P(t, T_{j+1})}. \qquad (9.92)$$

Under an ATSM, the price of a zero-coupon bond is given by

$$P(t, T) = \exp\left(\alpha(t, T) + \beta^{\mathrm{T}}(t, T) X_t \right), \qquad (9.93)$$

where $\mathbf{X}_t$ follows Equation 9.67, and $\alpha(t, T)$ and $\beta(t, T)$ satisfy Equation 9.71.

Schrager and Pelsser (2006) take an approach parallel to the swaption pricing under the market model with stochastic volatility approach (Wu, 2002). The price of the swaption is given by

$$V_0 = A_{m,n}(0) E^{\mathbb{Q}_s}\left[(R_{m,n}(T_m) - k)^+ \mid \mathcal{F}_0\right], \tag{9.94}$$

where $\mathbb{Q}_S$ stands for the forward swap measure, defined by

$$\left.\frac{d\mathbb{Q}_S}{d\mathbb{Q}}\right|_{\mathcal{F}_t} = \frac{A_{m,n}(t)/A_{m,n}(0)}{B(t)/B(0)}$$

$$= \frac{1}{A_{m,n}(0)} \sum_{j=m}^{n-1} \Delta T_j \frac{P(t, T_{j+1})}{B(t)} \triangleq m_s(t), \quad t \leq T_m. \tag{9.95}$$

Note that $P(t, T_{j+1})/B(t)$ is a $\mathbb{Q}$-martingale, following the process

$$d\left(\frac{P(t, T_{j+1})}{B(t)}\right) = \left(\frac{P(t, T_{j+1})}{B(t)}\right) \beta^{\mathrm{T}}(t, T_{j+1}) \mathbf{\Sigma} \sqrt{\mathbf{V}(t)} d\mathbf{W}_t. \tag{9.96}$$

The dynamics for the Radon–Nikodym derivative then is

$$dm_s(t) = \frac{1}{A_{m,n}(0)} \sum_{j=m}^{n-1} \Delta T_j d\left(\frac{P(t, T_{j+1})}{B(t)}\right)$$

$$= \frac{1}{A_{m,n}(0)} \left[\sum_{j=m}^{n-1} \Delta T_j \left(\frac{P(t, T_{j+1})}{B(t)}\right) \beta^{\mathrm{T}}(t, T_{j+1})\right] \mathbf{\Sigma}\sqrt{\mathbf{V}(t)} d\mathbf{W}_t, \tag{9.97}$$

and it follows that

$$\frac{dm_s(t)}{m_s(t)} = \frac{B(t)}{A_{m,n}(t)} \sum_{j=m}^{n-1} \Delta T_j d\left(\frac{P(t, T_{j+1})}{B(t)}\right)$$

$$= \left(\sum_{j=m}^{n-1} \Delta T_j \left(\frac{P(t, T_{j+1})}{A_{m,n}(t)}\right) \beta^{\mathrm{T}}(t, T_{j+1})\right) \mathbf{\Sigma}\sqrt{\mathbf{V}(t)} d\mathbf{W}_t$$

$$= \left(\sum_{j=m}^{n-1} \alpha_j \beta^{\mathrm{T}}(t, T_{j+1})\right) \mathbf{\Sigma}\sqrt{\mathbf{V}(t)} d\mathbf{W}_t. \tag{9.98}$$

Define a $\mathbb{Q}_S$-Brownian motion by

$$d\mathbf{W}_t^S = d\mathbf{W}_t - \left\langle d\mathbf{W}_t, \frac{dm_t}{m_t}\right\rangle$$

$$= d\mathbf{W}_t - \sqrt{\mathbf{V}(t)}\mathbf{\Sigma}^{\mathrm{T}}\left(\sum_{j=m}^{n-1} \alpha_j \beta(t, T_{j+1})\right) dt. \tag{9.99}$$

Then, under $\mathbb{Q}_S$, the process for $\mathbf{X}_t$ becomes

$$
d\mathbf{X}_t = \mathcal{K}(\boldsymbol{\theta} - \mathbf{X}_t)\,dt + \boldsymbol{\Sigma}\sqrt{\mathbf{V}(t)}\left(d\mathbf{W}_t^S + \sqrt{\mathbf{V}(t)}\boldsymbol{\Sigma}^T \right.
$$

$$
\left. \times \left(\sum_{j=m}^{n-1} \alpha_j(t)\boldsymbol{\beta}(t, T_{j+1}) \right) dt \right)
$$

$$
= \left(\mathcal{K}(\boldsymbol{\theta} - \mathbf{X}_t) + \boldsymbol{\Sigma} V(t)\boldsymbol{\Sigma}^T \left(\sum_{j=m}^{n-1} \alpha_j(t)\boldsymbol{\beta}(t, T_{j+1}) \right) \right) dt
$$

$$
+ \boldsymbol{\Sigma}\sqrt{\mathbf{V}(t)}\,d\mathbf{W}_t^S. \tag{9.100}
$$

The process for the swap rate is

$$
dR_{m,n}(t) = \text{drift} + \sum_{j=m}^{n} \frac{\partial R_{m,n}(t)}{\partial P(t, T_j)}\,dP(t, T_j)
$$

$$
= \text{drift} + \frac{dP(t, T_m)}{A_{m,n}(t)} - \frac{R_{m,n}(t)}{A_{m,n}(t)}\sum_{j=m}^{n-1} \Delta T_j\,dP(t, T_{j+1}) - \frac{dP(t, T_n)}{A_{m,n}(t)}
$$

$$
= \left(\sum_{j=m}^{n} \gamma_j(t)\boldsymbol{\beta}^T(t, T_j) \right) \boldsymbol{\Sigma}\sqrt{\mathbf{V}_t}\,d\mathbf{W}_t^S, \tag{9.101}
$$

where

$$
\gamma_m(t) = \frac{P(t, T_m)}{A_{m,n}(t)},
$$

$$
\gamma_j(t) = R_{m,n}(t)\alpha_j, \quad j = m+1, \dots, n-1,
$$

$$
\gamma_n(t) = \left(R_{m,n}(t) - \frac{1}{\Delta T_{n-1}} \right) \alpha_n.
$$

Consider the frozen coefficient approximation to the processes of $R_{m,n}(t)$ and $\mathbf{X}_t$:

$$
dR_{m,n}(t) = \left(\sum_{j=m}^{n} \gamma_j(0)\boldsymbol{\beta}^T(t, T_j) \right) \boldsymbol{\Sigma}\sqrt{\mathbf{V}_t}\,d\mathbf{W}_t^S,
$$

$$
d\mathbf{X}_t = \left(\mathcal{K}(\boldsymbol{\theta} - \mathbf{X}_t) + \boldsymbol{\Sigma} V(t)\boldsymbol{\Sigma}^T \left(\sum_{j=m}^{n-1} \alpha_j(0)\boldsymbol{\beta}(t, T_{j+1}) \right) \right) dt
$$

$$
+ \boldsymbol{\Sigma}\sqrt{\mathbf{V}(t)}\,d\mathbf{W}_t^S. \tag{9.102}
$$

For notational simplicity, we let

$$\beta_{m,n}^{\alpha}(t) = \sum_{j=m}^{n-1} \alpha_j(0)\boldsymbol{\beta}(t, T_{j+1}),$$

$$\beta_{m,n}^{\gamma}(t) = \sum_{j=m}^{n} \gamma_j(0)\boldsymbol{\beta}(t, T_j). \tag{9.103}$$

Noticing that

$$\boldsymbol{\Sigma} V(t)\boldsymbol{\Sigma}^{\mathrm{T}}\beta_{m,n}^{\alpha}(t)$$
$$= \boldsymbol{\Sigma}\left(\mathrm{diag}(a_i) + \mathrm{diag}(\mathbf{b}_i^{\mathrm{T}}\mathbf{X}_t)\right)\boldsymbol{\Sigma}^{\mathrm{T}}\beta_{m,n}^{\alpha}(t)$$
$$= \boldsymbol{\Sigma}\mathrm{diag}(a_i)\boldsymbol{\Sigma}^{\mathrm{T}}\beta_{m,n}^{\alpha}(t) + \boldsymbol{\Sigma}\mathrm{diag}(\mathbf{b}_i^{\mathrm{T}}\mathbf{X}_t)\boldsymbol{\Sigma}^{\mathrm{T}}\beta_{m,n}^{\alpha}(t)$$
$$= \boldsymbol{\Sigma} A(t)\boldsymbol{\Sigma}^{\mathrm{T}}\beta_{m,n}^{\alpha}(t) + \boldsymbol{\Sigma}\mathrm{diag}\left(\boldsymbol{\Sigma}^{\mathrm{T}}\beta_{m,n}^{\alpha}(t)\right)B^{\mathrm{T}}(t)\mathbf{X}_t$$
$$\triangleq \boldsymbol{\xi}_t + \boldsymbol{\Phi}_t\mathbf{X}_t, \tag{9.104}$$

we thus rewrite Equation 9.102 into

$$\mathrm{d}R_{m,n}(t) = (\beta_{m,n}^{\gamma}(t))^T\boldsymbol{\Sigma}\sqrt{\mathbf{V}_t}\,\mathrm{d}\mathbf{W}_t^S,$$
$$\mathrm{d}\mathbf{X}_t = (\mathcal{K}(\boldsymbol{\theta} - \mathbf{X}_t) + \boldsymbol{\xi}_t + \boldsymbol{\Phi}_t\mathbf{X}_t)\,\mathrm{d}t + \boldsymbol{\Sigma}\sqrt{V(t)}\,\mathrm{d}\mathbf{W}_t^S$$
$$= (\mathcal{K}\boldsymbol{\theta} + \boldsymbol{\xi}_t - (\mathcal{K} - \boldsymbol{\Phi}_t)\mathbf{X}_t)\,\mathrm{d}t + \boldsymbol{\Sigma}\sqrt{V(t)}\,\mathrm{d}\mathbf{W}_t^S. \tag{9.105}$$

Stacking the swap rate with the state variables:

$$\tilde{\mathbf{X}}_t = \begin{pmatrix} R_{m,n}(t) \\ \mathbf{X}_t \end{pmatrix},$$

and defining

$$\tilde{\mathcal{K}} = \begin{pmatrix} 0 \\ & \mathcal{K} - \boldsymbol{\Phi} \end{pmatrix}, \quad \tilde{\boldsymbol{\theta}} = \begin{pmatrix} 0 \\ (\mathcal{K} - \boldsymbol{\Phi})^{-1}(\mathcal{K}\boldsymbol{\theta} + \boldsymbol{\xi}_t) \end{pmatrix}, \quad \tilde{\boldsymbol{\Sigma}} = \begin{pmatrix} (\beta_{m,n}^{\gamma})^T \\ \boldsymbol{\Sigma} \end{pmatrix},$$

we finally end up with a succinct expression for the joint process

$$\mathrm{d}\tilde{\mathbf{X}}_t = \tilde{\mathcal{K}}\left(\tilde{\boldsymbol{\theta}} - \tilde{\mathbf{X}}_t\right)\mathrm{d}t + \tilde{\boldsymbol{\Sigma}}\sqrt{V(t)}\,\mathrm{d}\mathbf{W}_t^S. \tag{9.106}$$

Parallel to the proof of Lemma 9.1, we can show that the moment-generating function for $\tilde{\mathbf{X}}_t$ is

$$\varphi(\mathbf{u}; \tilde{\mathbf{X}}_t, t, T) = E^{\mathbb{Q}_S}\left[e^{\mathbf{u}^{\mathrm{T}}\tilde{\mathbf{X}}_T} \mid \mathcal{F}_t\right]$$
$$= \exp\left(\tilde{\alpha}_{\mathbf{u}}(t, T) + \tilde{\beta}_{\mathbf{u}}^{\mathrm{T}}(t, T)\tilde{\mathbf{X}}_t\right), \tag{9.107}$$

where $\tilde{\alpha}$ and $\tilde{\beta}$ satisfy

$$\left(\tilde{\beta}_{\mathbf{u}}\right)_t = \mathcal{K}^{\mathrm{T}}\tilde{\beta}_{\mathbf{u}} - \frac{1}{2}\tilde{\beta}_{\mathbf{u}}^{\mathrm{T}}H\tilde{\beta}_{\mathbf{u}}, \quad \tilde{\beta}_{\mathbf{u}}(T,T) = \mathbf{u}$$

$$(\tilde{\alpha}_{\mathbf{u}})_t = -\left(\tilde{\mathcal{K}}\tilde{\theta}\right)^{\mathrm{T}}\tilde{\beta}_{\mathbf{u}} - \frac{1}{2}\tilde{\beta}_{\mathbf{u}}^{\mathrm{T}}H_0\tilde{\beta}_{\mathbf{u}}, \quad \tilde{\alpha}_{\mathbf{u}}(T,T) = 0. \tag{9.108}$$

Here H_0 and H are defined in Equation 9.75. When we take, in particular, $\mathbf{u} = u\mathbf{e}_1$, where $\mathbf{e}_1 = (1, 0, \ldots, 0)^{\mathrm{T}}$, $\varphi(\mathbf{u}; \tilde{\mathbf{X}}_t, t, T)$ becomes the moment-generating function for $R_{m,n}(T)$ only. Denote

$$G(k) = E_t^{\mathbb{Q}_S}\left[(R_{m,n}(T) - k)^+\right], \quad \forall k. \tag{9.109}$$

The Laplace transform of $G(k)$ with respect to k is defined by

$$\begin{aligned}
\int_{-\infty}^{\infty} \mathrm{e}^{(a+iu)k} G(k)\, \mathrm{d}k &= \int_{-\infty}^{\infty} \mathrm{e}^{(a+iu)k} E_t^{\mathbb{Q}_S}\left[(R_{m,n}(T) - k)^+\right]\mathrm{d}k \\
&= E_t^{\mathbb{Q}_S}\left[\int_{-\infty}^{\infty} \mathrm{e}^{(a+iu)k}(R_{m,n}(T) - k)^+\, \mathrm{d}k\right] \\
&= E_t^{\mathbb{Q}_S}\left[\int_{-\infty}^{R_{m,n}(T)} \mathrm{e}^{(a+iu)k}(R_{m,n}(T) - k)^+\, \mathrm{d}k\right] \\
&= E_t^{\mathbb{Q}_S}\left[\frac{\mathrm{e}^{(a+iu)R_{m,n}(T)}}{(a+iu)^2}\right] \\
&= \frac{\varphi\left((a+iu)\mathbf{e}_1; \tilde{\mathbf{X}}_t, t, T\right)}{(a+iu)^2} \\
&\triangleq \psi(u), \tag{9.110}
\end{aligned}$$

where a is a positive constant. The value of the swaption can then be obtained from an inverse Laplace transform:

$$V_t = A_{m,n}(t)\frac{\mathrm{e}^{-ak}}{\pi}\int_0^{+\infty} \exp(-iuk)\, \psi(u)\, \mathrm{d}u. \tag{9.111}$$

Schrager and Pelsser (2006) implemented the above method with a two-factor CIR model of the form

$$\begin{aligned}
r_t &= \delta + X_1(t) + X_2(t), \\
\mathrm{d}X_i(t) &= \kappa_i(\theta_i - X_i(t))\, \mathrm{d}t + \sigma_i \sqrt{X_i(t)}\, \mathrm{d}W_i(t), \tag{9.112} \\
X_i(0) &= x_i, \quad i = 1, 2,
\end{aligned}$$

where $W_i(t)$, $i = 1, 2$ are independent Brownian motions under the risk-neutral

TABLE 9.1: Accuracy of the Approximation Method

Swap Tenor	Option Maturity			
	1		5	
ATMF				
1	25.34	9.5%	28.88	7.3%
	(−0.01)	(0.00%)	(−0.01)	(0.00%)
5	78.98	7.2%	90.61	5.3%
	(−0.02)	(0.00%)	(−0.05)	(0.00%)
10	99.82	5.6%	114.55	4.0%
	(−0.04)	(0.00%)	(−0.10)	(0.00%)
ITM				
1	101.05	(9.6%)	72.35	7.2%
	(0.03)	(0.06%)	(0.08)	(−0.04%)
5	411.03	(7.1%)	293.35	5.1%
	(0.06)	(0.08%)	(0.26)	(0.05%)
10	675.49	(5.1%)	487.83	3.6%
	(0.02)	(0.29%)	(0.24)	(0.07%)
OTM				
1	2.07	9.4%	8.43	7.3%
	(−0.05)	(−0.05%)	(−0.10)	(−0.04%)
5	2.27	7.3%	15.74	5.5%
	(−0.13)	−0.07%	(−0.41)	(−0.04%)
10	0.65	5.7%	9.73	4.2%
	(−0.10)	(−0.10%)	(−0.71)	(−0.07%)

Reprinted from Schrager DF and Pelsser AJ, 2006. *Mathematical Finance* 16(4): 673–694. With permission. The ITM and OTM are set at the levels of 85% and 115% relative to the ATM swap rate.

measure, $\mathbb{Q}$, and the coefficients and initial conditions are taken as

$$\kappa_1 = 0.2, \quad \kappa_2 = 0.2,$$
$$\theta_1 = 0.03, \quad \theta_2 = 0.01,$$
$$\sigma_1 = 0.03, \quad \sigma_2 = 0.01,$$
$$\delta = 0.02,$$
$$x_1 = 0.04, \quad x_2 = 0.02.$$

For comparison, Schrager and Pelsser computed swaption prices by both the transformation method and the standard Monte Carlo simulation method. The pricing results are presented in Table 9.1, where prices (in basis points) and implied Black's volatilities are listed. Also provided in the table are differences in prices and differences in implied volatilities obtained by the two numerical methods. From the results, we can see that the approximation method is very accurate. The entire calculation takes about 0.14 s. In general, swaption pricing under ATSMs can be implemented almost instantly, which is another advantage of the model.

9.7   Notes

Whenever feasible, users of ATSMs price options by the transformation method. Numerically, this is realized through FFT. When it comes to model calibration, however, ATSMs are not easy to handle, particularly when the coefficients of the models are time-dependent. Note that for a reasonably good fit of both interest-rate caps and swaptions, we need to assume time dependency of the model coefficients. As a matter of fact, calibration of affine models driven by the popular square-root processes is an outstanding challenge in financial engineering.

To price non-vanilla options, we may have to resort to Monte Carlo simulations. The numerical simulation of the square-root processes has attracted interest in recent years. An easy solution is to treat the square-root process as an addition of a number of squared Gaussian processes, so the time stepping of the square-root process is achieved by advancing those Gaussian processes (Roger, 1995). Several delicate new approaches were proposed recently by Andersen (2007).

Recent progress in the area of affine models is represented by the affine LIBOR model in Keller-Ressel et al. (2013). This model follows in the footsteps of the forward price model, but it guarantees non-negative interest rates by choosing positive driving processes, whose moment-generating function is used to define forward prices. Under the model, caplets and swaptions can be priced in closed-form – through fast Fourier transform. Yet for the swaptions, it encounters the curse of dimensionality: the layers of Fourier transforms increase with the underlying swap tenor, which limits its application.

Exercises

1. Prove that if, in a one-factor ATSM, X_t follows a Gaussian process, then the affine model reduces to the Hull–White model.

2. Use the Laplace transform to price a put option on a zero-coupon bond under the ATSM (Hint: choose the dumping parameter properly).

3. The value of a call option can be expressed as

$$V_0(k) = P(0, T) \int_0^\infty q(s)(s - k)^+ ds,$$

where $q(s)$ is the density function of the state variable at the option's maturity, T. Let $a > 0$. Prove that the Laplace transform of the integral is

$$\int_{-\infty}^\infty e^{(a+iu)k} \left(\int_0^\infty q(s)(s - k)^+ ds \right) dk = \varphi(u) \times (a + iu)^{-2},$$

where $\varphi(u)$ and $(a+iu)^{-2}$ are the Laplace transforms of $q(s)$ and $(s-k)^+$, respectively.

4. Price an option on a zero-coupon bond with an equilibrium CIR model. The characteristics of the option are

    a. maturity of the option: two years;

    b. maturity of the bond: five years (so in two years, it will become a three-year maturity bond).

Assume, under the risk-neutral measure, $\kappa = 1, \theta = 0.05, \sigma = 0.1, \delta_0 = 0, \delta_1 = 1$, and $X_0 = \theta$. Price the option across strikes: $K = [0:0.1:1]$. You will need a numerical integration scheme to compute the transform inversions. Moreover, once you have obtained prices, calculate and plot Black's implied volatilities.

Chapter 10

Market Models with Stochastic Volatilities

After rigorous justifications were published in 1997, the market model (Brace, Gatarek and Musiela, 1997; Jamshidian, 1997; and Miltersen, Sandmann and Sondermann, 1997), which is based on lognormal assumption for forward rates, established itself as the benchmark model for interest rate derivatives. One of many virtues of the market model is that it justifies the use of Black's formula for caplet and swaption prices, which has long been a standard market practice. Black's formula establishes a relationship between option prices and local volatilities of the forward rates, and such a relationship has enabled fast calibration of the standard model (Wu, 2003). Nevertheless, the standard market model is also known for its limitation of only generating flat implied volatility curves, whereas the implied volatility curves observed in LIBOR markets often have the shape of a smile or downward skew. The implication on the model is that, after being calibrated to at-the-money options, it mis-prices off-the-money options. Because of the benchmark role of caps and swaptions in the fixed-income derivatives markets, there had been great interest in extending the standard model so as to fix the problem of mis-pricing or, speaking in terms of implied volatilities, to capture the smiles or skews.

It had become a major challenge to model volatility smiles in the context of the LIBOR market model for interest-rate derivatives. In modern literature of option pricing, the notion of smiles means non-flat curves of implied Black's volatilities. Very often, such curves look like skewed smiles, or smirks, but they can also take other shapes. Over the last two decades, various solutions were proposed and met various degrees of success. Without exception, all solutions were based on adopting at least one of the following features or risk factors, in addition to diffusions to the driving dynamics of the forward rate curve: level-dependent volatilities, displaced diffusions, stochastic volatilities, and/or jumps.

Let us comment on some representative works for smile modeling. Andersen and Andreasen (1998) were the first to adopt constant-elasticity-variance (CEV) dynamics for forward rates. On top of the CEV model, Andersen and Brotherton-Ratcliffe (2001) superimposed an independent square-root volatility process that serves to generate additional curvature to the otherwise monotonic volatility skews. The pricing of caps and swaptions under the CEV-type models is then done by asymptotic expansions. Andersen and Andreasen

(2002) also combined displaced diffusions with stochastic volatilities. Caplet pricing under such models can be achieved through Fourier transforms. In these models, there is no correlation between the forward rates and their stochastic volatilities, and the mechanism for volatility smiles or skews lies in the use of either the CEV dynamics or displaced diffusions. Around the same time, correlation-based models were also developed, represented by the stochastic alpha, beta and rho model (SABR) (Hagan et al., 2002) and the Heston's type extension of the market model by Wu and Zhang (2002). Under these models, the generation of volatility smiles were mainly dictated by the correlation between a forward rate and its stochastic volatility, and options can be priced through asymptotic expansion, Laplace transforms, or numerical PDE methods. To model time dependent skews, Piterbarg (2003) combines stochastic volatility with displaced diffusion for LIBOR modeling. In yet another line of research, Glasserman and Kou (2003) develop a comprehensive term structure model with the jump-diffusion dynamics. Under this model, approximate closed-form formulae for caplets and swaptions can be developed (Glasserman and Merener, 2003). The Glasserman and Kou (2003) model was later extended by Eberlein and Ožkan (2004), with jump-diffusion processes replaced by the general Lévy processes; the latter serve as a framework for a wide class of jump-diffusion processes. In the industry, however, it turned out that the SABR model, which combines the CEV dynamics for state variables with lognormal stochastic volatilities, achieved enormous success. The SABR model can capture various shapes of implied volatility curves, and its approximate closed-form formula for vanilla options in terms of the Black implied volatilities enables efficient calculation of Greeks, the sensitivity parameters. Empirical studies are also most supportive of the SABR model, see e.g., Wu (2012).

For various reasons, jumps in interest rates are much less frequent than jumps in the equity markets. It has become a popular belief in fixed-income derivatives markets that correlated stochastic volatilities should be the primary mechanism for the implied volatility smiles. In this chapter, we will limit our introduction to two representative models of stochastic volatility: the SABR model (Hagan et al., 2002) and the LIBOR version of the Heston's model by Wu and Zhang (2002).

10.1   SABR Model

Originally, SABR was not intended as a term structure model, instead, it is a model on the volatility smile of a single forward rate of swap rate, maturity by maturity. Taking the pricing of caplet of maturity T_j and strike

K_j for example, the SABR model takes the following form:

$$df_j(t) = \gamma_j(t)f_j^{\beta_j}(t)dW_{j+1}(t), \tag{10.1}$$

$$d\gamma_j(t) = \nu_j\gamma_j(t)dZ_t, \quad \gamma_j(0) = \alpha_j, \tag{10.2}$$

where $W_{j+1}(t)$ and Z_t are correlated Brownian motions under the T_{j+1}-forward measure:

$$dW_{j+1}(t)dZ_t = \rho_j dt,$$

with $0 < \beta_j < 1, \alpha_j, \nu_j > 0$, and $|\rho_j| \leq 1$. For a swaption, the corresponding model is

$$ds_{m,n}(t) = \gamma_{m,n}(t)s_{m,n}^{\beta_{m,n}}(t)dW_{m,n}(t), \tag{10.3}$$

$$d\gamma_{m,n}(t) = \nu_{m,n}\gamma_{m,n}(t)dZ_t, \quad \gamma_{m,n}(0) = \alpha_{m,n}(0), \tag{10.4}$$

where both $W_{m,n}(t)$ and Z_t are correlated Brownian motions under the forward swap measure corresponding to numeraire $A_{m,n}(t)$, such that

$$dW_{m,n}(t)dZ_t = \rho_{m,n}dt,$$

with $0 < \beta_{m,n} < 1, \alpha_{m,n}, \nu_{m,n} > 0$, and $|\rho_{m,n}| \leq 1$. The SABR forward rate model and the SABR swap rate model are actually inconsistent, but this has been harmless and thus ignored in the industry. A peculiar feature of the SABR model is the adoption of the lognormal dynamics for the stochastic volatility, which is not mean reverting, for the sake of simplicity of the model and closed-form solution of the option pricing. Note that the SABR model is the stochastic version of the CEV model with CEV parameter β_j, which reduces to the CEV model if $\nu_j = 0$. The parameter ν_j is referred to as the volatility of the volatility, and often called "vol of vol" for convenience.

Without loss of generality, let us introduce the pricing of the caplet. For notational simplicity, we will drop the index j throughout our presentation. A caplet then has a payoff $(f - K)^+$ at maturity T.

Except for the special cases of $\beta = 0$ and $\beta = 1$, no closed form expression for the probability distribution of f is known. Yet the caplet price can be solved approximately by means of asymptotic expansions in the parameter $\varepsilon = \nu^2 T$. For most option markets, including the interest rate markets, this parameter is typically small and the approximate solution can be quite accurate. The derivation of solution is actually tedious, and we refer readers to Hagen et al. (2002) for details. Another reason for skipping the derivation of Hagan et al.'s pricing formula is that there is a competing and more systematic approach based on the heat kernel expansion method (Henry-Labordère, 2005), which we will introduce later in Chapter 14 with the dual-curve models of interest rates.

In particular, the so-called closed-form approximate formula is not for dollar price, but for the implied Black's volatilities (Hagan and Lesniewski, 2008)[1]:

---

[1]This is not the original formula, but the enhanced version of the original one.

$$\sigma_{\text{imp}} = \nu \, \frac{\log\left(f_0/K\right)}{D\left(\zeta\right)} \times \left\{1 + \left[\frac{2\gamma_2 - \gamma_1^2 + 1/f_{\text{mid}}^2}{24}\left(\frac{\sigma_0 C\left(f_{\text{mid}}\right)}{\nu}\right)^2\right.\right.$$
$$\left.\left. + \frac{\rho\gamma_1}{4}\frac{\sigma_0 C\left(f_{\text{mid}}\right)}{\nu} + \frac{2 - 3\rho^2}{24}\right]\varepsilon\right\},$$

where, for clarity, we have set $C\left(f\right) = f^\beta$. The value f_{mid} denotes a midpoint between f_0 and K (chosen according to either the geometric average $\sqrt{f_0 K}$ or the arithmetic average $\left(f_0 + K\right)/2$). The function $D\left(\zeta\right)$ appearing in the formula above is

$$D\left(\zeta\right) = \log\left(\frac{\sqrt{1 - 2\rho\zeta + \zeta^2} + \zeta - \rho}{1 - \rho}\right),$$

for

$$\zeta = \frac{\nu}{\alpha}\int_K^{f_0}\frac{dx}{C\left(x\right)} = \frac{\nu}{\alpha\left(1 - \beta\right)}\left(f_0^{1-\beta} - K^{1-\beta}\right),$$

and the two remaining parameters are

$$\gamma_1 = \frac{C'\left(f_{\text{mid}}\right)}{C\left(f_{\text{mid}}\right)} = \frac{\beta}{f_{\text{mid}}},$$
$$\gamma_2 = \frac{C''\left(f_{\text{mid}}\right)}{C\left(f_{\text{mid}}\right)} = -\frac{\beta\left(1 - \beta\right)}{f_{\text{mid}}^2}.$$

The above formula for the implied Black's volatility is very easy to implement with computers, and it lends itself well to risk management of large portfolios of options in real time. But first, the formula is used to determine the model parameters by fitting to the quoted implied Black's volatilities of traded options, which is a small scale root-finding problem.

Note that when $K \to f_0$, the implied volatility formula above reduces to the so-called $\frac{0}{0}$ situation and becomes undefined. To fix the problem, we need to work out the limit as $K \to f_0$. When taking the limit, we have

$$D(\zeta) \to \zeta$$

for

$$\zeta = \frac{\nu}{\alpha}\tilde{f}^{-\beta}(f_0 - K), \quad \tilde{f} \text{ is between } f_0 \text{ and } K,$$

which then results in a limit of implied volatility for the ATM options:

$$\sigma_{\text{ATM}} = \alpha K^{1-\beta} \times \left\{1 + \left[\frac{2\gamma_2 - \gamma_1^2 + 1/K^2}{24}\left(\frac{\sigma_0 C\left(K\right)}{\nu}\right)^2\right.\right.$$
$$\left.\left. + \frac{\rho\gamma_1}{4}\frac{\sigma_0 C\left(K\right)}{\nu} + \frac{2 - 3\rho^2}{24}\right]\varepsilon\right\},$$

The formulas can be applied to price the swaption smile or skew.

In some situations, swap rates can behave more like normal random variables, when it may become more accurate to use the implied volatility formula for the normal Black's model. Hagan et al. (2002) have also provided the volatility formula for the normal Black's model:

$$
\sigma_{\text{imp}}^{\text{n}} = \nu \frac{f_0 - K}{D(\zeta)} \left\{ 1 + \left[\frac{2\gamma_2 - \gamma_1^2}{24} \left(\frac{\sigma_0 C(f_{\text{mid}})}{\nu} \right)^2 \right. \right.
$$
$$
\left. \left. + \frac{\rho\gamma_1}{4} \frac{\sigma_0 C(f_{\text{mid}})}{\nu} + \frac{2 - 3\rho^2}{24} \right] \varepsilon \right\}.
$$

When $K \to f_0$, the formula becomes

$$
\sigma_{\text{imp}}^{\text{n}} = \alpha f_0^\beta \left\{ 1 + \left[\frac{2\gamma_2 - \gamma_1^2}{24} \left(\frac{\sigma_0 C(f_0)}{\nu} \right)^2 \right. \right.
$$
$$
\left. \left. + \frac{\rho\gamma_1}{4} \frac{\sigma_0 C(f_0)}{\nu} + \frac{2 - 3\rho^2}{24} \right] \varepsilon \right\}.
$$

It is worth noting that the normal SABR implied volatility formula is generally more accurate than the lognormal SABR implied volatility formula.

The SABR model turned out to be a great success in derivative pricing and risk management across asset classes, beyond interest rate derivatives. It outperforms other models with several advantages:

1. The SABR model can be used to accurately fit a variety of shapes of the IV curves observed in the market, including even the hockey-stick shape (see Figure 10.1).

2. It is an effective means of managing the smile risk in markets where each asset has only a single exercise date (e.g., swaptions and caplet/floorlet).

3. Additionally, it can be used to hedge delta risks, vega risks, vonna risks and volga risks.

With the availability of the SABR formulae, the roles played by various parameters, such as β, α, ρ and ν, have also become transparent. Let us describe in detail below.

1. β, the elasticity constant.

   - β controls the backbone, known as the trace of the ATM implied volatilities, $\sigma_B(f, f)$ (see the dash lines of Figure 10.2).

   - With any specific choice of β, market smiles can generally be fit more or less equally well.

   - β can be estimated from the historical data of the "backbone," $\sigma_B(f, f)$.

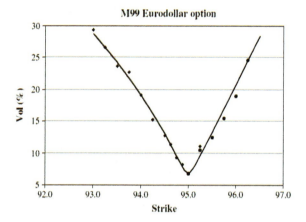

FIGURE 10.1: Implied volatility for the June 99 Eurodollar options, close-of-day values vs. predicted value (curve) by the SABR model. March 23, 1999, Bloomberg. (Courtesy of Applied. Math. Fin.)

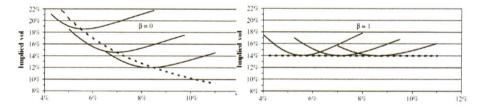

FIGURE 10.2: Backbone for different β. (Courtesy of Applied. Math. Fin.)

- A popular choice of β is 0.4 (it is claimed in Hagan et al. (2002) that for the JPY IR market, $\beta = 0$; yet for the USD IR market, $\beta = 1/2$).

2. α, the level of implied volatilities.

- α is calibrated to the level of ATM volatility.
- It is usually convenient to use σ_{ATM} to replace α in the parameter set.

3. ρ and ν, the correlation and the "vol of vol."

- ρ controls the smile and skew.
- ν controls the curvature of the smile/skew.
- The "vol of vol" ν is very big for short-dated options, and decreases as the time-to-exercise increases; where ρ starts near zero and becomes substantially more negative.

VOLATILITY OF VOLATILITY ν FOR EUROPEAN SWAPTIONS. ROWS ARE TIME–TO–EXERCISE; COLUMNS ARE TENOR OF THE UNDERLYING SWAP.

	1Y	2Y	3Y	4Y	5Y	7Y	10Y
1M	76.2%	75.4%	74.6%	74.1%	75.2%	73.7%	74.1%
3M	65.1%	62.0%	60.7%	60.1%	62.9%	59.7%	59.5%
6M	57.1%	52.6%	51.4%	50.8%	49.4%	50.4%	50.0%
1Y	59.8%	49.3%	47.1%	46.7%	46.0%	45.6%	44.7%
3Y	42.1%	39.1%	38.4%	38.4%	36.9%	38.0%	37.6%
5Y	33.4%	33.2%	33.1%	32.6%	31.3%	32.3%	32.2%
7Y	30.2%	29.2%	29.0%	28.2%	26.2%	27.2%	27.0%
10Y	26.7%	26.3%	26.0%	25.6%	24.8%	24.7%	24.5%

FIGURE 10.3: Fitted ν. (Courtesy of Applied. Math. Fin.)

The frequency of fitting for each parameter is different.

- Typically, α or σ_{ATM} are updated daily or every few hours.
- ρ and ν are re-fitted every month or as needed.
- β can stay unchanged for a long time.

Now, let us focus on the performance of the SABR model on swaptions in the IR market. There are the following observations.

- There is a weak dependence of the market skew/smile on the maturity of the underlying swaps.
- Both ρ and ν are fairly constant for each maturity.

Figures 10.3 and 10.4 are taken from Hagen et al. (2002), which shows the calibrated ν and ρ of swaptions for various option expiries and swap maturities.

MATRIX OF CORRELATIONS ρ BETWEEN THE UNDERLYING AND THE VOLATILITY FOR EUROPEAN SWAPTONS.

	1Y	2Y	3Y	4Y	5Y	7Y	10Y
1M	4.2%	-0.2%	-0.7%	-1.0%	-2.5%	-1.8%	-2.3%
3M	2.5%	-4.9%	-5.9%	-6.5%	-6.9%	-7.6%	-8.5%
6M	5.0%	-3.6%	-4.9%	-5.6%	-7.1%	-7.0%	-8.0%
1Y	-4.4%	-8.1%	-8.8%	-9.3%	-9.8%	-10.2%	-10.9%
3Y	-7.3%	-14.3%	-17.1%	-17.1%	-16.6%	-17.9%	-18.9%
5Y	-11.1%	-17.3%	-18.5%	-18.8%	-19.0%	-20.0%	-21.6%
7Y	-13.7%	-22.0%	-23.6%	-24.0%	-25.0%	-26.1%	-28.7%
10Y	-14.8%	-25.5%	-27.7%	-29.2%	-31.7%	-32.3%	-33.7%

FIGURE 10.4: Fitted ρ. (Courtesy of Applied. Math. Fin.)

We have the following additional remarks on the performance of the model.

1. In most markets, there is a strong smile for short-dated options that relaxes as the time-to-expiry increases.

2. Consequently, the "vol of vol" is larger for short-dated options and smaller for long-dated options, regardless of the particular underlying.

3. Correlation results are less clear: in some markets, a nearly flat skew for short-dated options develops into a strongly downward sloping skew for longer expiries.

4. In some markets, there is a strongly downward skew for all options maturities, while in other markets, the skew is close to zero for all maturities.

Next, we discuss the calculation of Greeks and hedging with the SABR model.

- Vega

The value of a call is

$$V_{call} = BS(f, K, \sigma_B(K, f), T).$$

The vega risk is given by

$$\frac{\partial V_{call}}{\partial \alpha} = \frac{\partial BS}{\partial \sigma_B} \frac{\partial \sigma_B(K, f; \alpha, \beta, \rho, \nu)}{\partial \alpha}.$$

It is traditional to scale vega so that it represents the change in value when the ATM volatility changes by a unit amount. So it follows that

$$vega = \frac{\partial V_{call}}{\partial \sigma_{ATM}} = \frac{\partial V_{call}}{\partial \sigma_B} \frac{\partial \sigma_B}{\partial \sigma_{ATM}} = \frac{\partial V_{call}}{\partial \sigma_B} \frac{\frac{\partial \sigma_B}{\partial \alpha}}{\frac{\partial \sigma_{ATM}}{\partial \alpha}}.$$

where $\sigma_{ATM} = \sigma_{f,f}$. According to Hagan et al. (2002), to the leading order, there are $\partial \sigma_B / \partial \alpha \approx \sigma_B / \alpha$ and $\partial \sigma_{ATM} / \partial \alpha \approx \sigma_{ATM} / \alpha$, hence we have the approximation

$$vega \approx \frac{\partial BS}{\partial \sigma_B} \cdot \frac{\sigma_B(K, f)}{\sigma_{ATM}(f)}.$$

Note that vega risks at different strikes are calculated by bumping the IV at each strike K, by an amount that is proportional to the IV, $\sigma_B(K, f)$, at that strike. So this is not a parallel shift, but a proportional shift of the volatility curve to calculate the total vega risk of a book of options.

- Vanna

The risk associated with the change in ρ is called vanna, which is calculated according to

$$vanna = \frac{\partial V_{call}}{\partial \rho} = \frac{\partial BS}{\partial \sigma_B} \cdot \frac{\partial \sigma_B(K, f; \alpha, \beta, \rho, \nu)}{\partial \rho}.$$

Vanna expresses the risk to the skew increasing.

- Volga

The volga (vol gamma) is the risk associated with the change in ν, the "vol of vol":

$$volga = \frac{\partial V_{call}}{\partial \nu} = \frac{\partial BS}{\partial \sigma_B} \cdot \frac{\partial \sigma_B(K, f; \alpha, \beta, \rho, \nu)}{\partial \nu},$$

and the volga expresses the risk to the smile becoming more pronounced. It is suggested (Hagan et al., 2002) to calculate both *vanna* and *volga* by using finite difference method. These risks, if unwanted, can be hedged by buying or selling away-from-the-money options.

- Delta

The delta risk is the risk associated with the underlying forward price f. It actually predicts change for a sideway movement of the volatility curve caused by the change in the forward price. The delta is calculated by

$$\Delta \equiv \frac{\partial V_{call}}{\partial f} = \frac{\partial BS}{\partial f} + \frac{\partial BS}{\partial \sigma_B} \frac{\partial \sigma_B(K, f; \alpha, \beta, \rho, \nu)}{\partial f}.$$

Here, the first term is the ordinary delta risk that can be calculated from the Black's model, and the second term is the SABR model's correction to the delta risk: Black vega risk times the predicted change in the IV caused by the change in the forward f.

Soon after its publication in 2002, the SABR model quickly gained popularity and has become the standard model for various derivatives markets, including

- interest rate derivative markets;

- equity option markets;

- fixed-income option markets;

- the foreign exchange derivatives markets;

- inflation option markets.

10.2 The Wu and Zhang (2001) Model

In this section, we present a genuine correlation-based model for LIBOR derivatives. We adopt a set-up similar to that of Andersen and Brotherton-Ratcliffe (2001), however, contrary to their approach, we exclude state dependent diffusion coefficients but include correlations between forward rates and a stochastic multiplier (hereafter *rate-multiplier correlations*). The model so developed can be regarded as the LIBOR term-structure version of the Heston's model (1993); the latter has been one of the most popular equity option models with stochastic volatility.[2] There are several additional reasons behind this extension. First, the time series data of interest rates suggests randomness of interest rate volatilities with mean reverting feature (Chen and Scott, 2001), and it is a popular belief that the stochastic volatilities are the primary factors behind the leptokurtic feature[3] of empirical interest rate distributions. With the Heston's type model, we can also capture the leptokurtic feature. Second, Heston's model establishes a direct correspondence between a downward skew to a negative correlation between the state variable and its stochastic volatility.[4] Third, among stochastic volatility models (see e.g., Lewis, 2000), the Heston's model carries good analytical tractability that renders exact closed-form pricing for equity options. Ironically, such a correlation has been deemed counterproductive for the Heston's type model in the context of LIBOR, as it causes dependence of volatility process on forward rates after we change from the risk neutral measure to any forward measure, which is a kind of circular dependence that spoils analytical tractability. A key question for developing a Heston's type model for LIBOR is whether it is possible to relegate such dependence without compromising the accuracy on derivative pricing.

The answer to the above question is positive. We have observed that under its corresponding forward swap measure, a swap-rate process retains the formalism of the Heston's model with, however, state-dependent coefficients for its volatility process. Yet, the time variability of those coefficients is rather small. This crucial observation has motivated us to get rid of the circular dependance through "freezing coefficients," to eventually obtain a "closed-form" formula for swaptions in terms of a fast Fourier transform (FFT).

Another focus of this section is the calibration of the LIBOR market model with square-root stochastic volatilities to cap prices. We try to determine the set of model parameters by optimally matching the implied caplet volatilities, not the prices themselves, in the least square sense. We identify the key parameters to be the magnitude of forward-rate volatilities, the correlation between the forward rates and stochastic volatilities, and the "vol of vol." Respectively, these three sets of parameters are responsible for the level, the skewness, and the curvature of the implied volatility surface. The calibration is achieved through a two-layer nested minimization procedure: the outer layer is for minimizing the total square error with respect to the "vol of vol," and the inner layer is for minimizing the square error in the implied volatilities for each individual maturity, with respect to *the magnitude of the forward rates* and *the correlations*. Moreover, the inner layer consists of decoupled bi-variable minimization problems, and thus can be solved instantly. In our

---

[2] A lognormal process whose volatility follows a square-root process (Cox et al. 1985).

[3] Higher peak and fatter tails than that of a normal distribution.

[4] Such a correspondence in fact also exists in other stochastic volatility models, e.g., Zhou (2003).

study, we have taken only state-independent parameters, and the procedure can be implemented in seconds. It is desirable to simultaneously calibrate to swaptions. However, there is the difficult issue of convergence that must be faced, and we thus choose to avoid the joint calibration of caps and swaptions.

The remaining part of this section is organized as follows. In this section we introduce the LIBOR market model with stochastic volatility. In Sections 10.3 and 10.4, we present the (approximate) closed-form pricing formulae for caplets and swaptions. In Section 10.5, we discuss the calibration strategy and demonstrate the performance of the strategy with market data. In Section 10.6, we make some notes. Finally, Section 10.7 is an Appendix where we give a proof of Proposition 2.2.

The notations for major variables and measures used in this section are as follows:

$B(t)$	-	the value of money market account at time t;
$P(t,T)$	-	time t value of zero-coupon bond of maturity T;
$A_{m,n}(t)$	-	time t value of the annuity, $\Delta T_j P(t,T_j), j = m+1,\ldots,$ n;
$f_j(t)$	-	the simple forward rate of maturity T_j;
$R_{m,n}(t)$	-	the prevailing swap rate for the term (T_m, T_n);
$V(t)$	-	the stochastic multiplier for forward-rate volatilities;
$\mathbb{Q}$	-	the risk neutral measure;
$\mathbb{Q}_j$	-	the forward measure of maturity T_j;
$\mathbb{Q}_S$	-	the forward swap measure corresponding to the numeraire of $A_{m,n}(t)$.

Our introduction of the market model with stochastic volatility starts with the usual lognormal dynamics of the Treasury zero-coupon bonds. Let $P(t,T)$ be the price of Treasury zero-coupon bond maturing at $T(\geq t)$ with par value \$1, and let $B(t)$ be the money market account under discrete compounding:

$$B(t) = \left(\prod_{j=0}^{\eta(t)-2} (1 + f_j(T_j)\Delta T_j) \right) \left(1 + f_{\eta(t)-1}(T_{\eta(t)-1})(t - T_{\eta(t)-1}) \right),$$

where $\Delta T_j = T_{j+1} - T_j$ and $\eta(t)$ is the smallest integer such that $T_{\eta(t)} \geq t$. Under a risk-neutral measure, it is typical to assume dynamics of lognormal martingale for the discount price of $P(t,T)$ as

$$d\left(\frac{P(t,T)}{B(t)} \right) = \left(\frac{P(t,T)}{B(t)} \right) \Sigma(t,T) \cdot d\mathbf{W}_t. \tag{10.5}$$

Here, $\Sigma(t,T)$ is the volatility vector of $P(t,T)$, and $\mathbf{W}_t$ is a finite dimensional Brownian motion under the risk-neutral measure, which we denote by

$\mathbb{Q}$, and "$\cdot$" is the usual vector product. The volatility function satisfies the boundedness condition, $E\left[\int_0^t \|\Sigma(s,T)\|^2 ds\right] < \infty, \forall t < T$.

Wu and Zhang (2002) adopt specifically a stochastic multiplier to the risk neutralized processes of the forward rates:

$$df_j(t) = f_j(t)\sqrt{V(t)}\gamma_j(t) \cdot \left[d\mathbf{W}_t - \sqrt{V(t)}\Sigma_{j+1}(t)dt\right],$$
$$dV(t) = \kappa(\theta - V(t))dt + \nu\sqrt{V(t)}dZ_t. \tag{10.6}$$

Here,

$$\Sigma(t, T_{j+1}) = -\sum_{k=\eta(t)}^{j} \frac{\Delta T_k f_k(t)}{1 + \Delta T_k f_k(t)}\gamma_k(t) + \Sigma(t, T_{\eta(t)}), \tag{10.7}$$

κ, θ and ν are time-dependent variables,[5] and Z_t is an additional 1-D Brownian motion under the risk-neutral measure. As a multi-factor model, the forward rates can be correlated such that

$$\text{Cov}_{jk}^i(t) = \gamma_j(t) \cdot \gamma_k(t), \qquad i \le j, k \le N, \quad 1 \le i \le N. \tag{10.8}$$

Adoption of a stochastic volatility in models similar to (10.6) was first considered by Chen and Scott (2001) and Andersen and Brotherton-Ratcliffe (2001). In these models, for the sake of analytical tractability, the stochastic volatility is kept independent of interest rates. As a distinct feature of the Wu and Zhang model, the correlations between the stochastic multiplier and forward rates are allowed, such as

$$E^Q\left[\left(\frac{\gamma_j(t)}{\|\gamma_j(t)\|} \cdot d\mathbf{W}_t\right) \cdot dZ_t\right] = \rho_j(t)dt, \quad \text{with} \quad |\rho_j(t)| \le 1. \tag{10.9}$$

Here, $\left(\frac{\gamma_j(t)}{\|\gamma_j(t)\|} \cdot d\mathbf{W}_t\right)$ is equivalent to (the differential of) a single Brownian motion that drives $f_j(t)$. The correlation coefficients, $\{\rho_j(t)\}$, will play an essential role to capture volatility smiles. For easy reference, we also call V a stochastic multiplier. Note that for the model above, Equation 10.7 is the no-arbitrage condition.

Mathematically, we can construct a market model where each component of the forward-rate volatility vector is associated with a stochastic multiplier.[6] While analytical or semi-analytical pricing of caplets and swaptions under such a model remains feasible, calibration will be extremely difficult, if not impossible. Technically, adopting a uniform volatility multiplier for all rates rather than one multiplier for each rate retains much greater analytical tractability, and it makes model calibration amenable.

---

[5]The distributional properties of $V(t)$ are described in Lemma 9.4.1.
[6]Trolle and Schwartz (2008) develop a HJM model with such a feature.

10.3 Pricing of Caplets

We now consider caplet pricing under the extended LIBOR model (10.6). A caplet is a call option on a forward rate. Assuming that the notional value of a caplet is one dollar, then the payoff of the caplet at T_{j+1} is

$$\Delta T_j (f_j(T_j) - K)^+ \overset{\triangle}{=} \Delta T_j \max\{f_j(T_j) - K, 0\}.$$

To price the caplet we choose $P(t, T_{j+1})$, in particular, to be the numeraire asset and let $\mathbb{Q}_{j+1}$ denote the corresponding forward measure (i.e. the martingale measure corresponding to numeraire $P(t, T_{j+1})$). The next proposition establishes the relationship between Brownian motions under the risk-neutral measure and under the T_{j+1} forward measure (Wu and Zhang, 2002).

Proposition 10.3.1. *Let $\mathbf{W}_t$ and Z_t be Brownian motions under $\mathbb{Q}$, then $\mathbf{W}_t^{j+1}$ and Z_t^{j+1}, defined by*

$$\begin{aligned}
d\mathbf{W}_t^{j+1} &= d\mathbf{W}_t - \sqrt{V(t)}\Sigma_{j+1}(t)dt, \\
dZ_t^{j+1} &= dZ_t + \xi_j(t)\sqrt{V(t)}dt,
\end{aligned} \tag{10.10}$$

are Brownian motions under $\mathbb{Q}_{j+1}$, where

$$\xi_j(t) = \sum_{k=\eta(t)}^{j} \frac{\Delta T_k f_k(t)\rho_k(t)\lambda_k(t)}{1 + \Delta T_k f_k(t)},$$

where $\lambda_k(t) = \|\gamma_k(t)\|$.   □

Proof: The Radon-Nikodym derivative of $\mathbb{Q}_{j+1}$ with respect to $\mathbb{Q}$ is

$$\begin{aligned}
\frac{d\mathbb{Q}_{j+1}}{d\mathbb{Q}} &= \frac{P(t, T_{j+1})/P(0, T_{j+1})}{B(t)} \\
&= e^{\int_0^t -\frac{1}{2}V(\tau)\Sigma_{j+1}^2(\tau)d\tau + \sqrt{V(\tau)}\Sigma_{j+1}\cdot d\mathbf{W}_t} \\
&\overset{\triangle}{=} m_{j+1}(t), \quad t \leq T_{j+1}.
\end{aligned}$$

Clearly, we have

$$dm_{j+1}(t) = m_{j+1}(t)\sqrt{V(t)}\Sigma_{j+1}(t) \cdot d\mathbf{W}_t.$$

Let $\langle \cdot, \cdot \rangle$ denote covariance. By the CMG change of measure theorem, we obtain the Brownian motions under $\mathbb{Q}_{j+1}$:

$$
\begin{aligned}
d\mathbf{W}_t^{j+1} &= d\mathbf{W}_t - \langle d\mathbf{W}_t, dm_{j+1}(t)/m_{j+1}(t)\rangle \\
&= d\mathbf{W}_t - \sqrt{V(t)}\Sigma_{j+1}(t)dt, \\
dZ_t^{j+1} &= dZ_t - \langle dZ_t, dm_{j+1}(t)/m_{j+1}(t)\rangle \\
&= dZ_t - \langle dZ_t, \sqrt{V(t)}\Sigma_{j+1}(t)\cdot d\mathbf{W}_t\rangle \\
&= dZ_t + \sqrt{V(t)}\sum_{k=1}^{j}\frac{\Delta T_k f_k(t)\lambda_k(t)}{1+\Delta T_k f_k(t)}\langle dZ_t, \frac{\gamma_k(t)}{\lambda_k(t)}\cdot d\mathbf{W}_t\rangle \\
&= dZ_t + \sqrt{V(t)}\sum_{k=1}^{j}\frac{\Delta T_k f_k(t)\lambda_k(t)}{1+\Delta T_k f_k(t)}\rho_k(t)dt \qquad \square
\end{aligned}
$$

In terms of $\mathbf{W}_t^{j+1}$ and Z_t^{j+1}, the extended market model (10.6) becomes

$$df_j(t) = f_j(t)\sqrt{V(t)}\gamma_j(t)\cdot d\mathbf{W}_t^{j+1}, \tag{10.11}$$

$$dV(t) = [\kappa\theta - (\kappa + \nu\xi_j(t))V(t)]\,dt + \nu\sqrt{V(t)}dZ_t^{j+1}. \tag{10.12}$$

In formalism, the multiplier process remains a square-root process under $\mathbb{Q}_{j+1}$. Yet part of the coefficients, $\xi_j(t)$, depends on forward rates, and such dependence prohibits analytical option valuation. The time variability of $\xi_j(t)$, however, is small. In fact, we can write

$$
\xi_j(t) = \sum_{k=1}^{j}\frac{\Delta T_k f_k(0)\rho_k(t)\lambda_k(t)}{1+\Delta T_k f_k(0)} + \frac{\rho_k(t)\lambda_k(t)\Delta T_k}{(1+\Delta T_k f_k(0))^2}(f_k(t) - f_k(0))
$$
$$
+ O\left(\rho_k(t)\lambda_k(t)\Delta T_k^2(f_k(t) - f_k(0))^2\right). \tag{10.13}
$$

In light of the martingale property, $E^{Q^{j+1}}[f_j(t)|\mathcal{F}_0] = f_j(0)$, we see that

$$
E^{Q^{j+1}}[\xi_j(t)|\mathcal{F}_0] = \sum_{k=1}^{j}\frac{\Delta T_k f_k(0)\rho_k(t)\lambda_k(t)}{1+\Delta T_k f_k(0)}
$$
$$
+ O(\rho_k(t)\lambda_k(t)\,Var(\Delta T_k f_k(t))),
$$
$$
Var(\xi_j(t)|\mathcal{F}_0) \approx (\rho_k(t)\lambda_k(t))^2\,Var(\Delta T_k f_k(t)).
$$

According to the model, $Var(\Delta T_k f_k(t)) \sim \Delta T_k f_k^2(t)\lambda_k^2(t)V(t)t$. Since $\Delta T_k f_k(t)$ is mostly under 5%, the expansion in (10.13) is dominated by the first term. Hence, to remove the dependence of $V(t)$ on $f_j(t)$'s, we choose to ignore higher order terms in (10.13) and consider the approximation

$$
\xi_j(t) \approx \sum_{k=1}^{j}\frac{\Delta T_k f_k(0)\rho_k(t)\lambda_k(t)}{1+\Delta T_k f_k(0)}. \tag{10.14}
$$

This is close to the technique of "freezing coefficients." For notational simplicity we denote

$$\tilde{\xi}_j(t) = 1 + \frac{\nu}{\kappa}\xi_j(t),$$

and thus retain a neat equation for the process of $V(t)$:

$$dV(t) = \kappa \left[\theta - \tilde{\xi}_j(t)V(t)\right]dt + \nu\sqrt{V(t)}dZ_t^{j+1}. \tag{10.15}$$

For the processes joined by (10.11) and (10.15), caplets can be priced along the approach pioneered by Heston (1993). According to the arbitrage pricing theory (APT) (Harrison and Pliska, 1979; Harrison and Krep, 1981), the price of the caplet on $f_j(T_j)$ can be expressed as

$$C_{let}(0) = P(0, T_{j+1})\Delta T_j E^{Q^{j+1}}\left[(f_j(T_j) - K)^+|\mathcal{F}_0\right]$$
$$= P(0, T_{j+1})\Delta T_j f_j(0)\left(E^{Q^{j+1}}\left[e^{X(T_j)}\mathbf{1}_{X(T_j)>k}|\mathcal{F}_0\right]\right.$$
$$\left. -e^k E_0^{Q^{j+1}}\left[\mathbf{1}_{X(T_j)>k}|\mathcal{F}_0\right]\right),$$

where $X(t) = \ln f_j(t)/f_j(0)$ and $k = \ln K/f_j(0)$. The two expectations above can be valuated using the moment generating function of $X(T_j)$, defined by

$$\phi(X(t), V(t), t; z) \triangleq E\left[e^{zX(T_j)}|\mathcal{F}_t\right], \quad z \in C.$$

In terms of $\phi_{T_j}(z) \triangleq \phi(0, V(0), 0; z)$, we have that (e.g., Kendall (1994) or more recently Duffie, Pan and Singleton (2000))

$$E^{Q^{j+1}}\left[\mathbf{1}_{X(T_j)>k}|\mathcal{F}_0\right] = \frac{\phi_{T_j}(0)}{2} + \frac{1}{\pi}\int_0^\infty \frac{\text{Im}\{e^{-iuk}\phi_{T_j}(iu)\}}{u}du,$$

$$E^{Q^{j+1}}\left[e^{X(T_j)}\mathbf{1}_{X(T_j)>k}|\mathcal{F}_0\right] = \frac{\phi_{T_j}(1)}{2} + \frac{1}{\pi}\int_0^\infty \frac{\text{Im}\{e^{-iuk}\phi_{T_j}(1+iu)\}}{u}du. \tag{10.16}$$

The integrals above can then be evaluated numerically. For later reference, we call this approach the Heston method.

When the Brownian motions $\mathbf{W}_t^{j+1}$ and Z_t^{j+1} are independent, the moment generating function can be directly worked out. In general, one can solve for $\phi(x, V, t; z)$ from the Kolmogorov backward equation corresponding to the joint processes:

$$\frac{\partial\phi}{\partial t} + \kappa(\theta - \tilde{\xi}_j V)\frac{\partial\phi}{\partial V} - \frac{1}{2}\lambda_j^2(t)V\frac{\partial\phi}{\partial x}$$
$$+ \frac{1}{2}\nu^2 V\frac{\partial^2\phi}{\partial V^2} + \nu\rho_j V\lambda_j(t)\frac{\partial^2\phi}{\partial V\partial x} + \frac{1}{2}\lambda_j^2(t)V\frac{\partial^2\phi}{\partial x^2} = 0, \tag{10.17}$$

subject to terminal condition

$$\phi(x, V, T_j; z) = e^{zx}. \tag{10.18}$$

It is known that the solution is of the form

$$\phi(x, V, t; z) = e^{A(t,z)+B(t,z)V+zx}, \tag{10.19}$$

where A and B satisfy the following equations

$$
\begin{aligned}
\frac{dA}{dt} + \kappa\theta B &= 0, \\
\frac{dB}{dt} + \frac{1}{2}\nu^2 B^2 + (\rho_j \nu \lambda_j z - \kappa \xi)B + \frac{1}{2}\lambda_j^2(z^2 - z) &= 0,
\end{aligned}
\tag{10.20}
$$

subject to terminal conditions

$$A(T_j, z) = 0, \quad B(T_j, z) = 0,$$

and A and B can be solved analytically for constant coefficients (Heston, 1993). The analytical solutions can be extended to the case of piecewise constant coefficients through recursions. In the statement of the next proposition, we have suppressed the time dependence of the coefficients for simplicity. The proof of the proposition is provided in this chapter's appendix for completeness.

Proposition 10.3.2. *Suppose that all coefficients of (10.20) are constants over time intervals* $T_{k-1} \le t < T_k$, $k = 1, 2, \ldots, j$, *and* $\nu^2 > 0$ *for all* t, *then* A *and* B *are given by the following recursive formulae*

$$
\left\{
\begin{aligned}
A(t, z) &= A(T_k, z) + a_0 \left\{ \left[u_k^+ + B(T_k, z)\right](T_k - t) \right. \\
&\qquad\qquad\qquad \left. - \frac{1}{b_2} \ln\left[\frac{u_k^- - u_k^+}{u_k^- - u_k^+ e^{d(T_k - t)}}\right]\right\}, \\
B(t, z) &= B(T_k, z) + u_k^- u_k^+ \frac{(1 - e^{d(T_k - t)})}{(u_k^- - u_k^+ e^{d(T_k - t)})}, \\
&\quad\text{for } T_{k-1} \le t < T_k, \quad k = j, j-1, \ldots, 1,
\end{aligned}
\right.
\tag{10.21}
$$

where,

$$d = \sqrt{b_1^2 - 4b_2 b_0}, \quad u_k^{\pm} = \frac{-b_1 \pm d}{2b_2} - B(\tau_k, z),$$

and

$$a_0 = \kappa\theta, \quad b_2 = \frac{1}{2}\nu^2, \quad b_1 = \rho_j \nu \lambda_j z - \kappa\xi, \quad b_0 = \frac{1}{2}\lambda_j^2(z^2 - z). \tag{10.22}$$

$\square$

10.4   Pricing of Swaptions

The equilibrium swap rate for a period (T_m, T_n) is defined by

$$R_{m,n}(t) = \frac{P(t, T_m) - P(t, T_n)}{A_{m,n}(t)},$$

where

$$A_{m,n}(t) = \sum_{j=m}^{n-1} \Delta T_j P(t, T_{j+1})$$

is an annuity. The payoff of a swaption on $R_{m,n}(T_m)$ can be expressed as

$$A_{m,n}(T_m) \cdot \max(R_{m,n}(T_m) - K, 0),$$

where K is the strike rate.

The swap rate can be regarded as the price of a tradable portfolio relative to the price of the annuity $A_{m,n}(t)$. This portfolio consists of one long T_m-maturity zero-coupon bond and one short T_n-maturity zero-coupon bond. According to arbitrage pricing theory (Harrison and Pliska, 1979; Harrison and Krep, 1981), the swap rate is a martingale under the measure corresponding to the numeraire $A_{m,n}(t)$. This measure is called the *forward swap measure* (Jamshidian, 1997) and is denoted by $\mathbb{Q}_S$ in this section. Similarly to pricing under a forward measure, we need to characterize the Brownian motions under the forward swap measure (Wu and Zhang, 2002).

Proposition 10.4.1. *Let $\mathbf{W}_t$ and Z_t be Brownian motions under $\mathbb{Q}$, then $\mathbf{W}_t^S$ and Z_t^S, defined by*

$$\begin{aligned} d\mathbf{W}_t^S &= d\mathbf{W}_t - \sqrt{V(t)}\Sigma_A(t)dt, \\ dZ_t^S &= dZ_t + \sqrt{V(t)}\xi^S(t)dt, \end{aligned} \tag{10.23}$$

are Brownian motions under $\mathbb{Q}_S$, where

$$\Sigma_A(t) = \sum_{j=m}^{n-1} \alpha_j \Sigma(t, T_{j+1}), \qquad \xi^S(t) = \sum_{j=m}^{n-1} \alpha_j \xi_j, \tag{10.24}$$

with weights

$$\alpha_j = \alpha_j(t) = \frac{\Delta T_j P(t, T_{j+1})}{A_{m,n}(t)}.$$

$\square$

Proof. Denote the forward swap measure by $\mathbb{Q}_S$. The Radon-Nikodym derivative for $\mathbb{Q}_S$ is

$$\frac{d\mathbb{Q}_S}{d\mathbb{Q}} = \frac{A_{m,n}(t)/A_{m,n}(0)}{B(t)}$$

$$= \frac{1}{A_{m,n}(0)} \sum_{j=m}^{n-1} \Delta T_j P(0, T_{j+1}) e^{\int_0^t -\frac{1}{2} V(\tau) \Sigma_{j+1}^2(\tau) d\tau + \sqrt{V(\tau)} \Sigma_{j+1} \cdot d\mathbf{W}_t}$$

$$\stackrel{\triangle}{=} m_S(t), \quad t \le T_m.$$

There is

$$dm_S(t) = \frac{1}{A_{m,n}(0)} \sum_{j=m}^{n-1} \Delta T_j P(0, T_{j+1}) e^{\int_0^t -\frac{1}{2} V(\tau) \Sigma_{j+1}^2(\tau) d\tau + \sqrt{V(\tau)} \Sigma_{j+1} \cdot d\mathbf{W}_t}$$

$$\sqrt{V(t)} \Sigma_{j+1}(t) \cdot d\mathbf{W}_t$$

$$= \frac{1}{A_{m,n}(0) B(t)} \sum_{j=m}^{n-1} \Delta T_j P(t, T_{j+1}) \sqrt{V(t)} \Sigma_{j+1}(t) \cdot d\mathbf{W}_t$$

$$= m_S(t) \sum_{j=m}^{n-1} \alpha_j \sqrt{V(t)} \Sigma_{j+1}(t) \cdot d\mathbf{W}_t.$$

It follows that

$$d\mathbf{W}_t^S = d\mathbf{W}_t - \langle d\mathbf{W}_t, dm_S(t)/m_S(t) \rangle$$

$$= d\mathbf{W}_t - \sqrt{V(t)} \sum \alpha_j \Sigma_{j+1}(t) dt,$$

$$= d\mathbf{W}_t - \sqrt{V(t)} \Sigma_A(t) dt,$$

$$dZ_t^S = dZ_t - \langle dZ_t, dm_S(t)/m_S(t) \rangle$$

$$= dZ_t - \langle dZ_t, \sqrt{V(t)} \sum \alpha_j \Sigma_{j+1}(t) \cdot d\mathbf{W}_t \rangle$$

$$= dZ_t + \sqrt{V(t)} \sum_{j=m}^{n-1} \alpha_j \sum_{k=1}^{j} \frac{\Delta T_k f_k(t) \lambda_k(t)}{1 + \Delta T_k f_k(t)} \langle dZ_t, \frac{\gamma_k(t)}{\lambda_k(t)} \cdot d\mathbf{W}_t \rangle$$

$$= dZ_t + \sqrt{V(t)} \sum_{j=m}^{n-1} \alpha_j \sum_{k=1}^{j} \frac{\Delta T_k f_k(t) \lambda_k(t)}{1 + \Delta T_k f_k(t)} \rho_k(t) dt$$

$$= dZ_t + \sqrt{V(t)} \sum_{j=m}^{n-1} \alpha_j \xi_j(t) dt$$

$$= dZ_t + \sqrt{V(t)} \xi^S(t) dt$$

$\square$

Using Ito's lemma, it can be shown that under the forward swap measure, the swap rate process becomes

$$dR_{m,n}(t) = \sqrt{V(t)} \sum_{j=m}^{n-1} \frac{\partial R_{m,n}(t)}{\partial f_j(t)} f_j(t)\gamma_j(t) \cdot d\mathbf{W}^S(t),$$

$$dV(t) = \kappa\left[\theta - \tilde{\xi}^S(t)V(t)\right]dt + \nu\sqrt{V(t)}dZ^S(t). \tag{10.25}$$

Here,

$$\tilde{\xi}^S(t) = 1 + \frac{\nu}{\kappa}\xi^S(t).$$

For the partial derivatives of the swap rate with respect to forward rates, we have:

Proposition 10.4.2. *Let*

$$R_{m,n}(t) = \sum_{k=m}^{n-1} \alpha_k f_k, \quad \alpha_k = \frac{\Delta T_k P(t, T_{k+1})}{A_{m,n}(t)},$$

there is

$$\frac{\partial R_{m,n}(t)}{\partial f_j(t)} = \alpha_j + \frac{\Delta T_j}{1 + \Delta T_j f_j(t)}\left[\sum_{l=m}^{j-1} \alpha_l(f_l - R_{m,n}(t))\right], \quad m \leq j \leq n-1 \quad \square$$

The proof can be found in Chapter 6.

Similar to swaption pricing under the standard market model (e.g., Sidennius, 2000; Andersen and Andreasen, 2000), we approximate the swap rate process by a lognormal process with however a stochastic volatility:

$$dR_{m,n}(t) = R_{m,n}(t)\sqrt{V(t)}\gamma_{m,n}(t) \cdot d\mathbf{W}^S(t), \quad 0 \leq t < T_m,$$

$$dV(t) = \kappa\left[\theta - \tilde{\xi}_0^S(t)V(t)\right]dt + \nu\sqrt{V(t)}dZ^S(t), \tag{10.26}$$

where

$$\xi_0^S(t) = \sum_{j=m}^{n-1} \alpha_j(0)\xi_j(t),$$

$$\gamma_{m,n}(t) = \sum_{j=m}^{n-1} w_j(0)\gamma_j(t), \quad w_j(t) = \frac{\partial R_{m,n}(t)}{\partial f_j}\frac{f_j(t)}{R_{m,n}(t)},$$

and $\rho^S = \sum_{j=m}^{n-1} w_j(0)\rho_j$.

In the above approximations, we have removed the dependence of $\xi_0^S(t)$ on forward rates through taking full advantage of the negligible time variability of $w_j(t)$ and $\alpha_j(t)$ (compared with that of forward rates). As a result, the approximate swap-rate process has moment generating function in closed form, and we thus retain the analytical tractability of the model under the forward swap measure. This is the key treatment in this section, which works well for market models with the square-root volatility dynamics, but may not work for those with general volatility dynamics. Note that when $n = m + 1$, $R_{m,m+1}(t) = f_m(t)$ and $A_{m,n}(t) = \Delta T_m P(t, T_{m+1})$, i.e., the swap rate reduces to a forward rate, and the swaption reduces to a caplet.[7] Theoretically, we can treat a caplet as a special case of swaptions.

Instead of following the Heston's approach for numerical pricing, we adopt a transformation method developed by Carr and Madan (1998). Under the forward swap measure, we have the following expression for swaption prices

$$SP(0) = A_{m,n}(0)R_{m,n}(0)E^{Q_s}\left[\left(e^{X(T_m)} - e^k\right)^+ |\mathcal{F}_0\right], \qquad (10.27)$$

where

$$X(T_m) = \ln R_{m,n}(T_m)/R_{m,n}(0) \quad \text{and} \quad k = \ln K/R_{m,n}(0).$$

The moment-generating function of $X(T_m)$, $\phi_{T_m}(z) \triangleq \phi(0, V(0), 0; z)$, which is characterized in (10.19) and derived in Proposition 10.3.2, with however $\{m, \rho^S, \lambda_{m,n}\}$ taking the place of $\{j, \rho_j, \lambda_j\}$, where $\lambda_{m,n} = \|\gamma_{m,n}\|$. Once we have obtained the moment generating function $\phi_{T_m}(z)$, we again follow Carr and Madan (1998)'s approach the valuation swaption. The first step is to treat the expectation in (13.40) as a function of strike:

$$G(k) = E^{Q_s}\left[\left(e^{X(T_m)} - e^k\right)^+ |\mathcal{F}_0\right].$$

Then, let $q(s)$ denote the density function of $X(T_m) = \ln R_{m,n}(T_m)/R_{m,n}(0)$, we write,

$$G(k) = \int_k^\infty (e^s - e^k)q(s)ds,$$

Note that $G(k)$ is not square integrable over $(-\infty, \infty)$ as it tends to 1 when k tends to $-\infty$. Take some constant $a > 0$ and consider the Laplace transform of the value function,

$$\psi(u) = \int_{-\infty}^\infty e^{(a+iu)k}G(k)dk = \int_{-\infty}^\infty \int_k^\infty e^{(a+iu)k}(e^s - e^k)q(s)dsdk$$

$$= \int_{-\infty}^\infty q(s) \int_{-\infty}^s (e^{s+ak} - e^{(1+a)k})e^{iuk}dkds$$

$$= \int_{-\infty}^\infty q(s)\left[\frac{e^{(a+1+iu)s}}{a+iu} - \frac{e^{(a+1+iu)s}}{a+1+iu}\right]ds$$

$$= \frac{\phi_{T_m}(1+a+iu)}{(a+iu)(1+a+iu)}.$$

---

[7]For convenience we have taken the same ΔT for both caps and swaptions. Note that in reality caps and swaptions can have different intervals between cash flows. In such case, we may take the smallest interval for ΔT.

Having obtained $\psi(u)$ in closed form, the swaption price follows from an inverse Fourier transform

$$G(k) = exp(-ak) \int_0^\infty e^{iuk} \psi(u) du. \tag{10.28}$$

The FFT implementation is described with details in chapter 9 and will not be repeated here.

Several remarks are in order. First, a rigorous error analysis of the lognormal approximation poses an open challenge. For the standard market model (without stochastic volatility), a paper by Brigo et al. (2004) studies the quality of the approximation using entropy distance, but an error estimation for option pricing remains beyond reach. The analysis may be applicable to the approximation (10.26) for the case of zero rate-multiplier correlation. In this chapter, we resort to numerical pricing comparisons in order to gauge the pricing accuracy of the FFT method. For the details of the FFT method, we refer to Section 9.1 and will not repeat here.

10.5   Model Calibration

Calibration is a procedure for determining the parameters of a model based on observed information of certain securities. This is a necessary step when the model is applied for production uses. For the market model with the square-root volatility process, specifically, we need to determine the following set of model coefficients or parameters, $\{\kappa, \theta, \nu; \gamma_j(t), \rho_j(t), j = 1, \ldots, N\}$, based on information on LIBOR, swap rates, and the prices of a set of caps, floors and swaptions. Computationally, this can be a challenging problem. It has been suggested to use some parameters estimated from time series data in order to reduce the scale (i.e., number of unknowns) of the problem. For instance, we may first estimate the multiplier process using the time series data of implied Black's volatilities of at-the-money caplets, as is suggested in Chen and Scott (2001). Once the process of $V(t)$ is specified, we can proceed to determine the pair of $\{\|\gamma_j\|, \rho_j\}$ through matching to the smile of T_j-maturity caplets. This will lead to a bi-variate optimization problem, which is easily manageable. Implicitly, taking a process estimated from time series data as a risk neutral process means that the related risk premium is treated as zero, which, at least in the case of stochastic volatility, is not justified (see for instance, Wu, 2004). Hence, using the estimated process may sometimes undermine the quality of calibration to an extent that is beyond acceptable.

In this section, we consider simultaneous calibration of processes of forward rates and the multiplier based on caplet smiles. When there is no stochastic volatility, i.e., $\nu \equiv 0$, we want $V \equiv 1$ and thus reduce the extended model to the standard market model. For this reason, we take $V_0 = \theta = 1$. The parameter κ will also be taken fixed, but its choice is less critical as the role of κ overlaps with that of ν. Note that either a smaller κ or a bigger ν will result in stronger effect of the stochastic volatility. After V_0, θ and κ are fixed a priori, the calibration problem reduces to the determination of ν and the pairs of $\|\gamma_j(t)\|$ and $\rho_j(t), j = 1, \ldots, N$.

We solve for ν and $\{\|\gamma_j(t)\|, \rho_j(t)\}_{j=1}^N$ through a two-layer nested minimization:

$$\min_{\nu} \left(\sum_{j=1}^N \min_{\|\gamma_j\|, \rho_j} \left(\sum_{k=1}^K (v_{k,j}^{(mk)} - v_{k,j}^{(md)})^2 \right) \right). \tag{10.29}$$

Here, $v_{k,j}^{(mk)}$ and $v_{k,j}^{(md)}$ are the implied caplet volatilities of the kth caplet of the jth cap, and the sup-indexes "mk" and "md" stand for market and model respectively. Hence, once a ν is taken, $\|\gamma_j^{(\nu)}\|, \rho_j^{(\nu)}$ are solved separately by matching to the implied volatility curve of T_j-maturity caplets, $j = 1, \ldots, N$. Note that there is a constraint, $|\rho_j| \leq 1$, on ρ_j. This constraint is easily removed by letting $\rho_j = \cos \theta_j$ and the pair of unknowns become $\|\gamma_j^{(\nu)}\|$ and $\theta_j^{(\nu)}$. In general, we can parameterize $\{\|\gamma_j\|, \rho_j\}_{j=1}^N$ and ν as time-dependent functions. Since we are calibrating only to caplets, we have the luxury to let $\{\|\gamma_j\|, \rho_j\}_{j=1}^N$ be time-independent as well. Our experiences with the use of time-dependent ν, however, are not very encouraging: the resulted ν can change dramatically, and sometimes we do not even achieve convergence due to numerical instability. For this reason, we limit ourselves to finding a constant ν.

One of the key treatments adopted in our solution procedure is to make the implied Black's volatility an explicit function of the state variable and the call-option value. This is achieved by interpolating the implied Black's volatility over a mesh of the state variables and the call-option values. Bivariate spline functions are used for the interpolation. Note that we are matching the implied volatilities through an iteration procedure, so there are many calculations of implied volatilities that employ a root-finding procedure. Making implied volatility an explicit function avoids the root-finding procedure and thus greatly speeds up the algorithm.

Let us comment on how to determine the components of the volatility vector, $\{\gamma_j\}$. It is known that once $\{\|\gamma_j\|\}$ are obtained, the determination of $\{\gamma_j\}$ is subject to forward-rate correlations. Given, for instance, historical forward-rate correlations, we can solve for $\{\gamma_j\}$ by matching model correlations to the historical correlations. Specifically, taking the rate-multiplier correlations into account, we can derive the following equation for $\{\gamma_j/\|\gamma_j\|\}$'s,

$$(1 - \rho_j \rho_k) \left(\frac{\gamma_j}{\|\gamma_j\|} \right) \cdot \left(\frac{\gamma_k}{\|\gamma_k\|} \right) + \rho_j \rho_k = C_{jk},$$

where C_{jk} is the historical correlation between the time series data of f_j and f_k, and it is assumed to be dependent on $T_j - T_k$. The existence of $\{\gamma_j/\|\gamma_j\|\}$ requires that the matrix with components

$$\frac{C_{jk} - \rho_j \rho_k}{1 - \rho_j \rho_k}, \quad i \leq j \wedge k, \tag{10.30}$$

be non-negative definite. Intuitively, (10.30) represents the correlation between the two forward rates after the factor of stochastic volatility is removed. An eigenvalue decomposition of the matrix with the elements given in (10.30) will produce $\{\gamma_j/\|\gamma_j\|\}$. For details we refer to Chapter 7.

For the implementation of the FFT method, we have to fix several additional parameters: A, b, N and α. Based on the condition $Ab = \pi N$, we take

$$A = \frac{3}{2}\sqrt{\pi N}, \quad \text{and} \quad b = \frac{2}{3}\sqrt{\pi N}. \tag{10.31}$$

By taking A and b in such a way, we ensure that both A and b increase when N increases, thus a finer division in both physical and frequency spaces is coupled with broader ranges. When $N = 80$, in particular, we have

$$A = 23.78, \quad \text{and} \quad b = 10.57.$$

The dampening parameter, α, is taken to be $\alpha = 0.5$, which has performed well. Note that a big α may induce numerical instability in the inverse Laplace transform.

Example: In the following example, we calibrate the market model with stochastic volatility with market data of July 3, 2002. The LIBOR and swap rates are shown in Table 10.1, and the cap data are listed in Table 10.2 (at the end of this section). The implied volatilities of caplets of various maturities are stripped from the cap data, and are displayed later in Figure 10.3. To calibrate to this set of data, we tried using $\kappa = 0.25$ and $\kappa = 1$, two numbers that correspond to weak and strong effects of mean reversion for the stochastic volatility respectively.

Let us first present the calibration results for the smaller kappa, $\kappa = 0.25$. We take initial guess for $\{\|\gamma_j\|, \theta_j, \nu\}$ as

$$\{0.35, \quad \frac{\pi}{2}, \quad 1\}.$$

Note that the guideline for selecting the initial guess for $\{\|\gamma_j\|\}$ is simply to get close to the level of implied caplet volatilities. After the calibration procedure, we obtain

$$\nu = 1.98,$$

while $\{\|\gamma_j\|, \rho_j = \cos\theta_j\}$, as functions of maturity, are shown in Figures 10.5 and 10.6. From Figure 10.6, we see that $\|\gamma_j\|$ decays steadily with increasing maturity, consistent with the lowering implied volatility curve for increasing maturity. The ρ's obtained by calibration, meanwhile, are not as steady but remain in a reasonable range.

The implied caplet volatility surface by the calibrated model is shown in Figure 10.7, alongside the implied volatility surface for the market prices of caplets. It can seen that, although the implied caplet volatility surface by the model is not as smooth, the overall agreement between the two surfaces is quite good. More detailed comparisons are made through Figure 10.8, where

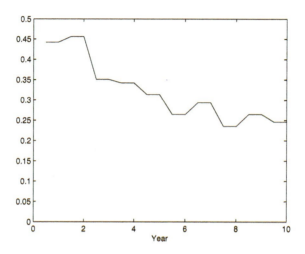

FIGURE 10.5: $\|\gamma_j\|$ versus T_j.

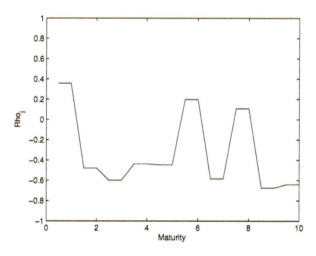

FIGURE 10.6: ρ_j versus T_j.

the quality of fitting of implied volatility curves for all maturities is displayed. The solid curves are the implied Black volatilities of the market values of the caplets, while the dotted curves are the implied Black volatilities of the caplets calculated from the calibrated model. One can see that the level and skewness of all curves are well matched, but the curvature of some curves is slightly missed, probably due to the use of a constant "vol of vol," ν.

It is time to justify the choice of A, b in (10.31) for $N = 80$. As demonstration, we show the Laplace transform of the 5-year caplets, $\eta_5(u)$, in Figures 10.9 and 10.10. One can see that both the real part and imaginary part are

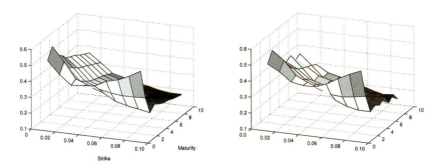

FIGURE 10.7: Implied caplet volatility surface, market (left) vs. model (right).

very small for $u > 5$. The Laplace transform of caplets of other maturities behaves similarly. Hence, our choice of $A = 23.78$, is thus already very conservative. Figures 10.9 and 10.10 may suggest the use of a smaller A, but, as is constrained by $Ab = \pi N$, a smaller A means a bigger b, which could cause greater errors in the calculation of option values, which is done by interpolation. We cannot reduce N either, as $\eta_5(u)$ is quite steep near the origin.

Next, we show the calibration results for the larger kappa, $\kappa = 1$. This kappa corresponds to a half life of mean reversion of $\ln(2)/\kappa = 0.7$ years, and it represents a greater strength of mean reversion and thus a weaker effect of stochastic volatility for longer time horizon. The initial values for $\{\|\gamma_j\|, \theta_j\}$, and ν are

$$\{0.35, \quad \pi/2, \quad 2\}.$$

Let us explain the results. The "vol of vol" by calibration is

$$\nu = 3.65,$$

and other results of calibration are shown in Figures 10.11 to 10.14 Figure 10.11 shows $\|\gamma_j\|$ versus the maturity T_j, while Figure 10.12 shows ρ_j versus the maturity T_j. The matching of implied volatility surfaces and curves is shown in Figures 10.13 and 10.14, from where we might say that the quality of calibration also looks quite good. These two calibration exercises also suggest that the result of a calibration depends on several input parameters. To gain some insight into the selection of those parameters, such as κ, we need to make more careful comparisons.

When compared with the implied volatility surface by calibration with $\kappa = 0.25$, the implied volatility surface for $\kappa = 1$ is smoother but flatter, a sign of a weak effect of stochastic volatility. In fact, we can see in Figure 10.12 that ρ_j drifts steadily toward -1 when maturity increases. This can be interpreted as that, in order to generate enough skewness, the model needs almost perfect negative correlations. The implication of this is that forward rates of long maturities would be almost perfectly correlated. Accordingly, $\|\gamma_j\|$ demonstrates a pattern known for calibrated standard LIBOR market

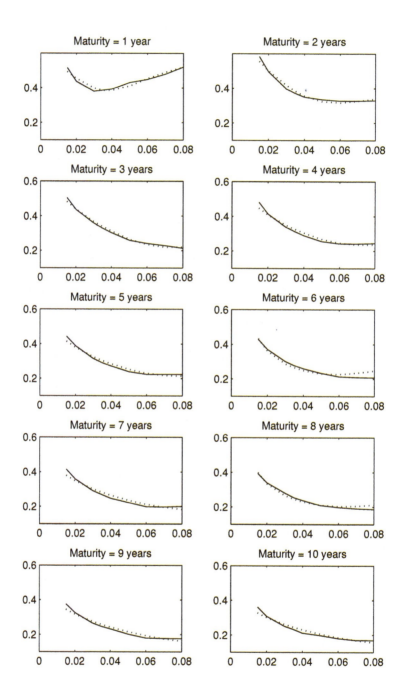

FIGURE 10.8: Implied volatility surface for the calibrated model, $\kappa = 0.25$.

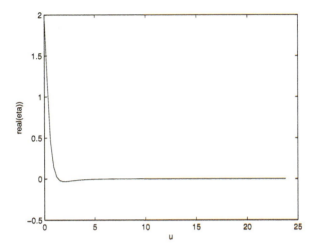

FIGURE 10.9: Real part of $\eta_T(u)$ for the 5-year caplets.

models: the volatilities of the forward rates first increases and then decreases, in an exponential way. Note that $\|\gamma_j\|$ again appears to be very close to the mean level of implied caplet volatilities of maturity T_j. One can imagine that for even greater maturities, the quality of calibration would become unacceptable due to the very weak stochastic volatility. Hence, it is not advisable to adopt a kappa as large as one.

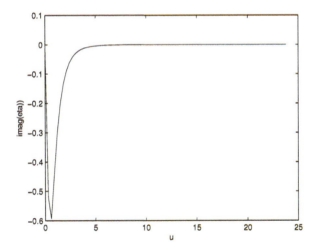

FIGURE 10.10: Imaginary part of $\eta_T(u)$ for the 5-year caplets.

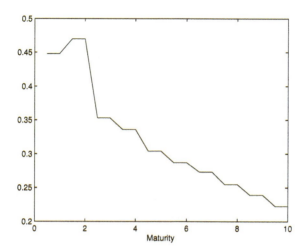

FIGURE 10.11: Real part of $\eta_T(u)$ for the 5-year caplets.

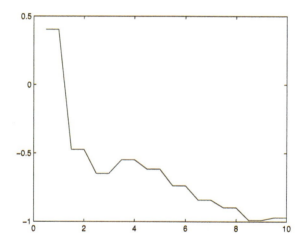

FIGURE 10.12: Imaginary part of $\eta_T(u)$ for the 5-year caplets.

10.6 Notes

After the 2008 financial crisis, LIBOR rates went lower swiftly, soon to the level of about 20 basis points and sometimes even lower, and had stayed in that level for years. To the SABR model, such a low level of interest rate

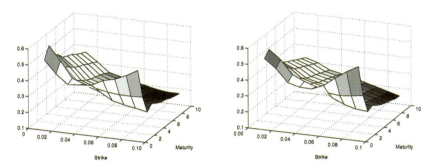

FIGURE 10.13: Implied volatility surface, market (left) vs. model (right).

TABLE 10.1: LIBOR and Swap Rates of July 3, 2002 (Bloomberg)

	Term	Rate
LIBOR	0.5	0.019463
	1	0.022425
Swap	2	0.031450
	3	0.037440
	4	0.041680
	5	0.044710
	6	0.047130
	7	0.049070
	8	0.050630
	9	0.051920
	10	0.052970
	15	0.056860
	20	0.058540
	30	0.059400

TABLE 10.2: Implied Cap Volatilities of July 3, 2002 (Bloomberg)

Maturity	Strikes								
	0.015	0.020	0.030	0.035	0.040	0.050	0.060	0.070	0.080
1y	51.70%	43.80%	38.00%	38.60%	39.30%	43.10%	44.90%	48.00%	52.10%
2y	55.00%	46.90%	39.50%	37.10%	35.00%	33.50%	32.70%	32.70%	33.00%
3y	52.30%	45.00%	37.30%	34.50%	32.20%	29.40%	28.30%	27.90%	27.70%
4y	50.60%	43.60%	35.90%	33.20%	30.90%	27.80%	26.40%	25.90%	25.70%
5y	48.50%	42.10%	34.60%	32.00%	29.80%	26.60%	25.00%	24.50%	24.20%
6y	46.90%	40.80%	33.50%	31.00%	28.90%	25.80%	24.00%	23.40%	23.00%
7y	45.60%	39.70%	32.60%	30.20%	28.10%	25.10%	23.10%	22.50%	22.20%
8y	44.30%	38.60%	31.80%	29.40%	27.40%	24.40%	22.50%	21.80%	21.40%
9y	43.10%	37.60%	31.00%	28.70%	26.80%	23.80%	21.80%	21.10%	20.70%
10y	42.00%	36.60%	30.20%	28.00%	26.10%	23.30%	21.30%	20.50%	20.10%

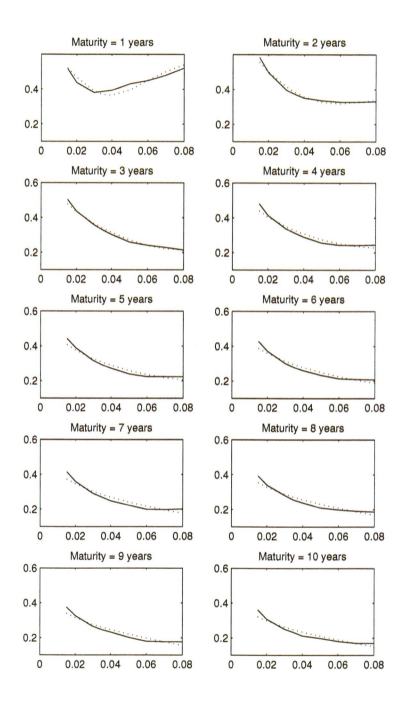

FIGURE 10.14: Implied volatility curves for the calibrated model, $\kappa = 1$.

brought a problem: the SABR solution implies negative densities for low strikes, so the option pricing for low strikes are inaccurate. One research focus over the past 10 years was to improve the solution accuracy of the SABR model for low-strike options as well as for long dated options. Some representative developments are Henry-Labordère (2008), Paulot (2009) and Antonov and Spector (2012). Henry-Labodère (2007), in particular, developed a term structure model for LIBOR based on the SABR model, so-called SABR-LMM model:

$$df_j(t) = V_t f_j^\beta(t)\gamma_j(t) \cdot [d\mathbf{W}_t - \Sigma(t, T_{j+1})dt],$$
$$dV_t = \nu V_t dZ_t, \quad V_0 = \alpha, \tag{10.32}$$

where all LIBOR rates share a single stochastic multiplier, similar to the Wu and Zhang (2002) model. Moreover, Henry-Labordère (2007) adopts the heat-kernel expansion method to solve for the transition density function for both LIBOR rates and swap rates, and achieved more accurate pricing results in terms of the implied Black's volatilities for both caplets and swaptions. A dual-curve version of SABR-LMM will be introduced in Chapter 14, where we will go through the details of the heat-kernel expansion method.

Appendix: Proof of Proposition 2.2

For clarity we let $\tau = T - t$ and $\lambda = \|\gamma_j\|$. Substituting the formal solution (10.19) to (10.17), we obtain the following equations for the undetermined coefficient:

$$\frac{dA}{d\tau} = a_0 B,$$
$$\frac{dB}{d\tau} = b_2 B^2 + b_1 B + b_0, \tag{10.33}$$

where

$$a_0 = \kappa\theta, \quad b_0 = \frac{1}{2}\lambda^2(z^2 - z), \quad b_1 = (\rho\epsilon\lambda z - \kappa\xi), \quad b_2 = \frac{1}{2}\epsilon^2.$$

Now consider (10.33) with constant coefficients and general initial conditions

$$A(0) = A_0, \qquad B(0) = B_0.$$

Since B is independent of A, it is solved first. In the special case when

$$b_2 B_0^2 + b_1 B_0 + b_0 = 0,$$

we have a easy solution

$$B(\tau) = B_0,$$
$$A(\tau) = A_0 + a_0 B_0 \tau. \tag{10.34}$$

Otherwise, let Y_1 be the solution to

$$b_2 Y^2 + b_1 Y + b_0 = 0.$$

Assume $b_2 \neq 0$, then

$$Y_1 = \frac{-b_1 \pm d}{2b_2}, \quad \text{with} \quad d = \sqrt{b_1^2 - 4b_0 b_2}. \qquad (10.35)$$

Without making any difference we take the "+" sign for Y_1. We then consider the difference between Y_1 and B:

$$Y_2 = B - Y_1.$$

Clearly, Y_2 satisfies

$$
\begin{aligned}
\frac{dY_2}{d\tau} &= \frac{d(Y_1 + Y_2)}{d\tau} \\
&= b_2 (Y_1 + Y_2)^2 + b_1 (Y_1 + Y_2) + b_0 \\
&= b_2 Y_2^2 + (2b_2 Y_1 + b_1) Y_2 \\
&= b_2 Y_2^2 + dY_2,
\end{aligned}
\qquad (10.36)
$$

with initial condition

$$Y_2(0) = B_0 - Y_1.$$

Note in the last equality of (10.36) we have used the equation (10.35). Equation (10.36) belongs to the class of Bernoulli equations which can be solved explicitly. One can verify that the solution is

$$Y_2 = \frac{d}{b_2} \frac{g e^{d\tau}}{(1 - g e^{d\tau})}, \quad \text{with} \quad g = \frac{\frac{-b_1 + d}{2b_2} - B_0}{\frac{-b_1 - d}{2b_2} - B_0}. \qquad (10.37)$$

It follows that

$$
\begin{aligned}
B(\tau) &= Y_1 + Y_2 \\
&= \frac{-b_1 + d}{2b_2} + \frac{d}{b_2} \frac{g e^{d\tau}}{(1 - g e^{d\tau})} \\
&= B_0 + \left(\frac{-b_1 + d}{2b_2} - B_0 \right) \frac{(1 - e^{d\tau})}{(1 - g e^{d\tau})}.
\end{aligned}
$$

Having obtained B, we integrate the first equation of (10.33) to get A:

$$
\begin{aligned}
A(\tau) &= A_0 + a_0 \int_0^\tau B(s)\, ds \\
&= A_0 + a_0 B_0 \tau + a_0 \left(\frac{-b_1 + d}{2b_2} - B_0 \right) \int_0^\tau \frac{1 - e^{d\tau}}{1 - g e^{d\tau}} d\tau \\
&= A_0 + a_0 B_0 \tau + a_0 \left(\frac{-b_1 + d}{2b_2} - B_0 \right) \left[\tau - \int_0^\tau \frac{(1 - g) e^{d\tau}}{1 - g e^{d\tau}} d\tau \right]
\end{aligned}
$$

$$= A_0 + a_0 \frac{(-b_1 + d)\tau}{2b_2} - a_0 \left(\frac{-b_1 + d}{2b_2} - B_0 \right) \int_1^{e^{d\tau}} \frac{(1-g)}{1-gu} du$$

$$= A_0 + a_0 \frac{(-b_1 + d)\tau}{2b_2} - a_0 \left(\frac{-b_1 + d}{2b_2} - B_0 \right) \frac{(g-1)}{g} \ln \left(\frac{1-ge^{d\tau}}{1-g} \right)$$

$$= A_0 + a_0 \left[\frac{(-b_1 + d)}{2b_2} \tau - \frac{1}{b_2} \ln \left(\frac{1-ge^{d\tau}}{1-g} \right) \right].$$

Letting

$$u^\pm = \frac{-b_1 \pm d}{2b_2} - B_0,$$

and then

$$A_0 = A(\tau_j, z),$$
$$B_0 = B(\tau_j, z),$$

and replacing τ by $\tau - \tau_j$, we arrive at (10.21). The solution $\phi(z)$ so obtained belongs to $\mathbb{C}^1$ and hence is a weak solution to (10.17). $\qquad\square$

Chapter 11

Lévy Market Model

The LIBOR market models we have studied so far are driven by diffusions, so LIBOR rates are a continuous function of time. Although less important today for interest-rate modeling than stochastic volatility as another kind of risk factor, jumps in derivatives modeling and pricing were not ignored. In the 1990s, interest rates often jumped following the routine meetings of the U.S. Federal Open Market Committee (FOMC). After stepping into the new millennium, FOMC had enhanced their dialogue and communications with the financial industry, such that surprises to the market due to the minutes of FOMC meetings have now become rare. Yet jumps in interest rates can be caused by other reasons, including political or economic events, see e.g., El-Jahel, Lindberg, and Perraudin (1997) and Johannes (2003). The best known example perhaps was the swift and repeated slashing of the U.S. Federal Fund rate right after the event of 9/11, when the U.S. Federal Reserve lowered the interest rates by 25 basis points consecutively for several days. From an empirical point of view, volatility smiles or skews are related to the so-called leptokurtic feature of distribution of the underlying variable, which means a higher peak and fatter tails than those of a normal distribution. A jump-diffusion process is well-known to have such a feature. The intention of this chapter is to take jump risks into the modeling and pricing of LIBOR derivatives. Since the Lévy processes is a broad class of stochastic processes for jumps and diffusion risks, we will devote this chapter to the Lévy LIBOR model for interest rate derivatives. In many ways, the Lévy LIBOR model can be regarded as a generalization to the Glasserman and Kou model (2003), a pioneering piece of work that models jumps using marked point processes.

11.1 Introduction to Lévy Processes

11.1.1 Infinite Divisibility

Definition 11.1.1. *A process* $\{X_t, t \geq 0\}$ *defined on a probability space* $(\Omega, \mathcal{F}, \mathbb{P})$ *is said to be a Lévy process if*

1. *The paths of X_t are right continuous with left limits.*

2. $X_0 = 0$ *almost surely.*

3. *For $0 \le s \le t$, $X_t - X_s$ is equal in distribution to X_{t-s}.*

4. *For $0 \le s \le t$, $X_t - X_s$ is independent of $\{X_u, u \le s\}$.*

Define the characteristic exponent of X_t by

$$\Psi_t(\theta) = -\ln \mathbb{E}\left[e^{i\theta X_t}\right], \quad \forall \theta \in \mathbb{R}.$$

For any $t > 0$, and any positive integer n, there is

$$X_t = X_{t/n} + (X_{2t/n} - X_{t/n}) + \cdots + (X_t - X_{(n-1)t/n}).$$

Due to Properties 3 and 4 of the Lévy process, we have

$$\Psi_m(\theta) = m\Psi_1(\theta) = n\Psi_{m/n}(\theta),$$

or

$$\Psi_{m/n}(\theta) = \frac{m}{n}\Psi_1(\theta).$$

So for any rational number t,

$$\Psi_t(\theta) = t\Psi_1(\theta).$$

For any irrational number t, we find a sequence of rational numbers $\{t_n\} \downarrow t$ as $n \to \infty$, then we have

$$\Psi_{t_n}(\theta) \to \Psi_t(\theta)$$

due to the right continuity of the Lévy process. So we conclude that any Lévy process has the property of

$$\Psi_t(\theta) = t\Psi_1(\theta)$$

for any number $t > 0$. With this equality we have just established that any infinitely divisible process has the characteristic exponent linear in t and, moreover,

Lévy processes $\subset$ infinitely divisible processes.

From now on we denote $\Psi(\theta) = \Psi_1(\theta)$.

11.1.2 Basic Examples of the Lévy Processes

11.1.2.1 Poisson Processes

Consider a random variable N with probability distribution

$$P(N = k) = \frac{e^{-\lambda}\lambda^k}{k!}$$

for some $\lambda > 0$. The characteristic function is

$$E\left[e^{i\theta N}\right] = \sum_{k \geq 0} e^{i\theta k} \frac{e^{-\lambda}\lambda^k}{k!} = e^{-\lambda(1-e^{i\theta})}.$$

The characteristic exponent is $\Psi(\theta) = \lambda(1 - e^{i\theta})$.

The Poisson process, $\{N_t : n \geq 0\}$, is a stationary process for which the distribution is Poisson with parameter λt, so that

$$P(N_t = k) = \frac{e^{-\lambda t}(\lambda t)^k}{k!}$$

$$\mathbb{E}\left[e^{i\theta N_t}\right] = e^{-\lambda t(1-e^{i\theta})} = e^{-t\Psi(\theta)}.$$

11.1.2.2 Compound Poisson Processes

Suppose $\{\xi_i : i \geq 1\}$ is a sequence of i.i.d. random variables, independent of N, with a common law F having no atom at zero. A compound process $\{X_t, t \geq 0\}$ is defined by

$$X_t = \sum_{i=1}^{N_t} \xi_i, \quad t \geq 0.$$

By conditioning on N_t, we have the characteristic function of X_t:

$$\mathbb{E}\left[e^{i\theta \sum_{i=1}^{N_t} \xi_i}\right] = \sum_{n \geq 0} \mathbb{E}\left[e^{i\theta \sum_{k=1}^{n} \xi_i}\right] e^{-\lambda t} \frac{(\lambda t)^n}{n!}$$

$$= \sum_{n \geq 0} \left(\int_{\mathbb{R}} e^{i\theta x} F(dx)\right)^n \frac{e^{-\lambda t}(\lambda t)^n}{n!}$$

$$= e^{-\lambda t} e^{\lambda t \int_{\mathbb{R}} e^{i\theta x} F(dx)}$$

$$= e^{-\lambda t \int_{\mathbb{R}} (1-e^{i\theta x}) F(dx)}.$$

The characteristic exponent of X_t is

$$\Psi_t(\theta) = \lambda t \int_{\mathbb{R}} (1 - e^{i\theta x}) F(dx).$$

By direct computations, we also have

$$E\left[X_t\right] = E\left[N_t\right] E\left[\xi_1\right] = \lambda t \int_{\mathbb{R}} x F(dx).$$

Note that when $F(dx) = \delta(x - 1)\,dx$, the compound Poisson process reduces to the Poisson process. Also, for $0 \le s < t < \infty$,

$$X_t = X_s + \sum_{i=N_s+1}^{N_t} \xi_i.$$

The last summation is an independent copy of X_{t-s}.

A compound Poisson with a drift is defined as

$$X_t = \sum_{i=1}^{N_t} \xi_i + ct, \quad t \ge 0,$$

with $c \in \mathbb{R}$. The characteristic exponent is

$$\Psi(\theta) = \lambda \int_{\mathbb{R}} (1 - e^{i\theta x}) F(dx) - ic\theta.$$

If we take $c = \lambda \int_{\mathbb{R}} x F(dx)$, then there is

$$\mathbb{E}\left[X_t\right] = 0,$$

which is called the *centered compound Poisson process*, with characteristic exponent

$$\Psi(\theta) = \lambda \int_{\mathbb{R}} (1 - e^{i\theta x} + i\theta x) F(dx).$$

11.1.2.3   Linear Brownian Motion

Take the probability law

$$\mu_{\sigma,a}(dx) = \frac{1}{\sigma\sqrt{2\pi}} e^{-(x+a)^2/2\sigma^2}\,dx$$

which is the normal distribution, $N(-a, \sigma^2)$, with characteristic function

$$\int_{\mathbb{R}} e^{i\theta x} \mu_{\sigma,a}(dx) = e^{-\frac{1}{2}\sigma^2\theta^2 - ia\theta},$$

so that characteristic component is

$$\Psi(\theta) = \frac{1}{2}\sigma^2\theta^2 + ia\theta,$$

which is that of a Brownian motion with a drift:

$$X_t = -at + \sigma W_t.$$

11.1.3 Introduction of the Jump Measure

Next, we introduce the jump measure to describe how jumps occur. Consider a compound Poisson process for which the arrival rate and jump size, ΔX, are described by the *jump measure*

$$\mu(dt, dx) = \text{no. of jumps over } [t, t + dt] \text{ such that } \Delta X \in [x, x + dx].$$

Define, in addition,

$$\nu(dx) = \text{expected no. of jumps over } [0, 1] \text{ such that } \Delta X \in [x, x + dx],$$

or simply

$$\nu(dx)dt = E\left[\mu(dt, dx)\right].$$

We call $\nu(dx)dt$ the compensator to $\mu(dt, dx)$. In literature, $\nu(dx)$ is more often called the Lévy measure of a Lévy process, which gives the intensity of jumps (i.e., expected number of jumps per unit time) of size in $[x, x + dx]$.

According to the definition,

$$\int_{[0,t] \times R} x\mu(ds, dx) \triangleq \int_0^t \int_R x\mu(ds, dx) = \sum_{0 \leq s \leq t, \Delta X_s \neq 0} \Delta X_s,$$

i.e., the aggregated jump size over $[0, t]$, while

$$\int_{[0,t] \times R} x\nu(dx)dt = E\left[\sum_{0 \leq s \leq t, \Delta X_s \neq 0} \Delta X_s \right],$$

i.e., the expected aggregated jump size over $[0, t]$. For the compensated jump measure,

$$\tilde{\mu}(dt, dx) = \mu(dt, dx) - \nu(dx)dt,$$

there is always

$$E\left[\int_{[0,t] \times R} f(x)\tilde{\mu}(dt, dx) \right] = 0$$

for $f(x)$ of a rather general class of functions.

11.1.4 Characteristic Exponents for General Lévy Processes

The next theorem characterizes a general Lévy process.

Theorem 11.1.1 (Lévy-Khintchine formula for Lévy processes). *Suppose $a \in \mathbb{R}$, $\sigma \geq 0$ and $\nu(x)$ is a measure on $\mathbb{R}\backslash\{0\}$ s.t. $\int_{\mathbb{R}} (1 \wedge x^2)\nu(dx) < \infty$. Define*

$$\Psi(\theta) = ia\theta + \frac{1}{2}\sigma^2\theta^2 + \int_{\mathbb{R}} \left(1 - e^{i\theta x} + i\theta x \mathbf{1}_{\{|x| < 1\}}\right)\nu(dx),$$

then $\Psi(\theta)$ is the characteristic exponent of a Lévy process. $\square$

Here, we need to make some necessary remarks. The condition

$$\int_{\mathbb{R}} (1 \wedge x^2)\nu(dx) < \infty$$

implies

1. $\nu(|x| > 1) < \infty$. It means that the number of big jumps (with size $|x| > 1$) is finite.

2. $\int_{-1}^{1} x^2\nu(dx) < \infty$. It means that infinite quadratic variation for small (as well as big) jumps is not allowed. Items 1 and 2 ensure that the integral in the characteristic exponent is finite.

3. There are two possibilities for $\nu(|x| < 1)$.

    (a) $\nu(|x| < 1) < \infty$. It means that the number of small jumps is finite as well. In such a case, the Lévy processes reduce to jump-diffusion processes.

    (b) $\nu(|x| < 1) = \infty$. It means that the number of small jumps is infinite. Item 2 however ensures that, $\forall \epsilon \in (0, 1)$,

$$\begin{cases} \nu(\epsilon < |x| < 1) < \infty, \\ \nu(|x| \le \epsilon) = \infty, \end{cases}$$

    suggesting that the smaller the jump size the greater the intensity, and the discontinuity in the path is predominantly made up of arbitrarily small jumps.

A Lévy process can be represented by three parameters, a, σ, and ν. We call (a, σ, ν) the Lévy triplet, where

a	—	the drift coefficient,
σ	—	the diffusion coefficient,
ν	—	the Lévy measure for jump distribution in unit time.

The Lévy triplets for the three preliminary processes of Section 11.2 are

Poisson	—	$(0, 0, \lambda\delta(x - 1))$,
Compound Poisson	—	$(0, 0, \lambda F(x))$,
Linear Brownian motion	—	$(a, \sigma, 0)$.

We now proceed to prove Theorem 11.1.1: given $\Psi(\theta)$, there exists a Lévy process with the same characteristic exponent. What follows below is the strategy of the proof.

The first step is to rewrite the characteristic exponent as

$$\Psi(\theta) = \left\{ ia\theta + \frac{1}{2}\sigma^2\theta^2 \right\}$$

$$+ \left\{ \nu(\mathbb{R}\backslash(-1,1)) \int_{|x|\geq 1} (1 - e^{i\theta x}) \frac{\nu(dx)}{\nu(\mathbb{R}\backslash(-1,1))} \right\}$$

$$+ \left\{ \int_{0<|x|<1} (1 - e^{i\theta x} + i\theta x)\nu(dx) \right\}$$

$$\triangleq \Psi^{(1)}(\theta) + \Psi^{(2)}(\theta) + \Psi^{(3)}(\theta).$$

In case of $\nu(\mathbb{R}\backslash(-1,1)) = 0$, $\Psi^{(2)}(\theta) = 0$. We have established the following relationship between a process and its characteristic component:

$$\Psi^{(1)}(\theta) \longleftrightarrow X_t^{(1)} = \sigma W_t - at,$$

$$\Psi^{(2)}(\theta) \longleftrightarrow X_t^{(2)} = \sum_{i=1}^{N_t} \xi_i, \ |\xi_i| > 1, \ t \geq 0,$$

while for $\Psi^{(3)}(\theta)$, we have

$$\Psi^{(3)}(\theta) = \int_{0<|x|<1} (1 - e^{i\theta x} + i\theta x)\nu(dx)$$

$$= \sum_{n \geq 0} \left\{ \lambda_n \int_{2^{-(n+1)} \leq |x| < 2^{-n}} (1 - e^{i\theta x} + i\theta x) F_n(dx) \right\}$$

$$\triangleq \sum_{n \geq 0} \Psi_n^{(3)}(\theta),$$

where

$$\lambda_n = \nu(2^{-(n+1)} \leq |x| < 2^{-n}),$$

$$F_n(dx) = \lambda_n^{-1}\nu(dx)\mathbf{1}_{\{2^{-(n+1)} \leq |x| < 2^{-n}\}}.$$

In case $\lambda_n = 0$, we let $\Psi_n^{(3)} = 0$. There is the relationship:

$$\Psi_n^{(3)}(\theta) \longleftrightarrow M_t^{(n)} = \sum_{i=1}^{N_t^{(n)}} \xi_i - \lambda_n t \int_{\mathbb{R}} x F_n(dx),$$

where $N_t^{(n)}$ has the intensity λ_n and ξ_i has the law $F_n(dx)$. Define, in addition,

$$\mu^{(k)}(dt, dx) = \mu(dt, dx)\, \mathbf{1}_{\{2^{-(k+1)} \leq |x| < 1\}},$$

$$\nu^{(k)}(dt, dx) = \nu(dt, dx)\, \mathbf{1}_{\{2^{-(k+1)} \leq |x| < 1\}},$$

then

$$X_t^{(3,k)} = \sum_{n=0}^{k} M_i^{(n)} = \int_0^t \int_{0<|x|<1} x \left(\mu^{(k)}(ds, dx) - 1_{|x|\le 1} \nu^{(k)}(dx)ds \right).$$

What is left to show are whether $X_t^{(3,k)}$

1. has a limit which has the characteristic exponent $\Psi^{(3)}$ as $k \to \infty$? and

2. the limiting function is right-continuous with left limit?

The answers to both questions are positive and are attributed to Lévy and Khintchine, yet the proof given here relies on the the Lévy-Ito decomposition (Ito, 1942; Lévy, 1954).

Theorem 11.1.2 (Lévy-Ito decomposition). *Given any $a \in \mathbb{R}$, $\sigma \ge 0$ and a measure ν on $\mathbb{R}\backslash\{0\}$ satisfying*

$$\int_{\mathbb{R}} (1 \wedge x^2)\nu(dx) < \infty,$$

then (a, σ, ν) is the Lévy triplet of a Lévy process given by

$$X_t = X_t^{(1)} + X_t^{(2)} + X_t^{(3)},$$

where $X^{(i)}$, $i = 1, 2, 3$, are independent processes such that

$$X_t^{(1)} = \sigma W_t - at,$$

$$X_t^{(2)} = \sum_{i=1}^{N_t} \xi_i, \quad |\xi_i| > 1,$$

$$X_t^{(3)} = \int_0^t \int_{0<|x|<1} x \left(\mu(ds, dx) - 1_{|x|\le 1} \nu(dx)ds \right),$$

and $X_t^{(3)}$ is a square integrable martingale with characteristic exponent $\Psi^{(3)}(\theta)$. $\square$

The proof is taken from Kyprianou (2008), and is put in this chapter's appendix for completeness. For notational simplification, in our interest rate modeling in the subsequent sections, we do not separate big jumps from small jumps, and we reduce the three components of a Lévy process into two, such that

$$X_t = X_t^{(1)} + X_t^{(2)},$$

with

$$X_t^{(1)} = \sigma W_t - \tilde{a}t,$$

$$X_t^{(2)} = \int_0^t \int_R x \left(\mu(ds, dx) - \nu(dx)ds \right).$$

and

$$\tilde{a} = a - \int_{|x|>1} x\nu(dx).$$

We can use the new set of triplets $(\tilde{a}, \sigma, \nu)$ to represent a Lévy process.

11.2 The Lévy HJM Model

We are now ready to develop the HJM model driven by Lévy processes. We begin with the $\mathbb{P}$ dynamics of zero-coupon bonds:

$$\frac{dP(t,T)}{P(t,T)} = m(t,T)dt + \Sigma(t,T) \cdot d\mathbf{W}_t$$

$$+ \int_R (H(t,x,T) - 1)(\mu_P - \nu_P)(dx, dt),$$

where $m(t,T)$ is the growth rate, Σ is the percentage volatility, $H(t,x,T) - 1$ is the percentage jump size, $\mu_P(x,t)$ is the jump measure and $\nu_P(x,t)$ is the compensator to $\mu_P(x,t)$, such that

$$E^P\left[\int_x^{x+dx} \mu_P(dx, dt)\right] = \int_x^{x+dx} \nu_P(dx, t)dt.$$

To be free of arbitrage, there must exist another measure equivalent to $\mathbb{P}$ under which the discount price of a zero-coupon bond for any maturity T is a martingale. We need the following theorem for finding the martingale measure.

Theorem 11.2.1. *(Girsanov) If $\mathbb{Q} \sim \mathbb{P}$, then there must exist $\mathbb{P}$-measurable function $\varphi_t(x)$ and $\psi(x,t)$ such that*

$$\int_0^t \|\varphi_s\|^2 ds < \infty, \quad \int_0^t \int_R x|\psi(x,s) - 1|\nu_P(dx)ds < \infty,$$

so that the Radon-Nikodyn derivative of $\mathbb{Q}$ with respect to $\mathbb{P}$ can be expressed as

$$\frac{d\mathbb{Q}}{d\mathbb{P}}\bigg|_{\mathcal{F}_t} = \exp\left(\int_0^t -\frac{1}{2}\|\varphi_s\|^2 ds + \varphi_s \cdot d\mathbf{W}_s\right.$$

$$+ \int_0^t \int_R [(1 - \psi(x,s)) + \ln\psi(x,s)]\nu_P(dx)ds \quad (11.1)$$

$$\left. + \int_0^t \int_R \ln\psi(x,s)(\mu_P - \nu_P)(dx, ds)\right) = \zeta_t,$$

or equivalently

$$d\zeta_t = \zeta_t \left(\varphi_t \cdot d\mathbf{W}_t + \int_0^t \int_R (\psi(x,s) - 1)(\mu_P - \nu_P)(dx, ds) \right).$$

Under $\mathbb{Q}$,

$$\tilde{\mathbf{W}}_t = \mathbf{W}_t - \int_0^t \varphi_s ds$$

is a Brownian motion and $(\tau_n, \Delta X(\tau_n))$ *have the Lévy measure*

$$\nu_Q(dx, t) = \psi(x, t)\nu_P(dx, t) \qquad \Box$$

Proof: The proof consists of two steps:

1. We first show that the results hold for the pair of

$$\mu_P^{(\epsilon)} = \mu_P \, 1_{\{|x| \ge \epsilon\}}, \quad \nu_P^{(\epsilon)} = \nu_P \, 1_{\{|x| \ge \epsilon\}}.$$

2. We then let $\epsilon \to 0$ and show that ζ_t converges.

Let

$$U_1(t) = \int_0^t -\frac{1}{2}\|\varphi_s\|^2 ds + \varphi_s \cdot d\mathbf{W}_s$$

$$U_2(t) = \int_0^t \int_R [(1 - \psi(x,s))]\,\nu_P(dx, ds) + \int_0^t \int_R \ln \psi(x,s)\mu_P(dx, ds).$$

The exponential of $U_1(t)$ is clearly a $\mathbb{P}$ martingale. For the exponential of $U_2(t)$ we have (by the Ito's lemma for jump-diffusion processes),

$$de^{U_2(t)} = e^{U_2(t-)} \left(dU_2^c(t) + \int_R [\psi(x,t) - 1]\,\mu_P(dx, dt) \right)$$

$$= e^{U_2(t-)} \int_R [\psi(x,t) - 1]\,(\mu_P - \nu_P)(dx, dt),$$

which shows that the $e^{U_2(t)}$ is also a $\mathbb{P}$ martingale. Here, $U_2^c(t)$ means the continuous component of $U_2(t)$. To prove the first result, we need to calculate the characteristic function of $\tilde{\mathbf{W}}_t$ under $\mathbb{Q}$. We have

$$E^Q \left[e^{iu\tilde{\mathbf{W}}_t} \right] = E^P \left[\zeta_t e^{iu\tilde{\mathbf{W}}_t} \right]$$

$$= E^P \left[e^{U_1} e^{iu\tilde{\mathbf{W}}_t} \right] E^P \left[e^{U_2} \right]$$

$$= E^P \left[\exp \left(\int_0^t -\frac{1}{2}\|\varphi_s + iu\|^2 ds + (\varphi_s + iu) \cdot d\mathbf{W}_s + \frac{1}{2}u^2 t \right) \right]$$

$$= \exp \left(\frac{1}{2}u^2 t \right),$$

which proves $\tilde{\mathbf{W}}_t$ is a $\mathbb{Q}$-Brownian motion. Meanwhile,

$$E^{\mathbb{Q}}\left[e^{iu\sum_{s\leq t}\Delta X_s}\right]$$

$$=E^{\mathbb{P}}\left[e^{iu\sum_{s\leq t}\Delta X_s}e^{\int_0^t\int_R(\nu_P^{(\epsilon)}-\nu_Q^{(\epsilon)})(dx,ds)+\ln\psi(x,s)\mu_P^{(\epsilon)}(dx,ds)}\right]$$

$$=e^{\int_0^t\int_R(\nu_P^{(\epsilon)}-\nu_Q^{(\epsilon)})(dx,ds)]}E^{\mathbb{P}}\left[e^{\int_0^t\int_R\ln(\psi(x,s)e^{iux})\mu_P^{(\epsilon)}(dx,ds)}\right]$$

$$=e^{\int_0^t\int_R(\nu_P^{(\epsilon)}-\nu_Q^{(\epsilon)})(dx,ds)}e^{\int_0^t\int_R(\psi(x,s)e^{iux}-1)\nu_P^{(\epsilon)}(dx,ds)}$$

$$=e^{\int_0^t\int_R(e^{iux}-1)\nu_Q^{(\epsilon)}(dx,ds)},$$

which shows that the jump component has Lévy measure $\nu_Q^{(\epsilon)}(dx,dt)$. Finally, given the regularity condition of ψ, we can easily show that U_2 is well defined and both the mean and variance of $U_2 - U_2^{(\epsilon)}$ converge to zero as $\epsilon \to 0$, which completes the proof. □

The absence of arbitrage implies (Harrison and Pliska, 1981) the existence of $\mathbb{Q} \sim \mathbb{P}$ such that under $\mathbb{Q}$, the discount prices of zero-coupon bonds are martingales. To look for φ and ψ, we rewrite the dynamics of zero-coupon bonds to

$$\frac{dP(t,T)}{P(t,T)} = (m(t,T) + \Sigma(t,T) \cdot \varphi_t)\,dt$$

$$+ \int_R (H(t,x,T)-1)(\psi-1)\nu_P(dx,dt) \quad (11.2)$$

$$+\Sigma(t,T)\cdot(d\mathbf{W}_t - \varphi_t dt) + \int_R (H(t,x,T)-1)(\mu-\nu_Q)(dx,dt).$$

Note that the last line of (11.2) has the $\mathbb{Q}$ expectation equal to zero. For $P(t,T)/B_t$ to be a $\mathbb{Q}$ martingale, there must be

$$m(t,T) + \Sigma(t,T) \cdot \varphi_t + \int_R (H(t,x,T)-1)(\psi-1)\nu_P(dx,t) = r_t,$$

where r_t is the risk-free short rate. Differentiating the above equation w.r.t. T, we then obtain an equivalent condition imposed on φ and ψ:

$$m_T(t,T) + \Sigma_T(t,T) \cdot \varphi_t + \int_R H_T(t,x,T)(\psi-1)\nu_P(dx,t) = 0. \quad (11.3)$$

There are usually non-unique solutions for the pair, unless additional additional criteria are imposed. An industrial approach to determine the pair is to calibrate the model to market prices of benchmark instruments, which is a challenging issue and will not be discussed here.

We now derive the $\mathbb{P}$ dynamics of the instantaneous forward rates. Based on (11.2), we obtain the dynamics of the log price of the zero-coupon bonds:

$$d\ln P(t,T) = \left(m(t,T) - \frac{1}{2}\|\Sigma(t,T)\|^2\right)dt$$

$$+ \int_R [(1 - H(t,x,T) + \ln H(t,x,T)]\,\nu_P(dx,t)dt \qquad (11.4)$$

$$+\Sigma(t,T)\cdot d\mathbf{W}_t + \int_R \ln H(t,x,T)(\mu - \nu_P)(dx,dt).$$

Differentiating (11.12) with respect to T and multiply the minus sign to the resulted equation, we then obtain

$$df(t,T) = (-m_T(t,T) + \Sigma_T(t,T)\cdot\Sigma(t,T)$$

$$+ \int_R \frac{H_T(t,x,T)\,[H(t,x,T) - 1]}{H(t,x,T)}\nu_P(dx,t)\Bigg) dt \qquad (11.5)$$

$$-\Sigma_T(t,T)\cdot d\mathbf{W}_t - \int_R \frac{H_T(t,x,T)}{H(t,x,T)}(\mu_P - \nu_P)(dx,dt).$$

Define the diffusion volatility and jump size for the forward rate by

$$\sigma(t,T) = -\Sigma_T(t,T),$$

$$h(t,x,T) = -\frac{H_T(t,x,T)}{H(t,x,T)} = -\frac{\partial \ln H(t,x,T)}{\partial T}. \qquad (11.6)$$

Once $\sigma(t,T)$ and $h(t,x,T)$ are specified, we can conversely define $\Sigma(t,T)$ and $H(t,x,T)$. In fact, there are

$$\Sigma(t,T) = -\int_t^T \sigma(t,s)ds,$$

and

$$\int_t^T h(t,x,s)ds = -\ln\frac{H(t,x,T)}{H(t,x,t)}.$$

At maturity, the bond price does not jump, meaning $H(t,x,t) = 1$. It then follows that

$$H(t,x,T) = e^{-\int_t^T h(t,x,s)ds}$$

and

$$H_T(t,x,T) = -h(t,x,T)e^{-\int_t^T h(t,x,s)ds}.$$

We now rewrite (11.15) into

$$df(t,T) = \alpha(t,T)dt + \sigma\cdot d\mathbf{W}_t + \int_R h(t,x,T)(\mu - \nu_P)(dx,dt), \qquad (11.7)$$

with

$$\alpha(t,T) = -m_T(t,T) - \sigma(t,T)\cdot\Sigma(t,T)$$

$$- \int_R h(t,x,T)\,[H(t,x,T) - 1]\,\nu_P(dx,t). \qquad (11.8)$$

By making use of (11.3), we further have

$$
\alpha(t,T) = -\sigma(t,T) \cdot \varphi_t - \int_R h(t,x,T) e^{-\int_t^T h(t,x,s)ds}(\psi - 1)\nu_P(dx,t)
$$

$$
- \sigma(t,T) \cdot \Sigma(t,T) - \int_R h(t,x,T) \left[e^{-\int_t^T h(t,x,s)ds} - 1 \right] \nu_P(dx,t) \tag{11.9}
$$

$$
= -\sigma \cdot (\Sigma + \varphi_t) - \int_R h(t,x,T)(e^{-\int_t^T h(t,x,s)ds}\psi(x,t) - 1)\nu_P(dx,t),
$$

which is the result first obtained by Eberline and Oxkan (2003) as the no-arbitrage condition. Given (11.9), we obtain the risk-neutral dynamics of the forward rates:

$$
\begin{aligned}
df(t,T) &= \sigma(t,T) \cdot d\left[\mathbf{W}_t - (\varphi_t + \Sigma(t,T))dt \right] \\
&+ \int_R h(t,x,T)(\mu - \psi(x,t)e^{-\int_t^T h(t,x,s)ds}\nu_P)(dx,dt) \\
&= \sigma(t,T) \cdot d\left[\tilde{\mathbf{W}}_t - \Sigma(t,T)dt \right] \\
&+ \int_R h(t,x,T)(\mu - e^{-\int_t^T h(t,x,s)ds}\nu_Q)(dx,dt)
\end{aligned} \tag{11.10}
$$

Next, we will show that $f(t,T)$ is a martingale under T forward measure, $\mathbb{Q}_T$, which is yet to be defined. For this purpose, we need to solve for the price of zero-coupon bonds under the $\mathbb{Q}$ measure. We already know that under the $\mathbb{Q}$ measure, the price dynamics is

$$
\frac{dP(t,T)}{P(t,T)} = r_t dt + \Sigma(t,T) \cdot d\tilde{\mathbf{W}}_t + \int_R (H(t,x,T) - 1)(\mu - \nu_Q)(dx,dt). \tag{11.11}
$$

The dynamics for the log price is then

$$
\begin{aligned}
d\ln P(t,T) &= \left(r_t + \int_R [1 - H(t,x,T) + \ln H(t,x,T)]\nu_Q(dx)dt \right. \\
&\left. - \frac{1}{2}\|\Sigma(t,T)\|^2 \right) dt + \Sigma(t,T) \cdot d\tilde{\mathbf{W}}_t + \int_R \ln H(t,x,T)(\mu - \nu_Q)(dx,dt),
\end{aligned} \tag{11.12}
$$

from which we can solve for the solution of zero-coupon bond prices:

$$
\begin{aligned}
P(t,T) &= P(0,T) \\
\exp &\left\{ \int_0^t \left(r_s + \int_R [1 - H(s,x,T) + \ln H(s,x,T)]\nu_Q(dx)dt \right.\right. \\
&\left. - \frac{1}{2}\|\Sigma(s,T)\|^2 \right) ds + \Sigma(s,T) \cdot d\tilde{\mathbf{W}}_s \\
&\left. + \int_R \ln H(s,x,T)(\mu - \nu_Q)(dx,ds) \right\}.
\end{aligned} \tag{11.13}
$$

With the price formula for the zero-coupon bond, we can define the Radon-Nikodyn derivative for the forward measure to be

$$
\begin{aligned}
\left.\frac{d\mathbb{Q}_T}{d\mathbb{Q}}\right|_{\mathcal{F}_t} &= \frac{P(t,T)}{P(0,T)} \Big/ \frac{B(t)}{B(0)} \\
&= \exp\bigg\{ \int_0^t \int_R [1 - H(s,x,T) + \ln H(s,x,T)]\,\nu_Q(dx) \\
&\quad - \frac{1}{2}\|\Sigma(s,T)\|^2 ds + \Sigma(s,T)\cdot d\tilde{\mathbf{W}}_s \\
&\quad + \int_R \ln H(s,x,T)(\mu - \nu_Q)(dx,ds) \bigg\} \overset{\triangle}{=} \zeta_t
\end{aligned}
\tag{11.14}
$$

According to the Girsanov Theorem, we now understand that

$$
\hat{\mathbf{W}}_t = \tilde{\mathbf{W}}_t - \int_0^t \Sigma(s,T)ds
$$

is a Q_T Brownian motion and

$$
\nu_{Q_T} = H(t,x,T)\nu_Q
$$

is the Lévy measure of jumps under $\mathbb{Q}_T$. The $\mathbb{Q}_T$ dynamics of the forward rate is thus

$$
df(t,T) = \sigma(t,T)\cdot d\hat{\mathbf{W}}_t + \int_R h(t,x,T)(\mu - \nu_{Q_T})(dx,dt),
\tag{11.15}
$$

which is a martingale under the T-forward measure.

11.3 Market Model under Lévy Processes

Recall that the forward rate for simple compounding is defined by

$$
f_{\Delta T}(t,T) = \frac{1}{\Delta T}\left(\frac{P(t,T)}{P(t,T+\Delta T)} - 1\right).
$$

By the Lévy-Ito's lemma,

$$
\begin{aligned}
d\left(\frac{P(t,T)}{P(t,T+\Delta T)}\right) &= \left(\frac{P(t-,T)}{P(t-,T+\Delta T)}\right) \\
&\quad \times \Bigg([(\Sigma(t,T) - \Sigma(t,T+\Delta T)] \cdot \left[d\tilde{\mathbf{W}}_t - \Sigma(t,T+\Delta T)dt\right] \\
&\quad + \int_R \left[1 - \frac{H(t,X,T)}{H(t,x,T+\Delta T)}\right](H(t,x,T+\Delta T) - 1)\nu_Q(dx,dt) \\
&\quad - \int_R \left[1 - \frac{H(t,x,T)}{H(t,x,T+\Delta T)}\right](\mu - \nu_Q)(dx,dt) \Bigg) \\
&= \left(\frac{P(t-,T)}{P(t-,T+\Delta T)}\right) \\
&\quad \times \Bigg([(\Sigma(t,T) - \Sigma(t,T+\Delta T)] \cdot \left[d\tilde{\mathbf{W}}_t - \Sigma(t,T+\Delta T)dt\right] \\
&\quad - \int_R \left[1 - \frac{H(t,x,T)}{H(t,x,T+\Delta T)}\right](\mu - H(t,x,T+\Delta T)\nu_Q)(dx,dt) \Bigg).
\end{aligned}
\tag{11.16}
$$

The dynamics of the simple forward rates then follows:

$$
\begin{aligned}
\frac{df_{\Delta T}(t,T)}{f_{\Delta T}(t-,T)} &= \frac{1+\Delta T f_{\Delta T}(t-,T)}{\Delta T f_{\Delta T}(t-,T)} \\
&\quad \times \Bigg([\Sigma(t,T) - \Sigma(t,T+\Delta T)] \cdot \left[d\tilde{\mathbf{W}}_t - \Sigma(t,T+\Delta T)dt\right] \\
&\quad + \int_R \left[\frac{H(t,x,T)}{H(t,x,T+\Delta T)} - 1\right](\mu - H(t,x,T+\Delta T)\nu_Q)(dx,dt) \Bigg)
\end{aligned}
\tag{11.17}
$$

where $\mathbf{W}_t^{(T+\Delta T)}$ is a $Q_{T+\Delta T}$ Brownian motion. Now define

$$
\begin{aligned}
\gamma(t,T) &= \frac{(1+\Delta T f_{\Delta T}(t-,T)}{\Delta T f_{\Delta T}(t-,T)}(\Sigma(t,T) - \Sigma(t,T+\Delta T)) \\
h_{\Delta T}(t,x,T) &= \frac{(1+\Delta T f_{\Delta T}(t-,T)}{\Delta T f_{\Delta T}(t-,T)}\left[\frac{H(t,x,T)}{H(t,x,T+\Delta T)} - 1\right],
\end{aligned}
\tag{11.18}
$$

then the dynamics of the simple forward rates under $\mathbb{Q}_{T+\Delta T}$ is simplified to

$$
\begin{aligned}
\frac{df_{\Delta T}(t,T)}{f_{\Delta T}(t-,T)} &= \gamma(t,T) \cdot (d\tilde{\mathbf{W}}_t - \Sigma(t,T+\Delta T)dt) \\
&\quad + \int_R h_{\Delta T}(s,x,T)(\mu - H(t,x,T+\Delta T)\nu_Q)(dx,dt).
\end{aligned}
\tag{11.19}
$$

In terms of $\gamma(t, T)$, $h_{\Delta T}$ and the forward rates, we can specify both $\Sigma(t, T + \Delta T)$ and $H(t, x, T + \Delta T)$ as follows:

$$\Sigma(t, T + \Delta T) = \Sigma(t, T) - \frac{\Delta T f_{\Delta T}(t-, T)}{(1 + \Delta T f_{\Delta T}(t-, T))} \gamma(t, T)$$

$$= \ldots$$

$$= \Sigma(t, t + \epsilon) - \sum_{k=0}^{\left[\frac{(T-t)}{\Delta T}\right]} \frac{\Delta T f_{\Delta T}(t-, T - k\Delta T)}{(1 + \Delta T f_{\Delta T}(t-, T - k\Delta T))} \gamma(t, T - k\Delta T),$$

$$(11.20)$$

and

$$H(t, x, T + \Delta T) = \frac{1 + \Delta T f_{\Delta T}(t-, T)}{1 + \Delta T f_{\Delta T}(t-, T)(1 + h_{\Delta T}(t, x, T))} H(t, x, T)$$

$$= \ldots$$

$$= \prod_{k=0}^{[(T-t)/\Delta T]} \frac{1 + \Delta T f_{\Delta T}(t-, T - k\Delta T)}{1 + \Delta T f_{\Delta T}(t-, T - k\Delta T)(1 + h_{\Delta T}(t, x, T - k\Delta T))}$$

$$\times H(t, x, t + \epsilon)$$

$$(11.21)$$

We can now define the Lévy market model with the spanning forward rates:

$$f_j(t) = \frac{1}{\Delta T_j} \left(\frac{P(t, T_j)}{P(t, T_{j+1})} - 1 \right), \quad j = \eta(t), \ldots,$$

where $\eta_t = \min\{j | T_j \geq t\}$. Then the dynamics of the forward rates under the risk-neutral measure are

$$\frac{df_j(t)}{f_j(t-)} = \gamma_j(t) \cdot \left[d\tilde{\mathbf{W}}_t - \Sigma(t, T_{j+1}) dt \right]$$

$$+ \int_R h_j(t, x)(\mu - H(t, x, T_{j+1}) \nu_Q)(dx, dt),$$

$$(11.22)$$

where

$$\Sigma(t, T_{j+1}) = - \sum_{k=\eta_t}^{j} \frac{\Delta T_k f_k(t-)}{1 + \Delta T_k f_k(t-)} \gamma_k(t),$$

$$H(t, x, T_{j+1}) = \prod_{k=\eta_t}^{j} \frac{1 + \Delta T_k f_k(t-)}{1 + \Delta T_k f_k(t-)(1 + h_j(t, x))}.$$

$$(11.23)$$

11.4   Caplet Pricing

Under the T_{j+1}-forward measure, the Lévy measure is state dependent, which destroys the analytical tractability of a Lévy process. To regain the analytical tractability, we get rid of the state dependence by the "freezing coefficient" treatment, such as

$$H(t, x, T_{j+1}) \approx \prod_{k=\eta_t}^{j} \frac{1 + \Delta T_k f_k(0)}{1 + \Delta T_k f_k(0)(1 + h_j(t, x))}. \tag{11.24}$$

The resulted approximating process for $f_j(t)$ will be a usual Lévy process, which thus has the moment generating function in closed form, which we denote by φ_f. With the moment generating function, we may price caplet using the transformation method, introduced earlier in Chapter 9.

We will instead take the approach of Raible (2000) for general options to price the caplet, because it is more stylized. Let $\mathcal{L}_V(u)$ denote the Laplace transform of a function V at $u \in \mathbb{C}$, such that

$$\mathcal{L}_V(u) = \int_{\mathbb{R}} e^{-ux} V(x) dx.$$

The value of the option is given by

$$\begin{aligned}
c_j(0) &= P(0, T_{j+1}) E_0^{Q_{j+1}} [g(f_j(T_j))] \\
&= P(0, T_{j+1}) \int_{\mathbb{R}^+} g(f_j(T_j)) dQ_{j+1} \\
&= P(0, T_{j+1}) \int_{\mathbb{R}^+} g(f_j(0)e^x) dQ_{j+1}(x) \\
&= P(0, T_{j+1}) \int_{\mathbb{R}^+} g(f_j(0)e^x) \rho(x) dx.
\end{aligned}$$

where $\rho(x)$ is the density function of $\ln f_j(T_j)$. Define $\pi(x) = g(e^{-x})$ and let $\zeta = -\ln f_j(0)$, and then we have

$$\frac{c_j(0)}{P(0, T_{j+1})} = \int_{\mathbb{R}} \pi(\zeta - x)\rho(x) dx = (\pi * \rho)(\zeta),$$

which is a convolution of π with ρ, multiplied by the discount factor. Applying Laplace transform on both sides of the above equation for $u \in \mathbb{C}$, we get

$$\begin{aligned}
\mathcal{L}_V(u) &= \int_{\mathbb{R}} e^{-ux} (\pi * \rho)(x) dx \\
&= \int_{\mathbb{R}} e^{-ux} \pi(x) dx \int_{\mathbb{R}} e^{-ux} \rho(x) dx \\
&= \mathcal{L}_\pi(u)\mathcal{L}_\rho(u).
\end{aligned}$$

Finally, we perform the inverse Laplace transform to obtain the forward price of the caplet:

$$\frac{c_j(0)}{P(0, T_{j+1})} = \frac{1}{2\pi} \int_R e^{\varsigma(R+iu)} \mathcal{L}_V(R+iu)du$$

$$= \frac{e^{\varsigma R}}{2\pi} \int_R e^{i\varsigma u} \mathcal{L}_\pi(R+iu)\mathcal{L}_\rho(R+iu)du$$

$$= \frac{e^{\varsigma R}}{2\pi} \int_R e^{i\varsigma u} \mathcal{L}_\pi(R+iu)\varphi_f(R+iu)du$$

The numerical implementation of the Laplace transform is described in Chapter 9 in detail.

11.5   Swaption Pricing

To derive the swap-rate dynamics, we make use of the swap rate–forward rate relationship:

$$R_{m,n}(t) = \sum_{j=m}^{n-1} \alpha_j(t)f_j(t), \tag{11.25}$$

where the weights are a function of $\{f_j(t)\}_{j=m}^{n-1}$ as well:

$$\alpha_j = \frac{\prod_{k=m}^j (1 + f_k(t))^{-1}}{\sum_{i=m}^{n-1} \prod_{k=m}^i (1 + f_k(t))^{-1}}.$$

As a result, we can write

$$R_{m,n}(t) = R_{m,n}(\{f_j(t)\}),$$

for $j = m, \ldots, n-1$. The payoff function of swaptions can be expressed as

$$V_T = A_{m,n}(T)(R_{m,n}(T) - K)^+. \tag{11.26}$$

Under the forward swap measure, $\mathbb{Q}_S$, we have the following expression for swaption value:

$$V_t = A_{m,n}(t)E_t^{\mathbb{Q}_S}\left[(R_{m,n}(T) - K)^+\right].$$

The approximate dynamics of the swap rate can be approximated as follows:

$$dR_{m,n}(t) = \left(\sum_{j=m}^{n-1} \frac{\partial R_{m,n}(t-)}{\partial f_j} f_j(t-)\gamma_j(t)\right) \cdot \left[d\mathbf{W}_t^{(Q)} - \Sigma_A(t)dt\right]$$

$$+ \int_R [R_{m,n}(\{f_j(t-)(1 + h_j(t))\}) - R_{m,n}(\{f_j(t-)\})] \, (\mu - \nu_{m,n})(dx, dt),$$

$$\tag{11.27}$$

where $\nu_{m,n}(dx,t)$ is the Lévy measure under the swap measure $\mathbb{Q}_S$, for which we have

Proposition 11.5.1 (Glasserman and Merener, 2003). *The Lévy measure of the swap rate is*

$$\nu_{m,n}(dx,t) = \sum_{j=m}^{n-1} \alpha_j(t-) \prod_{k=m}^{j} \frac{1+\Delta T_j f_j(t-)}{1+\Delta T_j f_j(t-)(1+h_j(t,x))} \nu(dx,t) \quad \square$$

Proof: The Radon-Nikodyn derivative of the forward swap measure is defined by

$$\left.\frac{d\mathbb{Q}_S}{d\mathbb{Q}}\right|_{\mathcal{F}_t} = \frac{A_{m,n}(t)}{A_{m,n}(0)} \bigg/ \frac{B(t)}{B(0)} = \zeta(t).$$

In terms of the forward rates, we have

$$\zeta(t) = \frac{1}{A_{m,n}(0)} \prod_{j=0}^{\eta(t)-1} \frac{1}{1+\Delta T_j f_j(T_j)} \sum_{k=m}^{n-1} \Delta T_k \prod_{j=\eta(t)}^{k} \frac{1}{1+\Delta T_j f_j(t)}. \quad (11.28)$$

The jump in percentage term is

$$\frac{\zeta(\tau)-\zeta(\tau-)}{\zeta(\tau-)} = \frac{\sum_{j=m}^{n-1}\Delta T_j \prod_{k=\eta(t)}^{j} \frac{1}{(1+\Delta T_k f_k(\tau-))(1+h_k(\tau,x))}}{\sum_{j=m}^{n-1}\Delta T_j \prod_{k=\eta(t)}^{j} \frac{1}{1+\Delta T_k f_k(\tau-)}} - 1$$

$$= \frac{\sum_{j=m}^{n-1}\Delta T_j \prod_{k=\eta(t)}^{j} \frac{1}{(1+\Delta T_k f_k(\tau-))} \prod_{k=\eta(t)}^{j} \frac{1+\Delta T_k f_k(\tau-)}{1+\Delta T_k f_k(\tau-)(1+h_j(\tau,x))}}{\sum_{j=m}^{n-1}\Delta T_j \prod_{k=\eta(t)}^{j} \frac{1}{1+\Delta T_k f_k(\tau-)}} - 1$$

$$= \sum_{j=m}^{n-1} \alpha_j(\tau-) \prod_{k=\eta(t)}^{j} \frac{1+\Delta T_k f_k(\tau-)}{1+\Delta T_k f_k(\tau-)(1+h_k(\tau,x))} - 1,$$

where we have made use of the definition of α_j. Using the notation of random measure, we have

$$\frac{d\zeta(t)}{\zeta(t-)} = \int_R \left(\sum_{j=m}^{n-1} \alpha_j(\tau-) \prod_{k=\eta(t)}^{j} \frac{1+\Delta T_k f_k(t-)}{1+\Delta T_k f_k(t-)(1+h_k(t,x))} - 1 \right)$$
$$\times (\mu_Q(dx,dt) - \nu_Q(dx,t)dt) + \dots d\mathbf{W}_t.$$

According to the Girsanov's theorem, we identify the Lévy measure under the forward swap measure to be

$$\nu_{m,n}(dx,t) = \left(\sum_{j=m}^{n-1} \alpha_j(t-) \prod_{k=\eta(t)}^{j} \frac{1+\Delta T_k f_k(t-)}{1+\Delta T_k f_k(t-)(1+h_k(t,x))} - 1 \right) \nu_Q \quad \square$$

To retain analytical tractability, we yet again adopt the following "freezing coefficient" treatment: we freeze the state variable $\{f_j(t-)\}$ at time zero, so as to have

$$
\begin{aligned}
\frac{dR_{m,n}(t)}{R_{m,n}(t-)} \approx & \left(\sum_{j=m}^{n-1} \frac{\partial R_{m,n}(0)}{\partial f_j} \frac{f_j(0)}{R_{m,n}(0)} \gamma_j(t) \right) \cdot \left[d\mathbf{W}_t^{(Q)} - \Sigma_S(t)dt \right] \\
& + \int_R \left[\frac{R_{m,n}(\{f_j(0)(1+h_j(t))\})}{R_{m,n}(\{f_j(0)\})} - 1 \right] (\mu - \nu_{m,n})(dx, dt)
\end{aligned}
\tag{11.29}
$$

and

$$
\nu_{m,n}(dx, t) \approx \sum_{j=m}^{n-1} \alpha_j(0) \prod_{k=m}^{j} \frac{1 + \Delta T_j f_j(0)}{1 + \Delta T_j f_j(0)(1 + h_j(t, x))} \nu(dx, t).
$$

Once the the swap rate process is approximated by a Lévy process, swaption pricing can be done through the method of Laplace transform, as is described in Chapter 9. This, however, is only one of the major computational methods for option pricing under the Lévy process, the other two major methods are the analytical approximation method and the numerical method for the so-called partial integral differential equations (PIDE). In the following section, we describe the analytical approximation method which utilizes the Merton formula (1976) under the jump-diffusion model for the underlying state variables.

11.6 Approximate Swaption Pricing via the Merton Formula

Assuming constant interest, diffusion, and hazard rates, the risk-neutral dynamics of the T-forward price dynamics is then

$$
\frac{dF_t}{F_{t-}} = -\lambda E[Y - 1] dt + \sigma dW_t + [Y - 1] dN_t,
\tag{11.30}
$$

where the jump distribution is given by

$$
Y = e^{\mu + \sigma_Y \epsilon}, \quad \epsilon \sim N(0, 1).
$$

It follows that

$$
E[Y] = e^{\mu + \frac{1}{2}\sigma_Y^2} = 1 + m.
$$

Let $X_t = \ln(F_t/F_0)$. Then X_t has the following risk-neutral dynamics:

$$
dX_t = \left[-\lambda m - \frac{1}{2}\sigma^2 \right] dt + \sigma dW_t + \ln Y dN_t.
$$

Conditional to jumps, we have the following expression for the log price:

$$X_t = \left(-\lambda m - \frac{1}{2}\sigma^2\right)t + \sum_{i=1}^{N_t} \ln Y_i + \sigma W_t.$$

The pricing of European options is conditional to $0, 1, \cdots, \infty$ jumps, respectively, so we have

$$C = P(0,T) \sum_{n=0}^{\infty} e^{-\lambda T} \frac{(\lambda T)^n}{n!} \cdot e^{-[\lambda m + \frac{n}{T}\ln(1+m)]T} C_B(F_0, K, T, r_n, \sigma_n)$$

$$= P(0,T) \sum_{n=0}^{\infty} e^{-\lambda(1+m)T} \frac{(\lambda T(1+m))^n}{n!} C_B(F_0, K, T, r_n, \sigma_n),$$

where C_B is the Black's call formula such that

$$C_B(F, K, T, r, \sigma) = F\Phi(d_1) - e^{-rT}K\Phi(d_2),$$

$$d_{1,2} = \frac{\ln(F/K) \pm (r + 1/2\sigma^2)T}{\sigma\sqrt{T}},$$

$$r_n = -\lambda m + \frac{n}{T}\ln(1+m),$$

$$\sigma_n^2 = \sigma^2 + \frac{n}{T}\sigma_Y^2.$$

To price a caplet, we approximate a forward rate dynamics by a process described by the Merton model (11.30), where we need to determine the mean and variance of the jump size Y, which is achieved by matching the first two moments of the jump size (Glasserman and Merener, 2003). For the Merton model (11.30), the first two moments of the jump size are

$$E\left[f_j(\tau) - f_j(\tau-)|\tau, f_j(\tau-)\right] = f_j(\tau-)m,$$

$$E\left[(f_j(\tau) - f_j(\tau-))^2|\tau, f_j(\tau-)\right] = f_j^2(\tau-)\left[e^{\sigma_j^2}(1+m)^2 - 2m - 1\right].$$

where σ_j^2 is the variance of jump size of $\ln f_j(\tau)$.

For the distribution of jumps under the T_{j+1}-forward measure, we need to calculate the integral

$$E^Q\left[H(t, x, T_{j+1})\right] \approx \int_{R^+} \prod_{k=\eta(t)}^{j} \frac{1 + \Delta T_k f_k(0)}{1 + \Delta T_k f_k(0)(1 + h_j(t,x))} \nu_Q(dx, t). \quad (11.31)$$

Then, we take

$$I_1 = E^{Q_{j+1}}\left[f_j(\tau) - f_j(\tau-)|\tau, f_j(\tau-)\right]$$

$$= f_j(\tau-)\int_{R^+} h_j(x, \tau)\frac{H(t, x, T_{j+1})\nu_Q(dx, t)}{E^Q\left[H(t, x, T_{j+1})\right]}. \quad (11.32)$$

The second moment is

$$I_2 = E^{Q_{j+1}} \left[(f_j(\tau) - f_j(\tau-))^2 | \tau, f_j(\tau-) \right]$$
$$\approx f_j^2(\tau-) \int_{R^+} h_j^2(x, \tau) \frac{H(t, x, T_{j+1}) \nu_Q(dx, t)}{E^Q \left[H(t, x, T_{j+1}) \right]}, \qquad (11.33)$$

Set the second moment of the Merton model to that of the approximating Lévy model, we obtain

$$\hat{m} = I_1,$$
$$e^{\sigma^2} (1 + \hat{m}_j)^2 - 2\hat{m}_j - 1 = I_2, \qquad (11.34)$$

for $\hat{m} = e^{\hat{\mu}_j + \hat{\sigma}_j^2/2} - 1$. With a bit more algebra, we arrive at

$$\hat{\mu}_j = \ln(1 + I_1) - \frac{\sigma_j^2}{2}, \quad \hat{\sigma}_j = \left[\ln \left(\frac{1 + 2I_1 + I_2}{(1 + I_1)^2} \right) \right]^{1/2}.$$

Once we have obtained $\hat{\mu}_j$ and $\hat{\sigma}_j$, we can apply the Merton formula to evaluate the caplet.

11.7   Notes

In this chapter we present the Lévy market model, taking a much simpler approach than the one taken by Eberlein and Özkan (2005). Our starting point is the exponential Lévy dynamics for the zero-coupon bond processes. The change of measure theorem by Girsanov is utilized in the derivation. The Lévy market model can be regarded as a framework for LIBOR market model with both jump and diffusion risk, which includes the famous model of Glasserman and Kou (2002) as a special case. As we have demonstrated in the chapter, under the Lévy market model caplets and swaptions can be priced approximately in closed form, see also Glasserman and Merener (2003) and Eberlein and Özkan (2005). A natural question to ask is whether there is a LIBOR market model that contains both stochastic volatility and jumps, the answer is positive. Jarrow, Li and Zhao (2003) have studied a generalized LIBOR market model that combines the constant elasticity variance (CEV) dynamics with stochastic volatility (Andersen and Andreasen, 2000) and jumps (Glasserman and Kou, 2002). Their empirical studies have the conclusions that the generalized model outperforms the other two models, but still the generalized model cannot fully capture volatility smile.

Appendix: The Lévy-Ito Decomposition

Definition 11.7.1. *Fix $T > 0$. Define $\mathcal{M}_T^2 = \mathcal{M}_T^2(\Sigma, \mathcal{F}, \mathcal{F}_t, \mathbb{P})$ to be the space of real-valued, zero mean right-continuous, square integrable $\mathbb{P}$-martingales with respect to $\mathcal{F}$ over the finite time period $[0, T]$.*

Note that any zero mean square integrable martingale with respect to $\{\mathcal{F}_t : t \geq 0\}$ has a right-continuous version belonging to $\mathcal{M}_T^2$.

We will show that $\mathcal{M}_T^2$ is a Hilbert space with respect to the inner product

$$\langle M, N \rangle = \mathbb{E}\left[M_T N_T \right]$$

for any M, $N \in \mathcal{M}_T^2$. It is obvious to see that, for any M, N, $Q \in \mathcal{M}_T^2$,

1. $\langle aM + bN, Q \rangle = a\langle M, Q \rangle + b\langle N, Q \rangle$ for any $a, b \in \mathbb{R}$.

2. $\langle M, N \rangle = \langle N, M \rangle$.

3. $\langle M, M \rangle \geq 0$.

4. When $\langle M, M \rangle = 0$, by Doob's maximal inequality,

$$\mathbb{E}\left[\sup_{0 \leq s \leq T} M_s^2 \right] \leq 4\mathbb{E}\left[M_T^2 \right] = 4\langle M, M \rangle = 0$$

so, $\sup_{0 \leq t \leq T} M_t = 0$ almost surely.

5. Finally, we need to show any Cauchy sequences have a limit in $\mathcal{M}_T^2$.

Let $\{M^{(n)} : n = 1, 2, \ldots\}$ be a Cauchy sequence in $\mathcal{M}_T^2$ such that

$$\|M^{(m)} - M^{(n)}\| = \left(\mathbb{E}\left[(M_T^{(m)} - M_T^{(n)})^2 \right] \right)^{1/2} \longrightarrow 0 \quad \text{as} \quad m, n \uparrow \infty.$$

Then, $\{M_T^{(n)} : n \geq 1\}$ must be a Cauchy sequence in the Hilbert space of zero mean, square integrable random variables defined on $(\Omega, \mathcal{F}_t, \mathbb{P})$, which is a subspace of $L^2(\Omega, \mathcal{F}_T, \mathbb{P})$, endowed with the inner product $\langle M, N \rangle = \mathbb{E}\left[MN \right]$. Hence there exists a limiting variable $M_T \in L^2(\Omega, \mathcal{F}_T, \mathbb{P})$ with zero mean such that

$$\left(\mathbb{E}\left[(M_T^{(n)} - M_T)^2 \right] \right)^2 \longrightarrow 0, \quad \text{as } n \uparrow \infty.$$

Define

$$M_t = \mathbb{E}\left[M_T | \mathcal{F}_t \right] \quad \forall t \in [0, T],$$

which is also right-continuous, then

$$\|M_t^{(n)} - M_t\| \longrightarrow 0 \quad \text{as } n \uparrow \infty.$$

M_t is $\mathcal{F}_t$-adaptive by definition and by Jensen's inequality,

$$\begin{aligned}
\mathbb{E}\left[M_t^2\right] &= \mathbb{E}\left[(\mathbb{E}\left[M_T|\mathcal{F}_t\right])^2\right] \\
&\leq \mathbb{E}\left[\mathbb{E}\left[M_T^2|\mathcal{F}_t\right]\right] \\
&= \mathbb{E}\left[M_T^2\right].
\end{aligned}$$

Hence $M \in \mathcal{M}_T^2$ and $\mathcal{M}_T^2$ is a Hilbert space.

Suppose that $\{\xi_i : i \geq 1\}$ is a sequence of i.i.d. random variable with a common law F (with no mass at the origin) and that $N = \{N_t : t \geq 0\}$ is a Poisson process with rate $\lambda > 0$, we have

Lemma 11.7.1. *Suppose that* $\displaystyle\int_{\mathbb{R}} |x|F(dx) < \infty.$

1. *The process* $M = \{M_t : t \geq 0\}$ *defined by*

$$M_t \triangleq \sum_{i=1}^{N_t} \xi_i - \lambda t \int_{\mathbb{R}} xF(dx)$$

   is a zero mean martingale with respect to its natural filtration.

2. *If, moreover,* $\displaystyle\int_{\mathbb{R}} x^2 F(dx) < \infty$, *then*

$$\mathbb{E}\left[M_t^2\right] = \lambda t \int_{\mathbb{R}} x^2 F(dx),$$

   so M is square integrable martingale.

Proof: The proof consists of two steps.

1. By definition, M has stationary and independent increments so it is a Lévy process. Define $\mathcal{F}_t = \sigma(M_s : s \leq t)$ then for $t \geq s \geq 0$,

$$\begin{aligned}
\mathbb{E}\left[M_t|\mathcal{F}_s\right] &= M_s + \mathbb{E}\left[M_t - M_s|\mathcal{F}_s\right] \\
&= M_s + \mathbb{E}\left[M_{t-s}\right].
\end{aligned}$$

What is left to show is that

$$\mathbb{E}\left[M_u\right] = 0 \quad \text{for all } u \geq 0.$$

In fact, $\forall u \geq 0$,

$$\begin{aligned}
\mathbb{E}\left[M_u\right] &= \mathbb{E}\left[\sum_{i=1}^{N_u} \xi_i - \lambda u \int_{\mathbb{R}} xF(dx)\right] \\
&= \lambda u \mathbb{E}\left[\xi_1\right] - \lambda u \int_{\mathbb{R}} xF(dx) = 0.
\end{aligned}$$

In addition

$$\mathbb{E}\left[|M_u|\right] \leq \mathbb{E}\left[\left|\sum_{i=1}^{N_u} \xi_i\right| + \lambda u \int_{\mathbb{R}} xF(dx)\right]$$

$$\leq \lambda u \mathbb{E}\left[|\xi_1|\right] + \lambda u \int_{\mathbb{R}} xF(dx)$$

$$= \lambda u \left(\int_{\mathbb{R}} (|x| + x)F(dx)\right) < \infty.$$

2. Using the i.i.d. property of $\{\xi_i, i \geq 1\}$, we have

$$\mathbb{E}\left[M_t^2\right] = \mathbb{E}\left[\left(\sum_{i=1}^{N_t} \xi_i\right)^2\right] - \lambda^2 t^2 \left(\int_{\mathbb{R}} xF(dx)\right)^2$$

$$= \mathbb{E}\left[\sum_{i=1}^{N_t} \xi_i^2\right] + \mathbb{E}\left[\sum_{i=1}^{N_t}\sum_{j=1}^{N_t} \mathbf{1}_{\{i \neq j\}}\xi_i\xi_j\right] - \lambda^2 t^2 \left(\int_{\mathbb{R}} xF(dx)\right)^2$$

$$= \lambda t \int_{\mathbb{R}} x^2 F(dx) + \mathbb{E}\left[N_t^2 - N_t\right] \left(\int_{\mathbb{R}} xF(dx)\right)^2$$

$$- \lambda^2 t^2 \left(\int_{\mathbb{R}} xF(dx)\right)^2$$

$$= \lambda t \int_{\mathbb{R}} x^2 F(dx) + \lambda^2 t^2 \left(\int_{\mathbb{R}} xF(dx)\right)^2 - \lambda^2 t^2 \left(\int_{\mathbb{R}} xF(dx)\right)^2$$

$$= \lambda t \int_{\mathbb{R}} x^2 F(dx) \qquad \qquad \Box$$

Recall that

$$\lambda_n = \nu\left(2^{-(n+1)} \leq |x| < 2^{-n}\right),$$

$$F_n(dx) = \lambda_n^{-1}\nu(dx)\mathbf{1}_{\{2^{-(n+1)} \leq |x| < 2^{-n}\}},$$

we now define

$$N^{(n)} = \{N_t^{(n)} : t \geq 0\} \quad - \quad \text{Poisson process with rate } \lambda_n,$$
$$\{\xi_i : i = 1, 2, \ldots\} \quad - \quad \text{i.i.d random variable with law } F_n,$$

and $M^{(n)} = \{M_t^{(n)} : t \geq 0\}$ such that

$$M_t^{(n)} = \sum_{i=1}^{N_t^{(n)}} \xi_i^{(n)} - \lambda_n t \int_{\mathbb{R}} xF_n(dx)$$

$$\mathcal{F}_t^{(n)} = \sigma(M_s^{(n)} : s \leq t) \quad \text{for } t \geq s \geq 0$$

Finally, we put $\{M^{(n)} : n \geq 1\}$ on the same probability space with respect to the common filtration

$$\mathcal{F}_t = \sigma \left(\bigcup_{n \geq 1} \mathcal{F}_t^{(n)} \right).$$

Theorem 11.7.1. *If*

$$\sum_{n \geq 1} \lambda_n \int_{\mathbb{R}} x^2 F_n(dx) < \infty,$$

then there is a Lévy process $X = \{X_t, t \geq 0\}$ which is also a square integrable martingale with characteristic exponent

$$\Psi(\theta) = \int_{\mathbb{R}} (1 - e^{i\theta x} + i\theta x) \sum_{n \geq 1} \lambda_n F_n(dx)$$

for all $\theta \in \mathbb{R}$, such that for each fixed $T > 0$,

$$\lim_{k \to \infty} \mathbb{E} \left[\sup_{t \leq T} \left(X_t - \sum_{n=1}^{k} M_t^{(n)} \right)^2 \right] = 0.$$

Proof. We first show that $\sum_{n=1}^{k} M^{(n)}$ is a square integrable martingale. In fact, due to independence and zero mean,

$$\mathbb{E} \left[\left(\sum_{n=1}^{k} M_t^{(n)} \right)^2 \right] = \sum_{n=1}^{k} \mathbb{E} \left[\left(M_t^{(n)} \right)^2 \right] = t \sum_{n=1}^{k} \lambda_n \int_{\mathbb{R}} x^2 F_n(dx) < \infty.$$

Fix $T > 0$. We now claim $X^{(k)} = \{X_t^{(k)}, 0 \leq t \leq T\}$ such that

$$X_t^{(k)} = \sum_{n=1}^{k} M_t^{(n)}$$

is a Cauchy sequence with respect to $\| \cdot \|$. Note that for $k \geq l$,

$$\|X^{(k)} - X^{(l)}\|^2 = \mathbb{E} \left[\left(X_T^{(k)} - X_T^{(l)} \right)^2 \right]$$

$$= T \sum_{n=l+1}^{k} \lambda_n \int_{\mathbb{R}} x^2 F_n(dx) \longrightarrow 0 \quad \text{as} \quad k, l \uparrow \infty.$$

Then, there is X_T in $(\Omega, \mathcal{F}_T, \mathbb{P})$ such that

$$\|X_T^{(k)} - X_T\| \longrightarrow 0 \quad \text{as} \quad k \uparrow \infty, \tag{11.35}$$

due to completeness of $L^2(\Omega, \mathcal{F}_t, \mathbb{P})$. Define

$$X_t = \mathbb{E}[X_T|\mathcal{F}_t] \qquad (11.36)$$

and $X = \{X_t, 0 \leq t \leq T\}$, then there is also

$$\|X^{(k)} - X\| \longrightarrow 0 \quad \text{as } k \uparrow \infty.$$

Moreover, thanks to the Doob's maximal inequality, we have

$$\lim_{k \uparrow \infty} \mathbb{E}\left[\sup_{0 \leq t \leq T}\left(X_t - X_t^{(k)}\right)^2\right] = 0.$$

The above limit also implies the convergence of (the finite dimensional) distribution. Consequently, since $X^{(k)}$ are Lévy processes,

$$\begin{aligned}
\mathbb{E}\left[e^{i\theta(X_t - X_s)}\right] &= \lim_{k \uparrow \infty} \mathbb{E}\left[e^{i\theta(X_t^{(k)} - X_s^{(k)})}\right] \\
&= \lim_{k \uparrow \infty} \mathbb{E}\left[e^{i\theta X_{t-s}^{(k)}}\right] \\
&= \mathbb{E}\left[e^{i\theta X_{t-s}}\right],
\end{aligned}$$

which shows that X has stationary and independent increments. Due to the condition of the theorem, we readily have

$$\begin{aligned}
\mathbb{E}\left[e^{i\theta X_t}\right] &= \lim_{k \uparrow \infty} \mathbb{E}\left[e^{i\theta X_t^{(k)}}\right] \\
&= \lim_{k \uparrow \infty} \prod_{n=1}^{k} \mathbb{E}\left[e^{i\theta M_t^{(n)}}\right] \\
&= \exp\left\{-\int_{\mathbb{R}}(1 - e^{i\theta x} + i\theta x)\sum_{n \geq 1}\lambda_n F_n(dx)\right\} \\
&= \Psi^{(3)}(\theta).
\end{aligned}$$

There are two more minor issues. The first is to show the right-continuity of X. This comes from the fact that the space of right continuity function over $[0, T]$ is a closed space, under the metric $d(f, g) = \sup_{0 \leq t \leq T}|f(t) - g(t)|$. The second is the dependence of X on T, which should be dismissed.

Suppose we index X by T, say X^T, using

$$\sup_n a_n^2 = \left(\sup_n |a_n|\right)^2,$$
$$\sup_n |a_n + b_n| \leq \sup_n |a_n| + \sup_n |b_n|,$$

and Minkowski's inequality, we have for $T_1 \leq T_2$,

$$\mathbb{E}\left[\sup_{t \leq T_1}\left(X_t^{T_1} - X_t^{T_2}\right)^2\right]^{1/2}$$

$$\leq \mathbb{E}\left[\left(\sup_{t \leq T_1}\left|X_t^{T_1} - X_t^{(k)}\right| + \sup_{t \leq T_1}\left|X_t^{T_2} - X_t^{(k)}\right|\right)^2\right]^{1/2}$$

$$\leq \mathbb{E}\left[\sup_{t \leq T_1}\left(X_t^{T_1} - X_t^{(k)}\right)^2\right]^{1/2} + \mathbb{E}\left[\sup_{t \leq T_1}\left(X_t^{T_2} - X_t^{(k)}\right)^2\right]^{1/2}$$

$$\longrightarrow 0 \quad \text{as} \quad k \uparrow \infty.$$

Note that $X_t^{T_1}$ and $X_t^{T_2}$ are defined as expectations of $X_{T_1}^{T_1}$ and $X_{T_2}^{T_2}$ in (11.36), and $X_{T_1}^{T_1}$ and $X_{T_2}^{T_2}$ are defined as the limits of $X_{T_1}^{(k)}$ and $X_{T_2}^{(k)}$. So $X_t^{T_1} = X_t^{T_2}$ for any $t \in [0, T_1]$, thus X_t does not depend on T.

The limit X established in the Theorem is just $X^{(3)}$, which has a countable number of discontinuities. $\qquad\square$

Chapter 12

Market Model for Inflation Derivatives Modeling

The government inflation-indexed bond first appeared in Finland in 1945, but it only began to be treated as an asset class after the first issuance of inflation-indexed bond, known as Gilt, by the UK government in 1981. Since then, more European governments began to issue inflation-protected sovereignty debts. The U.S. Treasury joined the ranks in 1997 with the issuance of Treasury Inflation Protected Securities (TIPS), which spurred the growth of the inflation derivatives securities.

Research on pricing models for inflation-rate derivatives has also become active since 1997 (Barone and Castagna, 1997; Bezooyen et al., 1997). A theoretical framework, the so-called "foreign currency analogy," first suggested by Hughton (1998) and established by Jarrow and Yildirim (2003), had became very influential. Under this framework, the real interest rate, defined as the difference between the nominal interest rate and the inflation rate, is treated as the interest rate of a foreign currency, while the Consumer Price Index (CPI) is treated as the exchange rate between the domestic and the foreign currencies. To price inflation derivatives, one needs to model the nominal (domestic) interest rate, the foreign (real) interest rate, and the exchange rate (CPI). A handy solution for modeling inflation derivatives is to adopt the Heath-Jarrow-Morton's (1992) framework for interest rates of two currencies, and bridge them using a lognormal exchange-rate process. Manning and Jones (2003) push to the limit of the analytical tractability of this approach and obtain price formulae for inflation caplets and floorlets. For general inflation derivatives, one resorts to Monte Carlo simulations.

Although elegant in theory, a Heath-Jarrow-Morton type model is inconvenient to use. The model takes the unobservable instantaneous (nominal and/or real) forward rates as state variables, while the payoffs of most inflation derivatives are written on CPI or simple compounding inflation rates, making the models hard to calibrate.

A number of different models aimed at more convenient pricing and hedging of inflation derivatives have been developed since 2003. Some researchers adopted normal or lognormal dynamics for certain observable inflation-related variables, for example, the CPI index (Belgrade and Benhamou, 2004a; Korn and Kruse, 2004), the forward price of real zero-coupon bonds (Kazziha, 1999; Mercurio, 2005), or inflation forward rates (Kenyon, 2008). Extensions to these models have also been developed so that risk factors other than diffusion, like stochastic volatility (Mercurio and Moreni, 2006 and 2009; Kruse, 2007; Kenyon, 2008) and jumps (Hinnerich, 2008), are incorporated. Other researchers adopted the square-root process of Cox, Ross and Ingersoll (1985) to model the spot inflation rate, in conjunction with a short-rate model for the nominal spot rate (see, e.g., Chen et al., 2006 and Falbo et al., 2009). Nonetheless, over the years practitioners have been using a model of their own: the so-called market model which is based on the displaced diffusion dynamics for simple inflation forward rates and has not been documented in the publicly available literature. There is also research studying various issues in inflation-rate modeling. Among others, Cairns (2000) considers inflation models where nominal interest rates are ensured positive; Chen et al. (2006) estimate inflation risk premium; and Belgrade and Benhamou (2004b) examine seasonality in inflation rates and manage to take the seasonality into account in their CPI-based model.

While important advancements were made over the years, a certain degree of disorder in the literature has also been created. Now there are at least three versions of "market models" (Beldgrade-Benhamou-Koehler, 2004; Mercurio-Moreni, 2006; the practitioners' model), and at least two different notions of "inflation forward rates," adopted respectively by models based on zero-coupon inflation-indexed swaps (ZCIIS) and models based on year-on-year inflation indexed swaps (YYIIS). When using a ZCIIS-based model to price derivatives on YYIIS, the technique of convexity adjustment is used to calculate the YYIIS swap rates.

With this chapter we hope to sort out the disorder of the field. We will redefine the notion of inflation forward rate as the fair rate for a forward contract on inflation rate, which will be shown to be replicable statically and thus is unique. We will then justify the lognormal martingale dynamics for displaced inflation forward rates, and thus rigorously rebuild the practitioners' model. In such a way, we uniquely define the market model for inflation rates. Moreover, we will establish a Heath-Jarrow-Morton (HJM) type equation for instantaneous inflation forward rates and, by also making use of the classic HJM equation for nominal forward rates, re-derive the HJM type equation for real forward rates as established by Jarrow and Yildirim (2003), along with a correction to drop the notion of "volatility of the Consumer Price Index" in modeling.

This chapter has several important implications. First, we show that the ZCIIS- and YYIIS-based market models are identical and the use of "convexity adjustment" is wrong and unnecessary. Second, we unify the closed-form pricing of inflation caplets, floorlets and swaptions with the Black formula for displaced-diffusion processes which allows us to quote these derivatives using the "implied Black's volatilities." Finally, we provide a proper platform for developing smile models.

The rest of the chapter is organized as follows. In Section 12.1, we introduce major inflation derivatives and highlight real zero-coupon bonds, which is part of our primitive state variables. In Section 12.2, we define the notion of forward inflation rates, rebuild the extended market model, and develop a Heath-Jarrow-Morton type model in terms of continuous compounding forward nominal and inflation rates. Section 12.3 is devoted to pricing major inflation-indexed derivatives under the market model, where we produce closed-form formulae for caps, floors and swaptions. In Section 12.4, we briefly discuss the comprehensive calibration of the market model, and demonstrate some calibration results with market data. In Section 12.5, we demonstrate the construction of smile models, in particular, using the SABR methodology.

12.1   CPI Index and Inflation Derivatives Market

Inflation-rate security markets have evolved steadily over the past twenty years. Among developed countries, the outstanding notional values of inflation-linked government bonds has grown from about 50 billion dollars in 1997 to over 3.2 trillion dollars in 2016.[1] There are inflation-linked securities in most major currencies, including the Pound sterling, Canadian dollar, yen and, of course, the Euro and the U.S. dollar. The global daily turnover was about $10 billion a day on average in 2009, which is largely dominated by Euro and dollar denominated securities. For more information on global inflation-indexed security markets, we refer readers to the Annual Report by Barclays Capital. By comparing the size of inflation markets to the sizes of LIBOR or credit markets, one has to conclude that the interest on inflation securities has been tepid in the past. Nonetheless, since the 2007-08 financial crisis, there has been concern for potential high inflation, which could possibly be caused by the expansionary monetary policy adopted across the globe (Jung, 2008).

The payoff functions of inflation-linked securities depend on inflation rates, which are defined as the percentage change rates of the CPI. The CPI represents an average price of a basket of services and goods, and the average price is compiled by official statistical agencies of central governments. The evolution of CPI indexes in both Europe and the United States is displayed in Figure 12.1, which shows a trend of steady increase.[2]

The inflation rate of a country is defined in terms of its CPI. Denote by $I(t)$ the CPI of time t, then the inflation rate over the time period $[t, T]$ is

---

[1]Source: Components of Barclays Universal Government Inflation-lined All Maturities Bond Index, 2016.

[2]Sources: U.S. Bureau of Labor Statistics (ftp://ftp.bls.gov/pub/special.requests/cpi /cpiai.txt) and European Central Bank (http://appsso.eurostat.ec.europa.eu/nui/show. do?dataset=prc_hicp_midx&lang=en).

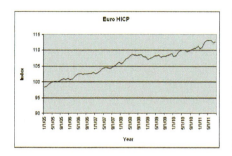
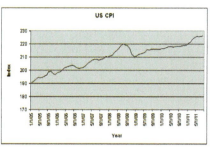

FIGURE 12.1: Consumer Price Indexes of the United States and the Euro zone.

defined as the percentage change of the index:

$$\hat{i}(t,T) = \frac{I(T)}{I(t)} - 1.$$

For purpose of comparison, we will more often use the annualized inflation rate,

$$i(t,T) = \frac{1}{T-t}\left(\frac{I(T)}{I(t)} - 1\right).$$

Suppose the limit of the annualized inflation rate exists for $T \downarrow t$, we obtain the so-called instantaneous inflation rate, $i(t)$, which will be used largely for mathematical and financial arguments instead of modeling. An important feature that distinguishes inflation rates from interest rates is that the former can be either positive or negative, while the latter should be positive in any normal circumstances otherwise we are in a situation of arbitrage.

Inflation-indexed bonds are mainly issued by central governments, while inflation-rate derivatives are offered and traded in the OTC markets by various financial institutions. The dollar-denominated inflation-linked securities have been predominately represented by TIPS, followed by ZCIIS and YYIIS. In recent years, caps, floors and swaptions on inflation rates have been gaining popularity. In the following subsections, we will describe at length these inflation-linked securities. In our modeling approach, ZCIIS will be taken as the underlying securities of the inflation derivatives markets and used for the construction of the "inflation forward rates." This is unlike the current practice of the markets which takes YYIIS as the underlying securities for derivatives pricing.

To understand the roles of the basic securities in model building, we need to set up the economy in mathematical terms. The uncertain economy is modeled by a filtered probability space $(\Omega, \mathcal{F}, \{\mathcal{F}_t\}_{t\in[0,\tau]}, Q)$ for some $\tau > 0$, where Q is the risk neutral probability measure under the uncertain economical environment, which can be defined in a usual way in an arbitrage-free market (Harrison and Krep, 1979; Harrison and Pliska, 1981), and the filtration $\{\mathcal{F}_t\}_{t\in[0,\tau]}$ is generated by a d-dimensional Q Brownian motion $\mathbf{Z} = \{\mathbf{Z}_t, t \geq 0\}$.

12.1.1 TIPS

TIPS are coupon bonds with fixed coupon rates but floating principals, and the latter is adjusted according to the inflation rate over the accrual period of a coupon payment. Note that typically there is a floor on the principal value of a TIPS, which is often the initial principal value. The existence of floors, as a matter of fact, turns TIPS into coupon bonds with embedded options. So strictly speaking, the pricing of TIPS should need a model.

Note that the CPI index is measured with a two-month lag. Yet this lagged index plays the role of the current index for the principal adjustments of TIPS and the payoff calculations of inflations derivatives. From a modeling point of view, lagging or not does not make a difference. With this understanding in mind, we will treat the lagged index as the current index throughout the paper.

12.1.2 ZCIIS

The ZCIIS is a swap contract between two parties with a single exchange of payments. Suppose that the contract was initiated at time t and will be expired at $T > t$, then the payment of one party equals to a notional value times the inflation rate over the contract period, i.e.

$$Not. \times \hat{i}(t, T),$$

while the counterparty makes a fixed payment in the amount

$$Not. \times \left((1 + K(t, T))^{T-t} - 1\right),$$

where $Not.$ stands for the notional value of the contract and $K(t, T)$ is the quote for the contract. Because the value of the ZCIIS is zero at initiation, ZCIIS directly renders the price of the so-called real discount bond, which pays inflation-adjusted principal:

$$P_R(t, T) = E^Q \left[e^{-\int_t^T r_s ds} \frac{I(T)}{I(t)} \middle| \mathcal{F}_t \right] = P(t, T)(1 + K)^{T-t}. \tag{12.1}$$

Here, $P(t, T)$ is the nominal discount factor from T back to t. For real zero-coupon bonds with the same maturity date T but an earlier issuance date, say, $T_0 < t$, the price is

$$P_R(t, T_0, T) = E^Q \left[e^{-\int_t^T r_s ds} \frac{I(T)}{I(T_0)} \middle| \mathcal{F}_t \right] = \frac{I(t)}{I(T_0)} P_R(t, T). \tag{12.2}$$

We emphasize here that $P_R(t, T_0, T)$, instead of $P_R(t, T)$, is treated as the time t price of a traded security. The latter can be considered as the initial price of a new security.

For modeling inflation-rate derivatives, we will take the term structure of real and nominal zero-coupon bonds, $P_R(t, T_0, T)$ and $P(t, T)$ for a fixed T_0 and all $T \geq t \geq T_0$, as model primitives. Note that $\{P_R(t, T), \forall T > t\}$ alone carries information on the term structure of real interest rates only. To see that, we make use the relationship between the instantaneous inflation rate and CPI:

$$\frac{I(T)}{I(T_0)} = e^{\int_{T_0}^{T} i(s) ds}. \tag{12.3}$$

Plugging (12.3) into (13.23) and making use of the Fisher's equation (Fisher, 1930; also see Cox, Ingersoll and Ross, 1985),

$$r(t) = R(t) + i(t), \tag{12.4}$$

where $R(t)$ is the real interest rate, we obtain

$$
\begin{aligned}
P_R(t, T) &= E^Q \left[e^{-\int_t^T (r_s - i(s)) ds} \Big| \mathcal{F}_t \right] \\
&= E^Q \left[e^{-\int_t^T R_s ds} \Big| \mathcal{F}_t \right].
\end{aligned}
\tag{12.5}
$$

According to (12.5), the real zero-coupon bond implies the discount factor associated to real interest rate. This is the reason why we use the subindex "R" for the price.

Note that the real interest rates are not good candidates for the state variable, because most inflation derivatives are written on inflation rates.

12.1.3 YYIIS

YYIIS are contracts to swap an annuity against a sequence of floating payments indexed to inflation rates over future periods. The fixed-leg payments of a YYIIS are $Not.\Delta \phi_i K, i = 1, 2, \ldots, N_x$, where $\Delta \phi_i$ is the year fractions between two consecutive payments, while the floating-leg payments are of the form

$$Not. \left(\frac{I(T_j)}{I(T_{j-1})} - 1 \right),$$

and are made at time $T_j, j = 1, 2, \ldots, N_f$. Note that the payment gaps $\Delta \phi_i = \phi_i - \phi_{i-1}$ and $\Delta T_j = T_j - T_{j-1}$ can be different, yet the terms for payment swaps are the same, i.e., $\sum_{i=1}^{N_x} \Delta \phi_i = \sum_{j=1}^{N_f} \Delta T_j$. The price of the YYIIS equals to the difference in values of the fixed and floating legs. The former can be calculated by discounting, yet the latter involves the evaluation of an expectation,

$$V_{float}^{(j)}(t) = Not.E^Q \left[e^{-\int_t^{T_j} r_s ds} \left(\frac{I(T_j)}{I(T_{j-1})} - 1 \right) \Big| \mathcal{F}_t \right].$$

We will show that, contrary to current practices, the theoretical pricing of the floating leg should be model independent.

12.1.4 Inflation Caps and Floors

An inflation cap is like a YYIIS with embedded optionality: with the same payment frequency, payments are made only when a netted cash flow to the payer (of the fixed leg) is positive, corresponding to cash flows of the following form to the cap holder

$$Not.\Delta T_i \left[\frac{1}{\Delta T_i} \left(\frac{I(T_i)}{I(T_{i-1})} - 1 \right) - K \right]^+ , i = 1, \ldots, N.$$

Accordingly, the cash flows of an inflation floor is

$$Not.\Delta T_i \left[K - \frac{1}{\Delta T_i} \left(\frac{I(T_i)}{I(T_{i-1})} - 1 \right) \right]^+ , i = 1, \ldots, N.$$

Apparently, the prices of caplets and floorlets depend on the variance of the future inflation rates, thus making their pricing model dependent.

12.1.5 Inflation Swaptions

An inflation swaption is an option to enter into a YYIIS swap in the future. At the maturity of the option, the holder of the option should enter into the underlying YYIIS if the option ends up in the money. The underlying security of the swaption is YYIIS. With the establishment of the theory of this chapter, the pricing of the underlying YYIIS will become model independent, which will consequently make the pricing of inflation swaptions easier.

12.2 Rebuilt Market Model and the New Paradigm

12.2.1 Inflation Discount Bonds and Inflation Forward Rates

The cash flows of several major inflation-indexed instruments, including the YYIIS and inflation caplets and floorlets, are expressed in term rates of simple inflation rates. For pricing and hedging purposes we need to define inflation forward rates as the fair rates for forward-rate agreements on inflation rates, parallel to the definition of nominal forward rates. We know that the prices of nominal zero-coupon bonds imply nominal forward rates. Not surprisingly, the prices of nominal zero-coupon bonds and real zero-coupon bonds jointly imply the inflation forward rates. To be precise, we introduce

Definition 1: The *discount bond associated to inflation rate* is defined by

$$P_I(t, T) \triangleq \frac{P(t, T)}{P_R(t, T)}. \tag{12.6}$$

Here, "$\triangleq$" means "being defined by."

Alternatively, with $P_I(t,T)$ and $P_R(t,T)$, we effectively factorize the nominal discount factor into real and inflation discount factors,

$$P(t,T) = P_R(t,T)P_I(t,T). \tag{12.7}$$

Note that neither $P_I(t,T)$ nor $P_R(t,T)$ is a price of a tradable security,[3] yet both are observable.

We define inflation forward rates as the *returns implied by the inflation discount bonds*.

Definition 2: The *inflation forward rate* for a future period $[T_1, T_2]$ seen at time $t \le T_2$ is defined by

$$f^{(I)}(t, T_1, T_2) \triangleq \frac{1}{(T_2 - T_1)} \left(\frac{P_I(t, T_1)}{P_I(t, T_2)} - 1 \right). \tag{12.8}$$

There is a slight problem with the above definition: the forward inflation rate is fixed at $t = T_2$, beyond the life of the T_1-maturity inflation bond, so we need to define $P_I(t, T_1)$ for $t > T_1$. In view of (12.2) and (12.7), we have

$$P_I(t, T_1) = \frac{I(t)}{I(T_0)} \frac{P(t, T_1)}{P_R(t, T_0, T_1)}. \tag{12.9}$$

The second ratio on the right-hand side of (12.9) is the relative price between two traded bonds with an identical maturity date, and thus its value beyond T_1 can be defined by constant extrapolation, yielding

$$P_I(t, T_1) = \frac{I(t)}{I(T_0)} \frac{I(T_0)}{I(T_1)} = \frac{I(t)}{I(T_1)}, \quad \forall t \ge T_1. \tag{12.10}$$

Given (12.10), we have the value of the forward rate at its fixing date to be

$$f^{(I)}(T_2, T_1, T_2) = \frac{1}{T_2 - T_1} \left(\frac{I(T_2)}{I(T_1)} - 1 \right), \tag{12.11}$$

so the inflation forward rate converges to inflation spot rate at maturity.

Next, we will argue that $f^{(I)}(t, T_1, T_2)$ so defined is the fair rate seen at time t for a forward contract on inflation rate over $[T_1, T_2]$. We rewrite (12.8) into

$$f^{(I)}(t, T_1, T_2) = \frac{1}{(T_2 - T_1)} \left(\frac{F_R(t, T_1, T_2)P(t, T_1)}{P(t, T_2)} - 1 \right), \tag{12.12}$$

where

$$F_R(t, T_1, T_2) \triangleq \frac{P_R(t, T_2)}{P_R(t, T_1)} = \frac{P_R(t, T_0, T_2)}{P_R(t, T_0, T_1)} \tag{12.13}$$

is the relative price of two tradable securities. The following result is the corner stone of our theory.

---

[3] $P_R(t,T)$ is treated as the price of a zero-coupon bond of a virtue "foreign currency" by Jarrow and Yildirim (2003).

Proposition 12.2.1. *Let $t \leq T_1 \leq T_2$. The T_1-forward price of a real bond with maturity T_2 seen at time t is $F_R(t, T_1, T_2)$.*

Proof: Do the following zero-net transactions.

1. At time $t \geq T_0$,

   (a) long the forward contract to buy $\frac{I(T_1)}{I(T_0)}$ dollars of T_2-maturity real bond at time T_1 with the unit price $F_R(t, T_1, T_2)$;

   (b) long one unit of T_1-maturity real bond at the price of $P_R(t, T_0, T_1)$;

   (c) short $\frac{P_R(t,T_0,T_1)}{P_R(t,T_0,T_2)}$ unit(s) of T_2-maturity real bond at the unit price of $P_R(t, T_0, T_2)$.

2. At time T_1, exercise the forward contract to buy the T_2-maturity real bond (that pays $I(T_2)/I(T_1)$) at the unit price $F_R(t, T_1, T_2)$, applying all proceeds from the T_1-maturity real bond.

3. At Time T_2, close out all positions.

The net profit or loss from the transactions is

$$P\&L = \left(\frac{1}{F_R(t, T_1, T_2)} - \frac{P_R(t, T_0, T_1)}{P_R(t, T_0, T_2)} \right) \frac{I(T_2)}{I(T_0)}. \tag{12.14}$$

For the absence of arbitrage, the forward price must be set equal to (12.13). □

Note that the forward contract used in the proof does not specify the number of units of the T_2-maturity real bond for purchasing, which is the only difference from a usual forward contract. Yet the seller can still perfectly hedge the forward contract.

In view of (12.12), we can treat $f^{(I)}(t, T_1, T_2)$ as the T_1-forward price for the payoff of $f^{(I)}(T_2, T_1, T_2)$ at T_2, and thus have proven

Proposition 12.2.2. *The inflation forward rate $f^{(I)}(t, T_1, T_2)$ is the unique arbitrage-free rate seen at the time t for a T_1-expiry forward contract on the inflation rate over the future period $[T_1, T_2]$.*

Proposition 12.2.2 should help to end the situation of the coexistence of multiple definitions of forward inflation rates. Note that our definition (12.8) coincides with one of the definitions of inflation forward rates, $Y_i(t)$, given in Mercurio and Moreni (2009).

12.2.2 The Compatibility Condition

We now proceed to the construction of dynamic models for inflation forward rates of both simple and instantaneous compounding. Under the risk neutral measure Q, $P(t,T)$ and $P_R(t,T_0,T)$ are assumed to follow the lognormal processes

$$dP(t,T) = P(t,T)\left(r_t dt + \Sigma(t,T) \cdot d\mathbf{Z}_t\right),$$
$$dP_R(t,T_0,T) = P_R(t,T_0,T)\left(r_t dt + \Sigma_R(t,T) \cdot d\mathbf{Z}_t\right), \qquad (12.15)$$

where r_t is the risk-free nominal (stochastic) interest rate, $\Sigma(t,T)$ and $\Sigma_R(t,T)$ are d-dimensional $\mathcal{F}_t$-adaptive volatility functions of $P(t,T)$ and $P_R(t,T_0,T)$, respectively,[4] and "$\cdot$" means scalar product. The volatility function are assumed sufficiently regular in t and T so that the SDE (12.15) admits a unique strong solution, and their partial derivatives with respect to T exist and have finite L_2 norms with respect to t. Moreover, the volatility functions must satisfy[5]

$$\Sigma(t,t) = \Sigma_R(t,t) = 0.$$

By making use of the dynamics of $P_R(t,T_0,T)$ and the dynamics of the CPI,

$$dI(t) = i(t)I(t)dt,$$

we can derive the dynamics of $P_R(t,T)$:

$$dP_R(t,T) = P_R(t,T)\left(((r_t - i(t))dt + \Sigma_R(t,T) \cdot d\mathbf{Z}_t\right). \qquad (12.16)$$

Being a T_1-forward price of a tradable security, $F(t,T_1,T_2)$ should be a lognormal martingale under the T_1-forward measure whose volatility is the difference between those of $P_R(t,T_0,T_2)$ and $P_R(t,T_0,T_1)$, i.e.,

$$\frac{dF_R(t,T_1,T_2)}{F_R(t,T_1,T_2)} = (\Sigma_R(t,T_2) - \Sigma_R(t,T_1)) \cdot (d\mathbf{Z}_t - \Sigma(t,T_1)dt). \qquad (12.17)$$

Note that $d\mathbf{Z}_t - \Sigma(t,T_1)dt$ is (the differential of) a Brownian motion under the T_1-forward measure, Q_{T_1}, defined by the Radon-Nikodym derivative

$$\left.\frac{dQ_{T_1}}{dQ}\right|_{\mathcal{F}_t} = \frac{P(t,T_1)}{B(t)P(0,T_1)},$$

where $B(t) = \exp(\int_0^t r_s ds)$ is the unit price of the money market account.

There is an important implication by (12.17). By Ito's lemma, we also have

$$\frac{dF_R(t,T_1,T_2)}{F_R(t,T_1,T_2)} = (\Sigma_R(t,T_2) - \Sigma_R(t,T_1)) \cdot (d\mathbf{Z}_t - \Sigma_R(t,T_1)dt). \qquad (12.18)$$

The coexistence of equations (12.17) and (12.18) poses a constraint on the volatility functions of the real zero-coupon bonds.

---

[4]It is not hard to see that the volatility of $P_R(t,T_0,T)$ does not depend on T_0.
[5]Note that both $P_R(t,t)$ and $I(t)$ have no volatility.

Proposition 12.2.3 (Compatibility condition). *For arbitrage pricing, the volatility functions of the real bonds must satisfy the following condition:*

$$(\Sigma_R(t, T_2) - \Sigma_R(t, T_1)) \cdot (\Sigma(t, T_1) - \Sigma_R(t, T_1)) = 0. \tag{12.19}$$

Its differential version is, by letting $T_2 \to T_1 = T$,

$$\dot{\Sigma}_R(t, T) \cdot \Sigma_I(t, T) = 0, \tag{12.20}$$

where the overhead dot means partial derivatives with respect to T, *the maturity.* □

Let us try to comprehend the compatibility condition. We know obviously that $\Sigma_I(t, T_1) \triangleq \Sigma(t, T_1) - \Sigma_R(t, T_1)$ is the percentage volatility of $P_I(t, T)$, while $\Sigma_R(t, T_2) - \Sigma_R(t, T_1)$ is the volatility of the real forward rate defined by

$$f_R(t, T_1, T_2) \triangleq \frac{1}{T_2 - T_1} \left(\frac{P_R(t, T_1)}{P_R(t, T_2)} - 1 \right).$$

Literally, (12.19) means that the price of inflation discount bond with maturity T_1 must be uncorrelated with real forward rates of any future period beyond T_1. This is reasonable and is not restrictive at all.

The differential version of the compatibility condition, (12.20), will be used later to derive an HJM type model for inflation rates.

12.2.3 Rebuilding the Market Model

For generality, we let $T = T_2$ and $\Delta T = T_2 - T_1$, then we can cast (12.12) into

$$f^{(I)}(t, T - \Delta T, T) + \frac{1}{\Delta T} = \frac{1}{\Delta T} \frac{F_R(t, T - \Delta T, T)P(t, T - \Delta T)}{P(t, T)}.$$

The dynamics of $f^{(I)}(t, T - \Delta T, T)$ follows readily from those of F_R and P's.

Proposition 12.2.4. *Under the risk neutral measure, the governing equation for the simple inflation forward rate is*

$$
\begin{aligned}
&d \left(f^{(I)}(t, T - \Delta T, T) + \frac{1}{\Delta T} \right) \\
&= \left(f^{(I)}(t, T - \Delta T, T) + \frac{1}{\Delta T} \right) \gamma^{(I)}(t, T) \cdot (d\mathbf{Z}_t - \Sigma(t, T)dt),
\end{aligned}
\tag{12.21}
$$

where

$$\gamma^{(I)}(t, T) = \Sigma_I(t, T - \Delta T) - \Sigma_I(t, T)$$

is the percentage volatility of the displaced inflation forward rate. □

In formalism, equation (12.21) is just the practitioners' model. Yet in applications, practitioners bootstrap the inflation forward rates from YYIIS and calibrate the model to inflation caps/floors for $\gamma^{(I)}(t,T)$. Let us present the market model for inflation rates in comprehensive terms. The state variables consist of two streams of spanning nominal forward rates (Brace et al., 1997; Miltersen et al., 1997; and Jamshidian, 1997) and forward inflation rates, $f_j(t) \triangleq f(t, T_j, T_{j+1})$ and $f_j^{(I)}(t) \triangleq f^{(I)}(t, T_{j-1}, T_j), j = 1, 2, \ldots, N$, that follow the following dynamics:

$$
\begin{cases}
df_j(t) = f_j(t)\gamma_j(t) \cdot (d\mathbf{Z}_t - \Sigma(t, T_{j+1})dt), \\
d\left(f_j^{(I)}(t) + \dfrac{1}{\Delta T_j}\right) = \left(f_j^{(I)}(t) + \dfrac{1}{\Delta T_j}\right)\gamma_j^{(I)}(t) \cdot (d\mathbf{Z}_t - \Sigma(t, T_j)dt),
\end{cases}
$$
(12.22)

where

$$
\Sigma(t, T_{j+1}) = - \sum_{k=\eta_t}^{j} \frac{\Delta T_{k+1}f_k(t)}{1 + \Delta T_{k+1}f_k(t)}\gamma_k(t)
$$

and $\eta_t = \min\{i|T_i > t\}$. The initial nominal and inflation forward rates are derived from prices of nominal and real discount bonds. We want to highlight here that $f_j^{(I)}(t)$ is also a martingale under its own "cash flow measure," i.e., the T_j-forward measure.

12.2.4 The New Paradigm

Analogously to the definition of nominal forward rates, we define the instantaneous inflation forward rates as

$$
f^{(I)}(t, T) \triangleq -\frac{\partial \ln P_I(t, T)}{\partial T}, \quad \forall T \geq t,
$$
(12.23)

or

$$
P_I(t, T) = e^{-\int_t^T f^{(I)}(t,s)ds}.
$$

By the Ito's lemma, we have

$$
\begin{aligned}
-d\ln P_I(t, T) &= d\ln\left(\frac{P_R(t, T)}{P(t, T)}\right) \\
&= -\left(i(t) + \frac{1}{2}\|\Sigma_I(t, T)\|^2\right)dt - \Sigma_I^T(t, T)(d\mathbf{Z}_t - \Sigma(t, T)dt).
\end{aligned}
$$
(12.24)

Differentiating the above equation with respect to T and making use of the compatibility condition (12.20), we then have

$$
df^{(I)}(t, T) = -\dot{\Sigma}_I \cdot (d\mathbf{Z}_t - \Sigma(t, T)dt).
$$
(12.25)

Equation (12.25) shows that $f^{(I)}(t, T)$ is a Q_T-martingale and its dynamics are fully specified by the volatilities of the nominal and inflation forward rates.

In an HJM context, the volatilities of nominal and inflation forward rates, $\sigma(t,T) = -\dot{\Sigma}(t,T)$ and $\sigma^{(I)}(t,T) = -\dot{\Sigma}_I(t,T)$, are first prescribed, and the volatilities of the zero-coupon bonds follow from

$$\Sigma(t,T) = -\int_t^T \sigma(t,s)ds \quad \text{and} \quad \Sigma_I(t,T) = -\int_t^T \sigma^{(I)}(t,s)ds.$$

Then, the extended HJM model with nominal and inflation forward rates is

$$\begin{cases} df(t,T) = \sigma(t,T) \cdot d\mathbf{Z}_t + \sigma(t,T) \cdot \left(\int_t^T \sigma(t,s)ds \right) dt, \\[3mm] df^{(I)}(t,T) = \sigma^{(I)}(t,T) \cdot d\mathbf{Z}_t + \sigma^{(I)}(t,T) \cdot \left(\int_t^T \sigma(t,s)ds \right) dt, \end{cases} \quad (12.26)$$

which takes the initial term structures of nominal and inflation forward rates as inputs.

If we treat (12.26) as a framework of no-arbitrage models, then the market model (12.21) fits in the framework with the volatility function

$$\sigma^{(I)}(t,T) = -\dot{\Sigma}_I(t,T) = \frac{\partial}{\partial T} \left(\sum_{k=0}^{\left[\frac{T-t}{\Delta T}\right]} \gamma^{(I)}(t, T - k\Delta T) \right),$$

where $[x]$ is the integer part of x.

12.2.5 Unifying the Jarrow-Yildirim Model

According to their definitions, nominal, inflation and real forward rates for continuous compounding satisfy the relationship

$$f_R(t,T) = f(t,T) - f^{(I)}(t,T).$$

Subtracting the two equations of (12.26) and applying the compatibility condition, (12.20), we then arrive at

$$df_R(t,T) = \sigma_R(t,T) \cdot d\mathbf{Z}_t + \sigma_R(t,T) \cdot \left(\int_t^T \sigma_R(t,s)ds \right) dt, \quad (12.27)$$

where

$$\sigma_R(t,T) = \sigma(t,T) - \sigma^{(I)}(t,T) = -\dot{\Sigma}_R(t,T).$$

In contrast, under our notations the equation established by Jarrow and Yildirim (2003) for the real forward rates is

$$df_R(t,T) = \sigma_R(t,T) \cdot d\mathbf{Z}_t + \sigma_R(t,T) \cdot \left(\int_t^T \sigma_R(t,s)ds - \sigma_I(t) \right) dt, \quad (12.28)$$

where $\sigma_I(t)$ is the volatility of the CPI index. Given that $\sigma_I(t) \equiv 0$, the two equations are identical.

Even if the CPI volatility were not zero, we can still re-derive the Jarrow and Yildirim model by recognizing that the volatility of $P_R(t, T_0, T)$ satisfies $\Sigma_R(t, t) = \sigma_I(t)$ and redoing the arguments. Based on the above analysis, we claim that the market model is consistent with the framework of the "foreign currency analogy."

In a two-currency economy, the exchange rate is usually stochastic with volatility. Having no volatility, the CPI behaves like a money market account instead of an exchange rate. Hence, we emphasize here that the inflation derivatives modeling is not completely analogous to cross-currency derivatives modeling.

12.3   Pricing Inflation Derivatives

We have established for the first time that simple inflation forward rates are lognormal martingales under respective forward measures. As a result, the current practices on pricing some inflation derivatives must undergo some changes.

12.3.1   YYIIS

The price of a YYIIS is the difference in value of the fixed leg and floating leg. While the fixed leg is priced as an annuity, the floating leg is priced by discounting the expectation of each piece of payment:

$$
\begin{aligned}
V_{float}^{(j)}(t) &= Not.P(t, T_j) E_t^{Q_j} \left[\left(\frac{I(T_j)}{I(T_{j-1})} - 1 \right) \right] \\
&= Not.\Delta T_j P(t, T_j) E_t^{Q_j} \left[f_j^{(I)}(T_j) \right] \qquad (12.29) \\
&= Not.\Delta T_j P(t, T_j) f_j^{(I)}(t),
\end{aligned}
$$

where we have made use of the martingale property of the inflation forward rates. The value of the floating leg is just a summation, and the value of the YYIIS is the difference between the values of the fixed and floating legs.

In the marketplace, YYIIS are treated as another set of securities parallel to ZCIIS, and the "inflation forward rates" implied by YYIIS and ZCIIS can be different. In existing literatures, pricing YYIIS using a ZCIIS-based model goes through the procedure of "convexity adjustment," which is shown to be flawed. Our theory, for the first time, suggests that such differences should create arbitrage opportunities.

12.3.2 Caps

In view of the displaced diffusion processes for simple forward inflation rates, we can price a caplet with $1 notional value straightforwardly as follows:

$$\Delta T_j E_t^Q \left[e^{-\int_t^{T_j} r_s ds} (f_j^{(I)}(T_j) - K)^+ \right]$$

$$=\Delta T_j P(t, T_j) E_t^{Q_j} \left[\left(\left(f_j^{(I)}(T_j) + \frac{1}{\Delta T_j} \right) - \left(K + \frac{1}{\Delta T_j} \right) \right)^+ \right] \quad (12.30)$$

$$=\Delta T_j P(t, T_j) \{ \mu_j(t) \Phi(d_1^{(j)}(t)) - \tilde{K}_j \Phi(d_2^{(j)}(t)) \},$$

where $\Phi(\cdot)$ is the standard normal accumulative distribution function, and

$$\mu_j(t) = f_j^{(I)}(t) + 1/\Delta T_j, \quad \tilde{K}_j = K + 1/\Delta T_j,$$

$$d_1^{(j)}(t) = \frac{\ln \mu_j / \tilde{K}_j + \frac{1}{2} \sigma_j^2(t)(T_j - t)}{\sigma_j(t) \sqrt{T_j - t}}, \quad d_2^{(j)}(t) = d_1^{(j)}(t) - \sigma_j(t) \sqrt{T_j - t},$$

with $\sigma_j(t)$ to be the mean volatility of $\ln(f_j^{(I)}(t) + \frac{1}{\Delta T_j})$:

$$\sigma_j^2(t) = \frac{1}{T_j - t} \int_t^{T_j} \|\gamma_j^{(I)}(s)\|^2 ds. \quad (12.31)$$

Equation (12.30) is the Black's formula for inflation caplet.

The inflation-indexed cap with maturity T_N and strike K is the sum of a series of inflation-indexed caplets with the cash flows at T_j for $j = 1, \cdots, N$. Denote by $\text{IICap}(t; N, K)$ the price of the inflation-indexed cap at time $t < T_1$. Based on (12.30), we have

$$\text{IICap}(t; N, K) = \sum_{j=1}^{N} \Delta T_j P(t, T_j) \{ \mu_j(t) \Phi(d_1^{(j)}(t)) - \tilde{K}_j \Phi(d_2^{(j)}(t)) \}. \quad (12.32)$$

Equation (12.30) is like an old bottle filled with new wine: the input inflation forward rates should be implied by ZCIIS instead of YYIIS. Given inflation caps of various maturities, we can consecutively bootstrap $\sigma_j(t)$, the "implied caplet volatilities" in either a parametric or a non-parametric way. With additional information on correlations between inflation rates of various maturities, we can determine $\gamma_j^{(I)}$, the volatility of inflation rates and thus fully specify the displace-diffusion dynamics for inflation forward rates. We may also include inflation swaption prices to the input set for determining $\gamma_j^{(I)}$'s.

12.3.3 Swaptions

The discussions on the pricing of inflation swaptions have been rare (Hinnerich, 2008). An inflation swaption is an option to enter into a YYIIS at the option's maturity. Without loss of generality, we consider here an underlying swap which has the same payment frequency for both fixed and floating legs. Similar to the situation of swaps on nominal interest rates, it is straightforward to show that the market prevailing inflation swap rate (that nullifies the value of a swap) is

$$S_{m,n}(t) = \frac{\sum_{i=m+1}^{n} \Delta T_i P(t, T_i) f_i^{(I)}(t)}{\sum_{i=m+1}^{n} \Delta T_i P(t, T_i)}. \tag{12.33}$$

The above expression can be recast into

$$S_{m,n}(t) + \frac{1}{\Delta T_{m,n}} = \sum_{i=m+1}^{n} w_i \mu_i(t), \tag{12.34}$$

where

$$w_i(t) = \frac{\Delta T_i P(t, T_i)}{A_{m,n}(t)}, \quad A_{m,n}(t) = \sum_{i=m+1}^{n} \Delta T_i P(t, T_i),$$

and

$$\frac{1}{\Delta T_{m,n}} = \sum_{i=m+1}^{n} w_i(t) \frac{1}{\Delta T_i}.$$

We have the following results on the dynamics of the swap rate.

Proposition 12.3.1. *The displaced forward swap rate* $S_{m,n}(t) + \frac{1}{\Delta T_{m,n}}$ *is a martingale under the measure* Q_S *corresponding to the numeraire* $A_{m,n}(t)$. *Moreover,*

$$d\left(S_{m,n}(t) + \frac{1}{\Delta T_{m,n}}\right) = \left(S_{m,n}(t) + \frac{1}{\Delta T_{m,n}}\right)$$
$$\times \sum_{i=m+1}^{n} \left[\alpha_i(t)\gamma_i^{(I)}(t) + (\alpha_i(t) - w_i(t))\Sigma_i(t)\right] \cdot d\mathbf{Z}_t^{(m,n)}, \tag{12.35}$$

where $d\mathbf{Z}_t^{(m,n)}$ *is a* Q_S-*Brownian motion,* $\Sigma_i(t) \stackrel{\triangle}{=} \Sigma(t, T_i)$, *and*

$$\alpha_i(t) = \frac{w_i(t)\mu_i(t)}{\sum_{j=m+1}^{n} w_j(t)\mu_j(t)}. \quad \square$$

Proof: According to (12.34),

$$S_{m,n}(t) + \frac{1}{\Delta T_{m,n}} = \sum_{i=m+1}^{n} w_i(t)\mu_i(t), \tag{12.36}$$

so the dynamics of the displaced swap rate will arise from, by Ito's lemma,

$$d\left(S_{m,n}(t) + \frac{1}{\Delta T_{m,n}}\right) = \sum_{i=m+1}^{n} \mu_i(t)dw_i(t) + w_i(t)d\mu_i(t) + dw_i(t)d\mu_i(t).$$

(12.37)

One can easily show that

$$dw_i(t) = w_i(t)(\Sigma_i(t) - \Sigma_A(t)) \cdot (d\mathbf{Z}_t - \Sigma_A(t)dt),$$

(12.38)

where $\Sigma_A(t) = \sum_{i=m+1}^{n} w_i \Sigma_i(t)$. Making use of (12.18) and (12.38), we obtain

$$d\left(\sum_{i=m+1}^{n} w_i(t)\mu_i(t)\right) = \sum_{i=m+1}^{n} w_i(t)\mu_i(t) \left[(\Sigma_i(t) - \Sigma_A(t)) \cdot (d\mathbf{Z}_t - \Sigma_A(t)dt)\right.$$

$$\left. + \gamma_i^{(I)}(t) \cdot (d\mathbf{Z}_t - \Sigma_i(t)dt) + \gamma_i^{(I)}(t) \cdot (\Sigma_i(t) - \Sigma_A(t))dt\right]$$

$$= \sum_{i=m+1}^{n} w_i(t)\mu_i(t) \left(\Sigma_i(t) - \Sigma_A(t) + \gamma_i^{(I)}(t)\right) \cdot (d\mathbf{Z}_t - \Sigma_A(t)dt)$$

$$= \left(\sum_{i=m+1}^{n} w_i(t)\mu_i(t)\right)$$

$$\times \left[\sum_{i=m+1}^{n} \alpha_i(t) \left(\gamma_i^{(I)}(t) + \Sigma_i(t)\right) - \Sigma_A(t)\right] \cdot (d\mathbf{Z}_t - \Sigma_A(t)dt),$$

which is (12.35).

Finally, we point out that $d\mathbf{Z}_t - \Sigma_A(t)dt$ is a Brownian motion under the martingale measure corresponding to the numeraire $A_{m,n}(t)$. Let Q_S denote this measure, then it is defined by the Radon-Nikodym derivative with the risk neutral measure by Q

$$\left.\frac{dQ_S}{dQ}\right|_{\mathcal{F}_t} = \frac{A_{m,n}(t)}{A_{m,n}(0)B(t)} = m_s(t) \quad \text{for} \quad t \leq T_n,$$

where $B(t)$ be the money market account under discrete compounding:

$$B(t) = \left(\prod_{j=0}^{\eta_t - 2} (1 + f_j(T_j)\Delta T_j)\right)(1 + f_{\eta_t - 1}(T_{\eta_t - 1})(t - T_{\eta_t - 1})),$$

and $\eta_t = \min\{j : t < T_j\}$. By Ito's lemma,

$$dm_s(t) = m_s(t)\Sigma_A(t) \cdot d\mathbf{Z}_t.$$

(12.39)

The Q_S Brownian motion corresponding to $\mathbf{Z}_t$ is defined by

$$d\mathbf{Z}_t^{(m,n)} = d\mathbf{Z}_t - \left\langle d\mathbf{Z}_t, \frac{dm_s(t)}{m_s(t)}\right\rangle$$

$$= d\mathbf{Z}_t - \Sigma_A(t)dt$$

(12.40)

□

The martingale property of the swap rate is easy to see because it is the relative value between the floating leg and the annuity, the numeraire asset of the measure $\mathbb{Q}_S$, both are tradable securities.

By appropriately freezing coefficients of (12.35), the displaced forward inflation swap rate $S_{m,n}(t) + \frac{1}{\Delta T_{m,n}}$ becomes a lognormal variable, and closed-form pricing of inflation swaptions will then follow. Consider a T_m-maturity swaption on the YYIIS over the period $[T_m, T_n]$ with strike K, we can derive its value as

$$V_t = A_{m,n}(t) \left[\left(S_{m,n}(t) + \frac{1}{\Delta T_{m,n}} \right) \Phi(d_1^{(m,n)}) - \tilde{K}_{m,n} \Phi(d_2^{(m,n)}) \right], \quad (12.41)$$

where

$$\tilde{K}_{m,n} = K + \frac{1}{\Delta T_{m,n}},$$

$$d_1^{(m,n)} = \frac{\ln\left(S_{m,n}(t) + 1/\Delta T_{m,n}\right)/\tilde{K}_{m,n} + \frac{1}{2}\sigma_{m,n}^2(t)(T_m - t)}{\sigma_{m,n}(t)\sqrt{T_m - t}},$$

$$d_2^{(m,n)} = d_1^{(m,n)} - \sigma_{m,n}(t)\sqrt{T_m - t},$$

$$\sigma_{m,n}(t) = \frac{1}{T_m - t} \int_t^{T_m} \left\| \sum_{i=m+1}^{n} \left[\alpha_i(t)\gamma_i^{(I)}(s) + (\alpha_i(t) - w_i(t))\Sigma(s, T_i) \right] \right\|^2 ds.$$

Equation (12.41) is the Black's formula for inflation swaptions.

Treatments of freezing coefficients similar to what we did to (12.35) are popular in the industry, and they are often very accurate in applications. A thorough analysis on the error estimation of such approximations, however, is still pending. For interesting insights about the magnitude of errors, we refer to Brigo et al. (2004).

The swaption formula, (12.41), implies a hedging strategy for the swaption. At any time t, the hedger should long $\Phi(d_1^{(m,n)})$ units of the underlying inflation swap for hedging. Proceeds from buying or selling the swap may go in or go out of a money market account. Besides, the swaption formula also contains the caplet formula, (12.30), as a special case when $n = m + 1$.

Finally we emphasize that, with the Black's formula (for displaced diffusion processes), inflation caps, floors and swaptions can be quoted using implied volatilities, regardless the sign (i.e., either positive or negative) of input inflation forward rates or swap rates.

12.4   Model Calibration

A comprehensive calibration of the inflation-rate model (13.22) means simultaneous determination of volatility vectors for inflation forward rates,

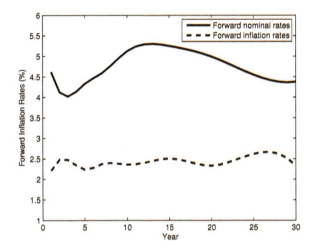

FIGURE 12.2: Term structure of the forward nominal rates and forward inflation rates.

based on market data of YYIIS, inflation caps and inflation swaptions. For non-parametric calibration, one can adopt the methodology for the calibration of the LIBOR market model developed by Wu (2003).

As demonstration, we have calibrated the two-factor market model to the prices of Euro ZCIIS and (part of the) inflation caps as of April 7, 2008,[6] and observed good performance. Figure 12.2 shows the term structures of inflation forward rates as well as nominal forward rates. Figure 12.3 shows the local volatility function obtained by calibrating the model to implied cap volatilities of various maturities but a fixed strike $K = 2\%$.

12.5 Smile Modeling

With the dynamics of displaced diffusions only, the market model cannot price volatility smiles in cap/floor markets. For that purpose we should extend or modify the current model in ways parallel to the extensions to the LIBOR market model, on which there is rich literature (see e.g., Brigo and Mercurio (2006) for an introductions of smile models). One quick solution for smiles modeling is to adopt the SABR (Hagen et al., 2002) dynamics for the expected

---

[6]For brevity the data are not presented here, which are however available upon request.

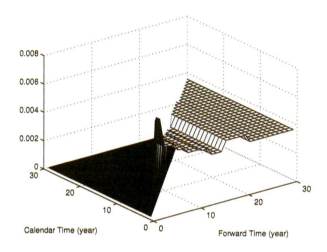

FIGURE 12.3: Calibrated local volatility surface, $\gamma_i^{(I)}(t)$.

displaced forward inflation rates, $\mu_j(t)$, and consider the following model:

$$
\begin{cases}
d\mu_j(t) = \mu_j^{\beta_j}(t)\gamma_j(t)dZ_t^j, \\
d\gamma_j(t) = \nu_j\gamma_j(t)dW_t^j, \quad \gamma_j(0) = \alpha,
\end{cases}
\tag{12.42}
$$

where β_j and ν_j are constants, and both Z_t^j and W_t^j are one-dimensional (correlated) Brownian motions under the T_j-forward measure. Mecurio and Mereni (2009) proposed and studied the above model for $\beta_j = 1$, and demonstrate a quality fitting of implied volatility smiles.

12.6   Notes

This chapter contains a number of achievements for inflation derivatives modeling. First, it properly defines the term structure of inflation forward rates, using arbitrage arguments, so that the definition is uniquely correct one. Second, it makes the pricing of YYIIS model free. Third, it justifies the Black's model of the industry that is based in displaced diffusion model for the forward inflation rates, so that inflation caplets/floorlets and swaptions can be quoted using "implied Black's volatilities." Fourth, it corrects the mistake of the dominant HJM type model that is based on "foreign currency analogy" (Jarrow and Yildirim, 2003). Lastly, based on the market model for inflation rate we can conveniently develop models to manage volatility smile risk in the inflation-rate derivatives.

Chapter 13

Market Model for Credit Derivatives

In the years before 2008, investors of credit markets had witnessed a rapid growth of liquidity in credit default swaps (CDS), options on the credit default swaps (or credit swaptions), tranches of collateralized debt obligations (CDO), as well as exotic portfolio of credit derivatives like bespoke single-tranche CDOs and CDOs of CDOs (so-called CDO^2). As the property market heated up in many areas of the United States, speculations in credit derivatives drove the property prices to a staggering level, eventually ending in a crash that almost brought down the U.S. banking system. Following the 2008 financial crisis, the credit market shrank considerably. Still, some of the derivatives, including CDS and CDO, survived the crisis. In this chapter we mostly address the pricing of CDS, and demonstrate that the model for CDS can be used for pricing CDO as well, with the adoption of copula, an additional mathematical tool, through the so-call bottom-up strategy.

Various pricing theories evolved with the pre-crisis markets. For the pricing of single-name swaptions, market practitioners and a series of researchers, including Schönbucher (2000), Arvanitis and Gregory (2001), Jamshidian (2002), and Hull and White (2002), among others, had led to the adoption of the Black's formula as the market standard. For the pricing of CDO tranches, where the modeling of dependent defaults is essential, there are survival-time copula models represented by Li (2000), Gregory and Laurent (2003), Andersen, Sidenius and Basu (2003) and Giesecke (2003); as well as the structural models represented by Zhou (2001) and Hull et al. (2005). The Gaussian copula model formulated by Gregory and Laurent (2003) and later improved by Andersen et al. (2003) was quite well received and has also become the market standard. Despite the successes, there are still major problems remaining. One fundamental problem, from a modeling point of view, is the detachment of the single-name credit derivatives market from the portfolio credit derivatives market, as there is no unified pricing framework for both markets yet. As a result, the notions of consistent pricing and hedging across the two markets are difficult to create. Although the Gaussian copula model can take the default probabilities estimated from the CDS markets as inputs, it has no capacity to utilize either spread dynamics or spread correlations observed in the CDS markets for CDO pricing.

There are also limitations with the popular portfolio credit derivatives models mentioned above. One of the major drawbacks of the survival-time copula approach for CDOs is that it does not constitute a proper dynamic model. The implication is that although these kind of models can be used to price CDO tranches, options on spread of either single name or portfolio cannot be priced under such models (e.g., exponential-copula model of Giesecke (2003)). With the Gaussian copula model, the market standard, default-time correlations are injected after the default times are mapped into normal random variables. There is, however, no economic intuition on how to input the correlations for these normal random variables. This problem is mitigated in the structural model of Hull et al. (2005), where a default is triggered when the value of a firm breaches a barrier (Black and Cox, 1976). However, structural models are difficult to use, as model users have to specify a number of model parameters, including drifts, volatility and correlations of the firm values, as well as default barriers. Given that firm values are not observable, a proper specification of the model is a daunting task by itself. In addition, firm-value based structural models do not naturally take the price information like credit spreads and implied spread volatilities from the single-name credit derivatives markets as inputs, hence the structural models do not ensure price consistency across the single-name and portfolio credit markets.

In this chapter, we intend to deliver a unified framework for pricing the CDS options and the CDOs. Motivated by the use of Black's formula for single-name swaptions, we develop a market model with forward hazard rates. Unlike usual structural models with firm values, the state variables of our market model are indirectly observable quantities. For the pricing of single-name credit derivatives, the new model is highly analogous to the LIBOR market model. In a very natural way, such a model can be applied to pricing CDO tranches using Monte Carlo simulations. Dynamically evolving CDS rates, CDS-rate correlations, and implied swaption volatilities, which are all observable in credit markets, can all be utilized for pricing the co-dependence of defaults of credit portfolios. In addition, the new model also has the capacity to price default-time correlation through the technique of survival-time copulas.

Our dynamic model is developed under the notion of arbitrage pricing. We define risky zero-coupon bonds for a credit name as cash flows backed by the coupons of corporate bonds. With such a definition, risky zero-coupon bonds, the basic building blocks of our model, become tradable. Consequently, forward spreads can be replicated with risky bonds, and the hedging strategies for of CDS and CDS options can then be readily defined. If liquidity and other market inefficiency are ignored, we can calibrate the market model to risky bonds and spread derivatives (CDS and CDS options), thus excluding arbitrage opportunities across these two markets. Survival-time copula, in principle, is the only input (for the model) that is unobservable in the credit market. When the volatilities of forward spreads of all names are set to zero, our market model reduces to the popular Gaussian copula model (Li, 2000; and Gregory and Laurent, 2003). Hence, our model can be regarded as a dynamic extension of the standard model for the pre-crisis markets.

This chapter is organized as follows. Sections 13.1 through 13.6 concern single-name credit derivatives. In Section 13.1, we present the definition of risky zero-coupon bonds and explain the intuition behind this definition. In Section 13.2, we define risky forward rates and forward spreads. In Section 13.3, we deliver swap-rate formulae for two types of credit swaps. Section 13.4 is devoted to the calculation of par CDS rates of both floating-rate and fixed-rate bonds. In Section 13.5, we demonstrate the estimation of implied survival rates and implied recovery rates. In Section 13.6, we present the market model for single-name swaptions. Through Section 13.1 to 13.6, we lay down the foundation for the pricing of portfolio credit derivatives. In Section 13.7, we then introduce CDOs, describe the Monte Carlo method for CDO pricing under the market model, and demonstrate the capacity of the model for accommodating spread correlations and default-time correlations. Pricing examples with standardized CDOs will be presented. Finally in Section 13.8, we provide some ending notes.

13.1 Pricing of Risky Bonds: A New Perspective

Our model is set in the filtered probability space, $(\Omega, (\mathcal{F}_t)_{t \geq 0}, \mathbb{Q})$, where the filtration satisfies the usual conditions of right-continuity and completeness, and $\mathbb{Q}$ is the risk-neutral measure. All stochastic processes are adapted to $(\mathcal{F}_t)_{t \geq 0}$. Without loss of generality, we model the default time as the first jump time of a Cox process with an intensity (or hazard rate) process, $\tilde{\lambda}(t)$. For technical convenience and clarity of presentation, we assume independence between credit spreads and U.S. Treasury yields, as well as independence between hazard rate and the recovery rate of the credit (with the understanding that this may be against some empirical findings (Duffee, 1998)). We also assume a single seniority across all bonds under the same credit name.

We begin with bond pricing. A defaultable coupon bond pays regular coupons until default or maturity, whichever comes first. In case of a default, the market convention is that a creditor will receive a final payment that consists of a fraction of both the principal and the last accrued interest. The schedule of the final payment varies. Without loss of generality, we assume that (1) the final payment is made at the next coupon date following default and (2) the last coupon accrues until the final payment date.[1] Let c be the coupon rate of a risky bond with term[2] $[T_m, T_n]$, τ be the default time and R_τ be the recovery rate at the default time. Then, the cash flow at

---

[1]The protection payment made only at a coupon date is a harmless idealization. There are other default payment schedules as well. For example, the last payment may occur at $\tau + 90$ days. Most payment schedules can be accommodated by adjusting the recovery rate.

[2]The coupon dates are $\{T_j\}_{j=m+1}^n$. The bond is said to be "forward starting" if $m > 0$.

$T_{j+1}, m \leq j \leq n - 1$ can be expressed as

$$\Delta T c \mathbf{1}_{\{\tau > T_{j+1}\}} + R_\tau (1 + \Delta T c) \mathbf{1}_{\{T_j < \tau \leq T_{j+1}\}},$$

where $\Delta T = 1/(\text{coupon frequency})$ and $\mathbf{1}_{\tau \in \Omega}$ is the indicator function that equals to 1 if $\tau \in \Omega$, or 0 otherwise. According to the arbitrage pricing theory (APT) (Harrison and Pliska, 1981), the bond is then priced as the risk-neutral expectation of discounted cash flows:

$$B^c(t) = \sum_{j=m}^{n-1} E_t^Q \left[\frac{B(t)}{B(T_{j+1})} \left\{ \Delta T_j c \mathbf{1}_{\{\tau > T_{j+1}\}} + R_\tau (1 + \Delta T_j c) \mathbf{1}_{\{T_j < \tau \leq T_{j+1}\}} \right\} \right]$$

$$+ E_t^Q \left[\frac{B(t)}{B(T_n)} \mathbf{1}_{\{\tau > T_n\}} \right]$$

$$= \sum_{j=m}^{n-1} P_{j+1}(t) \Delta T_j c E_t^{Q_{j+1}} \left[\mathbf{1}_{\{\tau > T_{j+1}\}} + R_\tau \mathbf{1}_{\{T_j < \tau \leq T_{j+1}\}} \right]$$

$$+ \sum_{j=m}^{n-1} P_{j+1}(t) E_t^{Q_{j+1}} \left[R_\tau \mathbf{1}_{\{T_j < \tau \leq T_{j+1}\}} \right] + P_n(t) E_t^{Q_n} \left[\mathbf{1}_{\{\tau > T_n\}} \right],$$

(13.1)

where, in addition,

$B(t)$	—	the money market account of value $\exp(\int_0^t r_s ds)$, where r_t is the risk-free spot rate;
$P_j(t)$	—	the time-t price of the risk-free zero-coupon bond of \$1 notional value maturing at T_j, equals to $E_t^Q [B(t)/B(T_j)]$;
ΔT_j	—	$T_{j+1} - T_j$, the length of the coupon interval $(T_j, T_{j+1}]$;
Q_j	—	the T_j forward measure;
$E_t^Q [\,\cdot\,]$	—	the expectation under Q conditional on $\mathcal{F}_t$.

The expression for payment upon a default, $R_\tau(1 + \Delta T_j c)$, conforms well with real-world practice: the compensation to the creditor is determined by the outstanding principal and the accrued interest of the defaulted bond, which are treated as in the same asset class, and future coupons are not taken into consideration (Schönbucher, 2004). The second line of (13.1) results from the changes of measures, followed by a regrouping of the PVs of the coupons and the principal. The payout at time T_{j+1} is priced by using the T_{j+1}-forward measure.

Parallel to the U.S. Treasury market, we introduce here the "C-strip" and the "P-strip" of risky zero-coupon bonds, which are backed separately by coupons and principals. It is not hard to see that the C-strip zero-coupon bonds should be defined as

$$
\begin{aligned}
\bar{P}_j(t) &= E_t^Q \left[\frac{B(t)}{B(T_j)} \left\{ 1_{\{\tau > T_j\}} + R_\tau 1_{\{T_{j-1} < \tau \le T_j\}} \right\} \right] \\
&= P_j(t) E_t^{Q_j} \left[1_{\{\tau > T_j\}} + R_\tau 1_{\{T_{j-1} < \tau \le T_j\}} \right] \overset{\triangle}{=} P_j(t) D_j(t),
\end{aligned}
\tag{13.2}
$$

while the "P-strip" zero-coupon bonds are defined as in the last line of equation (13.1). Here in (13.2), the "$\overset{\triangle}{=}$" means that a new variable $D_j(t)$ is defined through the equation. Unlike their Treasury counterparts, the risky zero-coupon bonds of the two strips have apparently different cash-flow structures. Given only the prices of risky coupon bonds, we cannot identify the PVs of zero-coupon bonds in either strip unless additional information is given or further assumptions are made.

In principle, zero-coupon bonds of the two strips can be generated through marketing the cash flows of the coupons and the principals separately. In reality, however, risky zero-coupon bonds are not traded and, as a result, direct price information is not available (with perhaps occasional exceptions in the Japanese market). Nonetheless, the prices of risky zero-coupon bonds of both strips can be backed out from associated coupon bond prices and additional information like the CDS rates. The notion of C-strip zero-coupon bonds, in particular, is important for the construction of our model.

13.2   Forward Spreads

To understand the product nature of CDS, we need to clarify the notion of "risky forward rates" adopted earlier by Schönbucher (2000) and Brigo (2005). A risky forward rate should be defined as the fair rate on a defaultable loan for a future period of time, say, $(T_j, T_{j+1}]$, that is backed by the coupon flows of defaultable bonds of an entity. If a default of the bond occurs before T_j, the loan ceases to exist. If a default occurs between T_j and T_{j+1}, then a recovery value proportional to the recovery rate of the bonds applies. Assume the notional of the loan to be \$1. According to the APT, the risky forward rate, denoted as $\hat{f}_j(t)$, must nullify the PV of the cash flows of the risky loan:

$$
\begin{aligned}
0 &= E_t^Q \left[\frac{B(t)}{B(T_j)} 1_{\{\tau > T_j\}} \right] \\
&\quad - E_t^Q \left[\frac{B(t)}{B(T_{j+1})} (1 + \Delta T_j \hat{f}_j(t))(1_{\{\tau > T_{j+1}\}} + R_\tau 1_{\{T_j < \tau \le T_{j+1}\}}) \right] \\
&= P_j(t) E_t^{Q_j} \left[1_{\{\tau > T_j\}} \right] - P_{j+1}(t)(1 + \Delta T_j \hat{f}_j(t)) D_{j+1}(t) \\
&\overset{\triangle}{=} P_j(t) \Lambda_j(t) - P_{j+1}(t)(1 + \Delta T_j \hat{f}_j(t)) D_{j+1}(t),
\end{aligned}
\tag{13.3}
$$

where $\Lambda_j(t)$ is the $\mathbb{Q}_j$ probability of survival until T_j, and it is equal to the $\mathbb{Q}$ probability of survival until T_j due to the independence between U.S. Treasury yields and the default probability of the entity. Equation (13.3) gives rise to

$$
\begin{aligned}
\hat{f}_j(t) &= \frac{1}{\Delta T_j}\left[\frac{P_j(t)}{P_{j+1}(t)}\frac{\Lambda_j(t)}{D_{j+1}(t)} - 1\right] \\
&= \frac{1}{\Delta T_j}\left[\frac{\bar{P}_j(t)}{\bar{P}_{j+1}(t)}\frac{\Lambda_j(t)}{D_j(t)} - 1\right].
\end{aligned}
\tag{13.4}
$$

The two lines in the equation above lead to two alternative expressions of $\hat{f}_j(t)$. The first expression is in terms of the LIBOR rates,

$$
\begin{aligned}
\hat{f}_j(t) &= \frac{1}{\Delta T}\left[\frac{P_j(t)}{P_{j+1}(t)} - 1\right] + \frac{1}{\Delta T}\frac{P_j(t)}{P_{j+1}(t)}\left(\frac{\Lambda_j(t)}{D_{j+1}(t)} - 1\right) \\
&= f_j(t) + \frac{1}{\Delta T}\frac{P_j(t)}{P_{j+1}(t)} \\
&\quad \times \left(\frac{E_t^{Q_j}\left[\mathbf{1}_{\{\tau > T_j\}}\right] - E_t^{Q_{j+1}}\left[\mathbf{1}_{\{\tau > T_{j+1}\}} + R\mathbf{1}_{\{T_j < \tau \leq T_{j+1}\}}\right]}{D_{j+1}(t)}\right) \\
&= f_j(t) + \frac{1}{\Delta T}\frac{P_j(t)}{P_{j+1}(t)}\left(\frac{E_t^{Q_{j+1}}\left[(1-R)\mathbf{1}_{\{T_j < \tau \leq T_{j+1}\}}\right]}{D_{j+1}(t)}\right) \\
&= f_j(t) + \frac{1 + \Delta T_j f_j(t)}{\Delta T}\left(\frac{E_t^{Q_{j+1}}\left[(1-R)\mathbf{1}_{\{T_j < \tau \leq T_{j+1}\}}\right]}{D_{j+1}(t)}\right),
\end{aligned}
\tag{13.5}
$$

where $f_j(t)$ is the default-free forward rate for the period (T_j, T_{j+1}) seen at time t, defined by

$$
f_j(t) = \frac{1}{\Delta T_j}\left(\frac{P_j(t)}{P_{j+1}(t)} - 1\right), \quad t \leq T_j.
$$

The second expression for the risky forward rate is

$$
\begin{aligned}
\hat{f}_j(t) &= \frac{1}{\Delta T}\left[\frac{\bar{P}_j(t)}{\bar{P}_{j+1}(t)} - 1\right] - \frac{1}{\Delta T}\frac{\bar{P}_j(t)}{\bar{P}_{j+1}(t)}\left[1 - \frac{\Lambda_j(t)}{D_j(t)}\right] \\
&= \bar{f}_j(t) - \frac{1 + \Delta T_j \bar{f}_j(t)}{\Delta T_j}\left(\frac{E_t^{Q_j}\left[R\mathbf{1}_{\{T_{j-1} < \tau \leq T_j\}}\right]}{D_j(t)}\right),
\end{aligned}
\tag{13.6}
$$

where

$$
\bar{f}_j(t) = \frac{1}{\Delta T_j}\left(\frac{\bar{P}_j(t)}{\bar{P}_{j+1}(t)} - 1\right), \quad t \leq T_j.
$$

is called the "defaultable effective forward rate" (Schönbucher, 2000). Note that $\bar{f}_j(t)$ should be understood as the effective rate of return over (T_j, T_{j+1}) provided that no default occurs until T_{j+1}. It is pointed out in Brigo (2004) that $\bar{f}_j(t)$ in general does not link directly to a financial contract. Putting (13.5) and (13.6) together, we have the order

$$f_j(t) \le \hat{f}_j(t) \le \bar{f}_j(t). \tag{13.7}$$

Note that $\hat{f}_j(t)$ achieves the upper bound and the lower bound when $R_\tau = 0$ and $R_\tau = 1$, respectively. The bounds on $\hat{f}_j(t)$ should be regarded as no-arbitrage constraints.

Here, we reiterate the insight of "non-separability" pointed out by Duffie and Singleton (1999), which means that the hazard rate and the loss rate cannot be simultaneously determined from bond prices alone. In view of (13.4), we can say that complete term structures of the survival probability and the recovery rate, $\Lambda_j(t)$ and $E_t^Q[R_\tau|T_{j-1} < \tau \le T_j]$, $j = 1, 2, \ldots$, can be uniquely determined from the term structures of $\hat{f}_j(t)$ and $\bar{f}_j(t)$. These in fact, as we soon shall demonstrate, are informatively equivalent to the term structures of risky-bond yields and CDS rates.

Intuitively, a "forward spread" is defined as the difference between a risky forward rate and its corresponding risk-free forward rate:

$$S_j(t) = \hat{f}_j(t) - f_j(t), \quad j = 1, 2, \ldots.$$

From equation (13.5), we obtain

$$
S_j(t) = (1 + \Delta T_j f_j(t)) \frac{E_t^{Q_{j+1}}\left[(1 - R_\tau)\mathbf{1}_{\{T_j < \tau \le T_{j+1}\}}\right]}{\Delta T_j D_{j+1}(t)} \tag{13.8}
$$
$$
\triangleq (1 + \Delta T_j f_j(t)) H_j(t).
$$

According to its definition, $H_j(t)$ can be interpreted as the "expected loss per risky dollar over $(T_j, T_{j+1}]$." Equation (13.8) can be rewritten as

$$1 + \Delta T_j \hat{f}_j(t) = (1 + \Delta T_j f_j(t))(1 + \Delta T_j H_j(t)), \quad j = 1, 2, \ldots.$$

By comparing definitions, we can say that $H_j(t)$ is the discrete-tenor version of the "mean loss rate" introduced in Duffie and Singleton (1999).

13.3 Two Kinds of Default Protection Swaps

A default protection swap consists of a fee leg (or premium leg) and a protection leg. Before the default of the reference entity, the protection buyer pays the protection seller a string of fees at regular time intervals. Upon default, the protection buyer either delivers the bond to the protection seller in exchange for par (so-called physical settlement), or receives from the protection seller a payment that is equal to the loss incurred (so-called cash settlement). In this section, we consider two kinds of swaps: swaps for fixed-rate bonds and swaps for floating-rate bonds. Without loss of generality, we assume the notional value of the bonds to be $1, and the coupon rate to be c or LIBOR, $f_j(T_j)$, respectively. For the swap on the fixed-rate bond, the protection payment is $(1 - R_\tau)(1 + \Delta Tc)$, while for the swap on the floating-rate bond, the protection payment is $(1 - R_\tau)(1 + \Delta T_j f_j(T_j))$. The payments simply reflect the loss to the bond holders in case of a default.[3] Note that the swaps of the second kind depend only on the default status of the reference entity.

We proceed to the determination of the fair rate for the default swap of the first kind. Let us denote a swap rate by $\bar{s}$. In the CDS markets, the contractual cash flows of the fee leg (for the protection of $1 notional) are typically

$$\bar{s} \Delta T_j \left[\mathbf{1}_{\{\tau > T_{j+1}\}} + \frac{(\tau - T_j)}{\Delta T_j} \mathbf{1}_{\{T_j < \tau \leq T_{j+1}\}} \right], \quad j = 1, 2, \dots,$$

i.e., in case of a default occurring between T_j and T_{j+1}, the protection buyer makes the final payment that is proportional to the time elapsed between the last fee payment and the default. From a financial engineering point of view, such a fee specification is troubling because the cash flow cannot be synthesized at the initiation of the swap contract. To make the swap contract replicable, we propose a slight modification to the fee leg: we let the cash flows be generated from the payouts of risky zero-coupon bonds:

$$\bar{s} \Delta T_j \left[\mathbf{1}_{\{\tau > T_{j+1}\}} + R_\tau \mathbf{1}_{\{T_j < \tau \leq T_{j+1}\}} \right].$$

This new definition only changes the last piece of fee payment after default. The value of the fee leg is now

$$PV_{fee} = \bar{s} \sum_{j=m}^{n-1} \Delta T_j P_{j+1}(t) E_t^{Q_{j+1}} \left[\mathbf{1}_{\{\tau > T_{j+1}\}} + R_\tau \mathbf{1}_{\{T_j < \tau \leq T_{j+1}\}} \right]$$

$$= \bar{s} \sum_{j=m}^{n-1} \Delta T_j \bar{P}_{j+1}(t). \tag{13.9}$$

Let

$$\bar{A}_{m,n}(t) := \sum_{j=m}^{n-1} \Delta T_j \bar{P}_{j+1}(t),$$

which is now a tradable annuity, analogous to the fixed leg of default-free swaps.

For a swap of the first kind, the cash flow of the protection seller can be written as

$$V_{prot} = (1 + \Delta Tc) \sum_{j=m}^{n-1} (1 - R_\tau) \mathbf{1}_{\{T_j \leq \tau \leq T_{j+1}\}}.$$

---

[3]The quantity $(1 + \Delta T_j f_j(T_j))$ can be regarded as "a constant dollar" seen at time T_j.

The PV of the protection payment is then

$$PV_{prot} = (1 + \Delta Tc) \sum_{j=m}^{n-1} P_{j+1}(t) E_t^{Q_{j+1}} \left[(1 - R_\tau) \mathbf{1}_{\{T_j \leq \tau \leq T_{j+1}\}} \right]. \quad (13.10)$$

By equating the fee leg to the protection leg and making use of (13.8), we obtain

$$\bar{s}_1 = (1 + \Delta Tc) \frac{\sum_{j=m}^{n-1} P_{j+1}(t) E_t^{Q_{j+1}} \left[(1 - R_\tau) \mathbf{1}_{\{T_j \leq \tau \leq T_{j+1}\}} \right]}{\bar{A}_{m,n}(t)}$$

$$= (1 + \Delta Tc) \sum_{j=m}^{n-1} \bar{\alpha}_j H_j(t), \quad (13.11)$$

where

$$\bar{\alpha}_j = \frac{\Delta T_j \bar{P}_{j+1}(t)}{\bar{A}_{m,n}(t)}.$$

Note that the case $c = 0$ corresponds to a prototypical default swap, which only depends on the default status of the reference entity and actually dominates the liquidity of single-name credit derivatives markets.

For credit default swaps of the second kind, the fair swap rate can be derived analogously as

$$\bar{s}_2 = \frac{\sum_{j=m}^{n-1} P_{j+1}(t)(1 + \Delta T_j f_j(t)) E_t^{Q_{j+1}} \left[(1 - R_\tau) \mathbf{1}_{\{T_j \leq \tau \leq T_{j+1}\}} \right]}{\bar{A}_{m,n}(t)}$$

$$= \sum_{j=m}^{n-1} \bar{\alpha}_j S_j(t). \quad (13.12)$$

Here, we have made use of the independence between credit spreads and the U.S. Treasury yields, as well as the martingale property of the forward rate: $E_t^{Q_{j+1}}[f_j(T_j)] = f_j(t)$. Note that a one-period CDS rate reduces to the forward spread.

Compared with the existing CDS rate formula (e.g., Schönbucher, 2004; Brigo, 2005), our definition simply states that a CDS rate is equal to the weighted average of credit spreads, which does not require the recovery rate as an input, and is analogous to the swap rate formula in LIBOR markets.

13.4   Par CDS Rates

A par CDS rate is a spread that, when added to a corresponding par rate of default-free bond, yields the coupon rate of a risky par bond. We will derive the par CDS rates for both floating rate bonds and fixed-rate bonds.

Typically, a defaultable floating-rate bond (which is also called a default-able floater) pays LIBOR plus a credit spread, denoted by s_F, until a default occurs, when a holder may obtain some recovered value of both principal and coupon. Hence, the cash flow of a defaultable floater at T_{j+1} can be written as

$$CF_{j+1} = (1 + \Delta T_j(f_j(T_j) + s_F)) \left(\mathbf{1}_{\{\tau > T_{j+1}\}} + R\mathbf{1}_{\{T_j < \tau \leq T_{j+1}\}}\right)$$
$$- \mathbf{1}_{\{\tau > T_{j+1}, j+1 < n\}}.$$

The question here is: if the floater is to be priced at par, what should the fair spread rate s_F be? To answer this question, we imagine that the holder of the floater is also "long" a protection swap of the second kind. Then, his/her cash flow at time T_{j+1} is

$$CF_{j+1} = \Delta T_j f_j(T_j) \mathbf{1}_{\{\tau > T_j\}} + \mathbf{1}_{\{T_j < \tau \leq T_{j+1}\}} + \mathbf{1}_{\{\tau > T_n, j+1 = n\}}$$
$$+ (s_F - \bar{s}_2)(\mathbf{1}_{\{\tau > T_j\}} + R\mathbf{1}_{\{T_{j-1} < \tau \leq T_j\}}).$$

$$(13.13)$$

The first line in (13.13) gives the cash flow of a rolling-forward CD that lasts until $T_{j_\tau} \wedge T_n$, where T_{j_τ} is the first fixing date after default, and such cash flows represent those of a par bond (that matures at $T_{j_\tau} \wedge T_n$). It then becomes clear that, for the defaultable floater to be priced at par, there must be

$$s_F = \bar{s}_2,$$

i.e., the default swap rate equals nothing else but the credit spread!

Given the clear relationship between a defaultable floater and a CDS, we then come up with the following hedging strategy for swaps of the second kind: once such a swap is written, the hedger goes "long" a default-free floater and goes "short" a defaultable floater, both at par. Then, the net cash flow at any fixing date is zero.

Next, we derive the par CDS rate for a corresponding fixed-rate bond with tenor $(T_m, T_n]$. We let s_X denote the par CDS rate for the fixed-rate bond. Then, s_X can be determined by equating the PVs of fixed-rate and floating-rate par bonds:

$$(R_{m,n}(t) + s_X)\bar{A}_{m,n}(t) = \sum_{j=m}^{n-1} \Delta T_j \bar{P}_{j+1}(t)[f_j(t) + \bar{s}_2]$$
$$= \sum_{j=m}^{n-1} \Delta T_j \bar{P}_{j+1}(t)\hat{f}_j(t),$$

where the PVs of the principals have been canceled, and $R_{m,n}(t)$ represents the corresponding swap rate (i.e., par rate) in LIBOR markets, defined by

$$R_{m,n}(t) = \sum_{j=m}^{n-1} \alpha_j f_j(t),$$

It follows that

$$s_X = \sum_{j=m}^{n-1} \bar{\alpha}_j \hat{f}_j(t) - R_{m,n}(t)$$

$$= \bar{s}_2 + \sum_{j=m}^{n-1} (\bar{\alpha}_j - \alpha_j) f_j(t).$$

Hence, the so-called "credit spread" is different for floating-rate bonds and fixed-rate bonds, and a par CDS rate for fixed-rate bonds is close to $\bar{s}_2$ instead of $\bar{s}_1$ (when $c = 0$).

It can be verified that the short position of the default swap on the risky coupon bond with coupon rate c can be hedged by

1. being "short" the risky coupon bond,

2. being "long" a risk-free floater at par, and

3. being "long" $(c - \bar{R}_{m,n}(t))$ units of the risky annuity, $\bar{A}_{m,n}(t)$.

Here, $\bar{R}_{m,n}(t) = R_{m,n}(t) + s_X$ is the risky par yield.

13.5 Implied Survival Curve and Recovery-Rate Curve

In reality, neither $\{\hat{f}_j(t)\}$ nor $\{\bar{f}_j(t)\}$ is directly observable. For CDS pricing and other applications, it is more convenient to make use of the term structures of forward hazard rate and forward recovery rate. We define the forward hazard rate for $(T_j, T_{j+1}]$ seen at time t by

$$\lambda_j(t) = \frac{1}{\Delta T_j} \left(\frac{\Lambda_j(t)}{\Lambda_{j+1}(t)} - 1 \right), \quad \text{for } t \le T_j, \tag{13.14}$$

and the forward recovery rate for the same period by

$$R_j(t) = E_t^Q [R_\tau | T_j < \tau \le T_{j+1}], \quad \text{for } t \le T_j \text{ and } j \ge \eta(t), \tag{13.15}$$

where $\eta(t)$ is smallest integer such that $T_{\eta(t)} \ge t$. The survival probability, $\Lambda_j(t)$, relates to the hazard rates by

$$\Lambda_j(t) = (1 + (T_{\eta(t)} - t)\lambda_{\eta(t)-1}(T_{\eta(t)-1}))^{-1} \prod_{k=\eta(t)}^{j-1} (1 + \Delta T_k \lambda_k)^{-1}. \tag{13.16}$$

The standard market practice is to back out the survival probabilities from CDS of various maturities, assuming a constant recovery rate (of 40%). Instead of doing the same thing with our model, we consider simultaneously backing out the implied hazard rates and recovery rates from CDSs as well as corporate bond prices. We choose Citigroup as the credit name for a demonstration, as there is relatively richer credit information on this company. A snapshot of market quotations is provided in Tables 13.1 and 13.2, where the currency is U.S. dollars (USD), the CDS rates are for $c = 0$, and the bond prices are "clean." To build the risk-free discount curve in USD, we have used LIBOR rates up to one year, and swap rates from 2 to 20 years. The interest rate information is provided in Table 13.3. Figure 13.1 presents the forward-rate curve constructed using the yield data in Table 13.3.

Table 13.1: Citigroup CDS Rates (7/28/2005, Bloomberg)

Maturity	1Y	3Y	5Y	10Y
Rates	0.07%	0.13%	0.19%	0.33%

Table 13.2: Prices of Benchmark Citigroup Bonds (7/28/2005, Bloomberg)

Maturity	Frequency	Coupon	Price
22/2/2010	Semi-annual	4.125%	98.123
1/10/2010	Semi-annual	7.25%	114.563
7/5/2015	Semi-annual	4.875%	97.563
18/5/2010	Quarterly	US LIB+15bps	99
16/3/2012	Quarterly	US LIB+12.5bps	99.80
5/11/2014	Quarterly	US LIB+28bps	100.501

Table 13.3: USD yield data (7/28/2005, Bloomberg)

LIBOR	3 mth	3.6931%
	6 mth	3.8435%
	12 mth	4.1731%
Swap	2Y	4.3200%
	3Y	4.3840%
	4Y	4.4330%
	5Y	4.4690%
	7Y	4.5365%
	10Y	4.6290%
	12Y	4.6905%
	15Y	4.7630%
	20Y	4.8320%

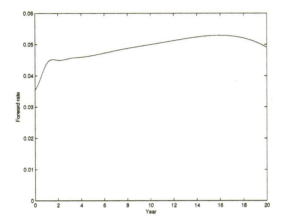

FIGURE 13.1: USD forward rates (7/28/2005).

We determine $\{\lambda_j, R_j\}$ through reproducing the CDS rates and the bond prices of Citigroup by the swap-rate formula (13.11) and the bond formula (13.1) respectively. Because the problem is under-determined, we have adopted cubic-spline and linear interpolation for the hazard rates and the recovery rates respectively, and imposed additional smoothness regularization. In our search algorithm, we took various initial guesses with $\lambda_j = 0$ and $0.0 \leq R_j \leq 0.6, \forall j$. Often but not always, the search ends up in one of the two solutions, depicted in Figures 13.2 to 13.4, depending on the closeness of the initial recovery rate to either $R_0 = 0.0$ or $R_0 = 0.4$. Existence of more than one solution reflects the ill-posedness nature of the calibration problem, particularly with regard to the determination of the recovery rate.

We remark here that the ill-posedness nature of the calibration problem is largely intrinsic. It is due to the insensitivity of the bond and CDS prices with respect to the change of the recovery rates,[4] particularly for a good credit name. As a consequence of the ill-posedness, implied recovery rates and default rates may depend on initial guesses, as is shown in Figure 13.2 and 13.3. To settle down to a single solution, additional regularizations (based on financial or mathematical considerations) are needed. On the other hand, the ill-posedness may not be a great concern. As is shown in Figure 13.4, risky forward rates and risky discount curve demonstrate noticeable stability. This is due to the complimentary effect between the hazard rates and the recovery rates in calibration.

By comparing the implied hazard rates with the CDS rates of Citigroup, we can say that the market quotations of CDS rates pretty much represent the risk-neutral hazard rates for zero recovery upon default. The quoted CDS spreads increase with time to maturity, and a risk-neutral hazard rate will also increase with its maturity.

---

[4]The insensitivity is also observed in Brigo (2005).

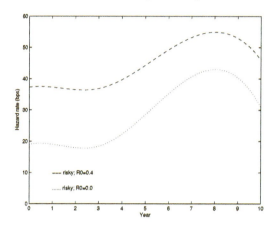

FIGURE 13.2: Implied hazard rates.

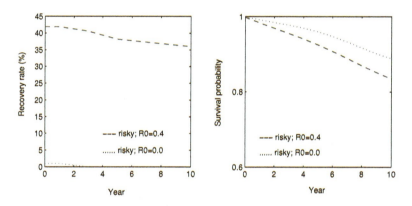

FIGURE 13.3: Implied default rates and recovery rates.

Next, we will demonstrate that when CDS rates are high, the calibration can become a well-posed problem. An example is with General Motors (GM). It had been a widely known credit event in July 2005, when the five-year CDS rate of GM shot up several hundred basis points, as shown in Table 13.4, and the yield spreads of its bonds also went up, see Table 13.5. When we calibrate the hazard rate and recovery rate of GM, we obtain almost identical results, see Figure 13.5 and Figure 13.6, regardless of the choice of initial guesses taken for calibration. As noted early, CDS rates essentially reflect the hazard rates. When the hazard rates are high, the CDS rates become sensitive to the recovery rate, thus making the calibration a well-posed problem.

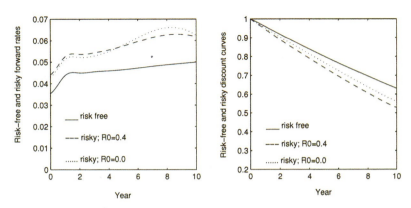

FIGURE 13.4: Forward rates and discount curves.

Table 13.4: GM CDS Rates (7/28/2005, Bloomberg)

Maturity	1Y	3Y	5Y	10Y
Rates	——	——	5.2117%	——

Table 13.5: Prices of Benchmark GM Bonds (7/28/2005, Bloomberg)

Maturity	Frequency	Coupon	Price
3/15/2006	Semi-annual	7.1%	101.405
5/1/2008	Semi-annual	6.375%	98.62
4/15/2016	Semi-annual	7.7%	90.41

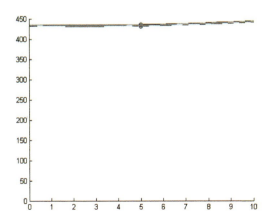

FIGURE 13.5: Implied hazard rates.

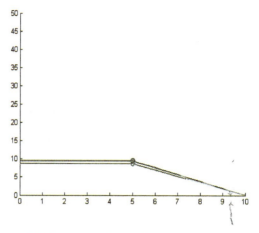

FIGURE 13.6: Implied recovery rates.

A remedy for the ill-posedness of the calibration can be the adoption of recovery swap quotes into the input set. A recovery swap has the following features.

- It is an agreement to swap a fixed recovery rate for a real recovery rate following a credit event.

- It is usually traded as zero-premium credit default swaps with the reference price set at the fixed recovery rate at certain notional value.

- When no default occurs, the recovery swap expires worthlessly.

Example: A dealer might quote a recovery swap in GM at 37/40. This means the dealer is prepared to sell a recovery swap with RS = 37% and buy at RS = 40%.

The liquidity of the recovery swaps rose during the 2011 European debt crisis. If we could take the recovery swaps as part of the input set, all we then need to do is to back out the implied hazard rates. Unfortunately, recovery swaps usually trade only on credits that are nearing default.

13.6   Credit Default Swaptions and an Extended Market Model

For either the protection buyer or the protection seller, an open position can be closed by either going "short" or going "long" of the same swap. Because the protection legs are exactly offset, the profit or loss for the pair of transactions comes from the difference of the fee legs, which is

$$(\bar{s}_{m,n}(t) - \bar{s}_{m,n}(0)) \sum_{j=m}^{n-1} \Delta T_j \bar{P}_{j+1}(t), \tag{13.17}$$

where $\bar{s}_{m,n}(t)$ is the prevailing CDS rate, either $\bar{s}_1$ or $\bar{s}_2$, seen at time $t \leq T$.

A credit swaption or CDS option, meanwhile, is a contract that gives its holder the right, not the obligation, to enter into a forward-starting swap at time $T \leq T_m$ with a predetermined swap rate, $\bar{s}^*$. Thus the profit/loss at maturity T of the swaption is

$$(\bar{s}_{m,n}(T) - \bar{s}^*)^+ \sum_{j=m}^{n-1} \Delta T_j \bar{P}_{j+1}(T). \tag{13.18}$$

To price the CDS option (or credit swaption), we need to make change of measures that are not equivalent but absolutely continuous.

Theorem 13.6.1 (Radon-Nikodyn). *Let* $\mathbb{P}$ *and* $\mathbb{Q}$ *be two probability measures on* $(\Omega, \mathcal{F}_{t\geq0})$ *with* $\mathbb{Q} \ll \mathbb{P}$ *($\mathbb{Q}$ absolutely continuous with respect to $\mathbb{P}$). Then there exists a unique $\mathbb{P}$-martingale L_t such that for all $t \geq 0$ and all $\mathcal{F}_t$-measurable and $\mathbb{Q}$-integrable random variable Y, we have*

$$E^{\mathbb{Q}}[Y] = E^{\mathbb{P}}[YL]. \tag{13.19}$$

Denote

$$L_t = \frac{d\mathbb{Q}}{d\mathbb{P}}\Big|_{\mathcal{F}_t}, \tag{13.20}$$

which satisfies

1. $L_t \geq 0$.

2. *Under* $\mathbb{Q}$, L_t *is almost surely bounded away from zero:* $\mathbb{Q}(\inf_{t\geq0} L_t > 0) = 1$.

3. *If* M_t *is a continuous from the right with left limit and adaptive, M_t is a $\mathbb{Q}$-martingale iff $L_t \times M_t$ is a $\mathbb{P}$-martingale. In particular, for $0 \leq t \leq T$:*

$$L_t E^{\mathbb{Q}}[M_T|\mathcal{F}_t] = E^{\mathbb{P}}[L_T M_T|\mathcal{F}_t].$$

4. *Every non-negative, integrable $\mathbb{P}$-martingale L with $L(0) = 1$ defined a measure $\mathbb{Q}$ by (13.20).* □

For a proof, we refer readers to Protter (1990).

Corollary 13.6.1. *Let* $A_t \geq 0$ *be a dividend-free asset and* $\mathbb{Q}$ *a spot martingale measure (corresponding to money market account $B(t)$ as numeraire). Define a new measure corresponding to A_t by*

$$L_t = \frac{d\mathbb{Q}_A}{d\mathbb{Q}}\Big|_{\mathcal{F}_t} = \frac{A_t}{A_0} \Big/ \frac{B(t)}{B(0)} 1_{A_0>0}$$

Then, the A_t-relative price of any default-related asset:

$$X_t^A = \frac{X_t}{A_t} 1_{A_t > 0}$$

with the property $1_{X_t > 0} = 1_{A_t > 0}$ is a $\mathbb{Q}_A$ martingale, i.e., for all $t \leq T$,

$$X_t^A = E^{\mathbb{Q}_A} \left[X_T^A | \mathcal{F}_t \right]. \tag{13.21}$$

Proof: When $L_t = 0$, $X_t^A = A_t = A_T = X_T^A = 0$, the equality holds automatically. When $L_t > 0$, there is also $A_t > 0$. It follows that

$$
\begin{aligned}
E^{\mathbb{Q}_A} \left[X_T^A | \mathcal{F}_t \right] &= E^{\mathbb{Q}} \left[\frac{L_T}{L_t} X_T^A | \mathcal{F}_t \right] \\
&= E^{\mathbb{Q}} \left[\frac{A_T}{A_t} \frac{B(t)}{B(T)} 1_{A_t > 0} \frac{X_T}{A_T} 1_{A_T > 0} | \mathcal{F}_t \right] \\
&= \frac{1_{A_t > 0}}{A_t} E^{\mathbb{Q}} \left[\frac{B(t)}{B(T)} X_T 1_{A_T > 0} | \mathcal{F}_t \right] \\
&= \frac{1_{A_t > 0}}{A_t} E^{\mathbb{Q}} \left[\frac{B(t)}{B(T)} X_T | \mathcal{F}_t \right] \\
&= \frac{X_t}{A_t} 1_{A_t > 0} = X_t^A.
\end{aligned}
$$

$\square$

Let $\bar{A}_{m,n}(0) \geq 0$, define the *default swap measure* $\bar{\mathbb{Q}}_S$, according to

$$\left. \frac{d\bar{\mathbb{Q}}_S}{d\mathbb{Q}} \right|_{\mathcal{F}_t} = \frac{\bar{A}_{m,n}(t)}{\bar{A}_{m,n}(0)} \Big/ \frac{B(t)}{B(0)} 1_{\bar{A}_{m,n}(0) > 0} = L_t, \tag{13.22}$$

then we have

Theorem 13.6.2 (Ho and Wu, 2008). *Under the new measure $\bar{\mathbb{Q}}_S$, there are*

1. *The $\bar{\mathbb{Q}}_S$-default probability is zero:*

$$E_t^{\bar{\mathbb{Q}}_S} \left[\mathbf{1}_{\{\tau < T\}} \right] = 0.$$

2. *The swap rate $\bar{s}_{m,n}(t)$ of either kind of swap is a martingale:*

$$E_t^{\bar{\mathbb{Q}}_S} \left[\bar{s}_{m,n}(T) \right] = \bar{s}_{m,n}(t), \quad \forall T \in (t, T_m).$$

$\square$

Proof:

1. By definition,

$$\bar{Q}_S\left[\tau \leq T\right] = E^{\bar{Q}_S}\left[1_{\tau \leq T}\right]$$

$$= E^Q\left[L_T 1_{\tau \leq T}\right]$$

$$= E^Q\left[\frac{\bar{A}_{m,n}(T)}{\bar{A}_{m,n}(0)}\frac{B(0)}{B(T)}1_{\tau > 0}1_{\tau \leq T}\right]$$

$$= \frac{1_{\tau > 0}}{\bar{A}_{m,n}(0)}E^Q\left[\frac{\bar{A}_{m,n}(T)}{B(T)}1_{\tau \leq T}\right] = 0.$$

2. The martingale property follows from taking $X_t = PV_{prot}(t)$ in the Corollary 13.6.1. □

The first result says that the probability of default under Q_S is zero, this is why Schonbucher (2004) call Q_S the survival measure.

In terms of the new measure, we can express the price of the default swaption at time $t \leq T$ as

$$C_t = E_t^Q\left[\frac{B(t)}{B(T)}\bar{A}_{m,n}(T)(\bar{s}_{m,n}(T) - \bar{s}^*)^+\right]$$

$$= \bar{A}_{m,n}(t)E_t^Q\left[\left(\frac{\bar{A}_{m,n}(T)}{\bar{A}_{m,n}(t)}\bigg/\frac{B(T)}{B(t)}\right)1_{\bar{A}_{m,n}(t)>0}(\bar{s}_{m,n}(T) - \bar{s}^*)^+\right] \quad (13.23)$$

$$= \bar{A}_{m,n}(t)E_t^{\bar{Q}_S}\left[(\bar{s}_{m,n}(T) - \bar{s}^*)^+\right]$$

A few comments must be noted here. $\bar{A}_{m,n}(t)$ is replicable until $t = \tau \wedge T_m$. The implication is that the price given by (13.23) is an arbitrage price. The default swap measure defined in (13.22) generalizes the definition of Schönbucher (2004), of which the annuity numeraire is non-tradable unless the recovery rate is zero. Second, in case of a default at $\tau < t$, $\bar{A}_{m,n}(t) = 0$, which implies that $\bar{Q}_S$ is absolutely continuous with respect to, but not equivalent to, Q. For swaption pricing, the lack of equivalence here is harmless. Note that Brigo (2005) provides an alternative derivation of the Black's formula that preserves the measure equivalence, based on a theory of restricted filtration (Jeanblanc and Rutkowski, 2000; Bielecki and Rutkowski, 2001). According to the interpretation of Schönbucher's survival measure by Bielecki and Rutkowski (2001) (section 15.2.2), these two approaches should be equivalent.

Based on the results of Proposition 7.1, we assume $\bar{s}_{m,n}(t)$ is a lognormal martingale under $\bar{Q}_S$:

$$d\bar{s}_{m,n}(t) = \bar{s}_{m,n}(t)\bar{\gamma}_{m,n} \, dW_t^S, \quad (13.24)$$

where W_t^S is a one-dimensional Brownian motion under $\bar{Q}_S$ and $\bar{\gamma}_{m,n}$ is the swap-rate volatility. The lognormality assumption, (13.26), leads readily to Black's formula for credit swaptions:

$$C = \bar{A}_{m,n}(t) \left[\bar{s}_{m,n}(t) N(d_1) - \bar{s}^* N(d_2) \right], \tag{13.25}$$

with

$$d_{1,2} = \frac{\ln(\bar{s}_{m,n}(t)/\bar{s}^*) \pm \frac{1}{2} \bar{\gamma}_{m,n}^2 (T - t)}{\bar{\gamma}_{m,n} \sqrt{T - t}}.$$

A hedging strategy follows from Black's formula (13.25): at time $t < T$, the hedger

- maintains $N(d_1)$ units of the credit default swap; and
- maintains $\bar{s}_{m,n}(t) N(d_1) - \bar{s}^* N(d_2)$ units of the annuity $\bar{A}_{m,n}(t)$.

Next we show that the above swaption pricing approach can be justified under the assumptions of lognormal dynamics for either the forward hazard rates or the mean loss rates, on top of the standard LIBOR market model for $f_j(t)$ (Brace, Gatarek and Musiela, 1997; Jamshidian, 1997; and Miltersen, Sandmann and Sondermann, 1997), with volatility γ_j, such that

$$\frac{df_j(t)}{f_j(t)} = \gamma_j(t) \cdot (d\mathbf{W}_t - \Sigma(t, T_{j+1}) dt), \tag{13.26}$$

where

$$\Sigma(t, T_{j+1}) = -\sum_{k=1}^{j} \frac{\Delta T_k f_k(t)}{1 + \Delta T_k f_k(t)} \gamma_k(t).$$

We begin with

Proposition 13.6.1. *Assume that the pre-default dynamic of the discrete hazard rates is lognormal:*

$$\frac{d\lambda_j(t)}{\lambda_j(t)} = \mu_j^\lambda(t) dt + \gamma_j^\lambda(t) \cdot d\mathbf{W}_t, \tag{13.27}$$

where $\mathbf{W}_t$ is a multi-dimensional Brownian motion under $\bar{\mathbb{Q}}_S$, then

$$\mu_j^\lambda(t) = -\gamma_j^\lambda(t) \cdot \sum_{k=\eta_t}^{j} \frac{\Delta T_k \lambda_k}{1 + \Delta T_k \lambda_k} \gamma_k^\lambda \tag{13.28}$$

□

Proof: According to the definition of discrete hazard rates, we have

$$\lambda_j(t) = \frac{1}{\Delta T_j} \left(\frac{\Lambda_j}{\Lambda_{j+1}} - 1 \right). \tag{13.29}$$

Assume the survival probabilities are lognormal as well:

$$d\Lambda_j(t) = \Lambda_j(t) \left\{ \Sigma_j^\Lambda \cdot d\mathbf{W}_t + \tilde{\lambda}_t dt \right\}.$$

where $\tilde{\lambda}_t$ is the instantaneous rate for default. It then follows that

$$
\begin{aligned}
d\lambda_j(t) &= \frac{1}{\Delta T_j} d\left(\frac{\Lambda_j}{\Lambda_{j+1}}\right) \\
&= \frac{1}{\Delta T_j}\left(\frac{\Lambda_j}{\Lambda_{j+1}}\right)\left[\Sigma_j^\Lambda - \Sigma_{j+1}^\Lambda\right]\cdot\left[d\mathbf{W}_t - \Sigma_{j+1}^\Lambda dt\right]
\end{aligned}
\tag{13.30}
$$

where σ_j^Λ is the percentage volatility of $\Lambda_j(t)$. Now rewrite the last equation into

$$
\frac{d\lambda_j(t)}{\lambda(t)} = \frac{1 + \Delta T_j\lambda_j(t)}{\Delta T_j\lambda_j(t)}\left[\Sigma_j^\Lambda - \Sigma_{j+1}^\Lambda\right]\cdot\left[d\mathbf{W}_t - \Sigma_{j+1}^\Lambda dt\right]
\tag{13.31}
$$

Now set

$$
\gamma_j^\lambda = \frac{1 + \Delta T_j\lambda_J(t)}{\Delta T_j\lambda_j(t)}\left[\Sigma_j^\Lambda - \Sigma_{j+1}^\Lambda\right].
\tag{13.32}
$$

Then there is

$$
\begin{aligned}
\Sigma_{j+1}^\Lambda &= \Sigma_j^\Lambda - \frac{\Delta T_j\lambda_j(t)}{\Delta 1 + T_j\lambda_j(t)}\gamma_j^\lambda \\
&\qquad\cdots \\
&= \Sigma_1^\Lambda - \sum_{k=1}^{j}\frac{\Delta T_k\lambda_k(t)}{\Delta 1 + T_k\lambda_j(t)}\gamma_k^\lambda
\end{aligned}
\tag{13.33}
$$

The drift term μ_j^λ is thus

$$
\mu_j^\lambda = -\gamma_j^\lambda\cdot\Sigma_{j+1}^\Lambda.
\tag{13.34}
$$

When there is no risk of imminent default, we can set $\Sigma_1^\Lambda = 0$. $\quad\square$

With $\{\lambda_j(t)\}$, we can construct $\{\Lambda_j(t)\}$ and $\{H_j(t)\}$. In fact, we have

$$
H_j(t) = \frac{(1 - R_j)\lambda_j}{1 + R_j\Delta T_j\lambda_j}.
\tag{13.35}
$$

Next, we proceed to the pricing of swaptions under the extended market model. We take the first kind of swaps as an example. The swap rate is given by (13.11). By Ito's lemma, we can derive an approximate swap-rate process as follows:

$$
\begin{aligned}
d\bar{s}_{m,n}(t) &= \sum_{j=m}^{n-1}\frac{\partial\bar{s}_{m,n}(t)}{\partial H_j}H_j(t)\gamma_j^H\cdot d\mathbf{W}_t^S \\
&= \bar{s}_{m,n}(t)\sum_{j=m}^{n-1}\frac{\partial\bar{s}_{m,n}(t)}{\partial H_j}\frac{H_j(t)}{\bar{s}_{m,n}(t)}\gamma_j^H\cdot d\mathbf{W}_t^S \\
&\approx \bar{s}_{m,n}(t)\sum_{j=m}^{n-1}\frac{\partial\bar{s}_{m,n}(0)}{\partial H_j}\frac{H_j(0)}{\bar{s}_{m,n}(0)}\gamma_j^H\cdot d\mathbf{W}_t^S \\
&= \bar{s}_{m,n}(t)\bar{\gamma}_{m,n}\cdot d\mathbf{W}_t^S,
\end{aligned}
\tag{13.36}
$$

where

$$\gamma_j^H = \frac{\gamma_j^\lambda}{1 + R_j \Delta T_j \lambda_j},$$

and

$$\bar{\gamma}_{m,n} \triangleq \sum_{j=m}^{n-1} \bar{\omega}_j \gamma_j^H, \quad \bar{\omega}_j \triangleq \frac{\partial \bar{s}_{m,n}(0)}{\partial H_j} \frac{H_j(0)}{\bar{s}_{m,n}(0)} \approx \bar{\alpha}_j \frac{H_j(0)}{\bar{s}_{m,n}(0)}, \tag{13.37}$$

and $\mathbf{W}_t^S$ is a multi-dimensional Brownian motion under $\bar{\mathbb{Q}}_S$, defined by

$$d\mathbf{W}_t^S = d\mathbf{W}_t - \sum_{j=m}^{n-1} \bar{\alpha}_j \bar{\Sigma}_{j+1}(t) dt,$$

and $\bar{\Sigma}_{j+1}$ is the volatility of $\bar{P}_{j+1}(t)$. The lognormal process for swap rates, (13.36), justifies the use of Black's formula, (13.25), for swaptions. Note that in the context of LIBOR market model, the approximations made in (13.36) and (13.37) are known to be accurate enough for application and have been justified with rigor (Brigo et al., 2004). The expressions in (13.36) describe the relation between forward spread volatilities and swap-rate volatilities, which can be used in practice to gauge the relative price richness/cheapness of a swaption. The model for spreads (13.27) can be calibrated to the implied volatilities of the default swaptions using the quadratic programming technology developed by Wu (2003) for market model calibrations.

Models more comprehensive than (13.27) can be developed by including other risk dynamics, like jumps, stochastic volatilities, and even correlations among multiple credit names (Eberlein et al., 2005). Such developments are largely parallel to existing extensions to the standard market model. Brigo (2005) and especially Schönbucher (2004) have made several extensions using swap rates as state variables.

13.7 Pricing of CDO Tranches under the Market Model

In this section, we explain how to apply the market model for pricing collateralized debt obligations (CDOs). A CDO is a way to restructure the cash flows of a portfolio of bonds (with various credit ratings). Tranches of ascending seniority are defined in terms of the percentages of the notional principal value of the portfolio, and losses are always allocated to the most junior tranche that is still alive. The spread of each tranche is just the premium of protection payments which, unlike the protection payments in a single-name CDS, are made periodically until all notional value is lost. Tranches are divided by *attachment points*. Take CDX IG for example, a standardized CDO of with the attachment points 0%, 3%, 7%, 10%, 15%, 30% and, of course, 100%. The equity tranche, which is the most junior tranche with attachment points 0% and 3%, will absorb the losses to the portfolio up to the first 3% and then cease to exist. Subsequent losses will then be borne by the next tranche with attachment points of 3% and 7%, which is called the mezzanine tranche. Losses to other senior tranches are determined similarly. For each tranche, the premium of protection is calculated based on the outstanding notional principal value. To express the remaining outstanding notional principals of the tranches, we introduce the following notations:

$\left[P_D^i, P_U^i\right]$	—	the attachment points for the ith tranche, in percentage;
$D^{(k)}(T_{j+1})$	—	the forward price of the outstanding notional value of the kth name at T_{j+1}, equal to $\mathbf{1}_{\{\tau>T_{j+1}\}} + \bar{R}\mathbf{1}_{\{T_j<\tau\leq T_{j+1}\}};$

Then, the total outstanding notional at T_{j+1} for the portfolio in percentage is

$$D^P(T_{j+1}) = \frac{1}{K}\sum_{k=1}^{K} D^{(k)}(T_{j+1}), \qquad (13.38)$$

where K be the number of credit names in the portfolio. The outstanding notional at T_{j+1} for the ith tranche can be expressed as

$$O_i(T_{j+1}) = \frac{1}{P_U^i - P_D^i}\left\{(D^P(T_{j+1}) - 1 + P_U^i)^+ \right.$$
$$\left. -(D^P(T_{j+1}) - 1 + P_D^i)^+\right\}. \qquad (13.39)$$

The loss to the ith tranche over (T_j, T_{j+1}) is thus

$$O_i(T_j) - O_i(T_{j+1}).$$

The expression (13.39) reiterates the fact that the outstanding notional for any tranche at any cash flow day can be regarded as a spread option on the total outstanding notional value of the portfolio backing the CDO.

Next, we consider the pricing of the premium rate on a tranche. Let s_i be the premium rate on the ith tranche; the value of the fee leg is then

$$PV_{fee} = s_i \sum_{j=m}^{n-1} \Delta T_j P_{j+1}(t) E_t^{Q_{j+1}}\left[O_i(T_{j+1})\right].$$

While the value of the protection leg is

$$PV_{prot} = \sum_{j=m}^{n-1} P_{j+1}(t) E_t^{Q_{j+1}}\left[O_i(T_j) - O_i(T_{j+1})\right].$$

The formula for the premium rate on the ith tranche is then

$$s_i = \frac{\sum_{j=m}^{n-1} P_{j+1}(t) E_t^{Q_{j+1}} \left[O_i(T_j) - O_i(T_{j+1}) \right]}{\sum_{j=m}^{n-1} \Delta T_j P_{j+1}(t) E_t^{Q_{j+1}} \left[O_i(T_{j+1}) \right]}.$$

In view of (13.39), we understand that the key to CDS rate calculation lies in the valuation of a sequence of call options of the form

$$E_t^{Q_{j+1}} \left[\left(D^P(T_{j+1}) - X \right)^+ \right], \quad j = m, \ldots, n-1. \tag{13.40}$$

We now consider the valuation of the above options by the Monte Carlo simulation method. In view of the definition of $D^P(T_{j+1})$ in (13.38), we need to simulate $D^{(k)}(T_{j+1})$, $k = 1, \ldots, K$. For simplicity, we assume constant recovery rate, $E^Q \left[R_\tau | T_{j-1} < \tau \leq T_j \right] = \bar{R} = constant$. We then can express $D^{(k)}(T_{j+1})$ as

$$D^{(k)}(T_{j+1}) = \mathbf{1}_{\{\tau > T_j\}} \left(\mathbf{1}_{\{u > \lambda_j^{(k)}(T_j) \Delta T_j\}} + \bar{R} \mathbf{1}_{\{u \leq \lambda_j^{(k)}(T_j) \Delta T_j\}} \right), \tag{13.41}$$

where u obeys the uniform distribution in $(0, 1)$, denoted by $U(0, 1; \Sigma_u)$, where Σ_u stands for the correlation matrix of u's. If taking a Gaussian copula for u's, we can proceed as follows.

1. Perform a Choleski decomposition of the input correlation matrix

$$\Sigma_g = AA^T. \tag{13.42}$$

2. Simulate $K(T_j)$ (which is the number of surviving firms at time T_j) independent standard normal random variables, $\{\tilde{\epsilon}_i\}_{i=1}^{K(T_j)}$.

3. Transform $\{\tilde{\epsilon}_i\}_{i=1}^{K(T_j)}$ to correlated normal random variables

$$\begin{pmatrix} \epsilon_1 \\ \vdots \\ \epsilon_{K(T_j)} \end{pmatrix} = A \begin{pmatrix} \tilde{\epsilon}_1 \\ \vdots \\ \tilde{\epsilon}_{K(T_j)} \end{pmatrix}.$$

4. Transform the normal random variables to uniform random variables

$$(u_1, \ldots, u_{K(T_j)}) = (N(\epsilon_1), \ldots, N(\epsilon_{K(T_j)})).$$

Furthermore, if we abide by the industrial convention to assume a uniform pairwise correlation such that $corr(\epsilon_k, \epsilon_j) = \rho > 0$ for any k and l, then steps (1) to (3) above are simplified into the calculations of

$$\epsilon_k = \sqrt{1 - \rho}\, \tilde{\epsilon}_k + \sqrt{\rho} \tilde{\epsilon}_c, \tag{13.43}$$

where $\tilde{\epsilon}_c$ and $\tilde{\epsilon}_k, k = 1, \ldots, K(T_j)$ are independent standard normal random variables.

The evolution of $\lambda_j^{(k)}(t)$ follows from the scheme of

$$\lambda_j^{(k)}(T_j) = \lambda^{(k)}(T_{j-1}) \exp\left((\mu_j^\lambda(T_{j-1}) - \frac{1}{2}\|\gamma_j^\lambda\|^2)\Delta T_j + \gamma_j^\lambda \Delta W^{(k)}\right). \quad (13.44)$$

Here, we have locally frozen $\mu_j^\lambda(T_{j-1})$, defined in (13.28). In the evolution of $\lambda_j^{(k)}(t)$, we can incorporate the correlations of the credit spreads observed in the single-name CDS market.

We are now ready to describe the algorithm for CDO pricing. The simulation of correlated defaults is the focus of the algorithm. Note that we need to input two set of correlations. The first set is for the correlations, Σ_H, of CDS rates, which are the state variables for the market model. This set of correlations can be observed from the market. The second set is for the correlations of default times, Σ_u, which is not quite observable and is dealt with using the technique of Gaussian copula. Let T the maturity of the CDO, ΔT the time interval for premium payments, $J = T/\Delta T$ be the maximal number of the premium payments, and M be the number of Monte Carlo simulation paths. We develop the following algorithm for pricing the options in (13.40).

```
/* Algorithm for pricing options on D_j^P(T_j), j = 1,...,J */

    For j = 1 : J
        V_j = 0
    end
    For m = 1 : M
        For k = 1 : K
            D^(k)(T_0) = 1
        end
        K(T_0) = K
        For j = 1 : J
            Generate {ΔW^(k)}_{k=1}^{K(T_{j-1})} ~ N(0, ΔTΣ_H)
            Generate {u_k}_{k=1}^{K(T_{j-1})} ~ U(0,1; Σ_u)
            Put D^P(T_j) = 0
            l = 0
            For k = 1 : K repeat
                If D^(k)(T_{j-1}) = 1, then
                    D^(k)(T_j) = 1
/* Simulate default over (T_{j-1}, T_j) for the k^{th} name */
                    l = l + 1
                    If u_l ≤ λ^(k)(T_{j-1})ΔT_j
                        D^(k)(T_j) = R̄
                        K(T_j) = K(T_j) - 1
                    end if
```

$$DP(T_j) \leftarrow DP(T_j) + D^{(k)}(T_j)$$

```
/* Simulate the hazard rate λ^(k)(T_j) according to the market
      model */
```

$$\lambda_j^{(k)}(T_j) = \lambda^{(k)}(T_{j-1}) \exp\left((\mu_j^\lambda(T_{j-1}) - \tfrac{1}{2}\|\gamma_j^\lambda\|^2)\Delta T_j + \gamma_j^\lambda \Delta W^{(k)}\right)$$

```
            end if
         end if
/* Calculate the payoff of the option */
         V_j ← V_j + (DP(T_j)/K - X)^+
      end
   end
/* Average payoff */
   For j = 1 : J
      V_j ← V_j/M
   end

/* The end of the algorithm */
```

One can see that the entire algorithm is rather easy to implement, and the computation time is about J times more than that of the Gaussian copula method of Li (2000).

Next, let us examine the ability of the model to back out correlations implied by various tranches of two standardized CDOs, namely, CDX IG and iTraxx IG. The quotes on August 24, 2004, are listed in Table 13.6.[5] The LIBOR and swap rates for the same day are listed in Table 13.7.

Table 13.6: Quotes on 8/24/2004

| | CDX IG Tranches | | | | | |
	0% to 3%	3% to 7%	7% to 10%	10% to 15%	15% to 30%	Index
5-year quotes	40.02	295.71	120.50	43.00	12.43	59.73
10-year quotes	58.17	632.00	301.00	154.00	49.50	81.00

| | iTraxx IG Tranches | | | | | |
	0% to 3%	3% to 6%	6% to 9%	9% to 12%	12% to 22%	Index
5-year quotes	24.10	127.50	54.00	32.50	18.00	37.79
10-year quotes	43.80	350.17	167.17	97.67	54.33	51.25

---

[5]The data are taken from Hull et al. (2005).

Table 13.7: USD Yield Data (8/24/2004, Bloomberg)

LIBOR	3 mth	1.760%
	6 mth	1.980%
	12 mth	2.311%
Swap	2Y	2.840%
	3Y	3.265%
	4Y	3.592%
	5Y	3.890%
	6Y	4.100%
	7Y	4.295%
	8Y	4.455%
	9Y	4.595%
	10Y	4.710%

Without loss of generality, we make a few reasonable simplifications in the handling of data. We assume that the curve of forward spreads is flat and equal to the index of the respective maturity, which implies that the CDS rate of any maturity is equal to the value of forward spreads. The CDS rate volatility is set at the constant level of either 50% or 100%, which represents the usual range of implied swaption volatilities (see Schöbucher (2004) and Brigo (2005)). The recovery rate is taken to be $\bar{R} = 40\%$, abiding to the industrial convention. Due to the lack of correlation data for CDS rates, we let the CDS rates and the Gaussian copula for default times share the same pairwise correlation. We take the number of paths to be $M = 10,000$, and the size of time stepping to be $\Delta t = 0.25$. The implied correlations of various tranches,[6] obtained through trial and error, are listed in Figures 13.7 and 13.8. The average accumulated default numbers for both CDOs, for a pairwise correlation of 20%, are shown in Figures 13.9 and 13.10.

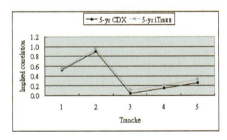
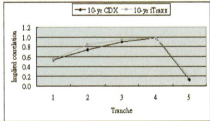

FIGURE 13.7: Implied correlations for $\|\gamma^\lambda\| = 0.5$.

Let us comment on the results. Under the market model, the implied correlation curves do not quite look like a smile. We have let $\|\gamma^H\|$, the CDS rates

---

[6]Note that the implied correlation for a tranche may not necessarily be unique.

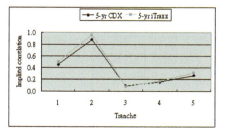
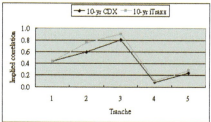

FIGURE 13.8: Implied correlations for $\|\gamma^\lambda\| = 1$.

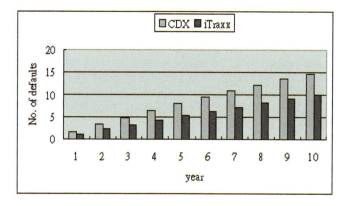

FIGURE 13.9: Accumulated no. of defaults for $\|\gamma^\lambda\| = 0.5$.

volatility, vary from 50% to 100%, and have witnessed a gradual deformation of the "smile curve." Remarkably, the "smile curves" of CDX IG and iTraxx IG stay close to each other, which is interesting but an explanation is not yet available. The histograms of average accumulated number of defaults look reasonable. The bigger numbers of defaults for CDX IG are consistent with the bigger numbers of spreads across all tranches. The algorithm is not optimized, but it has been very robust. The pricing of an entire CDO takes about 20 seconds in a PC with Intel Pentium 4 CPU (3.06GHz, 504MB RAM).

We tend to believe that the higher implied correlation for the mezzanine tranche is caused by the assumption of flat CDS rates across all maturities, which is against the fact that a CDS rate increases with maturity, as is seen in Table 13.1 for the CDS rates of Citigroup. If we bootstrap the forward spreads against the CDS rates, we should see that the forward spread curve has a bigger slope of increase than that of the CDS rate curve. The assumption of flat forward spread curve should have produced more defaults in the short horizon yet fewer defaults for the long horizon, which in turn can cause higher prices for junior tranches yet lower prices for senior tranches, as reflected by the level of implied correlations.

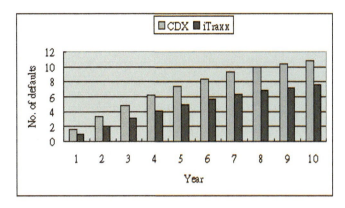

FIGURE 13.10: Accumulated no. of defaults for $\|\gamma^\lambda\| = 1$.

We want to point out here that we have tried to reproduce the CDO spreads without default correlation, by letting the random variables $\{u_k\}$ be independent. The results are no good, which thus supports a widely held opinion among market participants that the spread correlations are insufficient for pricing CDOs, and thus we also need default correlations.

13.8 Notes

Our objectives of this chapter are two fold. First, we want to achieve replication pricing of CDS and CDS options. Second, to put the pricing of credit instruments, including corporate bonds, CDS, CDS option and CDO under the same model with some properly chosen state variables. Both objectives are achieved, and the state variables eventually turn out to be the discrete hazard rates. Recently, Bloomberg LLP has started to publish the recovery rates together with their corresponding CDS rates. The availability of the recovery rates will make our market model with the discrete hazard rate easier to use, because extracting the implied recovery rates are often ill-posed problems.

Chapter 14

Dual-Curve SABR-LMM Market Model for Post-Crisis Interest Rate Derivatives Markets

In the absence of default risk, the forward rates of different tenors are all generated from a single discount curve, such that the forward-rate curves of different tenors are equivalent in the sense that from one we can infer the others. Before the 2007-2008 financial crisis, forward rates were considered free of default risk, and the modeling of interest-rate derivatives was carried out with a single LIBOR curve, typically the three-month curve. However, the situation changed during the financial crisis, when the default risks in LIBOR became significant and the classic relationship between the discount curve and the forward-rate curves broke down. To price derivatives based on the LIBOR of different tenors, the community of quantitative finance began to model the forward-rate curves of different tenors simultaneously, thereby creating the so-called multi-curve models.

It may be fair to say that multi-curve models developed thus far are largely adaptations or extensions of existing single-curve models to the multiple-curve setting. Thus, these extensions can be referred to using the model categories that existed before the financial crisis, namely, the short-rate models, instantaneous forward rate or HJM models, LIBOR market models, affine models, and potential models. Kijima, Tanaka, and Wong (2009) use the short rate approach to model discounting rates, LIBOR, and Treasury yields separately. Kenyon (2010) uses a Hull-White model for the short rate processes to drive the overnight-index-swap (OIS) and the LIBOR curves. Morini and Runggaldier (2014) model the risk-free short rate and the spreads between the term structures of LIBOR and the OIS rates using the Vasicek and Cox-Ingersoll-Ross processes. Henrard (2007, 2010) assumes that the spreads between the curves of different tenors are deterministic, and derives Gaussian Heath-Jarrow-Morton (HJM) formulas. Mercurio (2010a, 2010b) and Fujii, Shimada and Takahashi (2010, 2011) adapt the LIBOR market model such that the basis spreads for different tenors are modeled as different processes. Bianchetti (2010) makes use of the currency analogy and derives quanto convexity corrections to the valuation formulas under the standard LIBOR market model. A hybrid HJM-LIBOR market model is proposed in Moreni and Pallavicini (2010), where the HJM framework is used to obtain a model for multiple curves using a single family of Markov driving processes. Smile models have also been extended to the context of multiple curves. For LIBOR of different tenors, Bianchetti and Carlicch (2011) propose SABR type models, and Hoskinson (2014) develops the Heston type models.

The shift from single-curve modeling to multi-curve modeling was once hailed by some as a "model evolution" or "paradigm shift." However, with only a few exceptions, most multi-curve models are unable to accommodate either the basis swap curves or their dynamics; as the basis-swap curves appear smooth and monotonically decreasing in terms, a pattern cannot be retained under any multi-curve models that are developed through natural extensions of existing single-curve models. Luckily, this shortcoming in modeling can be overcome by potential methods and, in particular, affine models. Grbac et al. (2015) introduce a multiple-curve affine model that allows users to create an order among forward rates or swap rates of different tenors, so that basis spreads can be accommodated, in addition to the tractable dynamics and semi-analytic pricing formulas that the affine models already possess. Nguyen and Seifried (2015) develop a general class of multi-curve potential models that features positive stochastic basis spreads, positive term structures, and analytic pricing formulae for interest rate derivatives. The standard market practice, meanwhile, is to adopt the lognormal or the SABR-type of model for forward rates. Under these models, the pattern of basis curves can only be retained by taking perfect correlation among forward rates of different tenors, yet in doing so the basis curves will become frozen in time, which is not realistic.

The research behind this chapter is motivated by Bianchetti and Carlicchi's (2011) insights that basis spreads reflect the differences of the default-risk premiums in LIBOR rates of different tenors. Due to the periodic LIBOR panel review that intends to uphold the credit quality of penal banks, lending for a shorter term is subject to less counterparty risk than lending for a longer term. For example, lending monthly for three consecutive months is subject to less default risk than lending once for a three-month term. Preliminary results of our own studies suggest that the risk premiums in LIBOR are essentially proportional to the tenors or the duration of lending, which is identified as the cause of basis spreads. This finding allows us to rigorously characterize the spreads between LIBOR and OIS forward rates of different tenors, which we call the discrete loss rates. The discrete loss rates account for the premiums of the liquidity risk and the counterparty default risk, and the latter increases linearly with respect to the duration of lending. After confirming the role of the liquidity and counterparty default risk premiums in the generation of the basis spread curves, we then focus on the OIS forward rates and the discrete loss rates for modeling and pricing purposes.

In this chapter, we will take a somewhat conventional strategy to model LIBOR rates that will allow us to (1) accommodate the basis swap curves and (2) price the LIBOR derivatives across tenors. The strategy is to take the three-month tenor and superimpose a model of the term structure of the discrete loss rates, which account for the counterparty default[1] and liquidity risk premiums, on top of some popular models for the OIS forward rates. To price derivatives and/or to model the implied volatility smiles, we will adopt, in particular, the SABR dynamics (Hagan et al., 2002) for the expected discrete loss rates, and thus develop the so-called dual-curve version of the SABR-LMM model. Then, we will develop approximate pricing formulae for caplets/floorlets and swaptions under the dual-curve SABR-LMM model. Numerical pricing examples will be provided to demonstrate the accuracy of the approximate closed-form formulae.

Later, we show that the term structure of LIBOR of other tenors and their dynamics can be derived from those of the three-month LIBOR and the default risk premiums, with some additional handling of the volatility functions in order to retain analytical tractability of the developed model. Hence, we can also price derivatives on LIBOR of other tenors. It is a bit surprising that the pricing accuracy remain the same.

The chapter is organized as follows. In Section 14.1, we will first define the discrete loss rate, and then define LIBOR rates in terms of the OIS forward rates and the discrete loss rates, as well as the extension of the standard LIBOR market model. In Section 14.2, we introduce basis swaps. In Section 14.3, we introduce the dual-curve SABR-LMM model and develop the closed-form pricing of caplets under the approach of heat kernel expansion. In Section 14.4, we demonstrate the pricing accuracy for 3-month swaptions. In Section 14.5, we will price caplets and swaptions of other tenors, using the 6-month caplets and swaptions as examples. In Section 14.6, we will conclude with notes.

14.1 LIBOR Market Model under Default Risks

In the presence of counterparty credit (or default) risk, LIBOR needs to be redefined in terms of the prices of zero-coupon bonds as well as the survival probability of the counterparty. Subsequent modeling should be based on a proper definition of the LIBOR. For this purpose, we introduce the following notations:

---

[1]Subject to LIBOR panel review.

r_t	—	the risk-free spot rate at time t;
B_t	—	$= \exp(\int_0^t r_s ds)$, the time-t value of the default-free money market account;
$\mathbb{Q}$	—	the risk-neutral measure;
$P(t,T)$	—	$= B_t E_t^Q[B_T^{-1}]$, the risk-free discount factor for term $T - t$;
$\Delta T^{(i)}$	—	$= i/12$, the tenor in the fraction of a year for the floating leg of swaps, where i can take $1, 3, 6$ or 12;
$T_j^{(i)}$	—	$= j\Delta T^{(i)}$, the jth payment date for the floating leg with tenor $\Delta T^{(i)}$;
$\mathbb{Q}_j^{(i)}$	—	the $T_j^{(i)}$-forward measure;
$\Lambda(t,T)$	—	survival probability from t to T for a generic bank in the LIBOR panel, conditional on monthly review of the panel, with $\Lambda(t,t) = 1$;
τ	—	the time of a default by a generic bank in the LIBOR panel;
L	—	the loss rate of the principal value of risky zero-coupon bonds due to the issuer's default;
θ_m	—	the (slope of the) risk premium of counterparty default for lending beyond the mth month;
$f_j^{(i)}(t)$	—	the risk-free OIS forward rate for the period (T_j, T_{j+1}) with tenor $\Delta T_j^{(i)} = i/12$,
$D_{j+1}^{(i)}(t)$	—	the $T_{j+1}^{(i)}$-forward price of defautable zero-coupon bond issued by a generic panel bank;
$\hat{f}_j^{(i)}(t)$	—	the forward LIBOR rate for the period (T_j, T_{j+1}) with tenor $\Delta T_j^{(i)} = i/12$,
$H_j^{(i)}(t)$	—	the discrete loss rate for the period (T_j, T_{j+1}) with tenor $\Delta T_j^{(i)} = i/12$.

The slope of the risk premium for lending works as follows: if a lending is made by a panel bank for the period $\left(T_j^{(i)}, T_{j+1}^{(i)}\right]$ and thus the bank will not benefit from penal review over the period, then the default probability it faces over the period is defined to be

$$E_t^{Q_{j+1}^{(i)}}\left[1_{\{T_j^{(i)} < \tau \le T_{j+1}^{(i)}\}}\right] = 1 - \frac{\Lambda(t, T_{j+1}^{(i)})}{\Lambda(t, T_j^{(i)})} \frac{1}{\prod_{k=0}^{i-1}\left(1 + \frac{1}{12}k \times \theta_{i \times j + k}\right)}. \quad (14.1)$$

In the equation above, the lending bank adjusts the survival probability subject to monthly panel review, $\Lambda(t, T_{j+1}^{(i)})/\Lambda(t, T_j^{(i)})$, downward by a factor of $\prod_{k=0}^{i-1}(1 + \frac{1}{12}k \times \theta_{i \times j + k})^{-1}$, which accounts for the default risk of the borrowing bank. Note that the risk premium increases linearly over time or, in other words, is proportional to the duration of lending. Such a treatment of risk premiums is common in finance, see e.g., Tuckman (2002). Note that formula (14.1) is not affected by default(s) of panel banks prior to $T_j^{(i)}$. When such defaults happen, the defaulted banks will be removed from the panel and be replaced by other banks.

We define the forward LIBOR rate, $\hat{f}_j^{(i)}(t)$, through a forward loan. The forward loan is subject to default risk for the future period $(T_j^{(i)}, T_{j+1}^{(i)})$, which is defined by setting equal the present values (PVs) of lending and repaying legs:

$$
\begin{aligned}
0 &= E_t^Q \left[\frac{B_t}{B_{T_j^{(i)}}} \right] - E_t^Q \left[\frac{B_t}{B_{T_{j+1}^{(i)}}} \left(1 + \Delta T_j^{(i)} \hat{f}_j^{(i)}(t)\right) \left(1 - L1_{\{T_j^{(i)} < \tau \le T_{j+1}^{(i)}\}}\right) \right] \\
&= P(t, T_j^{(i)}) - \left(1 + \Delta T_j^{(i)} \hat{f}_j^{(i)}(t)\right) P(t, T_{j+1}^{(i)}) E_t^{Q_{j+1}^{(i)}} \left[1 - L1_{\{T_j^{(i)} < \tau \le T_{j+1}^{(i)}\}}\right] \\
&\stackrel{\triangle}{=} P(t, T_j^{(i)}) - \left(1 + \Delta T_j^{(i)} \hat{f}_j^{(i)}(t)\right) P(t, T_{j+1}^{(i)}) D_{j+1}^{(i)}(t).
\end{aligned}
$$

$$(14.2)$$

The last equation gives rise to the definition of the *forward LIBOR rate*:

$$
\hat{f}_j^{(i)}(t) = \frac{1}{\Delta T_j^{(i)}} \left(\frac{P(t, T_j^{(i)})}{P(t, T_{j+1}^{(i)}) D_{j+1}^{(i)}(t)} - 1 \right). \tag{14.3}
$$

In contrast to the risk-free OIS forward rates for the same future period,

$$
f_j^{(i)}(t) = \frac{1}{\Delta T_j^{(i)}} \left(\frac{P(t, T_j^{(i)})}{P(t, T_{j+1}^{(i)})} - 1 \right),
$$

we can tell that it is the risky $T_{j+1}^{(i)}$-maturity zero-coupon bond that makes a difference.

The forward LIBOR rate so-defined is also the fair rate for a forward rate agreement (FRA). Such a fair rate nullifies the value of the FRA. To emphasize, we state

Proposition 14.1.1. *The FRA rate for the term* $(T_j^{(i)}, T_{j+1}^{(i)})$ *is* $\hat{f}_j^{(i)}(t)$.

Proof: Let f_X be the FRA rate for $(T_j^{(i)}, T_{j+1}^{(i)}]$, then

$$
\begin{aligned}
0 &= E_t^Q \left[e^{-\int_t^{T_j^{(i)}} r_s ds} \frac{\Delta T_j^{(i)} (\hat{f}_j^{(i)}(T_j^{(i)}) - f_X)}{1 + \Delta T_j^{(i)} \hat{f}_j^{(i)}(T_j^{(i)})} \right] \\
&= E_t^Q \left[e^{-\int_t^{T_j^{(i)}} r_s ds} \left(1 - \frac{1 + \Delta T_j^{(i)} f_X}{1 + \Delta T_j^{(i)} \hat{f}_j^{(i)}(T_j^{(i)})}\right) \right] \\
&= E_t^Q \left[e^{-\int_t^{T_j^{(i)}} r_s ds} \left(1 - (1 + \Delta T_j^{(i)} f_X) P(T_j^{(i)}, T_{j+1}^{(i)}) D_{j+1}^{(i)}(T_j^{(i)})\right) \right] \\
&= P(t, T_j^{(i)}) - (1 + \Delta T_j^{(i)} f_X) P(t, T_{j+1}^{(i)}) D_{j+1}^{(i)}(t).
\end{aligned}
$$

$$(14.4)$$

Solve for f_X, we then obtain $f_X = \hat{f}_j^{(i)}(t)$.   □

Next, we look into the relationship between the (risky) forward LIBOR and the risk-free OIS forward rate. From their definitions, we have

$$
\begin{aligned}
\hat{f}_j^{(i)}(t) &= \frac{1}{\Delta T_j^{(i)}} \left[\frac{P(t, T_j^{(i)})}{P(t, T_{j+1}^{(i)})} - 1 + \frac{P(t, T_j^{(i)})}{P(t, T_{j+1}^{(i)})} \left(\frac{1}{D_{j+1}^{(i)}(t)} - 1 \right) \right] \\
&= \frac{1}{\Delta T_j^{(i)}} \left[\frac{P(t, T_j^{(i)})}{P(t, T_{j+1}^{(i)})} - 1 + \frac{P(t, T_j^{(i)})}{P(t, T_{j+1}^{(i)})} \left(\frac{1 - D_{j+1}^{(i)}(t)}{D_{j+1}^{(i)}(t)} \right) \right] \\
&= f_j^{(i)}(t) + \left(1 + \Delta T_j^{(i)} f_j^{(i)}(t) \right) \frac{E_t^{Q_{j+1}^{(i)}} \left[L 1_{\{T_j^{(i)} < \tau \leq T_{j+1}^{(i)}\}} \right]}{\Delta T_j^{(i)} D_{j+1}^{(i)}(t)}.
\end{aligned}
\tag{14.5}
$$

We now introduce the variable

$$
H_j^{(i)}(t) = \frac{1}{\Delta T_j^{(i)}} \left(\frac{1}{D_{j+1}^{(i)}(t)} - 1 \right) = \frac{E_t^{Q_{j+1}^{(i)}} \left[L 1_{\{T_j^{(i)} < \tau \leq T_{j+1}^{(i)}\}} \right]}{\Delta T_j^{(i)} D_{j+1}^{(i)}(t)},
\tag{14.6}
$$

then there is

$$
\hat{f}_j^{(i)}(t) = f_j^{(i)}(t) + \left(1 + \Delta T_j^{(i)} f_j^{(i)}(t) \right) H_j^{(i)}(t),
\tag{14.7}
$$

or

$$
1 + \Delta T_j^{(i)} \hat{f}_j^{(i)}(t) = \left(1 + \Delta T_j^{(i)} f_j^{(i)}(t) \right) \left(1 + \Delta T_j^{(i)} H_j^{(i)}(t) \right).
\tag{14.8}
$$

Apparently there is $H_j^{(i)}(t) \geq 0$ by definition, thus yielding the order

$$
\hat{f}_j^{(i)}(t) \geq f_j^{(i)}(t).
$$

Note that as $H_j^{(i)}(t)$ is the ratio of expected value of loss over the market value of the risky zero-coupon bond, we therefore call $H_j^{(i)}(t)$ the *discrete loss rate* over $(T_j^{(i)}, T_{j+1}^{(i)}]$.

Obviously, in the absence of counterparty default risk, $\hat{f}_j^{(i)}$ reduces to $f_j^{(i)}$, and the latter is known to be the fair FRA rate in a market free of credit risk.

Note that $\{\hat{f}_j^{(i)}(t), j = 1, 2, \ldots, 30/\Delta T^{(i)}\}$ is the forward LIBOR curve for the tenor $\Delta T^{(i)}$. In the absence of counterparty default risks, the term structures of forward rates of different tenors are linked by the same risk-free discount curve such that from the curve of one tenor, we can derive the curves of other tenors. As a result, it had previously sufficed to model the forward-rate term structure of a particular tenor, typically the three-month tenor due to liquidity consideration. However, such a link was lost during the 2008 financial crisis. One consequence is that the term structure of forward rates are now modeled individually in today's markets, tenor by tenor, which has been coined the term of multiple-curve models. By examining our definition of LIBOR (14.6–14.7), we realize that LIBOR curves of different tenors can be constructed using the risk-free discount curve and, in addition, the default risk premium of a generic panel bank applied in a linear fashion according to the duration of lending. In other words, with information of default risk premiums, we will be able to construct the term structure of LIBOR for any tenors from that of a particular tenor. The implication is that the simultaneous modeling of LIBOR curves for different tenors becomes unnecessary, and it is sufficient to model only one of the LIBOR curves.

We therefore propose a model based on the dynamics of risk-free OIS curve, $\{f_j^{(i)}(t)\}$, and the discrete loss rate curve, $\{H_j^{(i)}(t)\}$, of a particular tenor. Judging by liquidity, we will again take the three-month tenor, corresponding to $i = 3$. Once the dynamics for the two term structures are established, we can derive the dynamics of the LIBOR as well as the discrete loss rates of other tenors. Hence, the dual-curve LMM model can be used to price securities across tenors. The first equation of the dual-curve market model remains to be the standard LIBOR market model for the OIS forward rates:

$$df_j^{(i)}(t) = f_j^{(i)}(t)\gamma_j^{(i)}(t) \cdot \left(d\mathbf{W}_t - \Sigma_{j+1}^{(i)}(t)dt\right), \qquad (14.9)$$

where $\mathbf{W}_t$ is the n-dimensional Brownian motion under the risk-neutral measure $\mathbb{Q}$, $\gamma_j^{(i)}(t)$ is the volatility vector of the LIBOR rate, $\Sigma_{j+1}^{(i)}(t)$ is the volatility vector of $P(t, T_{j+1}^{(i)})$, which satisfies (Brace et al., 1997; Miltersen et al., 1997; Jamshidian, 1997)

$$\Sigma_{j+1}^{(i)}(t) = -\sum_{k=1}^{j} \frac{\Delta T_k f_k^{(i)}(t)}{1 + \Delta T_k^{(i)} f_k^{(i)}(t)} \gamma_j^{(i)}(t). \qquad (14.10)$$

The second equation for the LIBOR market model is a stochastic differential equation (SDE) that describes the dynamics of discrete loss rates, $H_j(t)$, which is given in the following proposition.

Proposition 14.1.2. *Assume that $H_j^{(i)}(t)$ is also a lognormal variable under the risk-neutral measure, then it satisfies*

$$\frac{dH_j^{(i)}(t)}{H_j^{(i)}(t)} = \gamma_j^{(i,H)}(t) \cdot \left[d\mathbf{W}_t - \left(\Sigma_{j+1}^{(i)} + \Sigma_{j+1}^{(i,D)}\right) dt\right], \qquad (14.11)$$

where $\gamma_j^{(i,H)}(t)$ is the volatility vector of the discrete loss rate, and

$$\Sigma_j^{(i,D)}(t) = -\frac{\Delta T_j^{(i)} H_j^{(i)}(t)}{1 + \Delta T_j^{(i)} H_j^{(i)}(t)} \gamma_j^{(i,H)}(t), \qquad (14.12)$$

where $\Sigma_j^{(i,D)}(t)$ is the percentage volatility of $D_{j+1}(t)$. $\qquad \square$

Proof: Since $D_{j+1}^{(i)}(t)$ is a positive martingale under the $\mathbb{Q}_{j+1}$ forward measure, in formalism, we can express its dynamics as

$$dD_{j+1}^{(i)}(t) = D_{j+1}^{(i)}(t)\Sigma_{j+1}^{(i,D)}(t) \cdot (d\mathbf{W}_t - \Sigma_{j+1}^{(i)}(t)dt),$$

where $\Sigma_{j+1}^{(i,D)}(t)$ is an $\mathcal{F}_t$-adapted and bounded random variable. For notational simplicity, we sometimes omit the argument t in $D_{j+1}^{(i)}(t)$, $\Sigma_{j+1}^{(i,D)}(t)$, as well as $\Sigma_{j+1}^{(i)}(t)$ during derivations. One can verify that $P(t, T_{j+1}^{(i)})D_{j+1}^{(i)}(t)/B_t$ is a martingale under the risk-neutral measure. To derive the dynamics of $H_j(t)$, we need the dynamics of $1/D_{j+1}^{(i)}(t)$:

$$d\left(\frac{1}{D_{j+1}^{(i)}(t)}\right) = -\frac{dD_{j+1}^{(i)}}{(D_{j+1}^{(i)})^2} + \frac{(dD_{j+1}^{(i)})^2}{(D_{j+1}^{(i)})^3}$$

$$= -\left(\frac{1}{D_{j+1}^{(i)}}\right)\Sigma_{j+1}^{(i,D)} \cdot \left[d\mathbf{W}_t - \left(\Sigma_{j+1}^{(i)} + \Sigma_{j+1}^{(i,D)}\right)dt\right]. \tag{14.13}$$

It then follows that

$$dH_j^{(i)}(t) = \frac{1}{\Delta T_j^{(i)}}d\left(\frac{1}{D_{j+1}^{(i)}(t)}\right)$$

$$= -\frac{1 + \Delta T_j^{(i)} H_j^{(i)}(t)}{\Delta T_j^{(i)}}\Sigma_{j+1}^{(i,D)} \cdot \left[d\mathbf{W}_t - \left(\Sigma_{j+1}^{(i)} + \Sigma_{j+1}^{(i,D)}\right)dt\right],$$

or

$$\frac{dH_j^{(i)}(t)}{H_j^{(i)}(t)} = -\frac{1 + \Delta T_j^{(i)} H_j^{(i)}(t)}{\Delta T_j^{(i)} H_j^{(i)}(t)}\Sigma_{j+1}^{(i,D)} \cdot \left[d\mathbf{W}_t - \left(\Sigma_{j+1}^{(i)} + \Sigma_{j+1}^{(i,D)}\right)dt\right],$$

Define

$$\gamma_j^{(i,H)}(t) = -\frac{1 + \Delta T_j^{(i)} H_j^{(i)}(t)}{\Delta T_j^{(i)} H_j^{(i)}(t)}\Sigma_{j+1}^{(i,D)}(t).$$

Suppose that $\gamma_j^{(i,H)}(t)$ is independent of $H_j^{(i)}(t)$, then $H_j^{(i)}(t)$ is a "lognormal" variable with dynamics

$$\frac{dH_j^{(i)}(t)}{H_j^{(i)}(t)} = \gamma_j^{(i,H)}(t) \cdot \left[d\mathbf{W}_t - \left(\Sigma_{j+1}^{(i)} + \Sigma_{j+1}^{(i,D)}\right)dt\right],$$

where

$$\Sigma_j^{(i,D)}(t) = -\frac{\Delta T_j^{(i)} H_j^{(i)}(t)}{1 + \Delta T_j^{(i)} H_j^{(i)}(t)}\gamma_j^{(i,H)}(t).$$

We can show that a unique solution exists globally. $\qquad\Box$

Together, equations (14.9–14.12) constitute the first version of the dual-curve LIBOR market model for the post-crisis markets. Yet, before adopting any models, we should take into consideration the observability of the state variables. As we shall demonstrate, the discrete loss rates (as well as the LIBOR rates) are not observable in the swap markets, and it will be a model-based exercise to extract them. As such, the discrete loss rate may not be a good candidate for modeling or pricing purposes, and will be later replaced by an alternative variable which can be extracted free of a model.

14.2 Swaps and Basis Swaps

We now consider the pricing of fully collateralized swaps. Let $(T_m, T_n) = (T_m^{(12)}, T_n^{(12)}) = (m, n)$ be the swap period in years, and $s_{m,n}^{(i)}$ be the swap rate with tenor $\Delta T_j^{(i)} = i/12$ year for the floating leg, for i taking 1, 3, 6 and 12. As usual, the swap rate is determined by equating the PVs of the cash flows of the floating leg to the fixed leg, and thus we have the following expression for the swap rate:

$$s_{m,n}^{(i)}(t) = \frac{\sum_{j=m/\Delta T^{(i)}}^{n/\Delta T^{(i)}-1} \Delta T_j^{(i)} P(t, T_{j+1}^{(i)}) E_t^{Q_{j+1}^{(i)}} \left[\hat{f}_j^{(i)}(T_j^{(i)}) \right]}{\sum_{k=m}^{n-1} \Delta T_k^{(12)} P(t, T_{k+1}^{(12)})}. \tag{14.14}$$

Because $H_j^{(i)}$ is not a martingale under $Q_{j+1}^{(i)}$, so is not $\hat{f}_j^{(i)}(t)$. Hence, in general, there is

$$E_t^{Q_{j+1}^{(i)}} \left[\hat{f}_j^{(i)}(T_j^{(i)}) \right] \neq \hat{f}_j^{(i)}(t).$$

Consequently, there is also $s_{m,n}^{(i)}(t) \neq s_{m,n}^{(j)}(t)$ for $i \neq j$. As we shall see next, the conversion between $E_t^{Q_{j+1}^{(i)}} \left[\hat{f}_j^{(i)}(T_j^{(i)}) \right]$ and $\hat{f}_j^{(i)}(t)$ is model dependent. For example, under the first dual-curve LIBOR market model, there is

$$
E_t^{Q_{j+1}^{(i)}}\left[\hat{f}_j^{(i)}(T_j^{(i)})\right] = \frac{1}{\Delta T^{(i)}} E_t^{Q_{j+1}^{(i)}}\left[\left(\frac{P(T_j^{(i)}, T_j^{(i)})}{P(T_j^{(i)}, T_{j+1}^{(i)}) D_{j+1}^{(i)}(T_j^{(i)})} - 1\right)\right]
$$

$$
= \frac{1}{\Delta T^{(i)}}\left(\frac{P(t, T_j^{(i)})}{P(t, T_{j+1}^{(i)})} E_t^{Q_{j+1}^{(i)}}\left[\frac{1}{D_{j+1}^{(i)}(T_j^{(i)})}\right] - 1\right)
$$

$$
= \frac{1}{\Delta T^{(i)}}\left(\frac{P(t, T_j^{(i)})}{P(t, T_{j+1}^{(i)})} \frac{1}{D_{j+1}^{(i)}(t)} E_t^{Q_{j+1}^{(i)}}\left[e^{\int_t^{T_j^{(i)}} \|\Sigma^{i,D}(s,T_{j+1}^{(i)})\|^2 ds}\right] - 1\right)
$$

$$
= \hat{f}_j^{(i)}(t) E_t^{Q_{j+1}^{(i)}}\left[e^{\int_t^{T_j^{(i)}} \|\Sigma^{i,D}(s,T_{j+1}^{(i)})\|^2 ds}\right] + \frac{E_t^{Q_{j+1}^{(i)}}\left[e^{\int_t^{T_j^{(i)}} \|\Sigma^{i,D}(s,T_{j+1}^{(i)})\|^2 ds}\right] - 1}{\Delta T^{(i)}}.
$$

$$(14.15)$$

Here, we have made use of the independence assumption between the market risk and the default risk, as well as also equation (14.13). Similar to the precrisis LIBOR market, we can obtain $E_t^{Q_{j+1}^{(i)}}\left[\hat{f}_j^{(i)}(T_j^{(i)})\right]$, but not $\hat{f}_j^{(i)}(t)$, by bootstrapping using swap rates. So we may consider the former as an indirectly observable variable that is model independent.

A basis swap is a swap to exchange the interest payments calculated according to LIBOR of two different tenors. In today's market,[2] there is $s_{m,n}^{(i)}(t) < s_{m,n}^{(j)}(t)$ for $i < j$. To prevent arbitrage, a spread equal to the difference of the swap rates must be added to the payments of LIBOR with a shorter tenor. When we plot the spreads against the terms of basis swaps, we obtain the basis swap (or spread) curves (or simply the basis curves).

We now highlight an important feature of our dual-curve setup: it can produce the basis spread curves. Put it differently, the risk premiums, $\{\theta_m\}$, can be calibrated to the basis spread curves of the market. In fact, we may take the exponential functional form for the (slope of the) default risk premium,

$$\theta_j = a\exp(-b \times j \times \Delta T^{(1)}),$$

and determine a and b by calibrating to the basis spread curves. For example, for the data of December 16, 2015, the calibrated results are $a = 37.1$ bps and $b = 3.4341$; the calibrated risk premiums and the basis curves are displayed in Figure 14.1. It is interesting that we are able to achieve a quality fitting of the basis curves even with a rather simple parametrization of (the slope of) the risk premiums.

The fact that $E_t^{Q_{j+1}^{(i)}}\left[\hat{f}_j^{(i)}(T_j^{(i)})\right]$ can be bootstrapped from swap rate curves brings about a candidate alternative to $H_j^{(i)}(t)$ for modeling purposes.

---

[2]After the 2008 financial crisis, swaps are almost all colleteralized.

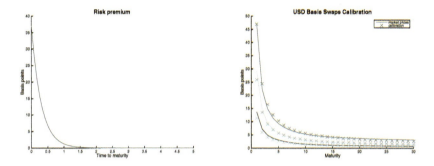

FIGURE 14.1: Slope of the risk premiums (left) and basis curves (right).

According to (14.8), there is

$$1 + \Delta T_j^{(i)} \hat{f}_j^{(i)}(T_j^{(i)}) = \left(1 + \Delta T_j^{(i)} f_j^{(i)}(T_j^{(i)})\right) \left(1 + \Delta T_j^{(i)} H_j^{(i)}(T_j^{(i)})\right).$$

Taking expectation of above equation conditional on $\mathcal{F}_t$ under the $T_{j+1}^{(i)}$-forward measure and define the expected LIBOR rate and expected loss rate as

$$L_j^{(i)}(t) = E_t^{Q_{j+1}^{(i)}} \left[\hat{f}_j^{(i)}(T_j^{(i)})\right], \quad h_j^{(i)}(t) = E_t^{Q_{j+1}^{(i)}} \left[H_j^{(i)}(T_j^{(i)})\right], \qquad (14.16)$$

then we arrive at

$$1 + \Delta T_j L_j^{(i)}(t) = \left(1 + \Delta T_j f_j^{(i)}(t)\right) \left(1 + \Delta T_j h_j^{(i)}(t)\right). \qquad (14.17)$$

As both $L_j^{(i)}(t)$ and $f_j^{(i)}(t)$ are indirectly observable in the swap markets, so is $h_j^{(i)}$. By definition, all three variables are martingales under the $T_{j+1}^{(i)}$-forward measure, which will yield great advantages for interest rate option modeling. For caplet pricing alone, we might consider modeling $L_j^{(i)}(t)$ for the four choices of i, but in doing so we will slip back to multiple-curve modeling and become unable to model the basis curves (see e.g., Mercurio, 2010a; Mercurio, 2010b). For this reason, we will model $f_j^{(i)}(t)$ and $h_j^{(i)}(t)$ simultaneously for a particular i. By utilizing the information of default risk premiums of a generic panel bank, we will be able to construct the term structures of LIBOR of other tenors and derive their evolution dynamics.

14.3 Option Pricing Using Heat Kernel Expansion

With the pricing of volatility smiles in mind, we adopt a natural extension of the existing SABR-LMM model (Henry-Labordère, 2007) to the dual-curve setting: we use the same CEV dynamics for both the OIS forward rate and the expected loss rate, and let them share the same stochastic volatility factor. The motivation for this choice is to take advantage of the analytical tractability of the SABR model which, as we shall see, can indeed be carried over to the dual-curve version of the model. Specifically, the transition density function of the dual-curve SABR model is the heat kernel of the 3-D Poincaré space, which can be studied in a way parallel to that for the 2-D Poincaré space. Since LIBOR of different tenors are now linked by the default and liquidity risk premium, we only need to model one LIBOR curve (of a particular tenor) through modeling the corresponding OIS forward rates curve and the expected loss rate curve. Due to liquidity consideration, we will take the three-month LIBOR for modeling, i.e., taking $i = 3$. For notational simplicity, we will drop the superscript $^{(i)}$ from both rates. Then, under the risk-neutral measure, the dual-curve SABR-LMM model can be prescribed as[3]

$$\begin{cases} df_j(t) = V(t)\gamma_j(t)f_j^\beta(t)\left(d\hat{W}_j(t) - \Sigma(t, T_{j+1})dt\right), \\ dh_j(t) = V(t)\sigma_j(t)h_j^\beta(t)d\tilde{W}_j(t), \\ dV(t) = \nu V(t)dZ(t), \quad V(0) = \alpha, \end{cases} \qquad (14.18)$$

where

$$\Sigma(t, T_{j+1}) = -V(t)\sum_{k=\eta(t)}^{j} \frac{\Delta T_k f_k^\beta(t)\rho_{kj}}{1 + \Delta T_k f_k(t)}\gamma_k(t), \qquad (14.19)$$

$\hat{W}_j, \tilde{W}_j$ and Z are Brownian motions under the the risk-neutral measure, and $\rho_{kj} = \langle d\hat{W}_k, d\hat{W}_j\rangle/dt$. Note that $\hat{W}_j(t)$ and $\tilde{W}_j(t)$ are mutually independent, by assumption yet Z_t can correlate with the other two Brownian motions.

Under the dual-curve SABR-LMM model, we consider the pricing of a generic caplet with payoff

$$\Delta T_j(L_j(T_j) - K)^+$$

at time T_{j+1}. According to the general pricing principle, we know the result is

$$Caplet(t, K) = \Delta T_j P(t, T_{j+1})E_t^{Q_{j+1}}\left[(L_j(T_j) - K)^+\right].$$

We will calculate the expectation above by following the approach of Henry-Labordère (2005), which consists of following three steps:

1. To derive an asymptotic value of the transition probability density using the heat kernel expansion of Henry-Labordère (2005).

2. To evaluate the volatility function for the corresponding local volatility model (LVM) for the LIBOR rate.

---

[3]With the pricing of swaption in mind, we use the same stochastic volatility factor for the OIS forward and expected loss rates of different indexes.

3. To derive the Black's implied volatilities from the LVM volatility function.

The technicalities of the remaining part of the chapter is largely attributed to Cui (2018), my Ph.D. student and research collaborator.

14.3.1 Derivation of the Heat Kernel

14.3.1.1 General Heat Kernel Expansion Formulae

Consider a set of diffusion processes:

$$dz_j = \mu^j(\mathbf{z}, t)dt + \sigma^j(\mathbf{z}, t)dW_j, \quad j = 1, 2, \ldots, N,$$

where any pair of Brownian motions can be correlated: $dW_i dW_j = \rho_{ij} dt$. Denote

$$g^{ij} = \sigma^i \sigma^j \rho_{ij},$$

then the matrix (g^{ij}) is symmetric and positive definite. The transition probability density function is known to satisfy the Kolmogorov backward equation:

$$\frac{\partial p}{\partial t} + \frac{\partial p}{\partial z_j} \mu^j + \frac{1}{2} \frac{\partial^2 p}{\partial z_i \partial z_j} g^{ij} = 0, \tag{14.20}$$

with terminal condition

$$p(T, \mathbf{z}; T, \mathbf{z}') = \delta(\mathbf{z} - \mathbf{z}').$$

Here in (14.20), we have adopted the Einstein's convention for summation. Note that $(g_{ij}(\mathbf{z})) \stackrel{\triangle}{=} (g^{ij}(\mathbf{z}))^{-1}$ is also a positive definite matrix, and Equation (14.20) can be treated as a heat equation on a Riemannian manifold with metric $(g_{ij}(\mathbf{z}))$. Let $\tau = T - t$, and then equation (14.20) becomes

$$\frac{\partial p}{\partial \tau} = \frac{\partial p}{\partial z_j} \mu^j + \frac{1}{2} \frac{\partial^2 p}{\partial z_i \partial z_j} g^{ij}, \tag{14.21}$$

with the initial condition

$$p(0, \mathbf{z}; 0, \mathbf{z}') = \delta(\mathbf{z} - \mathbf{z}'). \tag{14.22}$$

To present the heat kernel expansion formulae, we need to introduce the following notions or functions:

1. The geodesic distance between $\mathbf{z}$ and $\mathbf{z}'$ with respect to the Riemannian metric $(g_{ij}(\mathbf{z}))$ is $d(\mathbf{z}, \mathbf{z}')$, which is yet to be specified.

2. The *Synge world function* with respect to the Riemannian metric $(g_{ij}(\mathbf{z}))$:

$$\sigma_g(\mathbf{z}, \mathbf{z}') \stackrel{\triangle}{=} \frac{1}{2} d^2(\mathbf{z}, \mathbf{z}').$$

3. The *Van Vleck-Morette determinant with respect to* $(g_{ij}(\mathbf{z}))$:

$$\Delta(\mathbf{z}, \mathbf{z}') \triangleq \det\left(-\frac{\partial^2 \sigma_g(\mathbf{z}, \mathbf{z}')}{\partial z_i \partial z'_j}\right) \Big/ \sqrt{g(\mathbf{z})\, g(\mathbf{z}')},$$

where $g(\mathbf{z}) = det(g_{ij}(\mathbf{z}))$.

4. The Abelian connection:

$$\mathcal{A}_i = g_{ij}\mathcal{A}^j, \quad i = 1, \ldots, n,$$

where

$$\mathcal{A}^i = \frac{1}{2}\left(\mu^i - \frac{\partial_j(\sqrt{g}\, g^{ij})}{\sqrt{g}}\right).$$

5. The parallel transport along the geodesic curve between $\mathbf{z}$ and $\mathbf{z}'$ on the manifold, $\mathcal{C}(\mathbf{z}, \mathbf{z}')$:

$$\mathcal{P}(\mathbf{z}, \mathbf{z}') = \exp\left(-\int_{\mathcal{C}(\mathbf{z}, \mathbf{z}')} \mathcal{A}_i dz_i\right).$$

With the above functions/variables, we can present the heat kernel expansion of Equations (14.21–14.22) as follows.

Theorem 14.3.1 (Henry-Labordère, 2005). *The solution to (14.21–14.22) has the following expansion in τ:*

$$p(\tau, \mathbf{z}'|\mathbf{z}) = \frac{\sqrt{g(\mathbf{z}')}}{(4\pi\tau)^{n/2}} \sqrt{\Delta(\mathbf{z}, \mathbf{z}')}\mathcal{P}(\mathbf{z}, \mathbf{z}') \exp\left(-\frac{\sigma_g(\mathbf{z}, \mathbf{z}')}{2\tau}\right) \sum_{i=0}^{\infty} a_i(\mathbf{z}, \mathbf{z}')\tau^i,$$
$$(14.23)$$

for

$$a_0(\mathbf{z}, \mathbf{z}') = 1,$$
$$a_1(\mathbf{z}, \mathbf{z}') = \frac{1}{6}R + Q - \frac{1}{4}g_{ij}\mathcal{G}^{ij},$$

where

$$\mathcal{G}^{ij} = \frac{\partial}{\partial t}g^{ij}(t, \mathbf{z})|_{t=0},$$
$$Q = g^{ij}(\mathcal{A}_i\mathcal{A}_j - \mu_j\mathcal{A}_i - \partial_j\mathcal{A}_i),$$
$$R = g^{ij}R_{ij}, \quad R_{jl} = R^i_{jil}, \quad R^i_{jkl} = \partial_l\Gamma^i_{jk},$$

and Γ^i_{jk} is the so-called Christoffel symbol, defined by

$$\Gamma^i_{jk} = g^{il}(\partial_j g_{kl} + \partial_k g_{lk} - \partial_l g_{ik}).$$

$\square$

14.3.1.2 Heat Kernel Expansion for the Dual-Curve SABR-LMM Model

We will work with the T_{j+1}-forward measure, under which the dual-curve SABR model becomes

$$
\begin{cases}
df_j(t) = V(t)f_j^{\beta}(t)\gamma_j(t)dW_j(t), \\
dh_j(t) = V(t)h_j^{\beta}(t)\sigma_j(t)d\tilde{W}_j(t), \\
dV(t) = \mu_V(t, f_j, V)dt + \nu V(t)dZ(t),
\end{cases}
\tag{14.24}
$$

where W_j, $\tilde{W}_j$ and Z are Brownian motions under T_{j+1}-forward measure and

$$
\mu_V = -\nu V^2 \sum_{k=1}^{j} \frac{\Delta T_k \gamma_k(t) f_k^{\beta}(t) \rho_{kV}}{1 + \Delta T_k f_k(t)},
$$

for $\rho_{kV} = \langle dW_k, dZ \rangle / dt$. Note that μ_V is state dependent. To retain the analytical tractability of the model, we get rid of the state dependence by freezing the time of the state variables in μ_V, such as

$$
\mu_V \approx -\nu V^2 \sum_{k=1}^{j} \frac{\Delta T_k \gamma_k(t) f_k^{\beta}(0) \rho_{kV}}{1 + \Delta T_k f_k(0)}.
$$

After such a freezing-time treatment, the dynamics of course becomes an approximation to the original one. For notational simplicity, we further drop the index j for $f_j(t)$ and $h_j(t)$, and write $f(t) = f_t$ and $h(t) = h_t$ interchangeably. Then, the heat kernel of (14.24) satisfies

$$
\begin{aligned}
\frac{\partial p}{\partial \tau} = {} & \frac{\partial p}{\partial V}\mu_V + \frac{1}{2}\frac{\partial^2 p}{\partial f^2}f^{2\beta}\gamma^2 V^2 + \frac{1}{2}\frac{\partial^2 p}{\partial h^2}h^{2\beta}\sigma^2 V^2 \\
& + \frac{\partial^2 p}{\partial V \partial f}f^{\beta}\gamma\nu\rho V^2 + \frac{\partial^2 p}{\partial V \partial h}h^{\beta}\sigma\nu\bar{\rho}V^2 + \frac{1}{2}\frac{\partial^2 p}{\partial V^2}\nu V^2,
\end{aligned}
\tag{14.25}
$$

where ρ and $\bar{\rho}$ are the correlation between f and V and between h and V, respectively. With a time scaling $\tau' = \frac{\nu^2}{2}\tau$, Equation (14.25) becomes

$$
\begin{aligned}
\frac{\partial p}{\partial \tau'} = {} & \frac{2}{\nu^2}\left(\frac{\partial p}{\partial V}\mu^V + \frac{1}{2}\frac{\partial^2 p}{\partial f^2}f^{2\beta}\gamma^2 V^2 + \frac{1}{2}\frac{\partial^2 p}{\partial h^2}h^{2\beta}\sigma^2 V^2 \right. \\
& \left. + \frac{\partial^2 p}{\partial V \partial f}f^{\beta}\gamma\nu\rho V^2 + \frac{\partial^2 p}{\partial V \partial h}h^{\beta}\sigma\nu\bar{\rho}V^2 + \frac{1}{2}\frac{\partial^2 p}{\partial V^2}\nu V^2 \right).
\end{aligned}
\tag{14.26}
$$

For subsequent analysis, we write $\phi_f = \gamma(0)$ and $\phi_h = \sigma(0)$, and make a change of variables such as

$$
q_1 = \int_{f_0}^{f} \frac{1}{\phi_f f^{\beta}}df, \quad q_2 = \int_{h_0}^{h} \frac{1}{\phi_h h^{\beta}}dh, \quad \text{and} \quad q_3 = \frac{V}{\nu}.
\tag{14.27}
$$

Then, we obtain the heat kernel expansion by a direct application of Theorem 14.3.1.

Proposition 14.3.1. *The heat kernel expansion of the backward Kolmogorov equation is given by (14.23) with the following inputs:*

$$\sqrt{g} = \frac{\nu^2 \sqrt{(\det(\boldsymbol{\rho}^{-1})}}{f^\beta h^\beta \phi_f \phi_h V^3},$$

$$d = \cosh^{-1}\left(1 + \frac{(\mathbf{q} - \mathbf{q}')^T \boldsymbol{\rho}^{-1}(\mathbf{q} - \mathbf{q}')}{2q_3 q_3'}\right),$$

$$\Delta = \left(\frac{d}{\sinh(d)}\right)^3,$$

$$\ln \mathcal{P} = \int_{C(\mathbf{x},\mathbf{x}_0)} -\mathcal{A}_x dx \approx \sum_j \int_{x_j(0)}^{x_j} -\mathcal{A}_{x_j}^0 dx_j$$

$$= -\int_{f(0)}^f \mathcal{A}_f^0 df - \int_{h(0)}^h \mathcal{A}_h^0 dh - \int_{V(0)}^V \mathcal{A}_V^0 dV,$$

$$\begin{pmatrix} \mathcal{A}_f \\ \mathcal{A}_h \\ \mathcal{A}_V \end{pmatrix} = \begin{pmatrix} -\frac{\rho^{11}\beta}{2f} - \frac{\rho^{12}\beta h^{\beta-1}\phi_h}{2f^\beta \phi_f} + \frac{\rho^{13}\mu^V}{V^2 f^\beta \phi_f \nu^2} \\ -\frac{\rho^{12}\beta f^{\beta-1}\phi_f}{2h^\beta \phi_h} - \frac{\rho^{22}\beta}{2h} + \frac{\rho^{23}\mu^V}{V^2 h^\beta \phi_h \nu^2} \\ -\frac{\rho^{13}\beta f^{\beta-1}\phi_f}{2\nu} - \frac{\rho^{23}\beta h^{\beta-1}\phi_h}{2\nu} + \frac{1}{2V} + \frac{\rho^{33}\mu^V}{V^2 \nu^2} \end{pmatrix},$$

where

$$\boldsymbol{\rho} = \begin{pmatrix} 1 & 0 & \rho \\ 0 & 1 & \bar{\rho} \\ \rho & \bar{\rho} & 1 \end{pmatrix}$$

is the correlation matrix for the Brownian motions. $\qquad\square$

Proof: We start with equation (14.26). According to the equation, the inverse metric matrix at time 0 is

$$(g^{ij})_{3\times 3} = \begin{pmatrix} \frac{V^2}{\nu^2} f^{2\beta} \phi_f^2 & 0 & \frac{V^2}{\nu} f^\beta \phi_f \rho \\ 0 & \frac{V^2}{\nu^2} h^{2\beta} \phi_h^2 & \frac{V^2}{\nu} h^\beta \phi_h \bar{\rho} \\ \frac{V^2}{\nu} f^\beta \phi_f \rho & \frac{V^2}{\nu} h^\beta \phi_h \bar{\rho} & V^2 \end{pmatrix}.$$

Adopt the change of variables (14.27), then the inverse metric matrix becomes

$$(g^{ij})_{3\times 3}^{T_1} = \begin{pmatrix} \frac{V^2}{\nu^2} & 0 & \frac{V^2}{\nu}\rho \\ 0 & \frac{V^2}{\nu^2} & \frac{V^2}{\nu}\bar{\rho} \\ \frac{V^2}{\nu}\rho & \frac{V^2}{\nu}\bar{\rho} & \frac{V^2}{\nu^2} \end{pmatrix} = \frac{V^2}{\nu^2}\begin{pmatrix} 1 & 0 & \rho \\ 0 & 1 & \bar{\rho} \\ \rho & \bar{\rho} & 1 \end{pmatrix} \triangleq \frac{V^2}{\nu^2}\boldsymbol{\rho}.$$

Assume that the correlation matrix $\boldsymbol{\rho}$ is of full rank, and then we perform the Cholesky decomposition on $\boldsymbol{\rho}$:

$$\boldsymbol{\rho} = BB^T,$$

where B is an upper triangle matrix. Now, we adopt the second transformation

$$x = B^{-1}q,$$

then the inverse metric matrix becomes

$$(g^{ij})^{T_2}_{3\times 3} = \frac{V^2}{\nu^2} \begin{pmatrix} 1 & 0 & 0 \\ 0 & 1 & 0 \\ 0 & 0 & 1 \end{pmatrix} = x_3^2 I_3,$$

where we have used the fact that $(B^{-1})_{33} = 1$.

After the two consecutive transformations, the metric matrix becomes

$$(g_{ij})^{T_2}_{3\times 3} = \left((g^{ij})^{T_2}_{3\times 3} \right)^{-1} = \frac{1}{x_3^2} I_3,$$

which corresponds to the Poincaré space H^3.

– The geodesic distance $d(\mathbf{x}, \mathbf{x}')$

The geodesic distance on H^3 between the two points $(\mathbf{x}, \mathbf{x}')$ is (Henry-Labordère, 2008)

$$d(\mathbf{x}, \mathbf{x}') = \cosh^{-1}(1 + \frac{(\mathbf{x} - \mathbf{x}')^T (\mathbf{x} - \mathbf{x}')}{2x_3 x_3'})$$

$$= \cosh^{-1}(1 + \frac{(\mathbf{q} - \mathbf{q}')^T (B^{-1})^T B^{-1} (\mathbf{q} - \mathbf{q}')}{2q_3 q_3'})$$

$$= \cosh^{-1}(1 + \frac{(\mathbf{q} - \mathbf{q}')^T \rho^{-1} (\mathbf{q} - \mathbf{q}')}{2q_3' q_3}).$$

– Van Vleck-Morette determinant $\Delta(\mathbf{x}, \mathbf{x}')$

The Van Vleck-Morette determinant on H^3 is

$$\Delta(\mathbf{x}, \mathbf{x}') = \left(\frac{d(\mathbf{x}, \mathbf{x}')}{\sinh(d(\mathbf{x}, \mathbf{x}'))} \right)^3.$$

– The determinant of metric matrix g

In terms of the new coordinates x, the determinant of metric matrix is

$$\sqrt{g} = \frac{1}{x_3^3},$$

while the determinant of metric matrix with respect to old coordinates (f, h, V),

$$\sqrt{g}|_{(f,h,V)} = \frac{\sqrt{g}|_x}{\det(J)} = \frac{\nu^2 \sqrt{(\det(\rho^{-1})}}{f^\beta h^\beta \phi_f \phi_h V^3},$$

will be used in calculation of transition probability density function.

– The parallel transport $\mathcal{P}(\mathbf{x}, \mathbf{x}')$

The parallel transport satisfies

$$\ln \mathcal{P} = \int_{C(\mathbf{x}, \mathbf{x}_0)} -\mathcal{A}_x dx \approx \sum_j \int_{x_j(0)}^{x_j} -\mathcal{A}_{x_j}^0 dx_j$$

$$= -\int_{f(0)}^{f} \mathcal{A}_f^0 df - \int_{h(0)}^{h} \mathcal{A}_h^0 dh - \int_{V(0)}^{V} \mathcal{A}_V^0 dV,$$

where $C(x)$ is the geodesic curve in Poincaré space, and we approximate the integration along this geodesic by piecewise linear integration and freeze the state-dependent variables inside integrand by their initial values. Similar approximation could be found in Henry-Labordère (2008). The accuracy of such approximation seems not a concern. The one-form $\mathcal{A}^i$s are computed by their definition

$$\mathcal{A}^i = \frac{1}{2}(\mu^i - g^{-1/2}\partial_j(g^{1/2}g^{ij})), \qquad i = f, h, V.$$

For f,

$$\mathcal{A}^f = \frac{1}{2}(\mu^f - g^{-1/2}\partial_j(g^{1/2}g^{fj}))$$

$$= -\frac{1}{2}g^{-1/2}\left[\partial_f(g^{1/2}g^{ff}) + \partial_h(g^{1/2}g^{fh}) + \partial_V(g^{1/2}g^{fV})\right]$$

$$= -\frac{1}{2}g^{-1/2}\left[\partial_f\left(\frac{\sqrt{(\det(\boldsymbol{\rho}^{-1})}f^\beta \phi_f}{Vh^\beta \phi_h}\right) + \partial_V\left(\frac{\nu\sqrt{(\det(\boldsymbol{\rho}^{-1})}\rho}{Vh^\beta \phi_h}\right)\right]$$

$$= -\frac{\beta V^2 f^{2\beta-1}\phi_f^2}{2\nu^2} + \frac{V\rho f^\beta \phi_f}{2\nu}.$$

For h,

$$\mathcal{A}^h = -\frac{\beta V^2 h^{2\beta-1}\phi_h^2}{2\nu^2} + \frac{V\bar{\rho}h^\beta \phi_h}{2\nu}.$$

For V,

$$\mathcal{A}^V = \frac{1}{2}\left(\frac{2}{\nu^2}\mu^V - g^{-1/2}\partial_j(g^{1/2}g^{Vj})\right)$$

$$= \frac{\mu^V}{\nu^2} - \frac{1}{2}g^{-1/2}\partial_V(g^{1/2}g^{VV})$$

$$= \frac{\mu^V}{\nu^2} + \frac{1}{2}V.$$

Therefore, by definition

$$\mathcal{A}_i = g_{ij}\mathcal{A}^j,$$

$$\begin{pmatrix} \mathcal{A}_f \\ \mathcal{A}_h \\ \mathcal{A}_V \end{pmatrix} = \begin{pmatrix} \frac{\nu}{Vf^\beta \phi_f} & & \\ & \frac{\nu}{Vh^\beta \phi_h} & \\ & & \frac{1}{V} \end{pmatrix} \rho^{-1} \begin{pmatrix} \frac{\nu}{Vf^\beta \phi_f} & & \\ & \frac{\nu}{Vh^\beta \phi_h} & \\ & & \frac{1}{V} \end{pmatrix} \begin{pmatrix} \mathcal{A}^f \\ \mathcal{A}^h \\ \mathcal{A}^V \end{pmatrix}$$

$$= \begin{pmatrix} -\frac{\rho^{11}\beta}{2f} - \frac{\rho^{12}\beta h^{\beta-1}\phi_h}{2f^\beta \phi_f} + \frac{\rho^{13}\mu^V}{V^2 f^\beta \phi_f \nu^2} \\ -\frac{\rho^{12}\beta f^{\beta-1}\phi_f}{2h^\beta \phi_h} - \frac{\rho^{22}\beta}{2h} + \frac{\rho^{23}\mu^V}{V^2 h^\beta \phi_h \nu^2} \\ -\frac{\rho^{13}\beta f^{\beta-1}\phi_f}{2\nu} - \frac{\rho^{23}\beta h^{\beta-1}\phi_h}{2\nu} + \frac{1}{2V} + \frac{\rho^{33}\mu^V}{V^2 \nu^2} \end{pmatrix}.$$

Inserting relevant terms to the heat kernel expansion above and keeping up to the first-order terms, we then obtain an approximate expression of the heat kernel. □

14.3.2 Calculating the Volatility for Local Volatility Model

14.3.2.1 Calculation of the Local Volatility Function

Suppose that X_t is a real-valued one-dimensional Ito's process starting at $X_0 = 0$ with dynamics:

$$dX_t = a(t,\omega)dt + \mathbf{b}(t,\omega) \cdot d\mathbf{W}_t,$$

where $\mathbf{W}_t$ is a k-dimensional Brownian motion on the probability space $(\Omega, \mathcal{F}, P)$ and $\omega \in \Omega$ denotes dependence on some state variables. Gyöngy (1986) proved that there exists another stochastic process, $\tilde{X}_t$, which is a solution of the stochastic differential equation

$$d\tilde{X}_t = \tilde{a}(t, \tilde{X}_t)dt + \tilde{b}(t, \tilde{X}_t)d\tilde{W}_t,$$

with a scalar Brownian motion $\tilde{W}_t$, and non-random coefficients $\tilde{a}$ and $\tilde{b}$ defined by

$$\tilde{a}(t,x) = E\left[a(t,\omega)|X_t = x\right],$$
$$\tilde{b}^2(t,x) = E\left[\mathbf{b} \cdot \mathbf{b}(t,\omega)|X_t = x\right],$$

such that X_t and $\tilde{X}_t$ have the same marginal probability distribution for every $t > 0$.

Next, we will derive the local volatility function of L based on the Gyöngy's results. Equation (14.17) defines the expected value of a LIBOR rate, from which we obtain the dynamics of L_t as follows:

$$dL_t = (1 + \Delta T h_t)\gamma(t)f_t^\beta V_t dW + (1 + \Delta T f_t)\sigma(t)h_t^\beta V_t d\bar{W}.$$

According to the relationship between stochastic variance and local variance,

$$\sigma_{LV}^2(K,t) = E\left[\gamma^2(t)(1 + \Delta T h_t)^2 f_t^{2\beta} V_t^2 \right.$$
$$\left. + \sigma^2(t)(1 + \Delta T f_t)^2 h_t^{2\beta} V_t^2 \,\middle|\, L_t = K\right]. \tag{14.28}$$

Let $\mathbf{x}_t = (f_t, h_t, V_t)$ and let

$$D(\mathbf{x}_t) = \gamma^2(t)(1 + \Delta T h_t)^2 f_t^{2\beta} V_t^2 + \sigma^2(t)(1 + \Delta T f_t)^2 h_t^{2\beta} V_t^2,$$

and then the calculation of the local volatility boils down to the calculations of two triple integrals (in numerator and denominator respectively):

$$E\left[D(\mathbf{x}_t)|L_t = K\right] = \frac{\int_{R_+^3} D(\mathbf{x}_t) p(\tau, \mathbf{x}_t|\mathbf{x}_0)\delta(L_t = K) df_t dh_t dV_t}{\int_{R_+^3} p(\tau, \mathbf{x}_t|\mathbf{x}_0)\delta(L_t = K) df_t dh_t dV_t}, \qquad (14.29)$$

where $\delta(L_t = K)$ is the Dirac delta function, and $\{L_t = K\}$ implies $\{h_t = (K - f_t)/(1 + \Delta\tau f_t) \triangleq h(f_t)\}$. Let

$$A = \frac{1}{4\tau'} = \frac{1}{2\nu^2\tau},$$

$$E(\mathbf{x}) = \frac{\sqrt{g}}{(4\pi\tau')^{3/2}} \sqrt{\Delta}P(1 + a_1\tau') = \frac{1}{(\frac{\nu^2}{2})^{\frac{3}{2}}} \frac{\sqrt{g}}{(4\pi\tau)^{3/2}} \sqrt{\Delta}P(1 + \frac{\nu^2}{2}a_1\tau),$$

$$F(\mathbf{x}) = D \times E,$$

$$G(\mathbf{x}) = d^2(\mathbf{x}, \mathbf{x_0}),$$

then (14.29) can be rewritten into

$$E\left[D(\mathbf{x})|L_t = K\right] = \frac{\int_{R_+^3} F(\mathbf{x}) \exp(-AG(\mathbf{x}))\delta(L_t = K) d\mathbf{x}}{\int_{R_+^3} E(\mathbf{x}) \exp(-AG(\mathbf{x}))\delta(L_t = K) d\mathbf{x}}$$

$$= \frac{\int_{R_+^2} F(\mathbf{z}) \exp(-AG(\mathbf{z})) d\mathbf{z}}{\int_{R_+^2} E(\mathbf{z}) \exp(-AG(\mathbf{z})) d\mathbf{z}},$$

where $\mathbf{z}_t = (f_t, V_t)$ and, with a slight abuse of notations, we write

$$E(\mathbf{z}_t) = E(\mathbf{x}_t|h_t = h(f_t)),$$
$$F(\mathbf{z}_t) = F(\mathbf{x}_t|h_t = h(f_t)),$$
$$G(\mathbf{z}_t) = G(\mathbf{x}_t|h_t = h(f_t)).$$

In order to work out the integrals in approximate closed forms, we will use saddle point approximation method for the integrals. This method is intended for integrands that contain the factor $\exp(-AG(\mathbf{z}))$ for a large $A > 0$, when the mass of the whole integrand concentrates around the saddle point(s) of $G(\mathbf{z})$. In Figure 14.2, we show the 2-D plot of a typical transition probability density function (pdf), where one can see the concentration of the mass of the pdf for $\tau = 1$ year.

Proposition 14.3.2. *For very large $A > 0$ and strictly concave function $G(\mathbf{z}) \in \mathbf{C}(\mathbb{R}^N)$, the integral of the form*

$$I(A) = \int_{\Omega} F(\mathbf{z}) \exp(-AG(\mathbf{z})) d\mathbf{z}$$

has the following approximation:

$$I(A) = \frac{(\sqrt{2\pi})^N \sqrt{|\Sigma^{\mathbf{z}}|} F(\mathbf{z}^*) \exp(-AG(\mathbf{z}^*))}{A^{N/2}} \left[1 + \frac{1}{A} \left(\frac{1}{2} \sum_{i,j}^{N} \frac{F_{ij}(\mathbf{z}^*)}{F(\mathbf{z}^*)} \Sigma_{ij}^{z} \right. \right.$$

$$- \frac{1}{6} \sum_{i,j,k,l=1}^{N} \left(\frac{G_{ijk}(\mathbf{z}^*) F_l(\mathbf{z}^*)}{F(\mathbf{z}^*)} \right) (\Sigma_{ij}^z \Sigma_{kl}^z + \Sigma_{ik}^z \Sigma_{jl}^z + \Sigma_{il}^z \Sigma_{jk}^z)$$

$$\left. \left. + \frac{1}{A} H(G(\mathbf{z}), \Sigma^{\mathbf{z}}) \right) + O(A^{-\frac{3}{2}}) \right],$$

$$(14.30)$$

where $\mathbf{z}^*$ is the saddle point of $G(\mathbf{z})$, $(\Sigma^{\mathbf{z}})^{-1}$ is the Hessian matrix of $G(\mathbf{z})$ evaluated at $\mathbf{z}^*$, and

$$H(G(\mathbf{z}), \Sigma^{\mathbf{z}}) = \int_{\mathbb{R}} \frac{d\mathbf{y}}{\sqrt{|\Sigma^{\mathbf{z}}|}} \exp\left(-\frac{1}{2} \mathbf{y}^T (\Sigma^{\mathbf{z}})^{-1} \mathbf{y} \right)$$

$$\left[-\frac{1}{24} G_{ijkl}(\mathbf{z}^*) y^i y^j y^k y^l + \frac{1}{72} G_{i_1 j_1 k_1}(\mathbf{z}^*) G_{i_2 j_2 k_2}(\mathbf{z}^*) y^{i_1} y^{j_1} y^{k_1} y^{i_2} y^{j_2} y^{k_2} \right],$$

which is independent of $F(\mathbf{z})$. □

Proof: The integral

$$I(A) = \int_{\Omega} F(\mathbf{z}) \exp(-AG(\mathbf{z})) d\mathbf{z} \qquad (14.31)$$

will be calculated using the saddle-point method. Let $\mathbf{z}^* \in \Omega$ be the saddle point of $G(\mathbf{z})$. We first make a change of variable:

$$\mathbf{z} - \mathbf{z}^* = \frac{\mathbf{y}}{\sqrt{A}}.$$

Perform Taylor expansions of $F(z)$ and $G(z)$ around $\mathbf{z}^*$ till the second- and the fourth-order terms, respectively, and we have

$$F(\mathbf{z}) = F(\mathbf{z}^*) + F_i(\mathbf{z}^*) \frac{y^i}{\sqrt{A}} + \frac{1}{2} F_{ij}(\mathbf{z}^*) \frac{y^i y^j}{A} + O(A^{-\frac{3}{2}})$$

$$G(\mathbf{z}) = G(\mathbf{z}^*) + \frac{1}{2} G_{ij}(\mathbf{z}^*) \frac{y^i y^j}{A} + \frac{1}{6} G_{ijk}(\mathbf{z}^*) \frac{y^i y^j y^k}{(\sqrt{A})^3}$$

$$+ \frac{1}{24} G_{ijkl}(z^*) \frac{y^i y^j y^k y^l}{A^2} + O(A^{-\frac{5}{2}})$$

Here the first-order term of g vanishes because $\mathbf{z}^*$ is a saddle point. The integrand is then approximated by

$$F(\mathbf{z})\exp(-AG(\mathbf{z})) = F(\mathbf{z}^*)\left[1 + \frac{F_i(\mathbf{z}^*)}{F(\mathbf{z}^*)}\frac{y^i}{\sqrt{A}} + \frac{1}{2}\frac{F_{ij}(\mathbf{z}^*)}{F(\mathbf{z}^*)}\frac{y^i y^j}{A} + O(A^{-\frac{3}{2}})\right]$$

$$\times \exp(-AG(\mathbf{z}^*))\exp\left(-\frac{1}{2}G_{ij}(\mathbf{z}^*)y^i y^j\right) \times \left(1 - \frac{1}{6}G_{ijk}(\mathbf{z}^*)\frac{y^i y^j y^k}{\sqrt{A}}\right.$$

$$\left. -\frac{1}{24}G_{ijkl}(\mathbf{z}^*)\frac{y^i y^j y^k y^l}{A} + \frac{1}{72}G_{i_1 j_1 k_1}G_{i_2 j_2 k_2}\frac{y^{i_1}y^{j_1}y^{k_1}y^{i_2}y^{j_2}y^{k_2}}{A} + O(A^{-\frac{3}{2}})\right)$$

Plug in the right-hand side of the above equation into the integral (14.31), and we have

$$I(A) = \left(\frac{1}{\sqrt{A}}\right)^N F(\mathbf{z}^*)\exp(-AG(\mathbf{z}^*)) \times \int_{R^N}\exp\left(-\frac{1}{2}\mathbf{y}^T\Sigma_h^{-1}\mathbf{y}\right)$$

$$\left[1 + \frac{1}{2}\frac{F_{ij}}{F}\frac{y^i y^j}{A} - \left(\frac{1}{6}\frac{G_{ijk}F_l}{AF} + \frac{1}{24}\frac{G_{ijkl}}{A}\right)y^i y^j y^k y^l \right. \tag{14.32}$$

$$\left. + \frac{1}{72}\frac{G_{i_1 j_1 k_1}G_{i_2 j_2 k_2}}{A}y^{i_1}y^{j_1}y^{k_1}y^{i_2}y^{j_2}y^{k_2} + O(A^{-\frac{3}{2}})\right]d\mathbf{y},$$

where $(\Sigma^{\mathbf{z}})^{-1} = (G_{ij}(\mathbf{z}^*))$ denotes the Hessian of $G(\mathbf{z})$ at $\mathbf{z}^*$.

Clearly, the remaining job is to calculate the second and the fourth moments of the multivariate normal distribution $\mathcal{N}(0, \Sigma^z)$. Immediately, we obtain

$$\int_{R^N}\exp\left(-\frac{1}{2}\mathbf{y}^T(\Sigma^{\mathbf{z}})^{-1}\mathbf{y}\right)y^i y^j d\mathbf{y} = (\sqrt{2\pi})^N\sqrt{|\Sigma^{\mathbf{z}}|}\Sigma_{ij}^{\mathbf{z}}.$$

According to the Isserlis theorem (Isserlis, 1916), we also have

$$\int_{R^N}\exp\left(-\frac{1}{2}\mathbf{y}^T(\Sigma^{\mathbf{z}})^{-1}\mathbf{y}\right)y^i y^j y^k y^l d\mathbf{y}$$

$$= (\sqrt{2\pi})^N\sqrt{|\Sigma^{\mathbf{z}}|}\left[\Sigma_{ij}^{\mathbf{z}}\Sigma_{kl}^{\mathbf{z}} + \Sigma_{ik}^{\mathbf{z}}\Sigma_{jl}^{\mathbf{z}} + \Sigma_{il}^{\mathbf{z}}\Sigma_{jk}^{\mathbf{z}}\right]$$

Putting the above results back into (14.32), we finally arrive at (14.30). □

Let us make a remark here. Because $H(G(\mathbf{z}), \Sigma^{\mathbf{z}})$ is independent of $F(\mathbf{z})$, and the same saddle point is shared by both the numerator and the denominator, $H(G(\mathbf{z}), \Sigma^{\mathbf{z}})$ will thus be canceled out in the calculation of the local volatility function. Its evaluation therefore is unnecessary.

The Hessian $(\Sigma^{\mathbf{z}})^{-1}$ can be calculated from the Hessian of $G(\mathbf{x})$, denoted as $(\Sigma^{\mathbf{x}})^{-1} = \frac{\partial^2 G(\mathbf{x})}{\partial x_i \partial x_j}$. For our application, we interpret $L_t = K$ in the form of

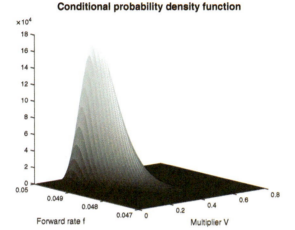

Conditional probability density function

FIGURE 14.2: Conditional transition pdf $p(1, \mathbf{z}|\mathbf{z_0})$ under $L(\mathbf{z}) = K$.

$x_2 = \alpha(x_1)$, and define

$$
T(x) = \begin{pmatrix}
1 & 0 & 0 & \cdots & 0 \\
\frac{d\alpha(x_1)}{dx_1} & 0 & 0 & \cdots & 0 \\
0 & 1 & 0 & \ddots & 0 \\
\vdots & 0 & \ddots & \ddots & \vdots \\
\vdots & \vdots & \ddots & \ddots & 0 \\
0 & 0 & \cdots & 0 & 1
\end{pmatrix}_{N \times (N-1)}
$$

Then

$$
(\Sigma^z)^{-1} = T^T(x)(\Sigma^x)^{-1}T(x).
$$

By applying the last theorem to calculate the conditional expectations in (14.29), we obtain

Corollary 14.3.1. *There are*

$$
E\left[D(\mathbf{x}_t)|L_t = K\right] = D(\mathbf{x}_t^*) \left[1 + \tau \left(\nu^2 \sum_{i,j=1}^{2} \left(\frac{D_{ij}(x^*)}{D(x^*)} \right. \right. \right.
$$
$$
+ \frac{D_j(x^*)}{2D(x^*)} \partial_i(\ln g \Delta \mathcal{P}^2) + \frac{D_i(x^*)}{2D(x^*)} \partial_j(\ln g \Delta \mathcal{P}^2) \bigg) \Sigma_{ij}^z \tag{14.33}
$$
$$
\left. \left. - \frac{\nu^2}{3} \sum_{i,j,k,l=1}^{2} G_{ijk}(\Sigma_{ij}^z \Sigma_{kl}^z + \Sigma_{ik}^z \Sigma_{jl}^z + \Sigma_{il}^z \Sigma_{jk}^z) \frac{D_l(x^*)}{D(x^*)} \right) \right].
$$

Proof: By the saddle point method,

$$
\int_{R_+^2} F(\mathbf{z}) \exp(-AG(\mathbf{z}))d\mathbf{z} \approx \frac{(\sqrt{2\pi})^2 \sqrt{|\Sigma^z|} F(\mathbf{z}^*) \exp(-AG(\mathbf{z}^*))}{A^{3/2}}
$$

$$
\times \left[1 + \frac{1}{A} \left(\frac{1}{2} \sum_{i,j=1}^{2} \frac{F_{ij}(\mathbf{z}^*)}{F(\mathbf{z}^*)} \Sigma_{ij}^z \right. \right.
$$

$$
\left. \left. - \frac{1}{6} \sum_{i,j,k,l=1}^{2} G_{ijk}(\Sigma_{ij}^z \Sigma_{kl}^z + \Sigma_{ik}^z \Sigma_{jl}^z + \Sigma_{il}^z \Sigma_{jk}^z) \frac{F_l(\mathbf{z}^*)}{F(\mathbf{z}^*)} + H(G, \Sigma^z) \right) \right]
$$

and

$$
\int_{R_+^2} E(\mathbf{z}) \exp(-AG(\mathbf{z}))d\mathbf{z} \approx \frac{(\sqrt{2\pi})^2 \sqrt{|\Sigma^z|} E(\mathbf{z}^*) \exp(-AG(\mathbf{z}^*))}{A^{3/2}}
$$

$$
\times \left[1 + \frac{1}{A} \left(\frac{1}{2} \sum_{i,j=1}^{2} \frac{E_{ij}(\mathbf{z}^*)}{E(\mathbf{z}^*)} \Sigma_{ij}^z \right. \right.
$$

$$
\left. \left. - \frac{1}{6} \sum_{i,j,k,l=1}^{2} G_{ijk}(\Sigma_{ij}^z \Sigma_{kl}^z + \Sigma_{ik}^z \Sigma_{jl}^z + \Sigma_{il}^z \Sigma_{jk}^z) \frac{E_l(\mathbf{z}^*)}{E(\mathbf{z}^*)} + H(G, \Sigma^z) \right) \right],
$$

the conditional expectation is

$$
E\left[D(\mathbf{x})|L_t = K\right]
$$

$$
= \frac{F(\mathbf{z}^*)}{E(\mathbf{z}^*)} \left[1 + \frac{1}{A} \left(\frac{1}{2} \sum_{i,j=1}^{2} \left(\frac{F_{ij}(\mathbf{z}^*)}{F(\mathbf{z}^*)} - \frac{E_{ij}(\mathbf{z}^*)}{E(\mathbf{z}^*)} \right) \Sigma_{ij}^z \right. \right.
$$

$$
\left. \left. - \frac{1}{6} \sum_{i,j,k,l=1}^{2} G_{ijkl}(\Sigma_{ij}^z \Sigma_{kl}^z + \Sigma_{ik}^z \Sigma_{jl}^z + \Sigma_{il}^z \Sigma_{jk}^z) \left(\frac{F_l(\mathbf{z}^*)}{F(\mathbf{z}^*)} - \frac{E_l(\mathbf{z}^*)}{E(\mathbf{z}^*)} \right) \right) \right]
$$

$$
= D(\mathbf{z}^*) \left[1 + \frac{1}{A} \left(\frac{1}{2} \sum_{i,j=1}^{2} \left(\frac{D_{ij}(\mathbf{z}^*)}{D(\mathbf{z}^*)} + \frac{D_j(\mathbf{z}^*)E_i(\mathbf{z}^*)}{D(\mathbf{z}^*)E(\mathbf{z}^*)} + \frac{D_i(\mathbf{z}^*)E_j(\mathbf{z}^*)}{D(\mathbf{z}^*)E(\mathbf{z}^*)} \right) \Sigma_{ij}^z \right. \right.
$$

$$
\left. \left. - \frac{1}{6} \sum_{i,j,k,l=1}^{2} G_{ijk}(\Sigma_{ij}^z \Sigma_{kl}^z + \Sigma_{ik}^z \Sigma_{jl}^z + \Sigma_{il}^z \Sigma_{jk}^z) \frac{D_l(\mathbf{z}^*)}{D(\mathbf{z}^*)} \right) \right],
$$

where the partial derivatives of $D(\mathbf{z}^*)$ can be calculated literally, and

$$E_i(\mathbf{x}) = \partial_i \left(\frac{\sqrt{g}}{(4\pi\tau)^{3/2}} \sqrt{\Delta}\mathcal{P}(1 + a_1\tau) \right),$$

$$= \partial_i \left(\frac{\sqrt{g}}{(4\pi\tau)^{3/2}} \sqrt{\Delta}\mathcal{P} \right)(1 + a_1\tau) + \frac{\sqrt{g}}{(4\pi\tau)^{3/2}} \sqrt{\Delta}\mathcal{P}(\partial_i a_1)\tau,$$

so we have

$$\frac{E_i}{E} = \frac{\partial_i(\frac{\sqrt{g}}{4\pi\tau}\sqrt{\Delta}\mathcal{P})}{\frac{\sqrt{g}}{4\pi\tau}\sqrt{\Delta}\mathcal{P}}(1 + a_1\tau)(1 - a_1\tau) + O(\tau)$$

$$= \frac{1}{2}\partial_i(\ln g\Delta\mathcal{P}^2) + O(\tau).$$

Substituting $2\tau\nu^2$ for $\frac{1}{A}$, we then arrive at the conclusions of the corollary. $\square$

14.3.2.2 Calculation of the Saddle Point

Finally, we address the calculation of the saddle point. The saddle point $\mathbf{x}^*$ is the point that minimizes the distance $d(\mathbf{x}, \mathbf{x}')$ subject to $L(\mathbf{x}) = K$, i.e.,

$$\min \quad \cosh^{-1}\left(1 + \frac{(\mathbf{q} - \mathbf{q}')^T \rho^{-1}(\mathbf{q} - \mathbf{q}')}{2q_3 q_3'} \right),$$

$$\text{s.t.} \quad L(q_1, q_2) = K,$$

where $\mathbf{q}' = (0 \quad 0 \quad V_0/\nu)^T$ is the value vector at time 0. As the function $\cosh^{-1}$ is strictly increasing, the original optimization is reduced to

$$\min \quad \frac{(\mathbf{q} - \mathbf{q}')^T \rho^{-1}(\mathbf{q} - \mathbf{q}')}{2q_3 q_3'}, \tag{14.34}$$

$$\text{s.t.} \quad L(q_1, q_2) = K.$$

This optimization problem (14.34) involves a convex function (see the proof of Proposition 14.3.3 next) and a nonlinear equality constraint, which is usually costly to solve. Yet we have noticed that the constraint has nothing to do with q_3. As a matter of fact, once the q_3 is fixed, problem (14.34) is reduced to a quadratic programming problem with a nonlinear equality constraint, which can be solved efficiently by solving the Karush-Kuhn-Tucker (KTT) optimality conditions (see e.g., Boyd and Vandenberghe, 2004). Figure 14.3 shows the objective function in (14.34) subject to the constraint. Since the objective function subject to the constraint is convex, we apply the following coordinate descent algorithm (Wright, 2015) to calculate the saddle point:

Algorithm for the saddle point under constraint: Choose a $q_3^{(0)}$ and for $k = 0, 1, \ldots$ repeat

1. For fixed $q_3^{(k)}$, solve the quadratic programming (QP) problem with a nonlinear constraint, and denote the optimal solution by $q_1^{(k)}$ and $q_2^{(k)}$;

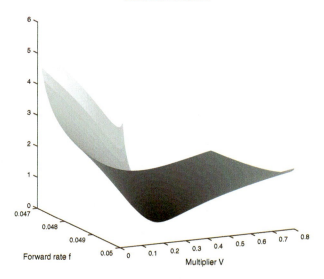

FIGURE 14.3: Geodesic distance under the constraint $L(\mathbf{z}) = K$.

2. Update q_3 according to the scheme

$$
q_3^{(k+1)} = \sqrt{\dfrac{\sum_{i,j=1}^{2} \rho^{ij} q_i^{(k)} q_j^{(k)} - 2 \sum_{i=1}^{2} \rho^{i3} q_i^{(k)} q_3' + \rho^{33}(q_3')^2}{\rho^{33}}},
$$

for $q_3' = V_0/\nu$.

3. Go back to step 1 with the updated $q_3^{(k+1)}$, and solve the QP again for $q_i^{(k+1)}, i = 1$ and 2 until $||\mathbf{q}^{(k+1)} - \mathbf{q}^{(k)}||_2$ is small enough.

Next, we will prove

Proposition 14.3.3. *The above algorithm converges to a global minimum.*

Proof: Denote the objective function of problem (14.34) as

$$
F(\mathbf{q}) = \frac{(\mathbf{q} - \mathbf{q}')^T \rho^{-1} (\mathbf{q} - \mathbf{q}')}{2 q_3 q_3'}.
$$

We first show that $F(\mathbf{q})$ is a convex function globally by showing the positive definiteness of the Hessian matrix:

$$
\begin{aligned}
\mathcal{H} &= (\partial_{ij} F)_{3\times 3} \\
&= \begin{pmatrix}
\dfrac{\rho^{11}}{q_3 q_3^0} & \dfrac{\rho^{12}}{q_3 q_3^0} & \dfrac{\rho^{13} q_3^0 - \rho^{11} q_1 - \rho^{12} q_2}{q_3^2 q_3^0} \\[2ex]
\dfrac{\rho^{12}}{q_3 q_3^0} & \dfrac{\rho^{22}}{q_3 q_3^0} & \dfrac{\rho^{23} q_3^0 - \rho^{12} q_1 - \rho^{22} q_2}{q_3^2 q_3^0} \\[2ex]
\dfrac{\rho^{13} q_3^0 - \rho^{11} q_1 - \rho^{12} q_2}{q_3^2 q_3^0} & \dfrac{\rho^{23} q_3^0 - \rho^{12} q_1 - \rho^{22} q_2}{q_3^2 q_3^0} & \partial_{33} F
\end{pmatrix}
\end{aligned}
$$

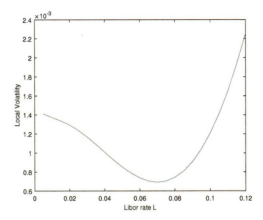

FIGURE 14.4: Local volatility function $C(L, 0)$.

where

$$\partial_{33}F = \frac{\rho^{11}q_1^2 + 2\rho^{12}q_1q_2 + \rho^{22}q_2^2 - 2\rho^{13}q_1q_3^0 - 2\rho^{23}q_2q_3^0 + \rho^{33}(q_3^0)^2}{q_3^3 q_3^0}.$$

Although it looks complex, matrix $\mathcal{H}$ can be transformed into the correlation matrix through two consecutive congruent transformations:

$$C_2^T C_1^T \mathcal{H} C_1 C_2 = \boldsymbol{\rho},$$

which of course is a positive definite matrix, with

$$C_1 = \begin{pmatrix} 1 & 0 & \frac{q_1}{q_3} \\ 0 & 1 & \frac{q_2}{q_3} \\ 0 & 0 & 1 \end{pmatrix} \quad \text{and} \quad C_2 = \begin{pmatrix} \sqrt{q_3 q_3'} & 0 & 0 \\ 0 & \sqrt{q_3 q_3'} & 0 \\ 0 & 0 & \sqrt{\frac{q_3^3}{q_3'}} \end{pmatrix}.$$

It is well known that congruent transformations preserve positive definiteness of a matrix. Due to the convexity of F, the coordinate descent algorithm is guaranteed to converge to its global minimum. $\qquad\square$

Figure 14.4 displays $\sigma_{LV}(L, 0)$, the local volatility function for L at time zero, of the dual-curve model obtained through the heat kernel expansion method. For notational simplicity, we write $C(L, 0) = \sigma_{LV}(L, 0)$.

14.3.3 Calculation of the Implied Black's Volatility

After calculating the local volatility function, $C(L_t, t)$ for L_t, we obtain the following approximation to the Black's implied volatility by combining the results of Henry-Labordère (2005) and Paulot (2015):

$$\sigma_{BS}(K,T) = \frac{\nu \left| \ln(\frac{K}{L_0}) \right|}{d^*(L_0,K)} \left[1 + \frac{C^2(L,0)T}{24} \left(2\frac{C''(L,0)}{C(L,0)} - \left(\frac{C'(L,0)}{C(L,0)}\right)^2 \right. \right.$$

$$\left. \left. + \frac{1}{L^2} + 12\frac{\partial_t C(L,t)|_{t=0}}{C^3(L,0)} \right) \right] \Bigg|_{L=\frac{L_0+K}{2}}, \qquad (14.35)$$

where $d^*(L_0, K)$ denotes the geodesic distance between the initial point $\mathbf{x}' = (f_0 \ h_0 \ V_0)^T$ and saddle point $\mathbf{x}^*$ solved in section 14.3.2.2.

Note that the zeroth-order term in the above formula is the same as that of Paulot (2015) because the geodesic distance here is calculated with the scaled time τ', thus containing a factor $\nu/\sqrt{2}$.

14.3.4 Numerical Results for 3M Caplets

The inputs for the model, which include the parameters for the dynamics of the OIS risk-free forward rates, the expected loss rates, and the stochastic multiplier, together with the initial terms structures of the OIS forward rates and the expected loss rates, are described in

Parameter set 1:
- Risk-free forward rates:

$$f_j^{(3)}(0) = \log(a+bj), f_0^{(3)}(0) = 5\%, f_{80}(0) = 9\%;$$
$$\sigma_j(t) = 0.17 + 0.002(T_{j-1}^{(3)} - t);$$

- Expected loss rates:

$$h_j^{(3)} = 0.0015 \exp(-0.4j\Delta T^{(3)});$$
$$\gamma_j(t) = 0.05 + 0.002(T_{j-1}^{(3)} - t);$$

- Stochastic multiplier:

$$\alpha = 0.25, \nu = 0.3, \beta = 0.5, \rho_{Vf} = -0.5, \rho_{Vh} = 0.5.$$

The correlation structure of f and h are

$$\rho_{ij} = \rho_L + (1 - \rho_L) \exp\left(-\left(\delta_A - \delta_B \min \left[T_{i-1}^{(3)}, T_{j-1}^{(3)} \right] \right) \left| T_{i-1}^{(3)} - T_{j-1}^{(3)} \right| \right),$$

for $\rho_L = 0.4, \delta_A = 0.5$ and $\delta_B = 0.01$. We will adopt this correlation structure for other numerical examples as well.

In Figure 14.5, we display the prices of caplets, in terms of the implied Black's volatilities, for various maturities and strikes. While the crosses represent the results by the analytical approximation method, the circles represent the results by the Monte Carlo simulation method. As one can see, for maturities up to nine years, the two sets of implied volatilities differ by no more than one basis point. Yet for the ten-year maturity, the errors can exceed one basis point, which is the situation when we may have to take into account the higher-order terms in the heat kernel expansion method.

14.4 Pricing 3M Swaptions

Let us now consider the pricing of swaptions on vanilla swaps, for which the floating leg pays quarterly while the fixed leg pays semiannually. For a T_m-year maturity swaption with underlying swap over the period (T_m, T_n) and strike rate K, the payoff function can be written as

$$A_{m,n}^{(6)}(T_m)\left(s_{m,n}^{(3)}(T_m) - K\right)^+,$$

where $A_{m,n}^{(6)}(t)$ is the fixed-leg annuity, which is to be taken as the numeraire of the forward swap measure Q_S. Then, the value of the swaption is given by

$$\text{Swaption}(t, K; T_m, T_n) = A_{m,n}^{(6)}(t)E_t^{Q_S}\left[(s - K)^+\right].$$

The implied Black's volatility σ_S is defined to be the solution to the equation

$$\text{Swaption}(t, K; T_m, T_n) = A_{m,n}^{(6)}(t)\text{Black}\left(K, s_{m,n}^{(3)}(t), \sigma_S\sqrt{T_m - t}\right),$$

where "Black" stands for the Black's formula.

The strategy to derive the asymptotic implied Black's volatility is very similar to that for caplets, with the exception of the very first step: we need to derive the dynamic of the swap rate under the dual-curve SABR-LMM model, with that we will then proceed to calculate the local volatility function, and then finally plug in the local volatility function into formulae (14.33) and (14.35). For brevity, we will only provide the parts that are different from the strategy for caplet pricing.

14.4.1 Dynamics of the State Variables

The pricing of the swaptions with quarterly floating legs is based on model (14.24). Under the forward swap measure corresponding to the numeraire:

$$A_{m,n}^{(6)}(t) = \sum_{j=m}^{n-1} \Delta T^{(6)} P(t, T_{j+1}^{(6)}) = \sum_{j=m}^{n-1} \Delta T^{(6)} P(t, T_{2(j+1)}^{(3)}), \qquad (14.36)$$

the dynamics of the state variables will change, as is described in the next proposition.

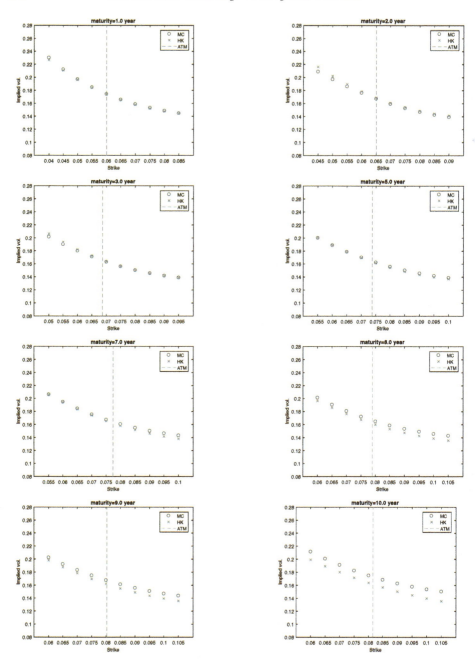

FIGURE 14.5: 3M caplet's implied Black volatilities using parameter set 1.

Proposition 14.4.1. *Under the forward swap measure that corresponds to the numeraire $A_{m,n}^{(6)}(t)$, the dynamics of dual-curve SABR-LMM model becomes*

$$
\begin{cases}
df_j(t) & = V^2 f_j^\beta \gamma_j \left[\Sigma_{j+1}^{(m,n)} - \Sigma_{j+1} \right] dt + V f_j^\beta \gamma_j dW_j^{Q_s}, \\
dh_j(t) & = V h_j^\beta \sigma_j d\bar{W}_j^{Q_s}, \\
dV(t) & = \nu V^2 \Sigma_V^{(m,n)} dt + \nu V(t) dZ^{Q_s},
\end{cases}
\tag{14.37}
$$

where $W_j^{Q_s}$, $\bar{W}_j^{Q_s}$ and Z^{Q_s} are Brownian motions under the forward swap measure and

$$
\Sigma_{j+1}^{(m,n)} = - \sum_{j=m}^{n-1} \alpha_j \sum_{k=\eta(t)}^{2j+1} \frac{\Delta T_k \gamma_k f_k^\beta \rho_{kj}}{1 + \Delta T_k f_k},
$$

$$
\Sigma_V^{(m,n)} = - \sum_{j=m}^{n-1} \alpha_j \sum_{k=\eta(t)}^{2j+1} \frac{\Delta T_k \gamma_k f_k^\beta \rho_{kV}}{1 + \Delta T_k f_k},
$$

$$
\Sigma_{j+1} = - \sum_{k=\eta(t)}^{2j+1} \frac{\Delta T_k \gamma_k f_k^\beta \rho_{kj}}{1 + \Delta T_k f_k},
$$

$$
\alpha_j(t) = \frac{\Delta T_j P(t, T_{j+1})}{A_{m,n}(t)}.
$$

Also, the correlation structure of the model is

$$
\rho = \begin{pmatrix} \rho_{N_s \times N_s} & 0 & \rho_{fV} \\ 0 & \bar{\rho}_{N_s \times N_s} & \bar{\rho}_{hV} \\ \rho_{fV}^T & \bar{\rho}_{hV}^T & 1 \end{pmatrix}_{(2N_s+1) \times (2N_s+1)}.
$$

$\square$

Proof: This proof again needs the independence between the market risk and the credit risk. The numeraire of the forward swap measure is changed to be $A_{m,n}(t)$, so the Radon- Nikodym derivative is

$$
\begin{aligned}
m_s(t) = \frac{dQ^s}{dQ}\bigg|_{\mathcal{F}_t} & = \frac{A_{m,n}(t)}{B_t} \bigg/ \frac{A_{m,n}(0)}{B_0} \\
& = \frac{1}{A_{m,n}(0)} \sum_{j=2m}^{2n-1} \frac{\Delta T^{(6)} P(t, T_{2(j+1)}^{(3)})}{B_t} \\
& = \frac{1}{A_{m,n}(0)} \sum_{j=2m}^{2n-1} \Delta T^{(6)} P(0, T_{2(j+1)}^{(3)}) m_{2(j+1)}(t),
\end{aligned}
$$

where $m_{2(j+1)}$ is Radon Nikodym derivative for forward measures, i.e.,

$$
m_{2(j+1)}(t) \triangleq \frac{dQ_{2(j+1)}}{dQ}\bigg|_{\mathcal{F}_t} = \frac{P(t, T_{2(j+1)}^{(3)})}{B_t} \bigg/ \frac{P(0, T_{2(j+1)})}{B_0},
$$

and

$$\frac{dm_{2(j+1)}(t)}{m_{2(j+1)}(t)} = -\sum_{k=2\eta(t)}^{2j+1} \frac{\Delta T_k^{(3)} \gamma_k f_k^\beta V}{1 + \Delta T_k^{(3)} f_k} dW_k^Q.$$

Then,

$$\frac{dm_s(t)}{m_s(t)} = \left[\frac{1}{A_{m,n}(0)} \sum_{j=2m}^{2n-1} \Delta T^{(6)} P(0, T_{2(j+1)}^{(3)}) dm_{2(j+1)}\right] \Big/ m_s$$

$$= \left[\frac{1}{A_{m,n}(0)} \sum_{j=2m}^{2n-1} \Delta T^{(6)} P(0, T_{2(j+1)}^{(3)}) m_{2(j+1)}\right.$$

$$\left. \left(-\sum_{k=\eta(t)}^{2j+1} \frac{\Delta T_k^{(3)} \gamma_k f_k^\beta V dW_k^Q}{1 + \Delta T_k^{(3)} f_k}\right)\right] \Big/ m_s$$

$$= \text{drift} - \left[\frac{1}{A_{m,n}(0)} \sum_{j=2m}^{2n-1} \Delta T^{(6)} P(0, T_{2(j+1)}) m_{2(j+1)} \sum_{k=\eta(t)}^{2j+1} \frac{\Delta T_k \sigma_k f_k^\beta V dW_k^Q}{1 + \Delta T_k f_k}\right] \Big/ m_s$$

$$= -\left[\frac{1}{A_{m,n}(0)} \sum_{j=2m}^{2n-1} \Delta T^{(6)} \frac{P(t, T_{2(j+1)}^{(3)})}{B_t} \sum_{k=\eta(t)}^{2j+1} \frac{\Delta T_k^{(3)} \gamma_k f_k^\beta V dW_k^Q}{1 + \Delta T_k^{(3)} f_k}\right] \Big/ m_s$$

$$= -\frac{1}{A_{m,n}(0)} \sum_{j=2m}^{2n-1} \Delta T^{(6)} \frac{P(t, T_{2(j+1)}^{(3)})}{m_s B_t} \sum_{k=\eta(t)}^{2j+1} \frac{\Delta T_k^{(3)} \gamma_k f_k^\beta V dW_k^Q}{1 + \Delta T_k^{(3)} f_k}$$

$$= -\sum_{j=2m}^{2n-1} 2\alpha_{2j+1} \sum_{k=\eta(t)}^{2j+1} \frac{\Delta T_k^{(3)} \gamma_k f_k^\beta V dW_k^Q}{1 + \Delta T_k^{(3)} f_k},$$

where

$$\alpha_j(t) = \frac{\Delta T^{(3)} P(t, T_{j+1}^{(3)})}{A_{m,n}(t)}.$$

Therefore, by CMG's change of measure theorem,

$$
dW_j^{Q_s} = dW_j^{Q} - \langle dW_j^{Q}, \frac{dm_s(t)}{m_s(t)} \rangle
$$

$$
= dW_j^{Q} + \sum_{j=2m}^{2n-1} 2\alpha_{2j+1} \sum_{k=\eta(t)}^{2j+1} \frac{\Delta T_k^{(3)} \gamma_k f_k^\beta V \rho_{kj}}{1 + \Delta T_k f_k} dt,
$$

$$
d\bar{W}_j^{Q_s} = d\bar{W}_j^{Q} - 0,
$$

$$
dZ^{Q_s} = dZ^{Q} - \langle dZ^{Q}, \frac{dm_s(t)}{m_s(t)} \rangle
$$

$$
= dZ^{Q} + \sum_{j=2m}^{2n-1} 2\alpha_{2j+1} \sum_{k=\eta(t)}^{2j+1} \frac{\Delta T_k^{(3)} \gamma_k f_k^\beta V \rho_{kV}}{1 + \Delta T_k^{(3)} f_k} dt.
$$

Substituting all these back to original dynamics yields (14.37). □

14.4.1.1   Swap Rate Dynamics under the Forward Swap Measure

Next, we derive the dynamics of the swap rate under the forward swap measure.

Proposition 14.4.2. *Under the forward swap measure, the swap rate (of collaterized swap) is a martingale, satisfying*

$$
ds_{m,n}(t) = \sum_{j=2m}^{2n-1} \frac{\partial s_{m,n}}{\partial f_j} \gamma_j f_j^\beta V dW_j^{Q_s} + \sum_{j=2m}^{2n-1} \frac{\partial s_{m,n}}{\partial h_j} \gamma_j h_j^\beta V d\bar{W}_j^{Q_s},
$$

where the partial derivatives can be computed in closed-form:

$$
\frac{\partial s_{m,n}}{\partial f_j} = \left(\frac{\Delta T_j}{1 + \Delta T_j f_j} \right) \left[\sum_{k=2m}^{j-1} \alpha_k L_k - \left(\sum_{l=2m}^{2n-1} 2\alpha_{2l+1} \bar{\mathcal{H}}(2l+1-j) \right) s_{m,n} \right]
$$
$$
+ \alpha_j (1 + \Delta T_j h_j),
$$

$$
\frac{\partial s_{m,n}}{\partial h_j} = \alpha_j (1 + \Delta T_j f_j),
$$

with $\bar{\mathcal{H}}(x) = 1 - \mathcal{H}(x)$, and $\mathcal{H}$ is the Heaviside step function defined such that $\mathcal{H}(x) = 1$ for $x \geq 0$ and $\mathcal{H}(x) = 0$ otherwise. □

Proof: By definition, a swap rate is given by

$$
s_{m,n}(t) = \frac{\sum_{k=2m}^{2n-1} \Delta T_k P(t, T_{k+1}) L_k(t)}{A_{m,n}(t)} = \sum_{k=2m}^{2n-1} \alpha_k(t) L_k(t),
$$

where

$$
L_k(t) = f_k(t) + h_k(t) + \Delta T_k f_k(t) h_k(t).
$$

Then, by taking partial derivative, we obtain

$$\frac{\partial s_{m,n}}{\partial f_j} = \sum_{k=2m}^{2n-1} \frac{\partial \alpha_k}{\partial f_j} L_k + \alpha_j(1 + \Delta T_j h_j)$$

and

$$\frac{\partial s_{m,n}}{\partial h_j} = \alpha_j(1 + \Delta T_j f_j).$$

Similar to Wu and Zhang (2008), $\frac{\partial \alpha_k}{\partial f_j}$ also can be computed explicitly, i.e.,

$$\frac{\partial \alpha_k}{\partial f_j} = \frac{\Delta T_k \frac{\partial P(t,T_{(k+1)})}{\partial f_j} A_{m,n} - \Delta T_k P(t,T_{k+1}) \frac{\partial A_{m,n}}{\partial f_j}}{A_{m,n}^2}$$

$$= \frac{\Delta T_k}{A_{m,n}} \left(\frac{-\Delta T_j}{1 + \Delta T_j f_j} P(t,T_{k+1}) \mathcal{H}(k-j) \right) - \frac{\alpha_k}{A_{m,n}} \frac{\partial A_{m,n}}{\partial f_j}$$

$$= \frac{\Delta T_k}{A_{m,n}} \left(\frac{-\Delta T_j}{1 + \Delta T_j f_j} P(t,T_{k+1}) \mathcal{H}(k-j) \right)$$
$$- \frac{\alpha_k}{A_{m,n}} \sum_{l=2m}^{2n-1} \Delta T_l^{(6)} \left(\frac{-\Delta T_j}{1 + \Delta T_j f_j} P(t,T_{2(l+1)}) \mathcal{H}(2l+1-j) \right)$$

$$= \left(\frac{-\Delta T_j}{1 + \Delta T_j f_j} \right) \alpha_k \mathcal{H}(k-j)$$
$$- \left(\frac{-\Delta T_j}{1 + \Delta T_j f_j} \right) \alpha_k \sum_{l=2m}^{2n-1} \frac{\Delta T_l^{(6)} P(t,T_{2(l+1)})}{A_{m,n}} \mathcal{H}(2l+1-j)$$

$$= \left(\frac{-\Delta T_j}{1 + \Delta T_j f_j} \right) \alpha_k \left[\mathcal{H}(k-j) - \sum_{l=2m}^{2n-1} 2\alpha_{2l+1} \mathcal{H}(2l+1-j) \right],$$

and $\mathcal{H}$ is the Heaviside step function defined such that $\mathcal{H}(x) = 1$ for $x \geq 0$ and $\mathcal{H} = 0$ otherwise. Note that

$$\sum_{l=2m}^{2n-1} 2\alpha_{2l+1} = \sum_{l=2m}^{2n-1} \frac{\Delta T_l^{(6)} P(t,T_{2(l+1)}^{(3)})}{A_{m,n}} = 1$$

and denoting $\bar{\mathcal{H}} = 1 - \mathcal{H}$, we have

$$\frac{\partial \alpha_k}{\partial f_j} = \left(\frac{\Delta T_j}{1 + \Delta T_j f_j} \right) \alpha_k \left[1 - \mathcal{H}(k-j) - \sum_{l=2m}^{2n-1} 2\alpha_{2l+1} \bar{\mathcal{H}}(2l+1-j) \right].$$

Substituting $\frac{\partial \alpha_k}{\partial f_j}$ into $\frac{\partial s_{m,n}}{\partial f_j}$ yields

$$
\begin{aligned}
\frac{\partial s_{m,n}}{\partial f_j} &= \sum_{k=2m}^{2n-1} \left(\frac{\Delta T_j}{1 + \Delta T_j f_j} \right) \alpha_k \left[1 - \mathcal{H}(k - j) - \sum_{l=2m}^{2n-1} 2\alpha_{2l+1} \bar{\mathcal{H}}(2l + 1 - j) \right] L_k \\
&\quad + \alpha_j (1 + \Delta T_j h_j) \\
&= \left(\frac{\Delta T_j}{1 + \Delta T_j f_j} \right) \left[\sum_{k=2m}^{j-1} \alpha_k L_k - \left(\sum_{l=2m}^{2n-1} 2\alpha_{2l+1} \bar{\mathcal{H}}(2l + 1 - j) \right) s_{m,n} \right] \\
&\quad + \alpha_j (1 + \Delta T_j h_j).
\end{aligned}
$$

The proof is completed. □

14.4.2 Geometric Inputs

As the local variance function is the conditional expectation of stochastic variance, so we thus have the following local variance of the swap rate:

$$
\begin{aligned}
\sigma_{LV}^2(K,t) =& E^{Q_s} \left[\sum_{j,k=2m}^{2n-1} \sigma_j \sigma_k \rho_{jk} f_j^\beta f_k^\beta \frac{\partial s_{m,n}}{\partial f_j} \frac{\partial s_{m,n}}{\partial f_k} V^2 \right. \\
&\left. + \sum_{j,k=2m}^{2n-1} \gamma_j \gamma_k \bar{\rho}_{jk} f_j^\beta f_k^\beta \frac{\partial s_{m,n}}{\partial h_j} \frac{\partial s_{m,n}}{\partial h_k} V^2 \bigg|_{s_{m,n}(t) = K} \right] \\
=& \sum_{j,k=2m}^{2n-1} \sigma_j \sigma_k \rho_{jk} E^{Q_s} \left[f_j^\beta f_k^\beta \frac{\partial s_{m,n}}{\partial f_j} \frac{\partial s_{m,n}}{\partial f_k} V^2 \bigg|_{s_{m,n}(t) = K} \right] \\
&+ \sum_{j,k=2m}^{2n-1} \gamma_j \gamma_k \bar{\rho}_{jk} E^{Q_s} \left[h_j^\beta h_k^\beta \frac{\partial s_{m,n}}{\partial h_j} \frac{\partial s_{m,n}}{\partial h_k} V^2 \bigg|_{s_{m,n}(t) = K} \right].
\end{aligned}
$$

To evaluate the conditional expectation, we again use the heat kernel expansion and integration by the saddle point method.

14.4.2.1 Inputs Parameter for the Heat Kernel Expansion

1. Metric matrix

$$
(g^{ij})_{N \times N} = \begin{pmatrix} g_f^{ij} & 0 & g_{fV} \\ 0 & g_h^{ij} & g_{hV} \\ (g_{fV})^T & (g_{hV})^T & V^2 \end{pmatrix},
$$

where

$$(g_f^{ij})_{N_s \times N_s} =$$

$$\begin{pmatrix} \frac{V^2}{\nu^2} f_{2m}^{2\beta} \phi_{f_{2m}}^2 & \cdots & \frac{V^2}{\nu^2} f_{2m}^{\beta} f_{2n-1}^{\beta} \phi_{f_{2m}} \phi_{f_{2n-1}} \rho_{2m,2n-1} \\ \cdots & \cdots & \cdots \\ \frac{V^2}{\nu^2} f_{2m}^{\beta} f_{2n-1}^{\beta} \phi_{f_{2m}} \phi_{f_{2n-1}} \rho_{2m,2n-1} & \cdots & \frac{V^2}{\nu^2} f_{2n-1}^{2\beta} \phi_{f_{2n-1}}^2 \end{pmatrix}$$

$$(g_h^{ij})_{N_s \times N_s} =$$

$$\begin{pmatrix} \frac{V^2}{\nu^2} f_{2m}^{2\beta} \phi_{f_{2m}}^2 & \cdots & \frac{V^2}{\nu^2} f_{2m}^{\beta} f_{2n-1}^{\beta} \phi_{f_{2m}} \phi_{f_{2n-1}} \bar{\rho}_{2m,2n-1} \\ \cdots & \cdots & \cdots \\ \frac{V^2}{\nu^2} f_{42}^{\beta} f_{2n-1}^{\beta} \phi_{f_{2m}} \phi_{f_{2n-1}} \bar{\rho}_{2m,2n-1} & \cdots & \frac{V^2}{\nu^2} f_{2n-1}^{2\beta} \phi_{f_{2n-1}}^2 \end{pmatrix}$$

$$(g_{fV})_{N_s \times 1} = \left(\frac{V^2}{\nu} f_{2m}^{\beta} \phi_{f_{2m}} \rho_{2m,V} \quad \cdots \quad \frac{V^2}{\nu} f_{2n-1}^{\beta} \phi_{f_{2n-1}} \rho_{2n-1,V} \right)^T$$

$$(g_{hV})_{N_s \times 1} = \left(\frac{V^2}{\nu} h_{2m}^{\beta} \phi_{h_{2m}} \rho_{2m,V} \quad \cdots \quad \frac{V^2}{\nu} h_{2n-1}^{\beta} \phi_{h_{2n-1}} \rho_{2n-1,V} \right)^T .$$

2. Geodesic distance

$$d = \cosh^{-1}\left(1 + \frac{(\mathbf{q} - \mathbf{q'})^T \boldsymbol{\rho}^{-1}(\mathbf{q} - \mathbf{q'})}{2 q_N q_N'}\right),$$

where $\mathbf{q} = (q_1^f, ..., q_{N_s}^f, q_1^h, ..., q_{N_s}^h, q_V)^T$ and

$$q_j^f = \int_{f_j(0)}^{f_j} \frac{1}{\phi_{f_j} f_j'^{\beta}} df_j'^{\beta}$$

$$q_j^h = \int_{h_j(0)}^{h_j} \frac{1}{\phi_{h_j} h_j'^{\beta}} dh_j'^{\beta}$$

$$q_V = \frac{V}{\nu}.$$

3. Van Vleck-Morette Determinant on H^N:

$$\Delta = \left(\frac{d}{\sinh(d)} \right)^N .$$

4. Determinant of metric matrix g

$$\sqrt{g} = \frac{\nu^{2N_s} \sqrt{\det(\boldsymbol{\rho}^{-1})}}{V^N \prod_{l=1}^{N_s} f_l^{\beta} h_l^{\beta} \phi_{f_l} \phi_{h_l}} .$$

5. Parallel transport

$$\ln \mathcal{P} \approx -\sum_j \int_{f_j}^{f_j(0)} \mathcal{A}_{f_j} df_j - \sum_j \int_{h_j}^{h_j(0)} \mathcal{A}_{h_j} dh_j - \int_V^{V(0)} \mathcal{A}_V dV,$$

where the one forms are

$$\mathcal{A}^{f_k} = \frac{1}{2}\mu^{f_k} - \frac{\beta V^2 f_k^{2\beta-1}\phi_{f_k}^2}{2\nu^2} + \frac{(N-2)V\rho_{kV}f_k^\beta \phi_{f_k}}{2\nu},$$

$$\mathcal{A}^{h_k} = \frac{1}{2}\mu^{h_k} - \frac{\beta V^2 h_k^{2\beta-1}\phi_{h_k}^2}{2\nu^2} + \frac{(N-2)V\bar{\rho}_{kV}h_k^\beta \phi_{h_k}}{2\nu},$$

$$\mathcal{A}^V = \frac{1}{2}\mu^V + \frac{N-2}{2}V,$$

and

$$\mathcal{A}_i = g_{ij}\mathcal{A}^i.$$

14.4.3 Local Volatility Function of Swap Rates

We will compute the expectations of the following functions

$$C^{jk} = f_j^\beta f_k^\beta \frac{\partial s_{m,n}}{\partial f_j} \frac{\partial s_{m,n}}{\partial f_k} V^2 \quad \text{and} \quad D^{jk} = h_j^\beta h_k^\beta \frac{\partial s_{m,n}}{\partial h_j} \frac{\partial s_{m,n}}{\partial h_k} V^2.$$

Note that if we use saddle point integration directly on each of these functions, we will need to calculate the partial derivatives of $s_{m,n}$ w.r.t. all f and h for each function, since $s_{m,n}$ depend on all f_j and $h_j, j = 4m, 4n-1$. This will be a heavy burden for computations. To mitigate such a burden, we fix all $\frac{\partial s}{\partial f}$ and $\frac{\partial s}{\partial h}$ at the saddle point, i.e.,

$$C^{jk} \approx f_j^\beta f_k^\beta \frac{\partial s_{m,n}(\mathbf{x}^*)}{\partial f_j} \frac{\partial s_{m,n}(\mathbf{x}^*)}{\partial f_k} V^2,$$

$$D^{jk} \approx h_j^\beta h_k^\beta \frac{\partial s_{m,n}(\mathbf{x}^*)}{\partial h_j} \frac{\partial s_{m,n}(\mathbf{x}^*)}{\partial h_k} V^2,$$

This approximation is motivated by the low variances of $\frac{\partial s_{m,n}}{\partial f}$ and $\frac{\partial s_{m,n}}{\partial h}$. So now both C^{jk} and D^{jk} depend only on $\mathbb{S}_{jk}^f = (f_j, f_k, V)$ and $\mathbb{S}_{jk}^h = (h_j, h_k, V)$ respectively. Then the conditional expectation for the local volatility of swap rate is given by

$$E^{Q_s}\left[f_j^\beta f_k^\beta \frac{\partial s_{m,n}}{\partial f_j} \frac{\partial s_{m,n}}{\partial f_k} V^2 \,\middle|\, s_{m,n}(t) = K \right] = C^{jk}(\mathbf{x}^*)\left[1 + \tau\left(\nu^2 \right. \right.$$

$$\sum_{a,b\in\mathbb{S}_{jk}^f} \left(\frac{C_{ab}^{jk}(\mathbf{x}^*)}{C^{jk}(\mathbf{x}^*)} + \frac{C_a^{jk}(\mathbf{x}^*)}{2C^{jk}(\mathbf{x}^*)}\partial_b(\ln g\Delta\mathcal{P}^2) + \frac{C_b^{jk}(\mathbf{x}^*)}{2C^{jk}(\mathbf{x}^*)}\partial_a(\ln g\Delta\mathcal{P}^2) \right)\Sigma_{ab}$$

$$\left. \left. -\frac{\nu^2}{3}\sum_{a,b,c,d\in\mathbb{S}_{jk}^f} G_{abc}(\Sigma_{ab}\Sigma_{cd} + \Sigma_{ac}\Sigma_{bd} + \Sigma_{ad}\Sigma_{bc})\frac{C_d^{jk}(\mathbf{x}^*)}{C^{jk}(\mathbf{x}^*)} \right) \right],$$

and

$$E^{Q_s}\left[h_j^\beta h_k^\beta \frac{\partial s_{m,n}}{\partial h_j} \frac{\partial s_{m,n}}{\partial h_k} V^2 \bigg| s_{m,n}(t) = K\right] = D^{jk}(\mathbf{x}^*)\left[1 + \tau\left(\nu^2\right.\right.$$

$$\sum_{a,b\in\mathbb{S}_{jk}^h}\left(\frac{D_{ab}^{jk}(\mathbf{x}^*)}{D^{jk}(\mathbf{x}^*)} + \frac{D_a^{jk}(\mathbf{x}^*)}{2D^{jk}(\mathbf{x}^*)}\partial_b(\ln g\Delta\mathcal{P}^2) + \frac{D_b^{jk}(\mathbf{x}^*)}{2D^{jk}(\mathbf{x}^*)}\partial_a(\ln g\Delta\mathcal{P}^2)\right)\Sigma_{ab}$$

$$\left.\left.-\frac{\nu^2}{3}\sum_{a,b,c,d\in\mathbb{S}_{jk}^h} G_{abc}(\Sigma_{ab}\Sigma_{cd} + \Sigma_{ac}\Sigma_{bd} + \Sigma_{ad}\Sigma_{bc})\frac{D_d^{jk}(\mathbf{x}^*)}{D^{jk}(\mathbf{x}^*)}\right)\right],$$

where $G = d^2$, and $(\Sigma)^{-1}$ is the Hessian matrix of G.

14.4.4   Calculation of the Saddle Point

To find the saddle point for the kernel of the swap rate, we need to solve the following optimization problem:

$$\min \quad \cosh^{-1}(1 + \frac{(\mathbf{q'} - \mathbf{q})^T\boldsymbol{\rho}^{-1}(\mathbf{q'} - \mathbf{q})}{2q_N'q_N})$$

$$s.t. \quad S(q_1, ..., q_{N-1}) = K.$$

Following the same arguments as those in the case of caplet pricing, we fix q_N to solve the QP problem inside, and then update q_N according to the formula

$$q_N^{(k+1)} = \sqrt{\frac{\sum_{i,j}^{N-1}\rho^{ij}q_i^{(k)}q_j - 2\sum_i^{N-1}\rho^{iN}q_i^{(k)}q_N^{(k)} + \rho^{NN}(q_N^{(k)})}{\rho^{NN}}},$$

where $(\rho^{ij})_{N\times N}$ is the inverse matrix of $\boldsymbol{\rho}$. We will repeat the iterations until convergence.

14.4.4.1   Interpolation in High Dimensional Cases

From our numerical experience, this algorithm works well for $N \le 45$, which corresponds to the cases of $T_n \le 5$ years. For longer tenors, it is better that we use interpolation to achieve the approximation. Note that for cases of short maturities for the underlying swaps, the difference between the saddle points and the current point, $\mathbf{x}^* - \mathbf{x}_0$, calculated using the above algorithm, have demonstrated a clear smoothness as a function of j, for both forward rates f_j and h_j, and for various strikes, as is shown in Figure 14.6. Here we set $T_m = 1$ year and $T_n = 5$ years.

Therefore, motivated by these observations, we choose a subset of $\{j\}$ (with dimensions ≤ 45) from all f_j and h_j, find their saddle points, and then use interpolations to determine other saddle points between them. In our examples for 10-year term swaptions, we actually use linear interpolation. Encouragingly, this works well.

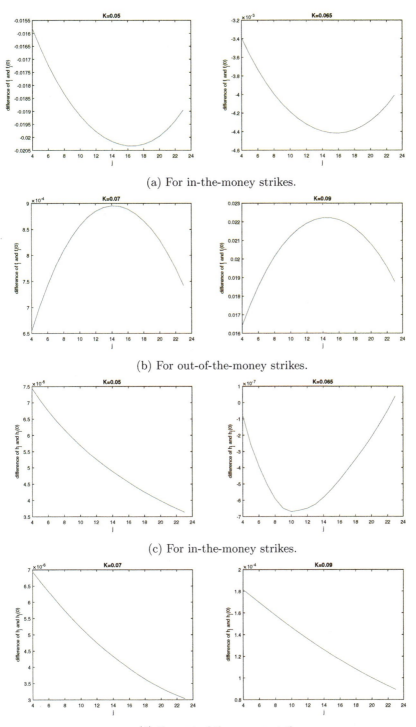

(a) For in-the-money strikes.

(b) For out-of-the-money strikes.

(c) For in-the-money strikes.

(d) For out-of-the-money strikes.

FIGURE 14.6: The differences, $\mathbf{x}^* - \mathbf{x}$.

14.4.5 Implied Black's Volatility

After having calculated the local volatility function for s_t, $C(s_t, t) = \sigma_{LV}(s_t, t)$, we use (14.35) to obtain the approximation to the Black's implied volatility.

14.4.6 Numerical Results for 3M Swaptions

The implied Black's volatilities for 3M swaptions using parameter set 1 are shown in Figure 14.7, which shows that the asymptotic implied volatility formula is of good accuracy as well.

We will try two more sets of parameters (see Figures 14.8 and 14.9).
Parameter set 2.

- Risk-free forward rates and volatilities:

$$f_j^{(4)}(0) = 6\% \quad \text{and} \quad \sigma_j(t) = 0.2.$$

- Expected loss rates and volatilities:

$$h_j^{(4)} = 0.0015 \exp(-0.4j\Delta T^{(4)}) \quad \text{and} \quad \gamma_j(t) = 0.2.$$

- Stochastic multiplier:

$$\alpha = 0.25, \quad \nu = 0.3, \quad \beta = 0.5, \quad \rho_{Vf} = -0.5, \quad \rho_{Vh} = 0.5.$$

Parameter set 3.

- Risk-free forward rates and volatilities:

$$f_j^{(4)}(0) = \log(a + bj), \quad f_0(0) = 1.5\%, \quad f_{20y}(0) = 5.5\%;$$
$$\sigma_j(t) = 0.17 - 0.002(T_{j-1}^{(4)} - t).$$

- Expected loss rates and volatilities:

$$h_j^{(4)} = 0.0015 \exp(-0.4j\Delta T^{(4)}); \qquad \gamma_j(t) = 0.05 + 0.002(T_{j-1}^{(4)} - t).$$

- Stochastic multiplier:

$$\alpha = 0.25, \quad \nu = 0.3, \quad \beta = 0.5, \quad \rho_{Vf} = -0.5, \quad \rho_{Vh} = 0.5.$$

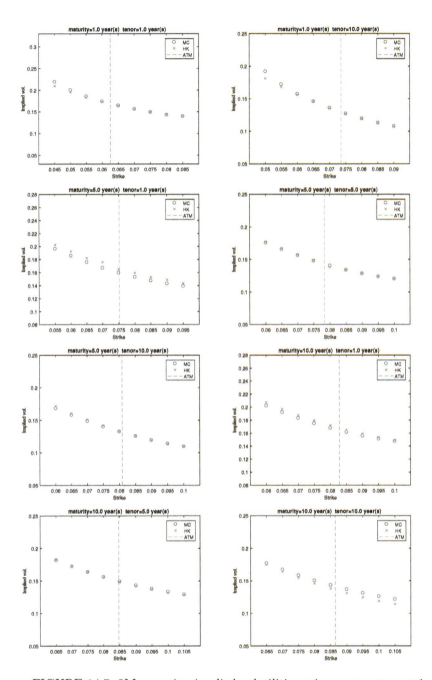

FIGURE 14.7: 3M swaption implied volatilities using parameter set 1.

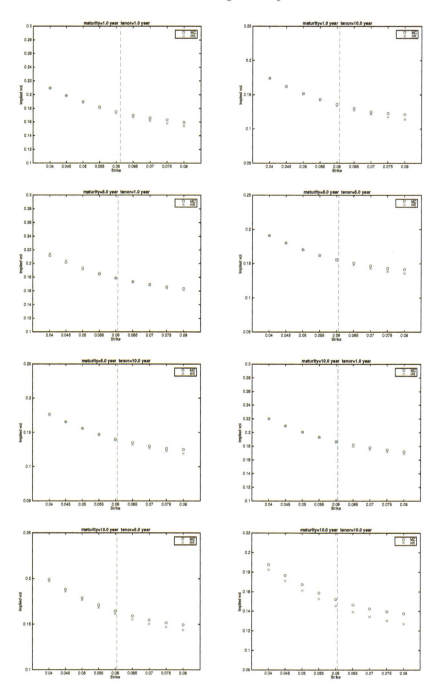

FIGURE 14.8: 3M swaption implied Black's volatilities using parameter set 2.

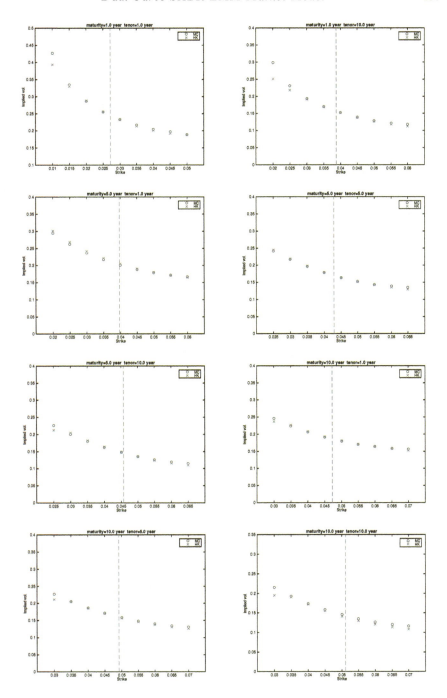

FIGURE 14.9: 3M swaption implied Black's volatilities using parameter set 3.

Interestingly, for 10-year maturity, the accuracy of swaptions is higher than that of the caplets.

14.5   Pricing Caps and Swaptions of Other Tenors

Since we have the dynamics of the 3M curves and a curve of risk premiums, we are able to derive the dynamics of forward-rate and swap-rate curves of other tenors, and eventually price the corresponding caplets and swaptions using similar methodologies. Without loss of generality, we take the caplets and swaptions of 6M tenor to demonstrate the procedure.

14.5.1   Linkage between 3M Rates and Rates of Other Tenors

14.5.1.1   The 6M Risk-Free OIS Rates

The OIS forward rate curve satisfies the traditional no-arbitrage relationship, i.e.,

$$1 + \Delta T^{(6)} f_k^{(6)}(t) = \left(1 + \Delta T^{(3)} f_{2k}^{(3)}(t)\right)\left(1 + \Delta T^{(3)} f_{2k+1}^{(3)}(t)\right),$$

meaning that the dynamics of 3M risk-free forward rates uniquely determine those of 6M risk-free forward rates.

14.5.1.2   The 6M Expected Loss Rates

Proposition 14.5.1. *By using the 3M expected loss rates and risk premiums, we can link up the 6M expected loss rates to the 3M expected loss rates as follows:*

$$h_k^{(6)} = \frac{L}{\Delta T^{(6)}}\left[1 - \left(1 - \frac{1}{L}\Delta T_{2k}^{(3)} h_{2k}^{(3)}(t)\right)\left(1 - \frac{1}{L}\Delta T_{2k+1}^{(3)} h_{2k+1}^{(3)}(t)\right)\Pi_k\right],$$

where

$$\Pi_k = \frac{\prod_{j=0}^{2}(1 + \frac{j}{12}\theta_{6k+j})\prod_{j=0}^{2}(1 + \frac{j}{12}\theta_{6k+3+j})}{\prod_{j=0}^{5}(1 + \frac{j}{12}\theta_{6k+j})}.$$

□

Proof: As the different tenors share the same default probability curve subject to the monthly panel review

$$\frac{\Lambda(t, T_{j+1}^{(6)})}{\Lambda(t, T_j^{(6)})} = \frac{\Lambda(t, T_{2j+2}^{(3)})}{\Lambda(t, T_{2j+1}^{(3)})} \cdot \frac{\Lambda(t, T_{2j+1}^{(3)})}{\Lambda(t, T_{2j}^{(3)})},$$

the following relationship between $D's$ holds for all $t \leq T_j^{(3)}$,

$$\frac{1}{L}(D_{j+1}^{(6)} - R) \prod_{k=0}^{5}(1 + \frac{k}{12}\theta_{6j+k}) = \frac{1}{L}(D_{2j+1}^{(3)} - R) \prod_{k=0}^{2}(1 + \frac{k}{12}\theta_{6j+k})$$

$$\times \frac{1}{L}(D_{2j+2}^{(3)} - R) \prod_{k=0}^{2}(1 + \frac{k}{12}\theta_{6j+3+k}(t)),$$

where, by definition

$$D_{j+1}^{(i)}(t) = E_t^{Q_{j+1}^{(i)}}\left[1 - L1_{\{T_j^{(i)} < \tau \leq T_{j+1}^{(i)}\}}\right]$$

$$= 1 - LE_t^{Q_{j+1}^{(i)}}\left[1_{\{T_j^{(i)} < \tau \leq T_{j+1}^{(i)}\}}\right]$$

$$= R + L\frac{\Lambda(t, T_{j+1}^{(i)})}{\Lambda(t, T_j^{(i)})}\frac{1}{\prod_{k=0}^{i-1}\left(1 + \frac{k}{12}\theta_{i \times j+k}\right)}.$$

Here, we have made use of the independence between default times and risk-free bond and $R = 1 - L$. When $t = T_j^{(6)} = T_{2j}^{(3)}$,

$$\frac{1}{L}\left(D_{j+1}^{(6)}(T_j^{(6)}) - R\right) \prod_{k=0}^{5}(1 + \frac{k}{12}\theta_{6j+k})$$

$$= \frac{1}{L}\left(D_{2j+1}^{(3)}(T_{2j}^{(3)}) - R\right) \prod_{k=0}^{2}(1 + \frac{k}{12}\theta_{6j+k})$$

$$\times \frac{1}{L}\left(D_{2j+2}^{(3)}(T_{2j}^{(3)}) - R\right) \prod_{k=0}^{2}(1 + \frac{k}{12}\theta_{6j+3+k}).$$

According to the martingale property of $D_{2j+2}^{(3)}(t)$,

$$D_{2j+2}^{(3)}(T_{2j}^{(3)}) = E^Q\left[D_{2j+2}^{(3)}(T_{2j+1}^{(3)})|\mathcal{F}_{T_{2j}^{(3)}}\right].$$

Together with the definition between H and D,

$$H_j^{(i)}(t) = \frac{1}{\Delta T_j^{(i)}}\left(\frac{1}{D_{j+1}^{(i)}(t)} - 1\right),$$

we have the relationship between $H's$ of different tenors at time $t = T_j^{(6)} = T_{2j}^{(3)}$,

$$\frac{1}{L}\left(\frac{1}{1 + \Delta T_j^{(6)} H_j^{(6)}(T_j^{(6)})} - R\right)\prod_{k=0}^{5}\left(1 + \frac{k}{12}\theta_{6j+k}\right)$$

$$= \frac{1}{L}\left(\frac{1}{1 + \Delta T_{2j}^{(3)} H_{2j}^{(3)}(T_{2j}^{(3)})} - R\right)\prod_{k=0}^{2}\left(1 + \frac{k}{12}\theta_{6j+k}\right)$$

$$\times \frac{1}{L}\left(E_{T_{2j}^{(3)}}^{Q}\left[\frac{1}{1 + \Delta T_{2j+1}^{(3)} H_{2j+1}^{(3)}(T_{2j+1}^{(3)})} - R\right]\right)\prod_{k=0}^{2}\left(1 + \frac{k}{12}\theta_{6j+3+k}\right).$$

Then, since $H's$ are relatively small in general, we approximate $\frac{1}{1+\Delta T \cdot H}$ by $1 - \Delta T \cdot H$,

$$\left(1 - \frac{1}{L}\Delta T_j^{(6)} H_j^{(6)}(T_j^{(6)})\right)\prod_{k=0}^{5}\left(1 + \frac{k}{12}\theta_{6j+k}\right)$$

$$= \left(1 - \frac{1}{L}\Delta T_{2j}^{(3)} H_{2j}^{(3)}(T_{2j}^{(3)})\right)\prod_{k=0}^{2}\left(1 + \frac{k}{12}\theta_{6j+k}\right)$$

$$\times \left(E_{T_{2j}^{(3)}}^{Q}\left[1 - \frac{1}{L}\Delta T_{2j+1}^{(3)} H_{2j+1}^{(3)}(T_{2j+1}^{(3)})\right]\right)\prod_{k=0}^{2}\left(1 + \frac{k}{12}\theta_{6j+3+k}\right) + O(H^2)$$

$$= \left(1 - \frac{1}{L}\Delta T_{2j}^{(3)} H_{2j}^{(3)}(T_{2j}^{(3)})\right)\prod_{k=0}^{2}\left(1 + \frac{k}{12}\theta_{6j+k}\right)$$

$$\times \left(1 - \frac{1}{L}\Delta T_{2j+1}^{(3)} E_{T_{2j}^{(3)}}^{Q}\left[H_{2j+1}^{(3)}(T_{2j+1}^{(3)})\right]\right)\prod_{k=0}^{2}\left(1 + \frac{k}{12}\theta_{6j+3+k}\right) + O(H^2).$$

We take conditional expectation $E^Q\left[\cdot|\mathcal{F}_t\right]$ on both sides and omit the $O(H^2)$ term, then

$$\left(1 - \frac{1}{L}\Delta T_j^{(6)} h_j^{(6)}(t)\right)\prod_{k=0}^{5}\left(1 + \frac{k}{12}\theta_{6j+k}\right)$$

$$= \left(1 - \frac{1}{L}\Delta T_{2j}^{(3)} h_{2j}^{(3)}(t)\right)\prod_{k=0}^{2}\left(1 + \frac{k}{12}\theta_{6j+k}\right)\left(1 - \frac{1}{L}\Delta T_{2j+1}^{(3)} h_{2j+1}^{(3)}(t)\right)$$

$$\prod_{k=0}^{2}\left(1 + \frac{k}{12}\theta_{6j+3+k}\right),$$

where the last step follows from the tower law.   □

In fact, we can have a much simpler approach to derive $h_k^{(i)}$. According to the definition of $H_j^{(i)}(t)$,

$$H_j^{(i)}(t) = \frac{1}{\Delta T_j^{(i)}} \left(\frac{1}{D_{j+1}^{(i)}(t)} - 1 \right),$$

there is

$$
\begin{aligned}
D_{j+1}^{(i)}(T_j^{(i)}) &= \frac{1}{1 + \Delta T_j^{(i)} H_j^{(i)}(T_j^{(i)})} \\
&= 1 - \Delta T_j^{(i)} H_j^{(i)}(T_j^{(i)}) + O(H_j^{(i)}(T_j^{(i)})^2).
\end{aligned}
\tag{14.38}
$$

Taking $Q_{j+1}^{(i)}$ expectation and making use the martingale property of $D_{j+1}^{(i)}(T_j^{(i)})$, we up with

$$
\begin{aligned}
D_{j+1}^{(i)}(t) &= 1 - \Delta T_j^{(i)} E^{Q_{j+1}^{(i)}} [H_j^{(i)}(T_j^{(i)})] + E^{Q_{j+1}^{(i)}}[O(H_j^{(i)}(T_j^{(i)}))^2] \\
&= 1 - \Delta T_j^{(i)} h_j^{(i)}(t) + E^{Q_{j+1}^{(i)}}[O(H_j^{(i)}(T_j^{(i)}))^2]
\end{aligned}
$$

which implies $h_j^{(i)}(t) = H_j^{(i)}(t) + E^{Q_{j+1}^{(i)}}[O(H_j^{(i)}(T_j^{(i)}))^2] \approx H_j^{(i)}(t)$. With such an approximation, the derivation of the expected loss rates becomes much simpler.

Similar approximations also apply to the expected loss rate of other tenors, making the generation of basis swap curve an easier task. We take the parameter set 3 for example. First, we use the dual-curve SABR-LMM model to simulate the 3M expected loss rates for one year, and then derive the 1M, 6M and 12M loss rates using the risk premiums shown in Figure 14.1, and finally calculate the differences of the expected loss rates to obtain the basis curves. The results are plotted in Figure 14.10. One can see the stylized pattern of the basis swap curve.

14.5.2 Dynamics of the 6M Risky LIBOR Rates

By model construction, the risky LIBOR rates are martingales under the corresponding forward measure, and the dynamics can be easily derived.

Proposition 14.5.2. *Dynamics of the 6M risky LIBOR under dual curve-SABR market model are given by*

$$
dL_k^{(6)} = \sum_{j=2k}^{2k+1} \frac{\partial L_k^{(6)}}{\partial f_j^{(3)}} \left[f_j^{(3)} \right]^\beta \gamma_j(t) V dW_j + \sum_{j=2k}^{2k+1} \frac{\partial L_k^{(6)}}{\partial h_j^{(3)}} \left[h_j^{(3)} \right]^\beta \sigma_j(t) V d\bar{W}_j,
$$

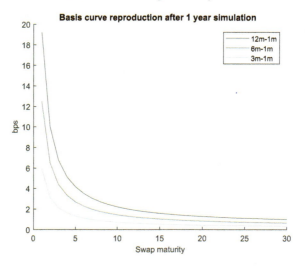

FIGURE 14.10: Simulated basis spread curves one year from now.

where

$$\frac{\partial L_k^{(6)}}{\partial f_{2k}^{(3)}} = \frac{1}{2}\left(1 + \Delta T^{(6)} h_k^{(6)}\right)\left(1 + \Delta T^{(3)} f_{2k+1}^{(3)}\right),$$

$$\frac{\partial L_k^{(6)}}{\partial f_{2k+1}^{(3)}} = \frac{1}{2}\left(1 + \Delta T^{(6)} h_k^{(6)}\right)\left(1 + \Delta T^{(3)} f_{2k}^{(3)}\right),$$

and

$$\frac{\partial L_k^{(6)}}{\partial h_{2k}^{(3)}} = \frac{1}{2}\left(1 + \Delta T^{(6)} f_k^{(6)}\right)\left(1 - \frac{1}{L}\Delta T^{(3)} h_{2k+1}^{(3)}\right)\Pi_k,$$

$$\frac{\partial L_k^{(6)}}{\partial h_{2k+1}^{(3)}} = \frac{1}{2}\left(1 + \Delta T^{(6)} f_k^{(6)}\right)\left(1 - \frac{1}{L}\Delta T^{(3)} h_{2k}^{(3)}\right)\Pi_k.$$

The proof is straightforward and is left to the readers.

14.5.3 Dynamics of the 6M Swap Rates

Similarly, we can also derive the dynamics of 6M swap rates under the corresponding forward swap measure.

Proposition 14.5.3. *Under the dual-curve SABR-LMM market model, the dynamics of the 6M swap rates is given by*

$$ds_{m,n}^{(6)}(t) = \sum_{j=4m}^{4n-1} \frac{\partial s_{m,n}^{(6)}}{\partial f_j^{(3)}} \gamma_j \left[f_j^{(3)} \right]^\beta V dW_j^{Q_s} + \sum_{j=4m}^{4n-1} \frac{\partial s_{m,n}^{(6)}}{\partial h_j^{(3)}} \sigma_j \left[h_j^{(3)} \right]^\beta V d\bar{W}_j^{Q_s},$$

where Q_s stands for the forward swap measure corresponding to numeraire $A_{m,n}^{(6)}$, and

$$\frac{\partial s_{m,n}^{(6)}}{\partial f_{2j}^{(3)}} = \left(\frac{\Delta T_j^{(3)}}{1 + \Delta T_j^{(3)} f_{2j}^{(3)}} \right) \left[\sum_{k=2m}^{j-1} \alpha_k^{(6)}(L_k^{(6)} - s_{m,n}^{(6)}) \right]$$
$$+ \frac{1}{2}\alpha_j^{(6)} \left(1 + \Delta T^{(6)} h_j^{(6)} \right) \left(1 + \Delta T^{(3)} f_{2j+1}^{(3)} \right),$$

$$\frac{\partial s_{m,n}^{(6)}}{\partial f_{2j+1}^{(3)}} = \left(\frac{\Delta T_j^{(3)}}{1 + \Delta T_j^{(3)} f_{2j}^{(3)}} \right) \left[\sum_{k=2m}^{j-1} \alpha_k^{(6)}(L_k^{(6)} - s_{m,n}^{(6)}) \right]$$
$$+ \frac{1}{2}\alpha_j^{(6)} \left(1 + \Delta T^{(6)} h_j^{(6)} \right) \left(1 + \Delta T^{(3)} f_{2j}^{(3)} \right),$$

$$\frac{\partial s_{m,n}^{(6)}}{\partial h_{2j}^{(3)}} = \frac{1}{2}\alpha_j^{(6)} \left(1 + \Delta T^{(6)} f_j^{(6)} \right) \left(1 - \frac{1}{L}\Delta T^{(3)} h_{2j+1}^{(3)} \right) \Pi_j,$$

$$\frac{\partial s_{m,n}^{(6)}}{\partial h_{2j+1}^{(3)}} = \frac{1}{2}\alpha_j^{(6)} \left(1 + \Delta T^{(6)} f_j^{(6)} \right) \left(1 - \frac{1}{L}\Delta T^{(3)} h_{2j}^{(3)} \right) \Pi_j,$$

for $j = 2m, 2m+1, ...2n$. $\qquad\qquad\square$

Proof: By the chain rule,

$$\frac{\partial s_{m,n}^{(6)}}{\partial f_{2j}^{(3)}} = \frac{\partial s_{m,n}^{(6)}}{\partial f_j^{(6)}} \cdot \frac{\partial f_j^{(6)}}{\partial f_{2j}^{(3)}}$$
$$= \left\{ \left(\frac{\Delta T_j^{(6)}}{1 + \Delta T_j^{(6)} f_j^{(6)}} \right) \left[\sum_{k=2m}^{j-1} \alpha_k^{(6)}(L_k^{(6)} - s_{m,n}^{(6)}) \right] + \alpha_j^{(6)}(1 + \Delta T^{(6)} h_j^{(6)}) \right\}$$
$$\times \frac{1}{2}(1 + \Delta T^{(3)} f_{2j+1}^{(3)}),$$

$$= \left(\frac{\Delta T_j^{(3)}}{1 + \Delta T_j^{(3)} f_{2j}^{(3)}} \right) \left[\sum_{k=2m}^{j-1} \alpha_k^{(6)}(L_k^{(6)} - s_{m,n}^{(6)}) \right]$$
$$+ \frac{1}{2}\alpha_j^{(6)}(1 + \Delta T^{(6)} h_j^{(6)})(1 + \Delta T^{(3)} f_{2j+1}^{(3)}),$$

$$\frac{\partial s_{m,n}^{(6)}}{\partial f_{2j+1}^{(3)}} = \left(\frac{\Delta T_j^{(3)}}{1 + \Delta T_j^{(3)} f_{2j}^{(3)}} \right) \left[\sum_{k=2m}^{j-1} \alpha_k^{(6)}(L_k^{(6)} - s_{m,n}^{(6)}) \right]$$
$$+ \frac{1}{2}\alpha_j^{(6)}(1 + \Delta T^{(6)} h_j^{(6)})(1 + \Delta T^{(3)} f_{2j}^{(3)}).$$

Similarly, we have

$$
\frac{\partial s_{m,n}^{(6)}}{\partial h_{2j}^{(3)}} = \frac{\partial s_{m,n}^{(6)}}{\partial h_j^{(6)}} \cdot \frac{\partial h_j^{(6)}}{\partial h_{2j}^{(3)}}
$$

$$
= \left[\alpha_j^{(6)} (1 + \Delta T^{(6)} f_j^{(6)}) \right] \cdot \frac{1}{2} (1 - \frac{1}{L} \Delta T^{(3)} h_{2j+1}^{(3)}) \Pi_j
$$

$$
= \frac{1}{2} \alpha_j^{(6)} \left(1 + \Delta T^{(6)} f_j^{(6)} \right) \left(1 - \frac{1}{L} \Delta T^{(3)} h_{2j+1}^{(3)} \right) \Pi_j,
$$

$$
\frac{\partial s_{m,n}^{(6)}}{\partial h_{2j+1}^{(3)}} = \frac{1}{2} \alpha_j^{(6)} \left(1 + \Delta T^{(6)} f_j^{(6)} \right) \left(1 - \frac{1}{L} \Delta T^{(3)} h_{2j}^{(3)} \right) \Pi_j.
$$

This completes the proof.   □

By following similar steps of the heat kernel expansion method, we can also calculate the implied Black's volatilities for caplets and swaptions with the 6M tenor and compare them with the implied Black's volatilities obtained from the Monte Carlo simulation method.

14.5.4   Numerical Results of 6M Caplets and Swaptions

Figure 14.11 shows the implied Black's volatilities for the 6M caplets using parameter set 1. Figures 14.12–14.14 show the implied Black's volatilities for the 6M swaptions using parameter sets 1–3. One can observe that the accuracy is quite good in general.

14.5.5   Model Calibration

Example 3: model calibration. We try our model to calibrate the 3M caplet volatilities of December 16, 2015, which are given in Appendix. The results is displayed in Figure 14.15. One can see that the overall accuracy of the calibration is good.

14.6   Notes

We have revealed the mechanism behind the basis spread curves, and constructed a dual-curve version of the SABR-LMM model for pricing LIBOR derivatives. We have chosen the approach of heat kernel expansion of Henry-Labordère (2005) to solve for the implied Black's volatilities, and show that the approach, although quite complex, works for the dual-curve SABR-LMM model. The disadvantage of the approach is that it seems to offer no convenience to calculating the Greeks, unlike the closed-form solution of Hagan

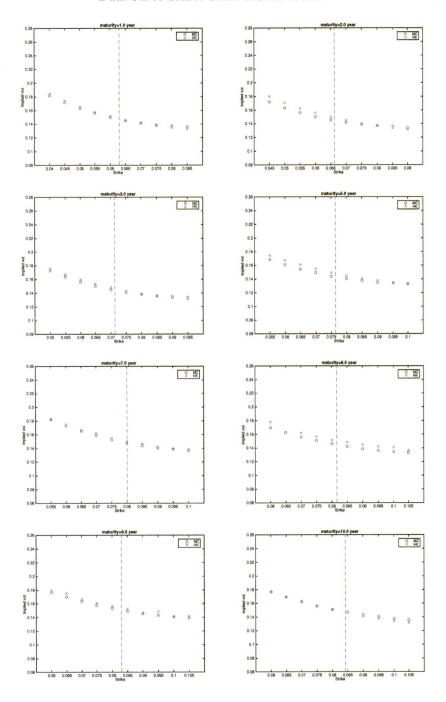

FIGURE 14.11: Implied Black's volatilities for the 6M caplets with parameters set 1.

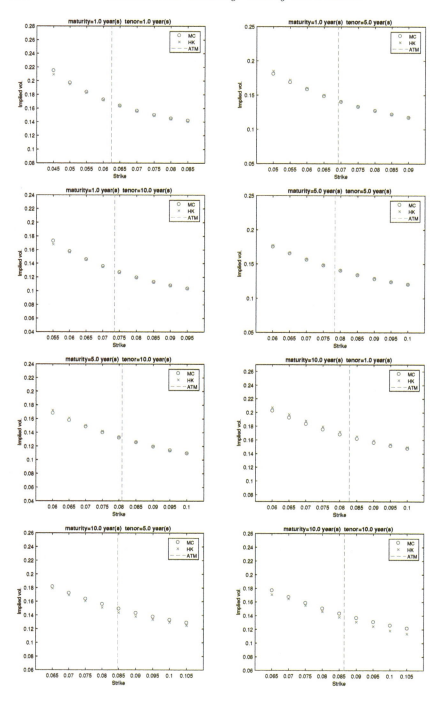

FIGURE 14.12: Implied Black's volatilities for the 6M swaption with parameters set 1.

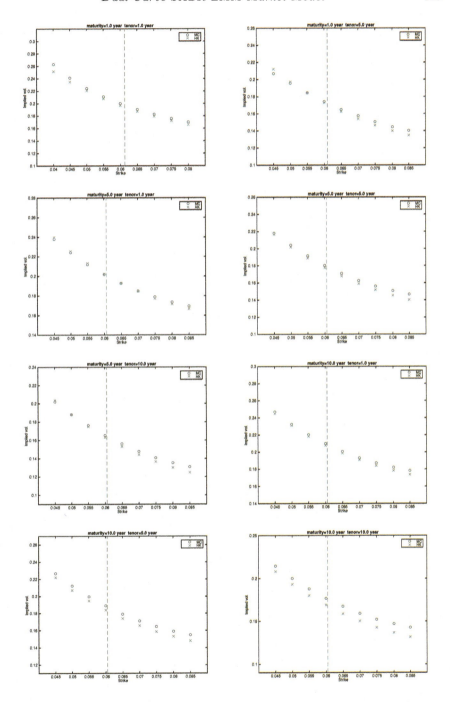

FIGURE 14.13: Implied Black's volatilities for the 6M swaption with parameters set 2.

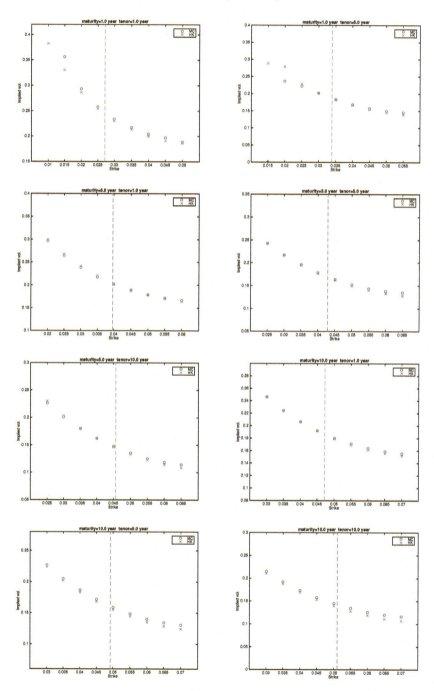

FIGURE 14.14: Implied Black's volatilities for the 6M swaption with parameter set 3.

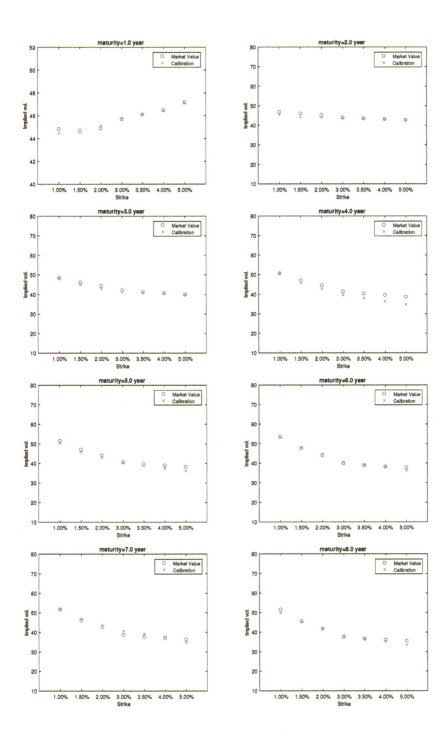

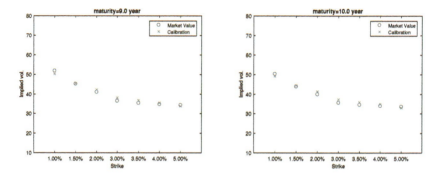

FIGURE 14.15: Calibration of caplet implied volatilities for 12/16/2015.

et al. (2002). To calculate the Greeks, we may have to rely on bumping. The dual-curve SABR-LMM itself is intuitive and financially plausible, so we think that more efforts are worthy to enhance this approach and on top of it develop simpler pricing formulas for caplets and swaptions.

Appendix: Implied Caplet Volatilities of 12/16/2015

Expiry	1.00%	1.50%	2.00%	3.00%	3.50%	4.00%	5.00%
1Yr	44.8	44.58	44.85	45.66	46.08	46.46	47.16
2Yr	46.7	45.84	45	43.75	43.32	42.98	42.51
3Yr	48.12	45.97	44.27	41.92	41.14	40.53	39.71
4Yr	50.45	46.84	44.36	41.2	40.2	39.47	38.52
5Yr	51.47	46.89	43.92	40.38	39.38	38.7	37.95
6Yr	53.4	47.5	43.81	39.7	38.65	38.02	37.53
7Yr	51.52	45.97	42.47	38.53	37.5	36.84	36.22
8Yr	51.35	45.36	41.61	37.48	36.46	35.86	35.41
9Yr	51.81	45.14	40.99	36.45	35.36	34.72	34.3
10Yr	50.22	43.83	39.85	35.55	34.52	33.94	33.58

Note: in basis points.

Chapter 15

xVA: Definition, Evaluation, and Risk Management

The 2008 financial crisis catalyzed many changes to the global derivatives market which, in particular, are seen through the proliferation of collaterals, the mandate to trade all derivatives through central counterparty clearing houses (CCP), and the emergence of various valuation adjustments, so-called xVA, in either pricing or accounting. The xVA contains a growing list of valuation adjustments (VAs): credit valuation adjustment (CVA), debit valuation adjustment (DVA), funding valuation adjustment (FVA), funding cost adjustment (FCA), collateral valuation adjustment (ColVA), margin valuation adjustment (MVA), capital valuation adjustment (KVA), and etc. Yet, except for CVA and DVA,[1] the other VAs have been controversial because, in addition to a lack of unanimous definition, their inclusion in pricing or booking causes price asymmetry and asset-liability asymmetry to the trading parties. In this chapter, we will redefine the notions of CVA, DVA and FVA and derive the formulae for the rest of xVA as the expected present values of excessive cash flows due to funding spreads under the risk-neutral pricing measure. We then identify the component of the xVA to be included into the fair price defined in IFRS 13 (i.e., exit price or market price, see FASB (2011)). In the end we will make our suggestions to managing the unhedgeable idiosyncratic risks behind the rest of xVA or xVA components.

The controversies started with the FVA, which is supposed to account for the funding costs for managing or manufacturing the cash flows of trades, including the funding costs for margins/collateral, capital as well as hedging. As early as 2012, a few major banks had started to adopt the notion and reported FVA as losses to their business, but such a practice has been criticized as a breach of the accounting principles which require booking with exit prices and maintaining asset-liability symmetry, see Hull and White (2012). Moreover, in a poll conducted by Risk.net in March 2015 (Sherif, 2015), about two thirds of quants believe that banks have overstated FVA losses and the current FVA model is wrong. As a matter of fact, there is not yet a market-wide consensus on issues like how to quantify the FVA, whether FVA is part of the fair-value pricing, or whether FVA is merely an accounting entry. For the remaining xVA items, we are facing more or less the same issues.

---

[1]DVA is the CVA of the counterparty.

To explain the motivations of our research, we need to get down to the FVA debate. Our starting point is the first as well as a prevalent FVA model of the market, see e.g. Cameron (2013). Let B stand for a bank and x_B be its funding spread for unsecured borrowing or lending.[2] If bank B has an uncollateralized derivative trade with the time-t value $V(t)$, then to bank B the FVA of the trade is defined to be

$$\text{FVA}_B = -E_0^Q \left[\int_0^{\tau \wedge T} x_B e^{-\int_0^t r_s ds} V(t) dt \right], \tag{15.1}$$

where τ is the time of the first bilateral default of the bank and its counterparty, T is the maturity of the derivative, r_t is the overnight risk-free rate,[3] and Q is the risk-neutral pricing measure. Note that the above formula arises from hedging an uncollateralized interest-rate swap using a fully collateralized swap, thus inducing either funding cost or funding benefit for posting or receiving cash collateral.[4] Formula (15.1) has been extended to a portfolio of trades with a single counterparty:

$$\text{FVA}_B = -E_0^Q \left[\int_0^{\tau} x_B e^{-\int_0^t r_s ds} \sum V_i(t) dt \right],$$

where the sum is over all trade with the counterparty, and $V_i(t)$ is the value of the ith trade; and to a portfolio of trades with multiple counterparties:

$$\text{FVA}_B = -E_0^Q \left[\int_0^{\tau_B} x_B e^{-\int_0^t r_s ds} \sum V_j(t) 1_{t < \tau_j} dt \right],$$

where the sum is over the netted trade values with all counterparties, and $V_j(t)$ is the netted value of all trades with counterparty j; and τ_j is the default time of counterparty j. Note that the FVA$_B$'s defined above are not the results of arbitrage pricing, so they may better be taken as metrics for the funding costs instead of valuation adjustments for fair-value pricing. Arguably these metrics make good sense for an uncollateralized swap and a swap portfolio, but it is not necessarily so for general derivatives or portfolios of derivatives, traded either in over-the-counter (OTC) markets or through CCPs.

To tackle funding costs in generality, a number of alternative models have been developed, including Brigo et al. (2011), Crépay (2011), Burgard and Kjaer (2011), Brigo et al. (2012), Bo and Capponi (2013), Lou (2015), Bichuch et al. (2015) and Li and Wu (2015). Under these models, FVA is defined as the expected present value of funding costs for posting/receiving collaterals as well as for hedging under the risk-neutral measure, and it is often modeled and evaluated together with CVA and DVA. Burgard and Kjaer (2011), in

---

[2]The funding spread can be asymmetrical for borrowing and for lending.

[3]The proxy for the overnight risk-free rate is the Federal funds rate for USD and EONIA rate for Euro.

[4]See Funding Valuation Adjustment (FVA), Part 1: A Primer, March 20, 2014, https://www.quantifisolutions.com.

particular, pioneer the arbitrage pricing of derivatives subject to market risks, counterparty default risks and funding costs/risks, and obtain the triad of valuation adjustments (i.e., CVA, DVA and FVA) under a consistent framework. As valuation adjustments to the otherwise no-default value of a derivative, CVA and DVA are well received, yet FVA is not: it makes fair-value pricing entity dependent such that two trading parties may no longer share the same "fair price" even if they share the same "risk-neutral pricing measure."

As an accounting entry, FVA is equally problematic. First, booking an FVA-adjusted price for a trade violates the rule of IFRS 13 (IASB, 2012) for using exit prices in fair-value accounting. Second, FVA booking for P&L accounting causes asset-liability asymmetry, unless a trading party willingly registers his counterparty's FVA loss/benefit as his FVA benefit/loss, which is not likely to happen. Nonetheless, as a measure of potential loss to their business, FVA is not ignored by major investment banks, and there have been efforts to accommodate the FVA without breaching the accounting principles. For examples, Albanese et al. (2015) and Andersen et al. (2016) advocate the reduction of FVA from a firm's equity value. A compromise, simultaneously occurring in both theory and practice, seems to arise in recent years. Hull and While (2014) have softened their stand against FVA and accepted the inclusion of the funding cost associated to the market funding liquidity risk premium into the fair value of a trade, as was first advocated by Morini and Prampolini (2010). In the marketplaces, there are increasing evidences that show banks have converged to using a uniform "market cost of funding" to price derivatives (Gregory, 2015).

Also controversial has been the DVA accounting adopted by some banks. When a bank's credit quality deteriorates, the value of its liabilities depreciates while its DVA appreciates for the same amount. As such, some banks went on to register the DVA appreciations as "profits," yet such paper profits have been disturbing and have drawn severe criticisms. As is argued in Castagna (2012), shareholders of the banks can hardly benefit from such "profits" unless the banks buy back related liabilities amid their financial distress, which is also not likely to happen. As a response to the criticisms, banks have recently excluded DVA from their Tier 1 equity capital and, according to a recent FASB revision,[5] started to register DVA under "other comprehensive incomes."

In this chapter, we will price general derivatives and quantify the xVA in two steps. First, we identify the risk-neutral pricing measure and then uncover the corresponding hedging or replicating strategies against the market risks and the counterparty default risks. Note that these strategies are unaffected by the funding spreads and are to be taken bilaterally by the counterparties. The cost of the replications in the absence of funding spreads is identified to be nothing else but the risk-neutral value of the derivative, and it can be decomposed into the no-default value and the bilateral CVAs. Second, we will figure out the additional costs caused by the funding spreads (for

---

[5]FASB News Release, January 5, 2016, www.fasb.org.

margins/collaterals, capitals as well as for hedging) and, following the market convention, define the FVA as the expected present values of the additional costs under the risk-neutral measure. Out of the FVA we further identify FCA, MVA, ColVA and KVA as components. Nonetheless, it is already known that pricing in unilateral funding risks will normally cause a price spread between the counterparties (e.g. Ruiz 2013, Wu 2015, Li and Wu 2015). As a major step forward, we will show that the market funding risk premium for unsecured lending and borrowing can be bilaterally priced into derivatives trades, without causing price asymmetry. This finding justifies the emerging market practice to charge a uniform market funding liquidity risk premium in derivatives trades.

As far as the management of funding risk is concerned, we suggest the adoption of existing popular risk metrics, like VaR or CVaR. Since the realized funding cost is a path-dependent random variable, the VaR or CVaR metric can be evaluated simultaneously with derivatives fair prices under Monte Carlo simulations. Note that FVA itself is not necessarily a good measure for risk management purposes, as it is an expectation of discounted funding costs under the risk-neutral measure,[6] which can be very different from the realized funding cost. Hence, either stacking up reserve or deducting Tier 1 capital according to the FVA number is not quite justified as proper risk management measure.

The pricing approach based on the bilateral replication has several advantages over other existing approaches. First, it is an extension to the classic unilateral replication pricing and it leads to the unique fair price that cannot be dominated by either party's trading strategy. Second, it does not take the pricing measure for granted and, instead, it treats the construction of pricing measure as a part of the pricing problem and identifies the replication strategies to be taken bilaterally by the counterparties. Third, the approach can be applied to xVA pricing under more general incomplete market models (e.g., models with stochastic interest rates, stochastic hazard rates, and jump risks), and the xVA formulations will remain quite the same. In addition, our xVA formulations can be conveniently adapted to pricing derivatives trades in CCP and in OTC markets, and the latter can be either collateralized or non-collateralized.

This chapter is organized as follows. In Section 15.1, we first describe the current conventions for posting margins, collaterals or risk capitals, then we present the bilateral replication for the fair value of a derivative in the absence of funding spread. In Section 15.2 we separate the P&L of a trade into the shortfall of payout replication and the funding cost, and present the xVA formulae. In particular, we show that the market funding risk premium can be bilaterally priced into a trade, without causing price asymmetry to the counterparties. In Section 15.3 we present an example of xVA calculation with an interest-rate swap. In Section 15.4 we conclude.

---

[6] Although it can be evaluated under the real-world measure as well.

15.1   Pricing through Bilateral Replications

15.1.1   Margin Accounts, Collaterals, and Capitals

We aim at developing a derivative pricing theory for both OTC markets and exchange markets. As of today, OTC derivative trades are largely collateralized[7] or CSA trades,[8] and collaterals are subject to periodic revisions. These are also the major features of CCP trades. As a result, we can tackle the pricing problem in the OTC market and exchange markets with the same approach.

Let us consider the pricing of a partially collateralized European option trade between two defaultable parties, B a bank and C a counterparty. If there is no default by either party until the maturity of the derivative, the party of liability will make a contractual payment to the counterparty. In case of a premature default, the party of exposure will seize the collateral, and will remain entitled to a fair share of recovery values. For subsequent discussions, we introduce the following notations.

Y_T	—	the contractual payoff of the derivative at maturity T;
τ_i	—	the default time of party i, for $i = B$ and C;
$I_i(t)$	—	≥ 0, the value of the initial margin (IM) posted by party i, $i = B$ and C;
$X_i(t)$	—	≥ 0, the value of the variable margin (VM) or collaterals posted by party i, $i = B$ and C;
$K_i(t)$	—	≥ 0, the value of risk capitals allocated to party i by shareholders, $i = B$ and C;
$V(t)$	—	the fair value of the derivative to B, the bank, s.t. $V(t) > 0$ — asset, $V(t) < 0$ — liability.

The margins, collaterals and capitals are typically in the form of cash or other Tier 1 capital assets, which cost fees to borrow and thus incurs funding costs. When one party is a clearinghouse, we may ignore its default risk and waive the requirements for margins/collaterals and capitals. Collaterals and variable margins may be lent out by the receiving party for its own funding purposes, which is called rehypothecation, and thus the party who poses funds may be rewarded with higher returns. The levels of margins or the amount of collaterals are revised periodically. In fact, the determination and maintenance of the margin levels or the collateral amount are specialized topics, which are addressed in details in e.g. Gregory (2015) or Green (2016).

---

[7]ISDA Margin Survey 2014, http://www2.isda.org/functional-areas/research/surveys/margin-surveys/.

[8]A collateralized trade is subject to Credit Support Annex (CSA), a legal document which regulates collateral for derivative transactions, and thus is called a CSA trade.

At the moment of the first bilateral default, $\tau = \tau_B \wedge \tau_C$, there may be a downward jump in the derivatives value. We let $V(\tau)$ denote the post-default value, such that $V(\tau) = V(\tau+)$, which has left limit and is right continuous.[9] For pricing purpose, we need to specify the post-default value of the derivative, by using either the Standardized Method or the Internal Model Method (IMM). The Standardized Method consists of a series of formulae set up by regulators (see BIPRU (Financial Conduct Authority, 2014)), and the IMM gives a firm the freedom of choice for advanced models, subject to regulator's approval. In our numerical demonstration, we will use the second method, under which the default settlement value can be conveniently described. Let $M(t)$ be the mark-to-market (MtM) value of the derivative, which typically is obtained through a dealer poll mechanism, and let $c_i(t) = I_i(t) + X_i(t)$ be the total value of margins/collaterals posted by party i, then the general bilateral close-out conditions upon the first bilateral default are

$$
\begin{aligned}
V(\tau = \tau_B) &= R_B[M(\tau_B) + c_B(\tau_B)]^- + [M(\tau_B) + c_B(\tau_B)]^+ - c_B(\tau_B), \\
V(\tau = \tau_C) &= R_C[M(\tau_C) - c_C(\tau_C)]^+ + [M(\tau_C) - c_C(\tau_C)]^- + c_C(\tau_C),
\end{aligned}
\tag{15.2}
$$

where R_B and R_C are the recovery rates for the losses of respective defaults, the sup indices "+" and "−" mean a floor and a cap to a function at the level of zero: $f^+(\tau) = \max\{f(\tau), 0\}$ and $f^-(\tau) = \min\{f(\tau), 0\}$. For notational simplicity we write $V(\tau_i)$ for $V(\tau = \tau_i)$, $i = B$ and C.

For both mathematical or notational simplicity, we make two non-essential assumptions: (1) the risk-free rate and default intensities are deterministic functions of time, and (2) the credit default swap (CDS) rates remain constants. Also, we assume there is no default at $t = 0$, the current moment.

15.1.2 Pricing in the Absence of Funding Cost

We will model the market with probability space $(\Omega, \mathcal{G}, \mathcal{G}_t, \mathbb{P})$, where $\mathbb{P}$ is the real-world measure, $\mathcal{G}_t$ is the filtration that represents all market information up to time t, such that $\mathcal{G}_t = \mathcal{F}_t \vee \mathcal{H}_t$, where $\mathcal{F}_t$ contains all market information except the default statuses of the trading parties, while $\mathcal{H}_t$ contains only the information of default statuses of the counterparties.

Without loss of generality, we consider the pricing of equity European options in a market where the following securities are traded: the money market account, shares or stocks, repurchasing agreements (repos), and CDS. The price dynamics of these securities are described below.

The balance of the money market account evolves according to

$$
dB_t = (r_t + x_m)B_t dt, \quad \text{with} \quad B_0 = 1,
$$

where x_m is the funding risk premium for the general market, so that $B_t = e^{\int_0^t (r_s + x_m)ds}$, which will be the numeraire asset for the risk-neutral measure to

---

[9]So-called a cádlág function.

be defined shortly. Note that the money market account is available for firms with negligible default probability. For other firms, repo type of transactions, i.e, collaterals, can be utilized to open the account. For notational simplicity we let $\hat{r}_t = r_t + x_m$.

The share price follows the usual lognormal dynamics:

$$dS_t = S_t \left[\mu_t dt + \sigma_t dW_t^{(P)} \right],$$

where $W_t^{(P)}$ is a one-dimensional Brownian motion under $\mathbb{P}$, μ_t is the expected return, and σ_t is the percentage volatility. Note that the lognormal dynamics is adopted in this chapter for the sake of simplicity, and more general dynamics can be considered with, of course, additional complexity.

The shares for hedging purpose will be "repoed in," such that the shares will be used as collaterals for borrowing. The instantaneous return from holding the repo is

$$dZ_S(t) = dS_t - (\hat{r}_t + \lambda_S - q_t)S_t dt$$
$$= S_t \left[(\mu_t - \hat{r}_t - \lambda_S + q_t)dt + \sigma_t dW_t^{(P)} \right],$$

where the second term on the RHS of the first equality is the cost of carry for the repo trade, with q_t being the dividend yield of the share and λ_S being the repo spread.[10] For a repo entered at time t, its value equals to zero: $Z_S(t) = 0$.

Let the notional value of the CDS be one dollar, then the instantaneous return of a CDS on the default of party i can be described by[11]

$$dU_i(t) = \hat{r}_t U_i(t)dt - s_i dt + L_i dJ_i^{(P)}, \quad i = B \text{ and } C,$$

where $s_i \geq 0$ is the annualized CDS rate, $J_i^{(P)}$ is a Poisson process that jumps from 0 to 1 upon the default of party i, the $\mathbb{P}$ intensity of the jump is $\lambda_i^{(P)}$, and $L_i \in [0,1]$ is the corresponding loss rate, assumed to be a constant for simplicity. Note that the above price dynamics holds true only if the CDS rate stays unchanged over time. Under our dynamical hedging strategy, we will always make use of the par CDS, which has zero value, when revising the hedge. For simplicity, we assume the Brownian motion and the Poisson processes are independent of one another.

According to the fundamental theory of asset pricing, there exists a measure, $\mathbb{Q}$, which is equivalent to $\mathbb{P}$ such that under $\mathbb{Q}$ the discounted prices of the repo and the CDS are martingales. For our asset price model, such a martingale measure is unique and is defined by the following Radon-Nikodym

---

[10]For simplicity, we have skipped discussing the details of haircuts in repo trades for long or short stocks, which can be easily accommodated by adjusting the repo spread.

[11]We ignore the funding costs for CDS, in order to isolate the funding costs for the derivative.

derivative:

$$
\left. \frac{d\mathbb{Q}}{d\mathbb{P}} \right|_{\mathcal{F}_t} = \frac{e^{-\int_0^t \gamma_s(u)dW_u^{(P)}}}{E_0^P\left[e^{-\int_0^t \gamma_s(u)dW_u^{(P)}}\right]} \frac{e^{\gamma_B J_B^{(P)}(t)}}{E_0^P\left[e^{\gamma_B J_B^{(P)}(t)}\right]} \frac{e^{\gamma_C J_C^{(P)}(t)}}{E_0^P\left[e^{\gamma_C J_C^{(P)}(t)}\right]}
$$

$$
= e^{\int_0^t -\frac{1}{2}\gamma_s^2(u)du + \gamma_s(u)dW_u^{(P)} + \gamma_B J_B^{(P)}(t) + \lambda_B^{(P)}t(1-e^{\gamma_B}) + \gamma_C J_C^{(P)}(t) + \lambda_C^{(P)}t(1-e^{\gamma_C})},
$$

$$(15.3)$$

with

$$
\gamma_S(t) = \frac{\mu_t - \hat{r}_t - \lambda_S + q_t}{\sigma_t},
$$

$$
\gamma_B = \ln \frac{s_B/L_B}{\lambda_B^{(P)}} \quad \text{and} \quad \gamma_C = \ln \frac{s_C/L_C}{\lambda_C^{(P)}}.
$$

Under $\mathbb{Q}$ the price dynamics of the repo and the CDS become

$$
\begin{aligned}
dZ_S(t) &= \sigma_t(t)S_t dW_t^{(Q)}, \\
dU_i(t) &= \hat{r}_t U_i(t)dt + L_i\left(dJ_i^{(Q)} - \lambda_i^{(Q)}dt\right), \quad i = B \text{ and } C,
\end{aligned}
$$

$$(15.4)$$

where $W_t^{(Q)}$ is a $\mathbb{Q}$-Brownian motion and $J_i^{(Q)}$ is a jump process with risk-neutral intensity $\lambda_i^{(Q)} = s_i/L_i$. Apparently, both $Z_S(t)$ and the discounted prices of $U_i(t)$ are $\mathbb{Q}$-martingales. We call $\mathbb{Q}$ the risk-neutral measure.

According to the fundamental theorem of asset pricing (Harrison and Pliska, 1981), we have the following result on the arbitrage-free valuation of the derivative, where we have used $E_t^Q[X]$ for $E^Q[X|\mathcal{G}_t]$ for notational simplicity.

Definition 15.1.1. *The risk-neutral valuation of the derivative to the counterparties is*

$$
V_f(0) = E_0^Q\left[e^{-\int_0^{\tau \wedge T} \hat{r}_s ds}V(\tau \wedge T)\right].
$$

$$(15.5)$$

When Y_T is $\mathcal{F}_T$-adapted, the risk-neutral valuation (15.5) contains the bilateral credit valuation adjustments (BCVA) as price components. In fact, there is

$$
\begin{aligned}
V_f(0) &= E_0^Q[1_{\{\tau > T\}}e^{-\int_0^T \hat{r}_s ds}Y_T + 1_{\{\tau \le T\}}e^{-\int_0^\tau \hat{r}_s ds}V_\tau] \\
&= E_0^Q[e^{-\int_0^T \hat{r}_s ds}Y_T] - E_0^Q[1_{\{\tau \le T\}}e^{-\int_0^T \hat{r}_s ds}Y_T] \\
&\qquad + E_0^Q[1_{\{\tau \le T\}}e^{-\int_0^\tau \hat{r}_s ds}V_\tau] \\
&= V_e(0) + E_0^Q[1_{\{\tau \le T\}}(e^{-\int_0^\tau \hat{r}_s ds}V_\tau - e^{-\int_0^T \hat{r}_s ds}Y_T)],
\end{aligned}
$$

$$(15.6)$$

where

$$
V_e(t) = E_t^Q[e^{-\int_t^T \hat{r}_s ds}Y_T]
$$

is the no-default value (NDV) of the derivative, which is defined in terms of the

$\mathcal{F}_T$-adapted payoff functions. Since Y_T is $\mathcal{F}_T$-adapted and thus independent of $\mathcal{H}_T$, we have, by using the tower law,

$$
\begin{aligned}
E_0^Q[1_{\{\tau \leq T\}} e^{-\int_0^T \hat{r}_s ds} Y_T] &= E_0^Q[1_{\{\tau \leq T\}} E^Q[e^{-\int_0^T \hat{r}_s ds} Y_T | \mathcal{G}_\tau]] \\
&= E_0^Q[1_{\{\tau \leq T\}} E^Q[e^{-\int_0^T \hat{r}_s ds} Y_T | \mathcal{F}_\tau]] \qquad (15.7) \\
&= E_0^Q[1_{\{\tau \leq T\}} e^{-\int_0^\tau \hat{r}_s ds} V_e(\tau)].
\end{aligned}
$$

By substituting (15.7) back to (15.6) and distinguishing between $\tau = \tau_B$ and $\tau = \tau_C$, we arrive at

Corollary 15.1.1. *When Y_T is $\mathcal{F}_T$-adapted, the risk-neutral valuation of the derivative has the following decomposition:*

$$V_f(0) = V_e(0) + CVA_B + CVA_C, \qquad (15.8)$$

where

$$
\begin{aligned}
CVA_B &= E_0^Q[1_{\{\tau = \tau_B \leq T\}} e^{-\int_0^{\tau_B} \hat{r}_s ds} (V(\tau_B) - V_e(\tau_B))], \\
CVA_C &= E_0^Q[1_{\{\tau = \tau_C \leq T\}} e^{-\int_0^{\tau_C} \hat{r}_s ds} (V(\tau_C) - V_e(\tau_C))]
\end{aligned}
\qquad (15.9)
$$

□

Being $\mathcal{F}_T$-adapted means Y_T is always definable, regardless the default statuses of the counterparties. In other words, a counterparty default will never cause the default of the underlying security. This is the case, for example, when the underlying is a stock index.

We gain two insights from formulae (15.8) and (15.9). First, the bilateral CVA is part of the risk-neutral valuation, which is symmetrical to the counterparties.[12] Second, the NDV of the derivative is the only right choice for the MtM value of the derivative when it comes to the calculation of LGD. Similar bilateral CVA formula are also seen in other literature, e.g., Brigo et al. (2012) and Bo and Capponi (2013). The main difference between our result and theirs is that our LGD is the difference between the post-default value and the no-default value, while the LGD of the other two papers is the percentage recovery of the pre-default value.

We have the following interpretations for NDV and CVAs.

Proposition 15.1.1. *In the absence of funding cost, $V_e(0)$ is the present value (PV) of cost to replicate the payoff, and CVA_i is the $\mathbb{Q}$-expected present value of the cost to replicate the LGD by party i.*

Proof: Without loss of generality, we first assume that the derivative is an asset to B and thus a liability to C. Then, C will replicate the derivative with a portfolio of repos and cash:

$$\Pi_C(t) = \delta_C Z_S(t) + \beta_C(t), \qquad (15.10)$$

---

[12] From the perspective of B, CVA_C is the CVA, while CVA_B is the DVA. The combined value of CVA_B and CVA_C is the bilateral CVA.

where $Z_S(t) = 0$,

$$\delta_C = \frac{\partial V_e(t)}{\partial S}.$$

and

$$\beta_C(t) = \beta_C(0)e^{\int_0^t \hat{r}_v dv} + \int_0^t e^{\int_u^t \hat{r}_v dv} \delta_C(u) dZ_S(u). \qquad (15.11)$$

To achieve perfect replication, $\beta_C(0)$ must be chosen so that

$$\beta_C(\tau \wedge T) = V_e(\tau \wedge T).$$

Taking $\mathbb{Q}$ expectation on both sides of the equation above after discounting and making use of the martingale property of $Z_S(t)$, we obtain

$$\beta_C(0) = E_0^Q \left[e^{-\int_0^{\tau \wedge T} \hat{r}_s ds} V_e(\tau \wedge T) \right] = V_e(0),$$

so that the cost to replicate the derivatives payout is just $V_e(0)$.

Party B, meanwhile, can hedge against his LGD using CDS. Note that our strategy to hedge against the counterparty credit risk (CCR) is based on the fact that, whenever liquid, CDS are the primary means for CVA desks to hedge against credit risks, see Green (2016).[13] The hedging portfolio of party B consists of CDS and cash:

$$\Pi_B(t) = \alpha_B U_C(t) + \beta_B(t),$$

where α_B, is the number of CDS, is given by

$$\alpha_B(t) = -\frac{V(\tau_C = t) - V_e(t)}{L_C}, \qquad (15.12)$$

and a debt in cash due to the CDS fee payment and interest accrual:

$$\beta_B(t) = \int_0^t e^{\int_s^t \hat{r}_u du} \alpha_B(s)(-s_C) ds.$$

The expected present value of the debt for hedging the LGD is

$$E_0^Q \left[e^{-\int_0^{\tau \wedge T} \hat{r}_s ds} \beta_B(\tau \wedge T) \right]$$

$$= E_0^Q \left[\int_0^T e^{-\int_0^t \hat{r}_s ds} \alpha_B(t)(-s_C) 1_{\{\tau > t\}} dt \right]$$

$$= E_0^Q \left[\int_0^T e^{-\int_0^t \hat{r}_s ds} (V(\tau_C = t) - V_e(t)) \lambda_C^{(Q)} e^{-\int_0^t (\lambda_B^{(Q)} + \lambda_C^{(Q)}) ds} dt \right] \qquad (15.13)$$

$$= E_0^Q [1_{\{\tau = \tau_C \leq T\}} e^{-\int_0^{\tau_C} \hat{r}_s ds} (V(\tau_C) - V_e(\tau_C))],$$

---

[13] Names without or with illiquid CDS can be directly mapped to liquid CDS of a name with similar profile.

which is CVA$_C$. Since the derivative is a liability to C, C does not hedge its counterparty risk, so that there is CVA$_B = 0$. It then follows that

$$E_0^Q \left[e^{-\int_0^{\tau \wedge T} \hat{r}_s ds} \beta_C(\tau \wedge T) \right] + E_0^Q \left[e^{-\int_0^{\tau \wedge T} \hat{r}_s ds} \beta_B(\tau \wedge T) \right]$$
$$= V_e(0) + \text{CVA}_C = V_f(0),$$

i.e., the fair price of the derivative is the replication cost of the payout minus the cost to the counterparty to hedge against the loss given default.

When the derivative is a liability to B, or the derivative can switch from an asset to a liability to B, C should be compensated with the present value of the cost to hedge against his LGD which, by symmetry, is CVA$_B$. Hence, in general, the fair value of a derivative to the trading parties is the cost to replicate the derivatives payout minus the bilateral costs to hedge against the loss given their counterparty's defaults.   □

We make three comments here. First, while $V_e(0)$ is the actual cost to replicate the derivatives payout, CVA_i is not, and it is the expected value of the cost to replicate the LGD. Second, in the absence of funding spreads, there is always

$$E_0^Q \left[e^{-\int_0^{\tau \wedge T} \hat{r}_s ds} \beta_i(\tau \wedge T) - \beta_i(0) \right] = 0, \quad \text{for } i = B \text{ and } C, \qquad (15.14)$$

regardless of what price is taken for the trade. Yet in the presence of funding spreads, the above equality will no longer hold, as we shall witness in the following section, thus giving rise to funding cost that can affect the P&L of the trade. Third, we will confirm later that the risk-neutral value is exactly the fair price for a derivatives trade between the counterparties, with or without idiosyncratic funding costs.

15.2   The Rise of Other xVA

We now take the funding spreads for margins, collaterals and capitals into account. Once B and C enter into a trade at time $t = 0$ for a value V_0 to B, the two parties start hedging, using repos and/or CDS, posting margins/collaterals and setting aside capital according to margin, collateral and capital rules. The hedging portfolios of the two parties can be expressed as

$$\Pi_B(t) = \delta_B Z_S(t) + \alpha_B U_C(t) + \beta_B(t),$$
$$\Pi_C(t) = \delta_C Z_S(t) + \alpha_C U_B(t) + \beta_C(t), \qquad (15.15)$$

where δ_i is the number of repo contracts held by party i, which can be

$$\delta_B = -\frac{\partial V_e(t)}{\partial S}, \quad \delta_C = \frac{\partial V_e(t)}{\partial S},$$

and α_i is the number of CDS contracts taken up by party i to hedge against LGD:

$$\alpha_B(t) = -\frac{V(\tau_C = t) - V_e(t)}{L_C}, \quad \alpha_C(t) = -\frac{V(\tau_B = t) - V_e(t)}{L_B},$$

and $\beta_i(t)$ is the total value of party i's cash in his savings account, with initial values

$$\beta_B(0) = \Pi_B(0) = -V_0, \quad \text{and} \quad \beta_C(0) = \Pi_C(0) = V_0.$$

Here, V_0 is the initial premium payment paid by or received by B, depending on whether it is positive or negative. Note that with the above setup, the funding costs for the initial premium of the derivative is also taken into account in evaluating the derivatives values to the trading parties.

The savings account serves a number of purposes:

1. Saving or unsecured borrowing.

2. To deposit or pay for the cash out of the revisions of IM, VM and capital.

3. To take the P&L for hedging.

Without loss of generality, let us focus on β_B, the saving account of party B. To describe the evolution of the balance of the savings account, we need the following notations:

x_B	—	the spread for unsecured borrowing or lending over $\hat{r}_t$ for party B,
$x_B^{(I)}$	—	the funding spreads over the $\hat{r}_t$ for the initial margin of B,
$x_B^{(X)}$	—	the funding spreads over the $\hat{r}_t$ for variable margin or collateral of B,
$\gamma_B^{(K)}$	—	the funding spreads over the $\hat{r}_t$ for capital of B (so-called net return of the capital to the shareholders),
$\phi_B(t)K_B(t)$	—	the portion of capital allocated to the savings account, of B,
$\delta_B dZ_S(t)$	—	the P&L for delta hedging using repos over $(t, t+dt)$, and
$\alpha_B dU_C(t)$	—	the P&L for hedging LGD of C over $(t, t+dt)$ using CDS.

Here, $\phi_B(t) \in [0,1]$, which represents the fraction of capital borrowed from shareholders being reallocated to the savings account. Then, the change of the savings account over the time interval $(t, t+dt)$ can be described as

$$d\beta_B(t) = \Big((\hat{r}_t + x_B)[\beta_B(t) + \phi_B(t)K_B(t)] - x_B^{(I)} I_B(t) - x_B^{(X)} X_B(t)$$
$$- [\gamma_B^{(K)} + \phi_B(t)\hat{r}_t]K_B(t)\Big) dt + \delta_B dZ_S(t) + \alpha_B dU_C(t), \tag{15.16}$$

for which we need to make additional elaborations.

1. The spread for the savings account can be nonlinear and asymmetric, such that

$$x_B(\beta_B(t)) = x_B^{(l)} 1_{\{\beta_B(t) \geq 0\}} + x_B^{(b)} 1_{\{\beta_B(t) < 0\}}.$$

Here, $x_B^{(b)} \geq 0$ is the default risk premium for party B, and $x_B^{(l)} \geq 0$ is the default risk premium of the borrower. In general, there is $x_B^{(l)} \neq x_B^{(b)}$.

2. Due to a partial allocation of capital to the savings account, the interest accrual of the savings account becomes $(\hat{r}_t + x_B)[\beta_B(t) + \phi_B(t) K_B(t)]dt$, in an expense of $r_t \phi_B(t) K_B(t) dt$ to the capital account.

3. In general, the funding spreads and the return on capital are positive, reflecting a simple reality that costs will be incurred for borrowing funds or capitals. Yet for VM or collaterals, there can be exceptions in case of rehypothecation, when the return from rehypothecation is higher than the borrowing cost, the corresponding spread turns negative, representing a funding benefit to the party who posts VM or collaterals.[14]

4. When the derivative is an asset, there will be a jump in the balance of the savings account upon counterparty default due to the the CDS payment for LGD.

We now analyze the P&L of the trade to party B. The present value of the P&L is simply the PV of B's hedged portfolio's value at the termination time of the option (upon either default or maturity):

$$P\&L = e^{-\int_0^{\tau \wedge T} \hat{r}_s ds} [\beta_B(\tau \wedge T) + V(\tau \wedge T)].$$

Due to perfect replication of the market risk and the possible LGD, there is the equality

$$V(\tau \wedge T) = V_e(\tau \wedge T) - \alpha_B L_C dJ_C^{(Q)}(\tau \wedge T)$$
$$= V_e(0)e^{\int_0^{t \wedge T} \hat{r}_v dv} - \int_0^{t \wedge T} e^{\int_u^{t \wedge T} \hat{r}_v dv} \delta_B(u) dZ_S(u)$$
$$- \alpha_B L_C dJ_C^{(Q)}(\tau \wedge T).$$

---

[14]It should be pointed out, however, the higher return is often due to the exposure to credit risk. Thus, for pricing purpose, one may have to adjust the spread properly to account for the credit risk.

We then rewrite the P&L into

$$
\begin{aligned}
P\&L = {} & \left[(\beta_B(0) + V_e(0)) - \int_0^{t \wedge T} e^{-\int_0^u \hat{r}_s ds} \alpha_B(u) s_C du \right] \\
& + \left[e^{-\int_0^{t \wedge T} \hat{r}_s ds} \beta_B(\tau \wedge T) - \beta_B(0) - \int_0^{t \wedge T} e^{-\int_0^u \hat{r}_s ds} \delta_B(u) dZ_S(u) \right. \\
& \left. - \int_0^{t \wedge T} e^{-\int_0^u \hat{r}_s ds} \alpha_B(u) dU_C(u) \right] \\
= {} & I + II.
\end{aligned}
$$

where I and II represent the terms enclosed by the two pairs of square brackets, respectively. We will show next that I represents the PV of the shortfall of payout replication, and II represents the PV of accumulative costs due to various funding spreads.

We first look at I. According to (15.13),

$$
\begin{aligned}
E_0^Q[I] &= \beta_B(0) + V_e(0) + CVA_C \\
&= \beta_B(0) + V_f(0) - CVA_B.
\end{aligned}
$$

If we take the trade price to be $V_0 = -\beta_B(0) = V_f(0)$, then there is

$$
E_0^Q[I] = -CVA_B,
$$

meaning that the debit valuation adjustment for party B is part of his P&L, and the DVA to B will result in a replication shortfall.

Next, we look at II. According to (15.16), we can derive that

$$
\begin{aligned}
d\left(e^{-\int_0^t \hat{r}_s ds} \beta_B(t) \right) = {} & e^{-\int_0^t \hat{r}_s ds} \left(x_B[\beta_B(t) + \phi_B(t) K_B(t)] - x_B^{(I)} I_B(t) \right. \\
& \left. - x_B^{(X)} X_B(t) - \gamma_B^{(K)} K_B(t) \right) dt \\
& + \delta_B e^{-\int_0^t \hat{r}_s ds} dZ_S(t) + \alpha_B e^{-\int_0^t \hat{r}_s ds} dU_C(t).
\end{aligned}
$$

Integrating the above equation from 0 to $t \wedge T$, we then have

$$
\begin{aligned}
& e^{-\int_0^{\tau \wedge T} \hat{r}_s ds} \beta_B(\tau \wedge T) - \beta_B(0) \\
& = \int_0^{\tau \wedge T} e^{-\int_0^t \hat{r}_s ds} \left(x_B[\beta_B(t) + \phi_B(t) K_B(t)] - x_B^{(I)} I_B(t) \right. \\
& \qquad \left. - x_B^{(X)} X_B(t) - \gamma_B^{(K)} K_B(t) \right) dt \\
& \quad + \int_0^{\tau \wedge T} \delta_B e^{-\int_0^t \hat{r}_s ds} dZ_S(t) + \int_0^{\tau \wedge T} \alpha_B e^{-\int_0^t \hat{r}_s ds} dU_C(t).
\end{aligned}
\tag{15.17}
$$

Substituting the right-hand side for $e^{-\int_0^{\tau \wedge T} r_s ds} \beta_B(\tau \wedge T) - \beta_B(0)$ in the expression of II, we arrive at

Proposition 15.2.1. *The realized funding cost to B is*

$$II = \int_0^{\tau \wedge T} e^{-\int_0^t \hat{r}_s ds} \left(x_B[\beta_B(t) + \phi_B(t)K_B(t)] - x_B^{(I)} I_B(t) \right.$$
$$\left. - x_B^{(X)} X_B(t) - \gamma_B^{(K)} K_B(t) \right) dt. \tag{15.18}$$

Note that II is a path-dependent random variable, except for the case of zero funding spreads for margins, collaterals and capitals, when there is $II \equiv 0$, regardless the price V_0 taken for the trade. Following the market convention, we define FVA as the expectation of the present value of realized funding cost under the risk-neutral measure.

Definition 15.2.1. *The FVA to party B is the expected value of excess cost due to the funding spreads for various funding transactions under the risk-neutral measure:*

$$FVA_B = E_0^Q \left[\int_0^{\tau \wedge T} e^{-\int_0^t \hat{r}_s ds} \left(x_B[\beta_B(t) + \phi_B(t)K_B(t)] - x_B^{(I)} I_B(t) \right. \right.$$
$$\left. \left. - x_B^{(X)} X_B(t) - \gamma_B^{(K)} K_B(t) \right) dt \right].$$

We comment here that the FVA so defined is not so useful because neither can it be accepted for valuation adjustment for trade price, nor can it serve as a sound measure for funding risks because it cannot be replicated. Yet the funding costs and risks behind it must be managed, which will be addressed shortly.

From the FVA definition we immediately have

Definition 15.2.2. *To party B, the funding valuation adjustment is given by*

$$FVA_B = FCA_B + MVA_B + ColVA_B + KVA_B, \tag{15.19}$$

with

$$FCA_B = E_0^Q \left[\int_0^{T \wedge \tau} x_B e^{-\int_0^t r_s ds} [\beta_B(t) + \phi_B(t)K_B(t)] dt \right],$$

$$MVA_B = - E_0^Q \left[\int_0^{T \wedge \tau} x_B^{(I)} e^{-\int_0^t r_s ds} I_B(t) dt \right],$$

$$ColVA_B = - E_0^Q \left[\int_0^{T \wedge \tau} x_B^{(X)} e^{-\int_0^t r_s ds} X_B(t) dt \right], \tag{15.20}$$

$$KVA_B = - E_0^Q \left[\int_0^{T \wedge \tau} \gamma_B^{(K)} e^{-\int_0^t r_s ds} K_B(t) dt \right],$$

and $\beta_B(t)$ evolving according to (15.16) □

We want to emphasize here that in general FCA and FVA can depend on V_0, the trade price of the derivative, through $\beta_B(0) = -V_0$. Hence, whenever necessary, we will write $\text{FCA}(-V_0)$ or $\text{FVA}(-V_0)$ to highlight such dependence.

Different choices of V_0 will lead to some major existing results on xVA. The first choice is $V_0 = V_e(0)$. When we ignore the funding spreads for collaterals and capitals and do not hedge against counterparty default, i.e., letting $x_B^{(I)} = x_B^{(X)} = \gamma_B^{(K)} = 0$, we will then have $\beta_B(t) = -V_e(t)$ under the diffusion dynamics for the underlying security due to perfect replication, and consequently have

$$\text{FVA}_B(-V_e(0)) = \text{FCA}_B(-V_e(0)) = -E_0^Q \left[\int_0^{\tau \wedge T} x_B e^{-\int_0^t \hat{r}_s ds} V_e(t) dt \right],$$

which is identical to (15.1), the FVA formula of the prevalent model.

The second choice of V_0 is the value of the derivative to a party, which is defined to be the value with which the risk-neutral expected return of the trade is the risk-free rate, $\hat{r}_t$. To party B, the value of the derivative, $V_0^{(B)}$, satisfies (Li and Wu, 2015)

$$V_0^{(B)} = V_f(0) + \text{FVA}(-V_0^{(B)}),$$

which is an implicit equation for $V_0^{(B)}$. By solving this equation we will then reproduce the xVA formulae of Green and Kenyon (2015). With additional simplifying assumptions, we can also reproduce the FCA formula of Piterbarg (2010) and the bilateral CVA and FCA formulae of Burgard and Kjaer (2011). It should be pointed out, however, the approach we have taken is much less restrictive. To derive the similar results, Burgard and Kjaer (2011) and Green and Kenyon (2015) need the "funding condition" (see Equation 6 of both papers), which is equivalent to requiring perfect unilateral replication of a derivative at all time until the first default or the maturity. Given the invalidation of "hedging own default using own bonds" (Castagna, 2012), we can see that such a "funding condition" holds only narrowly for derivatives receivables under the diffusion model for the underlying securities. In addition, our formulae have accommodated asymmetric funding spreads for unsecured borrowing and lending, which is not included in the other papers.

There is, however, normally a gap between the values to the counterparties. The value to party C, the counterparty, is implied by the equation

$$-V_0^{(C)} = -V_f(0) + \text{FVA}_C(V_0^{(C)}). \tag{15.21}$$

In Li and Wu (2015), $V_0^{(B)}$ and $V_0^{(C)}$ are called bid and ask prices, respectively. Note that the bid and ask prices so defined are subject to nonlinear pricing rules, and such nonlinearity occurs when one prices derivatives as the replication costs, as was observed in Pallavicini et al. (2011), Crépey (2011) and Bichuch et al. (2015). The bid and ask prices are in general different due to the idiosyncratic nature of funding costs. To see this, we let the derivative be an asset to B and assume there is no funding benefit but funding costs to both parties, then there will be $\text{FVA}_i \leq 0, i = B$ and C, thus yielding the order $V_0^{(B)} \leq V_0^{(C)}$. As has been shown by Li and Wu (2015), a trade that takes place at any price within the interval $[V_0^{(B)}, V_0^{(C)}]$ will be non-arbitrageable to both parties and, in addition, will make risk-neutral expected returns to both parties simultaneously lower than the risk-free rate.

To conclude, to derive a fair price which achieves symmetry in pricing and asset-liability symmetry in accounting, we have to ignore all idiosyncratic funding spreads, such that $x_i^{(b)} = x_i^{(l)} = x_i^{(I)} = x_i^{(X)} = \gamma_i^{(K)} = 0, i = B$ and C, and let the interest rate account for the market funding liquidity risk premium (Gregory, 2015), $\hat{r}_t = r_t + x_m$, and then perform risk-neutral valuation for the defaultable claim. The result is (15.5) or (15.8) if the payoff function is $\mathcal{F}_T$-adapted.

We finish this section with several more comments. First, for P&L accounting, we suggest to the booking of the idiosyncratic component(s) of the realized funding costs, which, to party B, is

$$
FC_B(t) = \int_0^t e^{\int_u^t \hat{r}_s ds} \left(x_B[\beta_B(u) + \phi_B(u)K_B(u)] - x_B^{(I)} I_B(u) \right.
$$
$$
\left. -x_B^{(X)} X_B(u) - \gamma_B^{(K)} K_B(u) \right) du \tag{15.22}
$$

for a trade entered at time $t = 0$. Actually, it is straightforward to show

$$
V_t + \Pi_B(t) = FC_B(t) + CVA_B(0)e^{\int_0^t \hat{r}_s ds}, \tag{15.23}
$$

which means that the P&L of the hedged portfolios is nothing else but the realized total (idiosyncratic) funding costs plus the accrued value of the initial credit valuation adjustment.

Second, more effective methodologies should be adopted for managing the funding risk. Over the years some major banks have been setting aside reserves based on the FVA numbers (Levine, 2014). In a similar spirit, some researchers have advocated the deduction of FVA from a bank's common equity Tier 1 (CET1) capital (Albanese et al., 2015; Andersen et al., 2016). It should be pointed out that, however, as an expected value of funding cost under the risk-neutral measure, an FVA number can be far from the funding costs actually realized to a desk or a firm. For better risk management, we suggest the adoption of other established risk metrics, like the VaR or CVaR, by making use of (15.16) and (15.22) through Monte Carlo simulations under the real-world measure. Note that with these metrics for funding risks we can also perform stress testing for various scenarios, including liquidity crunch.

Third, CVA is exposed to market risk also as it fluctuates with the share prices, and the CVA desks want to hedge the CVA dynamically. The hedging against share-price risk is based on the usual delta hedging, which is discussed in detail in Li and Wu (2015).

Finally, the xVA formulae (15.9) and (15.20) hold for general pricing models, including models with stochastic interest rates, stochastic hazard rates for defaults and stochastic funding spreads, and models with jumps and stochastic volatilities in their state variables.

15.3   Examples

For demonstration purposes, we consider the pricing the xVA of a bilateral interest-rate swap, which is subject to IM, VM and capital, from the viewpoint of B. We will take the approach of IMM in the specifications of IM, VM and capital. Without loss of generality, we only consider cash as the posted assets for both IM and VM. Note that, due to the IM, there can be over-collateralization for the swap trade from time to time, which can result in very small CVA and DVA.

The initial margin requirement for potential future exposure is

$$I_B(t) \quad \text{—} \quad \text{99\% VaR of the P\&L of the swap for 10-day horizon,}$$

and the initial margin account will accrue using the OIS rate.

The posting of VM is required only for the party of liability, and the amount will be revised periodically at a set of predetermined dates, $\{t_i\}$. The subsequent additions or reductions of VM are subject to (1) a threshold value, H, and (2) a minimum transfer amount, m, with $H \geq m \geq 0$. Essentially, a revision to VM is required only if (1) the liability exceeds H, and (2) the change of the liability (due to either appreciation or depreciation) exceeds m. Between two moments of margin revision, the VMs accrue interest with an interest rate no less than the OIS rate. To the counterparties, the VMs start with $X_i(t_0-) = 0, i = B$ and C, and are set according to the following scheme: for $i = 0, 1, \ldots,$

$$Adj_B(t_i) = |V_e(t_i)| - H - X_B(t_i-),$$
$$X_B(t_i+) = \left(X_B(t_i-) - 1_{\{V_e(t_i)<0\}} \, Adj_B(t_i) \, 1_{\{|Adj_B(t_i)|\geq m\}}\right)^+,$$

and

$$Adj_C(t_i) = |V_e(t_i)| - H - X_C(t_i-),$$
$$X_C(t_i+) = \left(X_C(t_i-) + 1_{\{V_e(t_i)>0\}} \, Adj_C(t_i) \, 1_{\{|Adj_C(t_i)|\geq m\}}\right)^+,$$

where t_i- and t_i+ denote the moment immediately before and after a revision. Upon a bilateral default, the default settlement value of the swap is described in (15.2), with $c_i(t) = I_i(t) + X_i(t)$ to be the total value of the collateral posted by party $i = B$ or C.

For simplicity, we consider only the capital requirement for market risk under IMM, which is

$$K_B(t) \quad - \quad \text{99\% VaR of the P\&L of the swap for 10-day horizon.}$$

We use no risk capital to stack up the savings account, meaning that $\phi_B = 0$ is taken.

We use the single-factor linear Gaussian model (LGM) (Hagan and Woodward, 1999), introduced in Section 4.11, for pricing purpose.

The LGM is naturally calibrated to the discount curve. The function $h(t)$ takes the form of (4.139) with $\kappa = 0.03$. Before pricing options, we normally calibrate α_t to vanilla interest-rate derivative like caps, floors or swaptions.

Our example is with an ATM receiver's swap of 20-year maturity, which receives fixed-rate interest annually and pays floating-rate interest semiannually according to 6M Euribor, out of a notional value of €10m. We make use of the actual 6M Euribor projection curve and the OIS curve of 02/05/2016, and calculate the 20-year ATM swap rate to be 1.135%. The local volatility function α_t is taken to be a piecewise linear function that is calibrated to the implied normal volatility matrix of swaptions of the same date, provided in Table 15.A in this chapter's appendix.

For calculating the xVA, we also need to specify the parameters for collateral and funding:

margin revision frequency	—	daily,
credit risk	—	$L_B = L_C = 0.6$,
		$\lambda_C = 1.0\%$, $s_B = 0\% : 0.25\% : 3\%$,
collateral	—	$H=€100000$, $m=€0$.

The funding spreads for the lending/borrowing through savings account, for collateral and for capital are

savings account	—	$x_B^{(l)} = 0$, $\quad x_B^{(b)} = \lambda_B L_B$,
IM	—	$x_B^{(I)} = \lambda_B L_B$,
VM	—	$x_B^{(X)} = 0$ (due to rehypothecation),
capital	—	$\gamma_B^{(K)} = 8\%$.

The xVA against the default rate of party B, the issuer of the swap, is shown in Figure 15.1. As one can see from the plot, CVA_B and CVA_C are all negligibly small due to over-collateralization, and ColVA is equal to zero due to rehypothecation (which eliminates the funding cost for VM). Compared with other xVA, KVA is significantly larger, apparently due to the 8% return rate on the capital demanded by the shareholders, which is significantly higher than other funding spreads. Finally, because higher hazard rate shortens expected default time of the issuer and consequently reduces the average time of accrual for the capital, the KVA in absolute term decreases when the hazard rate increases.

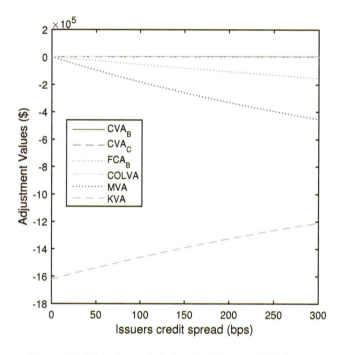

FIGURE 15.1: The xVA for the 20-year ATM swap.

15.4   Notes

As a major response to the 2007-08 financial crisis, xVA has been a focal point to regulators, practitioners, and researchers. This chapter provides a rather general yet simple framework to analyze and quantify xVA. We show that the CVA and DVA are expected value of costs to hedge against counterparty default, and FVA is the interest accrual of funding costs due to funding spreads. We have the conclusion that the fair price of a derivative is the classical risk-neutral value in a market where the rate for unsecured lending and borrowing is the risk-free rate plus the market funding liquidity risk premium, and such a value naturally contains the bilateral CVA. There are remaining issues we did not touch upon, such as, under the (IMM) or advanced method, how to set the levels of initial margin (IM), variable margin (VM) and capital level. The level of IM, VM and capital will have impacts on the bilateral CVA, and ultimately on derivatives price. Related to IM, VM and capital, there are also the issues of collateral management, spread sheet optimization, and management of xVA; these are the issues that are undergoing research and debate, and are beyond the scope of this book.

Appendix

Swap Maturity	1Y	2Y	3Y	4Y	5Y	7Y	10Y	15Y	20Y
Option Expiry									
1Y	0.003543	0.003889	0.004316	0.004802	0.00527	0.006123	0.006978	0.007557	0.007918
2Y	0.004732	0.00498	0.005279	0.005629	0.005957	0.006563	0.007273	0.007508	0.007752
3Y	0.00586	0.006018	0.006218	0.00642	0.006646	0.007066	0.007539	0.007663	0.007657
4Y	0.006539	0.006691	0.006784	0.006967	0.007112	0.007394	0.00773	0.007573	0.007489
5Y	0.007013	0.007117	0.007194	0.0073	0.007443	0.007648	0.00782	0.007505	0.007363
7Y	0.007519	0.007573	0.007612	0.007735	0.007807	0.007883	0.007698	0.007392	0.007117
10Y	0.007668	0.007642	0.007637	0.007882	0.007705	0.007869	0.007611	0.007116	0.006848
15Y	0.007492	0.007636	0.007657	0.007647	0.007394	0.007559	0.007199	0.006833	0.006442
20Y	0.007198	0.007364	0.007364	0.007294	0.007216	0.007141	0.006964	0.006419	0.005994

Table 15.A. Implied normal volatilities of swaptions

References

[1] Albanese, C, Andersen L and Iabichino S, 2015. FVA accounting, risk management and collateral trading. *Risk*, February, 64-69.

[2] Andersen L, 2007. Efficient simulation of the Heston stochastic volatility model. Working paper, Bank of America Securities.

[3] Andersen, L and Andreasen J, 2000. Volatility skews and extensions of the LIBOR market model. *Applied Mathematical Finance*, (7)1: 1–32.

[4] Andersen L and Andreasen J, 2002. Volatile Volatilities. *Risk*, 15: 163–168.

[5] Andersen L and Brotherton-Ratcliffe R, 2005. Extended LIBOR market models with stochastic volatility. *Journal of Computational Finance* 9(1): 1–40.

[6] Andersen, L, Duffie D and Song Y, 2016. Funding value adjustments. Working paper, http://ssrn.com/abstract=2746010.

[7] Andersen L, Sidenius J and Basu S, 2003. All your hedges in one basket. *RISK*, November, 67–72.

[8] Arrow K and Debreu G, 1954. Existence of an equilibrium for a economy. *Econometrica* 22: 265–290.

[9] Arvanitis A and Gregory J, 2001. Credit: the complete guide to pricing, hedging and risk management. Risk Books.

[10] Avellaneda M, Boyer-Olson D, Busca J and Fritz P, 2002. Reconstructing volatility. *Risk*, 87–91.

[11] Avellaneda M, Friedman C, Holmes R, and Samperi D, 1998. Calibrating Volatility Surfaces via Relative-Entropy Minimization. Working paper, New York University.

[12] Avellaneda M and Laurence P, 1999. *Quantitative Modeling of Derivative Securities: From Theory to Practice*. Chapman & Hall/CRC, Boca Raton, FL.

[13] Balduzzi P, Das S, Foresi S, and Sundaram R, 1996. A simple approach to three factor affine term structure models. *Journal of Fixed Income* 6: 43–53.

[14] Barone, E and Castagna, A, 1997. The information content of TIPS. Internal Report. SanPaolo IMI, Turin and Banca IMI, Milan.

[15] Baxter M and Rennie A, 1996. *Financial Calculus: An Introduction to Derivative Pricing*. Cambridge University Press, Cambridge.

[16] Belgrade N, and Benhamou E, 2004. Reconciling Year on Year and Zero Coupon Inflation Swap: A Market Model Approach. Preprint, CDC Ixis Capital Markets. Downloadable at: http://papers.ssrn.com/sol3/papers.cfm?abstract-id=583641.

[17] Belgrade N and Benhamou E, 2004. Smart modeling of the inflation market: taking into account the seasonality. Preprint, CDC Ixis Capital Markets.

[18] Belgrade N, Benhamou E and Koehler E, 2004. A Market Model for Inflation. Preprint, CDC Ixis Capital Markets. Downloadable at: http://papers.ssrn.com/sol3/papers.cfm?abstract-id=576081.

[19] Benhamou E, 2000. Pricing Convexity Adjustment with Wiener Chaos, London School of Economics working paper.

[20] Bianchetti M, 2010. Two curves, one price. *Risk* Magazine, August, 74–80.

[21] Bianchetti M and Carlicch M, 2011. Interest rates after the credit crunch: multiple-curve vanilla derivatives and SABR. Munich Personal RePEc Archive, paper 42248.

[22] Bichuch M, Capponi A and Sturm S, 2015. Arbitrage-free pricing of xVA - part I: framework and explicit examples. Working paper, arXiv:1501.05893[q-fin.PR].

[23] Bielecki T and Rutkowski M, 2001. *Credit Risk: Modeling, Valuation and Hedging*. Springer-Verlag.

[24] Björk T, Kabanov Y and Runggaldier W, 1997. Bond market structure in the presence of marked point processes. *Math. Fin.*, 7(2): 211–223.

[25] Black F and Cox J, 1976. Valuing corporate securities: some effects of bond indenture provision. *Journal of Finance* 31: 351–67

[26] Black F and Karasinski P, 1991. Bond and option pricing when short rates are lognormal. *Financial Analysts Journal* 47(4): 52–59.

[27] Black F and Scholes M, 1973. The pricing of options and corporate liabilities. *Journal of Political Economy* 81: 637–659.

[28] Black F, Derman E, and Toy W, 1990. A one-factor model of interest rates and its application to Treasury bond options. *Financial Analysts Journal*, January/February, 33–39.

[29] Börger R and van Heys J, 2008. Calibration of the Libor Market Model Using Correlations Implied by CMS Spread Options. Working paper, WGZ BANK, Germany.

[30] Boyd S and Vandenberghe L, 2004. *Convex Optimization*. Cambridge: Cambridge University Press.

[31] Boyle P, Broadie M and Glasserman P, 1997. Monte Carlo methods for security pricing. *Journal of Economic Dynamics and Control*, 21: 1267–1321.

[32] Brace A, Gatarek D, and Musiela M, 1997. The market model of interest rate dynamics. *Mathematical Finance* 7: 127–154.

[33] Brace A and Womersley R, 2000. Exact fit to the swaption volatility matrix using semi-definite programming. Working paper, National Australia Bank and University of New South Wales.

[34] Brigo D, 2004. Candidate market models and the calibrated CIR++ stochastic intensity model for credit default swap options and callable floaters. Proceedings of the 4th ICS Conference, Tokyo, March 18–19.

[35] Brigo D, 2005. Market models for CDS options and callable floaters. *Risk*, January, 89–94.

[36] Brigo D, Capponi A, Pallavicini A and Papatheodorou D, 2011. Collateral margining in arbitrage-free counterparty valuation adjustment including rehypotecation and netting. Working paper, http://ssrn.com/abstract=1744101.

[37] Brigo D and Liinev J, 2003. On the distributional distance between the Libor and the Swap market models. Working paper, Banca IMI and University of Ghent.

[38] Brigo D, Liinev J, Mercurio F and Rapisarda F, 2004. On the distributional distance between the lognormal LIBOR and Swap market models. Working paper, Banca IMI, Italy.

[39] Brigo D and Mercurio F, 2003. Analytical pricing of the smile in a forward LIBOR market model. *Quantitative Finance*, 3(1): 15–27.

[40] Brigo D and Mercurio F, 2006. *Interest Rate Models: Theory and Practice: With Smile, Inflation and Credit*, 2nd edition. Springer Finance, Berlin.

[41] Brotherton-Ratcliffe R and Iben B, 1993. Yield curve applications of swap products. In: RJ Schwartz and C Smith (eds), *Advanced Strategies in Financial Risk Management*, pp. 400–450. New York Institute of Finance, New York.

[42] Burgard C and Kjaer M, 2011. Funding cost adjustments for derivatives. *Asia Risk*, November, 63–67.

[43] Cairns A, 2000. A multifactor model for the term structure and inflation for long-term risk management with an extension to the equities market. Preprint. Heriot-Watt University, Edinburgh.

[44] Cameron M, 2013. The black art of FVA. *Risk*, April, 15–18.

[45] Carriere J, 1996. Valuation of early exercise price of options using simulations and nonparametric regression. *Insurance: Mathematics and Economics* 19: 19–30.

[46] Carr P and Madan D, 1998. Option valuation using the fast Fourier Transform. Working paper, Morgan Stanley and University of Maryland.

[47] Carr P and Wu L, 2008. Variance risk premiums. *Review of Financial Studies*, Advance Access published April 10.

[48] Castagna A, 2012. The impossibility of DVA replication. *Risk*, November.

[49] Chen RR, 1996. *Understanding and Managing Interest Rate Risks*. World Scientific, Singapore.

[50] Chen RR, Liu B and Cheng X, 2006. Pricing the Term Structure of Inflation Risk Premia: Theory and Evidence from TIPS. Working paper, Rutgers Business School.

[51] Chen RR and Scott L, 2001. Stochastic volatility and jumps in interest-rates: an empirical analysis. Working paper, Rutgers University and Morgan Stanley.

[52] Choy B, Dun T, and Schlogl E, 2004. Correlating market models. *Risk* 17(9): 124–129.

[53] Clement E, Lamberton D, and Protter P, 2002. An analysis of a least squares regression algorithm for American option pricing. *Finance and Stochastics*, 6:449–471.

[54] Cuchiero C, Fontana C, and Gnoatto A, 2017. Affine multiple yield curve models. Forthcoming in *Mathematical Finance*.

[55] Cox J, Ingersoll J, and Ross S, 1985. A theory of the term structure of interest rates. *Econometrica* 53(2): 385–407.

[56] Crépey S, 2011. A BSDE approach to counterparty risk under funding constraints. Available at grozny.maths.univ-evry.fr/pages perso/crepey.

[57] Dai Q and Singleton K, 2000. Specification analysis of affine term structure models. *Journal of Finance* 55: 1943–1978.

[58] Dodds S, 1998. Personal communication.

[59] Duffee G, 1998. The relation between Treasury yields and corporate bond yield spreads. *Journal of Finance* 53: 2225–2242.

[60] Duffie D and Kan R, 1996. A yield-factor model of interest rates. *Mathematical Finance* 6: 379–406.

[61] Duffie D and Singleton K, 1999. Modeling term structure of defaultable bonds. *Review of Financial Studies* 12: 687–720.

[62] Duffie D, Filipovic D, and Schachermayer W, 2002. Affine processes and applications in finance. Working paper, Stanford University.

[63] Duffie D, Pan J, and Singleton K, 2000. Transform analysis and asset pricing for affine jump diffusions. *Econometrica* 68: 1343–1376.

[64] Eberlein E, Kluge W, and Schönbucher P, 2005. The Lèvy LIBOR model with default risk. Working paper, University of Freiburg.

[65] Eberlein E and Özkan F, 2004. The Levy Libor model. Working paper, University of Freiburg.

[66] Fabozzi F, 2003. *Bond Markets, Analysis, and Strategies*, 5th edition. Prentice Hall: Upper Saddle River, NJ.

[67] FASB 2011. Accounting standards update. Financial Accounting Standards Board, 04.

[68] Falbo P, Paris F, and Pelizzari C, 2009. Pricing inflation-link bonds. *Quantitative Finance*, 1–15, iFirst.

[69] Feller W, 1971. *An Introduction to Probability Theory and Its Applications*, Vol. 2. Wiley, New York.

[70] Filipovic W, 2001. A general characterization of one factor affine term structure models. *Finance and Stochastics* 5(3): 389–412.

[71] Fisher I, 1930. *The Theory of Interest*. The Macmillan Company. ISBN13 978-0879918644.

[72] Flesaker B and Hughston L, 1996. Positive interest, *Risk Magazine*, January.

[73] Fujii M, Shimada Y, and Takahashi A, 2010. A note on construction of multiple swap curves with and without collateral. *FSA Research Review* 6: 139–157.

[74] Fujii M, Shimada Y, and Takahashi A, 2011. A market model of interest rates with dynamic basis spreads in the presence of collateral and multiple currencies. *Wilmott Journal* 54: 61–73.

[75] Giesecke K, 2003. A simple exponential model for dependent defaults. *Journal of Fixed Income* 13(3): 74–83.

[76] Glasserman P, 2003. *Monte Carlo Methods in Financial Engineering.* Springer, Berlin.

[77] Glasserman P and Kou S, 2003. The term structure of simple forward rates with jump risk. *Mathematical Finance* 13(3): 383–410.

[78] Glasserman P and Merrener N, 2003. Numerical solution of jump-diffusion LIBOR market models. *Journal of Computational Finance,* 7(1): 1–27.

[79] Glasserman P and Zhao X, 2000. Arbitrage-free discretization of log-normal forward LIBOR and swap rate models. *Finance and Stochastics,* 4:35–68.

[80] Grbac Z, Papapantoleon A, Schoenmakers J, and Skovmand D, 2015. Affine LIBOR models with multiple curves: theory, examples and calibration. Preprint, arXiv:1405.2450v4.

[81] Green A, 2016. *xVA: Credit, Funding and Capital Valuation Adjustments.* The Wiley Finance Series.

[82] Green A and Kenyon C, 2015. MVA: initial margin valuation adjustment by replication and regression. Working paper, arXiv:1405.0508[q-fin.PR].

[83] Green A, Kenyon C, and Dennis C, 2014. KVA: capital valuation adjustment. arXiv:1405.0515[q-fin.PR].

[84] Gregory J, 2015. *The xVA Challenge: Counterparty Credit Risk, Funding, Collateral and Capital,* 3rd edition, Wiley.

[85] Gregory J and Laurent JP, 2003. I will survive. *RISK,* June, 103–107.

[86] Gyöngy I, 1986. Mimicking the one-dimensional marginal distributions of processes having an ito differential. *Probability Theory and Related Fields,* 71(4): 501–516.

[87] Hagan P, Kumar D, Lesniewski A, and Woodward D, 2002. Managing smile risks. *Wilmott Magazine* (September): 84–108.

[88] Hagan P, 2003. Convexity conundrums: pricing CMS swaps, caps and floors. *Wilmott Magazine* (March): 38–44.

[89] Hagan P and West G, 2006. Interpolation methods for curve construction. *Applied Mathematical Finance,* 13(2), 89–129.

[90] Hagan P and Woodward D, 1999. Markov interest rate models. *Applied Mathematical Finance,* 6(4): 233–260.

[91] Harrison J and Kreps D, 1979. Martingales and arbitrage in multiperiod securities markets. *Journal of Economic Theory* 20: 381–408.

[92] Harrison J and Pliska S, 1981. Martingales and stochastic integrals in the theory of continuous trading. *Stochastic Processes and their Applications* 11: 215–260.

[93] Heath D, Jarrow R, and Morton A, 1992. Bond pricing and the term structure of interest rates: a new methodology. *Econometrica* 60: 77–105.

[94] Henrard M, 2007. The irony in the derivatives discounting. *Wilmott Journal* 2: 92–98.

[95] Henrard M, 2010. The irony in the derivatives discounting part II: the crisis. *Wilmott Journal* 2: 301–316.

[96] Henry-Labordère P, 2005. A general asymptotic implied volatility for stochastic volatility models April, available at http://ssrn.com/abstract=698601.

[97] Henry-Labordère P, 2007. Combining the SABR and LMM models. *Risk*, October, 102–107.

[98] Heston S, 1993. A closed-form solution for options with stochastic volatility with applications to bond and currency options. *The Review of Financial Studies*, (6)2: 327–343.

[99] Hinnerich M, 2008. Inflation indexed swaps and swaptions. *Journal of banking and Finance*, forthcoming.

[100] Ho S and Wu L, 2008. Arbitrage pricing of credit derivatives. *Credit Risk – Models, Derivatives and Management*, Chapter 22, 427–456. Financial Mathematics Series, Vol. 6, ed. W. Niklas. Chapman & Hall / CRC Boca Raton, London, New York.

[101] Ho T and Lee S, 1986. Term structure movements and pricing interest rate contingent claims. *Journal of Finance* 42: 1129–1142.

[102] Hoskinson R, 2014. A new approach to multiple curve Market Models of Interest Rates. Presentation, http://actuaries.asn.au/Library/Events/FSF/2014/ Hoskinson2f.pdf.

[103] Hubalek F and Kallsen J, 2005. Variance-Optimal Hedging and Markowitz-Efficient Portfolios for Multivariate Processes with Stationary Independent Increments with and without Constraints. Working Paper, TU Munchen.

[104] Hughston L, 1998. Inflation Derivatives. Working paper. Merrill Lynch.

[105] Hull J and White A, 1989. Pricing interest-rate derivative securities. Working paper, University of Toronto.

[106] Hull J and White A, 2003. The valuation of credit default swap options. Working paper, Rothman School of Management.

[107] Hull J and White A, 2012. The FVA debate. *Risk*, August, 25(7): 83–85.

[108] Hull J and White A, 2014. Valuing derivatives: funding value adjustment and fair value, *Financial Analysts Journal*. 70(3): 46–56.

[109] Hunt P and Kennedy J, 2000. *Financial Derivatives in Theory and Practice*. John Wiley & Sons.

[110] Hunter C, Jäckel P, and Joshi M, 2001. Cutting the drift. *Risk*, July.

[111] IASB, 2011. IFRS 13 fair value measurement. *International Accounting Standard Board*.

[112] Ikeda N and Watanabe S, 1989. *Stochastic Differential Equations and Diffusion Processes*, 2nd edition. North Holland-Kodansha, Amsterdam.

[113] Inui K and Kijima M, 1998. A Markovian framework in multi-factor Heath–Jarrow–Morton models. *Journal of Financial and Quantitative Analysis* 33(3): 423–440.

[114] Isserlis L, 1916. On certain probable errors and correlation coefficients of multiple frequency distributions with skew regression. *Biometrika*. 11: 185–190.

[115] Itô K, 1942. On stochastic processes, I. (Infinitely divisible laws of probability). *Jpn. J. Math.* 18: 261–301.

[116] Jamshidian F, 1989. An exact bond option pricing formula. *Journal of Finance* 44: 205–209.

[117] Jamshidian F, 1997. LIBOR and swap market models and measures. *Finance and Stochastics* 1(4): 293–330.

[118] Jamshidian F, 2002. Valuation of credit default swaps and swaptions. Preprint, NIB Capital Bank.

[119] Jarrow R, Li H, and Zhao F, 2003. Interest rate caps "smile" too! But can the LIBOR market model capture it? Working paper, Cornell University.

[120] Jarrow R and Yildirim Y, 2003. Pricing treasury inflation protected securities and related derivatives using an HJM model. *Journal of Financial and Quantitative Analysis*, (38)2: 409–430.

[121] Jeanblanc M and Rutkowski M, 2000. Default risk and hazard process. Mathematical Finance Bachelier Congress 2000, edited by Geman, Madan, Pliska and Vorst, Springer-Verlag.

[122] Jin Y and Glasserman P, 2001. Equilibrium positive interest rates: A unified view. *The Review of Financial Studies*, 14(1): 187–214.

[123] Johannes M, 2004. The statistical and economic role of jumps in continuous-time interest rate models. *The Journal of Finance*, 59(1): 227–260.

[124] Joshi M and Rebonato R, 2003. A stochastic-volatility, displaced-diffusion extension of the LIBOR market model. *Quantitative Finance*, (3)6: 458–469.

[125] Jung J, 2008. Real Growth. *RISK*, February.

[126] Karatzas I and Shreve S, 1991. *Brownian Motion and Stochastic Calculus*, 2nd edition. Springer, Berlin.

[127] Karlin S and Taylor H, 1981. *A Second Course in Stochastic Processes*. Academic Press, New York.

[128] Kazziha S, 1999. Interest Rate Models, Inflation-based Derivatives, Trigger Notes And Cross-Currency Swaptions. PhD Thesis, Imperial College of Science, Technology and Medicine. London.

[129] Keller-Ressel M, Papapantoleon A, and Teichmann J, 2013. The affine libor models. *Mathematical Finance*, 23(4): 627–658.

[130] Kendall M, 1994. *Advanced Theory of Statistics*, 6th edition. London: Edward Arnold; New York: Halsted Press.

[131] Kenyon C, 2008. Inflation is normal. *RISK*, July, 76–82.

[132] Kenyon C, 2010. Post-shock short-rate pricing, *Risk Magazine*, 23(11): 83–87.

[133] Kijima M, Tanaka K, and Wong T, 2009. A multi-quality model of interest rates. *Quantitative Finance*, 9(2): 133–145.

[134] Kloeden P and Platen E, 1992. Numerical solution of stochastic differential equations. Springer-Verlag.

[135] Korn R and Korn E, 2000. *Option Pricing and Portfolio Optimization: Modern Methods of Financial Mathematics*. American Mathematical Society, Providence, RI.

[136] Korn R and Kruse S, 2004. A simple model to value inflation-linked financial products, (in German), Blatter der DGVFM, XXVI (3): 351–367.

[137] Kruse S, 2007. Pricing of Inflation-Indexed Options under the Assumption of a Lognormal Inflation Index as well as under stochastic volatility. Working paper, University of Applied Sciences–Bonn, Germany.

[138] Kyprianou A, 2008. Lévy processes and continuous-state branching processes: part I. Lecture Notes, University of Bath.

[139] Lee R, 2004. Option pricing by transform methods: Extensions, unification, and error control. *Journal of Computational Finance* 7: 51–86.

[140] Leippold M and Wu L, 2002. Asset pricing under the quadratic class. *Journal of Financial and Quantitative Analysis* 37: 271–295.

[141] Levine M, 2014. It cost JPMorgan $1.5 billion to value its derivatives right. Bloomberg, https://www.bloomberg.com/view/articles /2014-01-15/it-cost-jpmorgan-1-5-billion-to-value-its-derivatives-right-draft.

[142] Lévy P, 1954. Théorie de l'addition des variables aléatoires, second edition. Gaulthier-Villars, Paris.

[143] Lewis A, 2000. *Option Valuation under Stochastic Volatility.* Finance Press, Newport Beach.

[144] Li D, 2000. On default correlations: a copula approach. *Journal of Fixed Income,* 9: 43–54.

[145] Li R, 2004. Option pricing by transform methods: extensions, unification, and error control. *Journal of Computational Finance,* 7(3):51–86.

[146] Li C and Wu L, 2015. CVA and FVA under Margining. *Studies in Economics and Finance,* 32(3): 298–321.

[147] Lichters R, Stamm R, and Gallagher D, 2015. *Modern Derivatives Pricing and Credit Exposure Analysis, Theory and Practice of CSA and xVA Pricing, Exposure Simulation and Backtesting,* Palgrave Macmillan.

[148] Litterman R and Scheinkman J, 1991. Common factors affecting bond returns. *Journal of Fixed Income,* 1(1): 54–63.

[149] Longstaff F and Schwartz E, 2001. Valuing American options by simulation: A simple least-squares approach. *The Review of Financial Studies,* 14:113–147.

[150] Lou W, 2015. Funding in option pricing: the Black-Scholes framework extended. *Risk,* April, 1–6.

[151] Macaulay F, 1938. Some theoretical problems suggested by the movement of interest rates, bond yields, and stock prices in the U.S. since 1856. National Bureau of Economic Research, New York.

[152] Manning S and Jones M, 2003. *Modeling inflation derivatives - a review.* The Royal Bank of Scotland Guide to Inflation-Linked Products. Risk.

[153] Mercurio F, 2005. Pricing inflation-indexed derivatives. *Quantitative Finance,* 5(3): 289–302.

[154] Mercurio F, 2010a. A LIBOR market model with a stochastic basis. *Risk* Magazine, December, 84–89.

[155] Mercurio F, 2010b. Interest rates and the credit crunch: new formulas and market models. Bloomberg Portfolio Research Paper No. 2010-01-FRONTIERS.

[156] Mercurio F and Moreni N, 2006. Inflation with a smile. *Risk* March, Vol. 19(3): 70–75.

[157] Mercurio F and Moreni N, 2009. Inflation modelling with SABR dynamics. *Risk* June, 106–111.

[158] Mercurio F and Pallavicini A, 2006. Swaption skews and convexity adjustment. *RISK* (August): 64–69.

[159] Merton RC, 1973. Theory of rational option pricing. *Bell Journal of Economics and Management Science* 4(Spring): 141–183.

[160] Merton RC, 1976. Option pricing when underlying stock returns are discontinuous. *Journal of Financial Economics*, 3: 125-144.

[161] Mikosch T, 1998. *Elementary Stochastic Calculus*. World Scientific, Singapore.

[162] Moreni N and Pallavicini A, 2010. Parsimonious HJM modelling for multiple yield-curve dynamics. Preprint, arXiv:1011.0828v1.

[163] Morini M and Prampolini A, 2010. Risky funding: a unified framework for counterparty and liquidity charges. Working paper, http://ssrn.com/abstract=1669930.

[164] Morini N, and Runggaldier W, 2014. On multicurve models for the term structure. In: *Nonlinear Economic Dynamics and Financial Modelling* (R. Dieci, X. Z. He and C. Hommes, eds.), 275-290. Berlin, Heidelberg, New York: Springer.

[165] Morton A, 1988. A class of stochastic differential equations arising in models for the evolution of bond prices. Technical Report, Cornell University.

[166] Musiela M and Rutkowski M, 1995. Continuous-time term structure models. Working paper, Rheinische Friedrich-Wilhelms-Universitat, Bonn.

[167] Nguyen T and Seifried F, 2015. The Multi-Curve Potential Model. Preprint, www.ssrn.com, abstract id=2502374.

[168] Oksendal B, 1992. *Stochastic Differential Equations: An Introduction with Applications*, 3rd edition. Springer, Berlin.

[169] Pallavicini A, Perini D, and Brigo D, 2011. Funding valuation adjustment: a consistent framework including CVA, DVA, collateral, netting rules and rehypothecation. Working paper, http://ssrn.com/abstract=1969114.

[170] Paulot L, 2015. Asymptotic Implied Volatility at the Second Order with Application to the SABR Model. P.K. Friz et al. (eds.), *Large Deviations and Asymptotic Methods in Finance*, Springer Proceedings in Mathematics & Statistics 110.

[171] Pedersen M, 1999. Bermudan Swaptions in the LIBOR market model. SimCorp Financial Research Working Paper.

[172] Pietersz R, Pelsser A, and Regenmortel M, 2004. Fast drift approximated pricing in the BGM model. *Journal of Computational Finance* 8(1): 93–124.

[173] Piterbarg V, 2003. A Stochastic Volatility Forward LIBOR Model with a Term Structure of Volatility Smiles. Working paper, Barclays Capital.

[174] Piterbarg V, 2010. Funding beyond discounting: collateral agreements and derivatives pricing. *Risk*, February, 97–102.

[175] Press W, Teukolsky S, Vetterling W, and Flannery B, 1992. *Numerical Recipes in C: The Art of Scientific Computing*, 2nd edition. Cambridge University Press, Cambridge.

[176] Raible S, 2000. Lévy process in Finance: Theory, Numerics, and Empirical Facts, Ph.D. thesis, University of Freiburg.

[177] Rebonato R, 1999. On the pricing implications of the joint lognormal assumption for the swaption and cap markets. *Journal of Computational Finance* 2(3): 57–76.

[178] Rebonato R, 2002. *Modern Pricing of Interest-Rate Derivative*. Princeton University Press, Princeton and Oxford.

[179] Ritchken P and Sankarasubramanian L, 1995. Volatility structures of forward and the dynamics of the term structure. *Mathematical Finance* 5: 55–72.

[180] Rockafellar RT, 1970. *Convex Analysis*. Princeton University Press, Princeton, NJ.

[181] Rogers L, 1997. The potential approach to the term structure of interest rates and foreign exchange rates. *Mathematical Finance*, 7: 157–176.

[182] Rudin W, 1976. *Principles of Mathematical Analysis*, 3rd edition. The McGraw-Hill Companies Inc., New York.

[183] Ruiz I, 2013. FVA demystified: CVA, DVA, FVA and their interaction (Part I). Working paper.

[184] Sato K, 1999. *Levy Processes and Infinitely Divisible Distributions*. Cambridge University Press, Cambridge.

[185] Schrager DF and Pelsser AJ, 2006. Pricing swaptions and coupon bond options in affine term structure models. *Mathematical Finance* 16(4): 673–694.

[186] Schönbucher P. (2000): A LIBOR market model with default risk. Working paper, Bonn University.

[187] Schönbucher P. (2004): A measure of survival. *Risk*, August, 79–85.

[188] Sherif N, 2015. FVA models overstate costs - Risk.net poll. *Risk.net*, May.

[189] Sidenius J, 2000. LIBOR market model in practice. *Journal of Computational Finance*, 3(3): 5–26.

[190] Stewart GW and Sun JG, 1990. *Matrix Perturbation Theory*. Academic Press, San Diego.

[191] Steel J, 2000. *Stochastic Calculus and Financial Applications*. Springer, Berlin.

[192] Trolle A and Schwartz E, 2008. A general stochastic volatility model for the pricing of interest rate derivatives. *Review of Financial Studies* Advance Access published online on April 28.

[193] Tuckman B, 2002. *Fixed Income Securities: Tools for Today's Markets*, 2nd edition, Wiley.

[194] Varadhan SRS, 1980. *Diffusion Processes and Partial Differential Equations*. Tata Institute Lectures, Springer, New York.

[195] Vasicek OA, 1977. An equilibrium characterization of the term structure. *Journal of Financial Economics* 5: 177–188.

[196] van Bezooyen J, Exley C, and Smith A, 1997. A market-based approach to valuing LPI liabilities. Downloadable at: http://www.gemstudy.com/DefinedBenefitPensionsDownloads.

[197] Wright S, 2015) Coordinate descent algorithms. *Mathematical Programming*, 151(1): 3–34.

[198] Wu L, 2002. LIBOR market model with stochastic volatility. Working paper, Hong Kong University of Science and Technology.

[199] Wu L, 2003. Fast at-the-money calibration of the LIBOR market model through Lagrange Multipliers. *Journal of Computational Finance*, 6(2): 39–77.

[200] Wu L, 2005. To recover or not to recover: that is not the question. Working paper, HKUST.

[201] Wu L, 2015. CVA and FVA to Derivatives Trades Collateralized by Cash. IJTAF, 18(5), 1550035.

[202] Wu L and Zhang F, 2008. Fast swaption pricing under the market model with a square-root volatility process. *Quantitative Finance*, 8(2): 163–180.

[203] Zhang Z-Y and Wu L, 2003. Optimal low-rank approximation of correlation Matrices. *Linear Algebra and Its Application*, 364: 161–187.

Bibliography

Abramowitz M and Stegun I, 1964. *Handbook of Mathematical Functions, with Formulas, Graphs and Mathematical Tables*. National Bureau of Standards, Washington, DC.

Andersen L and Andreasen J, 1998. Volatility skews and extensions of the LIBOR market model. Working paper, GE Financial Products.

Bachelier L, 1900. Theorie de la speculation. *Ann Ecole Norm Sup* 17(S.3): 21–86.

Bickel P and Doksum K, 1991. *Mathematical Statistics: Basic Ideas and Selected Topics*. Prentice-Hall, Englewood Cliffs, NJ.

Black F, 1976. The pricing of commodity contracts. *Journal of Financial Economics* 3: 167–179.

Breiman L, 1968. *Probability*. Addison-Wesley, Reading, MA.

Brown R, 1827. A brief description of microscopical observations (German translation). *Annals of Physics* 14(June and July): 294–313.

Carr P and Madan D, 1999. Option valuation using the fast Fourier transform. *Journal of Computational Finance* 3: 463–520.

Chen RR and Scott L, 1993. Maximum likelihood estimation for a multifactor equilibrium model of the term structure of interest rates. *Journal of Fixed Income* 3: 14–31.

Chen RR and Scott L, 1995. Interest rate options in multi-factor Cox–Ingersoll–Ross models of the term structure. *Journal of Derivatives* 3: 53–72.

Chung KI and Williams RJ, 1990. *Introduction to Stochastic Integration*, 2nd edition. Birkauser, Cambridge.

Cootner PH, 1964. *The Random Character of Stock Prices*. MIT Press, Cambridge, MA.

Cox J and Rubinstein M, 1985. *Option Theory*. Prentice-Hall, Englewood Cliffs, NJ.

Cox J, Ross S, and Rubinstein M, 1979. Option pricing: A simplified approach. *Journal of Financial Economics* 7: 229–264.

Cui S, 2018. Dual-Curve Term Structure Models for Post-Crisis Interest Tare Derivatives Markets. Ph.D. Thesis, the Hong Kong University of Science and Technology.

Doob J, 1953. *Stochastic Processes*. Wiley, New York.

Duffie D and Ken R, 1995. Multi-factor interest rate models. *Philosophical Transactions of the Royal Society, Series A* 317: 577–586.

Duffie D and Protter P, 1988. From discrete to continuous time finance: Weak convergence of the financial gain process. *Mathematical Finance* 1: 1–16.

Einstein A, 1956. *Investigation on the Theory of Brownian Movement*. Dover Publications, London.

Fabozzi F, 1997. *Fixed Income Mathematics: Analytical and Statistical Techniques*. Irwin, Chicago.

Figlewski S, Silber W, and Subrahmanyam M, 1990. *Financial Options: From Theory to Practice*. Business One Irwin, Homewood, IL.

Girsanov I, 1960. On transforming a certain class of stochastic processes by absolute substitution of measures. *Theory of Probability and Applications* 5: 285–301.

Hagan P and Lesniewski, A, 2008. LIBOR Market Model with SABR Style Stochastic Volatility. Working paper.

Heath D, Jarrow R, and Morton A, 1990a. Bond pricing and the term structure of interest rates: A discrete time approximation. *Journal of Financial and Quantitative Analysis* 25: 419–440.

Heath D, Jarrow R, and Morton A, 1990b. Contingent claim valuation with a random evolution of interest rates. *Review of Futures Markets* 9: 54–76.

Heston S, 1993. A closed-form solution for options with stochastic volatility with applications to bond and currency options. *The Review of Financial Studies*, (6)2: 327–343.

Hull J, 1998. *Introduction to Futures and Options Markets*. Prentice-Hall, Upper Saddle River, NJ.

Ito K, 1944. Stochastic integral. *Proceedings of the Imperial Academy* 20: 648–665.

Jarrow R, 1983. *Option Pricing*. R.D. Irwin, Homewood, IL.

Jarrow R, 1995. *Modeling Fixed-Income Securities and Interest-Rate Options*. McGraw-Hill, New York.

Jarrow R and Heath D, 1998. *Fixed-Income Securities and Interest Rates*. McGraw-Hill, New York.

Judge GG, Griffiths WE, Hill RC, Lutkepohl H, and Lee TC, 1988. *Introduction to the Theory and Practice of Econometrics*, 2nd edition. Wiley, New York.

Kapner KR and Marshall JF, 1994. *Understanding Swaps*. Wiley, New York.

Kloeden PE and Platen E, 1994. *Numerical Solution of SDE Through Computer Experiments*. Springer, Berlin.

Levy P, 1948a. Sur certains processus stochastiques Homogenge. *Composition Math* 7: 283–339.

Levy P, 1948b. *Processus Stochastiques et Mouvement Brownien*. Gauthier Villars, Pairs.

Li A, Ritchen P, and Sankarasubramanian L, 1995. Lattice models for pricing American interest rate claims. *Journal of Finance* 50: 719–737.

Longstaff F and Schwartz E, 1992a. Interest rate volatility and the term structure: a two-factor general equilibrium model. *Journal of Finance* 47: 1259–1282.

Longstaff F and Schwartz E, 1992b. A two-factor interest rate model and contingent claims valuation. *Journal of Fixed Income* 3: 16–23.

Miltersen K, Sandmann K, and Sondermann K, 1997. Closed-form solutions for term structure derivatives with lognormal interest rates. *Journal of Finance* 52: 409–430.

Musiela M and Rutkowski M, 1997. *Martingale Methods in Financial Modelling*. Springer Verlag, Berlin.

Oksendal B, 1995. *Stochastic Differential Equations, an Introduction with Applications*, 4th edition. Springer, Berlin.

Paley R and Wiener N, 1933. Notes on random functions. *Mathematische Zeitschrift* 37: 647–668.

Protter, P. 1990. *Stochastic Integration and Differential Equations*. Springer.

Rogers C, 1995. Which model for the term structure of interest rates should one use? In: M Davis, D Duffie, and I Karatzas (eds), *Mathematical Finance*, IMA, Vol. 65, pp. 93–116. Springer, New York.

Siegel DR and Siegel DF, 1990. *The Futures Markets: Arbitrage, Risk Management and Portfolio Strategies*. Probus, Chicago.

Sun G and Jagannathan R, 1998. An evaluation of multi-factor CIR models using LIBOR, swaps rates, and cap and swaptions prices. Working Paper.

Thorp E, 1969. Optimal gambling systems for favorable games. *Review of the International Statistical Institute* 37: 3.

Varadhan SRS, 1980b. *Lectures on Brownian Motion and Stochastic Differential Equations*. Tata Institute of Fundamental Research, Bombay.

Wong MA, 1991. *Trading and Investing in Bond Options*. Wiley, New York.

Wu T, 2012. Pricing and hedging the smile with SABR: Evidence from the interest rate caps market. *The Journal of Futures Markets*. Vol. 32, Bo. 8: 773–791.

Wu L and Zhang F, 2006. LIBOR market model with stochastic volatility. *Journal of Industrial and Management Optimization* 2(2): 199–227.

Index